WORLD HEALTH ORGANIZATION

INTERNATIONAL AGENCY FOR RESEARCH ON CANCER

IARC MONOGRAPHS

ON THE

EVALUATION OF CARCINOGENIC RISKS
TO HUMANS

Some Organic Solvents, Resin Monomers and Related Compounds,
Pigments and Occupational Exposures in Paint
Manufacture and Painting

VOLUME 47

This publication represents the views and expert opinions
of an IARC Working Group on the
Evaluation of Carcinogenic Risks to Humans
which met in Lyon,

18–25 October 1988

1989

IARC MONOGRAPHS

In 1969, the International Agency for Research on Cancer (IARC) initiated a programme on the evaluation of the carcinogenic risk of chemicals to humans involving the production of critically evaluated monographs on individual chemicals. In 1980, the programme was expanded to include the evaluation of the carcinogenic risk associated with exposures to complex mixtures.

The objective of the programme is to elaborate and publish in the form of monographs critical reviews of data on carcinogenicity for chemicals and complex mixtures to which humans are known to be exposed, and on specific occupational exposures, to evaluate these data in terms of human risk with the help of international working groups of experts in chemical carcinogenesis and related fields, and to indicate where additional research efforts are needed.

This project is supported by PHS Grant No. 5–UO1 CA33193–07 awarded by the US National Cancer Institute, Department of Health and Human Services. Additional support has been provided by the Commission of the European Communities since 1986.

CONTENTS

CONTENTS

CONTENTS

NOTE TO THE READER

The term 'carcinogenic risk' in the *IARC Monographs* series is taken to mean the probability that exposure to an agent will lead to cancer in humans.

Inclusion of an agent in the *Monographs* does not imply that it is a carcinogen, only that the published data have been examined. Equally, the fact that an agent has not yet been evaluated in a monograph does not mean that it is not carcinogenic.

The evaluations of carcinogenic risk are made by international working groups of independent scientists and are qualitative in nature. No recommendation is given for regulation or legislation.

Anyone who is aware of published data that may alter the evaluation of the carcinogenic risk of an agent to humans is encouraged to make this information available to the Unit of Carcinogen Identification and Evaluation, International Agency for Research on Cancer, 150 cours Albert Thomas, 69372 Lyon Cedex 08, France, in order that the agent may be considered for re-evaluation by a future Working Group.

Although every effort is made to prepare the monographs as accurately as possible, mistakes may occur. Readers are requested to communicate any errors to the Unit of Carcinogen Identification and Evaluation, so that corrections can be reported in future volumes.

IARC WORKING GROUP ON THE EVALUATION OF CARCINOGENIC RISKS TO HUMANS: SOME SOLVENTS, RESIN MONOMERS AND RELATED COMPOUNDS, PIGMENTS AND OCCUPATIONAL EXPOSURES IN PAINT MANUFACTURE AND PAINTING

Lyon, 18–25 October 1988

LIST OF PARTICIPANTS

Members

D. Anderson, The British Industrial Biological Research Association, Woodmansterne Road, Carshalton, Surrey SM5 4DS, UK

R.B. Beems, Department of Biological Toxicology, TNO–CIVO Institutes, PO Box 360, 3700 AJ Zeist, The Netherlands

V. Beral, Division of Sexually Transmitted Diseases, Centers for Disease Control, Atlanta, GA 30333, USA

M. Bignami, Higher Institute of Health, via Regina Elena 299, 00161 Rome, Italy

A. Brøgger, Department of Genetics, Institute for Cancer Research, Norwegian Radium Hospital, Montebello, 0310 Oslo 3, Norway

M.R. Elwell, Toxicologic Pathology, Toxicology Research and Testing Program, National Institute of Environmental Health Sciences, PO Box 12233, Research Triangle Park, NC 27709, USA

J. Fajen, National Institute for Occupational Safety and Health, Robert A. Taft Laboratories, 4676 Columbia Parkway, Cincinnati, OH 45226, USA

L. Fishbein, Environ Corporation, Counsel in Health and Environmental Science, The Flour Mill, 1000 Potomac Street NW, Washington DC 20007, USA

M. Gérin, Department of Occupational and Environmental Health, Faculty of Medicine, University of Montréal, CP 6128, Station A, Montréal, Québec H3C 3J7, Canada

P. Grandjean, Institute of Community Medicine, Odense University, J.B. Winsowsvej 19, 5000 Odense, Denmark

K. Hemminki, Institute of Occupational Health, Topeliuksenkatu 41 a A, 00250 Helsinki, Finland

P.T. Henderson, State University Limburg, Postbox 616, 6200 MD Maastricht, The Netherlands

C. Hogstedt, National Institute of Occupational Health, 171 84 Solna, Sweden (*Chairman*)

M. Ikeda, Department of Public Health, Kyoto University, Faculty of Medicine, Konoe-cho, Sakyo-ku, Kyoto 606, Japan (*Vice-Chairman*)

A.F. Karamysheva, Laboratory of Carcinogenic Substances, All-Union Cancer Research Centre, Kashirskoye Shosse 24, 115478 Moscow, USSR

R.J. Kavlock, Developmental Toxicology Division, Health Effects Research Laboratory, US Environmental Protection Agency, MD-67, Research Triangle Park, NC 27711, USA

R. Laib, Institute for Worker Physiology of the Dortmund University, Ardeystrasse 67, 4600 Dortmund 1, Federal Republic of Germany

L.S. Levy, Institute of Occupational Health, University of Birmingham, PO Box 363, Birmingham B15 2TT, UK

G. Matanoski, Department of Epidemiology, School of Hygiene and Public Health, 615 N Wolfe, Baltimore, MD 21205, USA

S. Skerfving, Department of Occupational and Environmental Medicine, University Hospital, 221 85 Lund, Sweden

Representatives/Observers

Representative of the National Cancer Institute

J. Rice, NCI-Frederick Cancer Research Facility, Building 538, Frederick, MD 21701, USA

Representative of Tracor Technology Resources, Inc.

S. Olin, Tracor Technology Resources, Inc., 1601 Research Boulevard, Rockville, MD 20850, USA

Representative of the Commission of the European Communities

H. Martin, Division of Industrial Health and Hygiene, Health and Safety Directorate, Commission of the European Communities, Bâtiment Jean Monnet, 2920 Luxembourg, Grand Duchy of Luxembourg

Chemical Manufacturers' Association

N. Krivanek, E.I. du Pont de Nemours & Company, Haskell Laboratories, PO Box 50, Newark, DE 19714, USA

CONCAWE

W. Glaap, German Exxon Chemical GmbH, Dompropst-Ketzer-strasse 7-9, 5000 Cologne 1, Federal Republic of Germany

European Chemical Industry Ecology and Toxicology Centre

M. Wooder, Rohm & Haas Co., European Operations, Chesterfield House, Bloomsbury Way, London WC1A 2TP, UK

Bygghälsan

A. Englund, Bygghälsan, Box 94, 182 11 Danderyd, Sweden

Secretariat

A. Aitio, Unit of Carcinogen Identification and Evaluation (*Officer in Charge of the Programme*)

J. Cabral, Unit of Mechanisms of Carcinogenesis

E. Cardis, Unit of Biostatistics Research and Informatics

B.-H. Chen, International Programme on Chemical Safety, World Health Organization, Geneva, Switzerland

J. Cheney, Editorial and Publishing Services

M. Friesen, Unit of Environmental Carcinogenesis and Host Factors

M.-J. Ghess, Unit of Carcinogen Identification and Evaluation

E. Heseltine, Lajarthe, 24290 Montignac, France

T. Kauppinen, Unit of Carcinogen Identification and Evaluation

J. Kaldor, Unit of Biostatistics Research and Informatics

M. Khlat, Unit of Descriptive Epidemiology

K. L'Abbé, Unit of Analytical Epidemiology

C. Malaveille, Unit of Environmental Carcinogenesis and Host Factors

D. Mietton, Unit of Carcinogen Identification and Evaluation

R. Montesano, Unit of Mechanisms of Carcinogenesis

I. O'Neill, Unit of Environmental Carcinogenesis and Host Factors

C. Partensky, Unit of Carcinogen Identification and Evaluation

I. Peterschmitt, Unit of Carcinogen Identification and Evaluation, Geneva, Switzerland

R. Saracci, Unit of Analytical Epidemiology

D. Shuker, Unit of Environmental Carcinogenesis and Host Factors

L. Shuker, Unit of Carcinogen Identification and Evaluation (*Secretary*)

L. Simonato, Unit of Analytical Epidemiology

L. Tomatis, Director

V. Turusov, Director's Office

J. Wilbourn, Unit of Carcinogen Identification and Evaluation

H. Yamasaki, Unit of Mechanisms of Carcinogenesis

Secretarial assistance

J. Cazeaux
M. Lézère
M. Mainaud
S. Reynaud

PREAMBLE

IARC MONOGRAPHS PROGRAMME ON THE EVALUATION OF CARCINOGENIC RISKS TO HUMANS[1]

PREAMBLE

1. BACKGROUND

In 1969, the International Agency for Research on Cancer (IARC) initiated a programme to evaluate the carcinogenic risk of chemicals to humans and to produce monographs on individual chemicals. The *Monographs* programme has since been expanded to include consideration of exposures to complex mixtures of chemicals (which occur, for example, in some occupations and as a result of human habits) and of exposures to other agents, such as radiation and viruses. With Supplement 6(1), the title of the series was modified from *IARC Monographs on the Evaluation of the Carcinogenic Risk of Chemicals to Humans* to *IARC Monographs on the Evaluation of Carcinogenic Risks to Humans*, in order to reflect the widened scope of the programme.

The criteria established in 1971 to evaluate carcinogenic risk to humans were adopted by the working groups whose deliberations resulted in the first 16 volumes of the *IARC Monographs* series. Those criteria were subsequently re-evaluated by working groups which met in 1977(2), 1978(3), 1979(4), 1982(5) and 1983(6). The present preamble was prepared by two working groups which met in September 1986 and January 1987, prior to the preparation of Supplement 7(7) to the *Monographs*.

2. OBJECTIVE AND SCOPE

The objective of the programme is to prepare, with the help of international working groups of experts, and to publish in the form of monographs, critical reviews and evaluations of evidence on the carcinogenicity of a wide range of agents to which humans are or may be exposed. The *Monographs* may also indicate where additional research efforts are needed.

The *Monographs* represent the first step in carcinogenic risk assessment, which involves examination of all relevant information in order to assess the strength of the available evi-

[1]This project is supported by PHS Grant No. 5 U01 CA33193–07 awarded by the US National Cancer Institute, Department of Health and Human Services, and with a subcontract to Tracor Technology Resources, Inc. Since 1986, this programme has also been supported by the Commission of the European Communities.

dence that, under certain conditions of exposure, an agent could alter the incidence of cancer in humans. The second step is quantitative risk estimation, which is not usually attempted in the *Monographs*. Detailed, quantitative evaluations of epidemiological data may be made in the *Monographs*, but without extrapolation beyond the range of the data available. Quantitative extrapolation from experimental data to the human situation is not undertaken.

These monographs may assist national and international authorities in making risk assessments and in formulating decisions concerning any necessary preventive measures. The evaluations of IARC working groups are scientific, qualitative judgements about the degree of evidence for carcinogenicity provided by the available data. These evaluations represent only one part of the body of information on which regulatory measures may be based. Other components of regulatory decisions may vary from one situation to another and from country to country, responding to different socioeconomic and national priorities. *Therefore, no recommendation is given with regard to regulation or legislation, which are the responsibility of individual governments and/or other international organizations.*

The *IARC Monographs* are recognized as an authoritative source of information on the carcinogenicity of chemicals and complex exposures. A users' survey, made in 1988, indicated that the *Monographs* are consulted by various agencies in 57 countries. Each volume is generally printed in 4000 copies for distribution to governments, regulatory bodies and interested scientists. The *Monographs* are also available *via* the Distribution and Sales Service of the World Health Organization.

3. SELECTION OF TOPICS FOR MONOGRAPHS

Topics are selected on the basis of two main criteria: (a) that they concern agents for which there is evidence of human exposure, and (b) there is some evidence or suspicion of carcinogenicity. The term agent is used to include individual chemical compounds, groups of chemical compounds, physical agents (such as radiation), biological factors (such as viruses) and mixtures of agents such as occur in occupational exposures and as a result of personal and cultural habits (like smoking and dietary practices). Chemical analogues and compounds with biological or physical characteristics similar to those of suspected carcinogens may also be considered, even in the absence of data on carcinogenicity.

The scientific literature is surveyed for published data relevant to an assessment of carcinogenicity; the IARC surveys of chemicals being tested for carcinogenicity(8) and directories of on-going research in cancer epidemiology(9) often indicate those agents that may be scheduled for future meetings. An ad-hoc working group convened by IARC in 1984 gave recommendations as to which chemicals and exposures to complex mixtures should be evaluated in the *IARC Monographs* series(10).

As significant new data on subjects on which monographs have already been prepared become available, re-evaluations are made at subsequent meetings, and revised monographs are published.

4.   DATA FOR MONOGRAPHS

The *Monographs* do not necessarily cite all the literature concerning the subject of an evaluation. Only those data considered by the Working Group to be relevant to making the evaluation are included.

With regard to biological and epidemiological data, only reports that have been published or accepted for publication in the openly available scientific literature are reviewed by the working groups. In certain instances, government agency reports that have undergone peer review and are widely available are considered. Exceptions may be made on an ad-hoc basis to include unpublished reports that are in their final form and publicly available, if their inclusion is considered pertinent to making a final evaluation (see p. 27 *et seq.*). In the sections on chemical and physical properties and on production, use, occurrence and analysis, unpublished sources of information may be used.

5.   THE WORKING GROUP

Reviews and evaluations are formulated by a working group of experts. The tasks of this group are five-fold: (i) to ascertain that all appropriate data have been collected; (ii) to select the data relevant for the evaluation on the basis of scientific merit; (iii) to prepare accurate summaries of the data to enable the reader to follow the reasoning of the Working Group; (iv) to evaluate the results of experimental and epidemiological studies; and (v) to make an overall evaluation of the carcinogenicity of the agent to humans.

Working Group participants who contributed to the consideration and evaluation of the agents within a particular volume are listed, with their addresses, at the beginning of each publication. Each participant who is a member of a working group serves as an individual scientist and not as a representative of any organization, government or industry. In addition, representatives from national and international agencies and industrial associations are invited as observers.

6.   WORKING PROCEDURES

Approximately one year in advance of a meeting of a working group, the topics of the monographs are announced and participants are selected by IARC staff in consultation with other experts. Subsequently, relevant biological and epidemiological data are collected by IARC from recognized sources of information on carcinogenesis, including data storage and retrieval systems such as CAS ONLINE, MEDLINE and TOXLINE, including EMIC and ETIC for data on genetic and related effects and teratogenicity, respectively.

The major collection of data and the preparation of first drafts of the sections on chemical and physical properties, on production and use, on occurrence, and on analysis are carried out under a separate contract funded by the US National Cancer Institute. Efforts are made to supplement this information with data from other national and international sources. Representatives from industrial associations may assist in the preparation of sections on production and use.

Production and trade data are obtained from governmental and trade publications and, in some cases, by direct contact with industries. Separate production data on some substances may not be available because their publication could disclose confidential information. Information on uses is usually obtained from published sources but is often complemented by direct contact with manufacturers.

Six months before the meeting, reference material is sent to experts, or is used by IARC staff, to prepare sections for the first drafts of monographs. The complete first drafts are compiled by IARC staff and sent, prior to the meeting, to all participants of the Working Group for review.

The Working Group meets in Lyon for seven to eight days to discuss and finalize the texts of the monographs and to formulate the evaluations. After the meeting, the master copy of each monograph is verified by consulting the original literature, edited and prepared for publication. The aim is to publish monographs within nine months of the Working Group meeting.

7. EXPOSURE DATA

Sections that indicate the extent of past and present human exposure, the sources of exposure, the persons most likely to be exposed and the factors that contribute to exposure to the subject of the monograph are included at the beginning of the monograph.

Most monographs on individual chemicals or complex mixtures include sections on chemical and physical data, and production, use, occurrence and analysis. In other monographs, for example on physical agents, biological factors, occupational exposures and cultural habits, other sections may be included, such as: historical perspectives, description of an industry or habit, exposures in the work place or chemistry of the complex mixture.

The Chemical Abstracts Services Registry Number, the latest Chemical Abstracts Primary Name and the IUPAC Systematic Name are recorded. Other synonyms are given, but the list is not necessarily comprehensive.

Information on chemical and physical properties and, in particular, data relevant to identification, occurrence and biological activity are included. A separate description of technical products gives relevant specifications and includes available information on composition and impurities and a list of trade names, which may not be comprehensive. Some of the trade names may be those of mixtures in which the substance being evaluated is only one of the ingredients.

The dates of first synthesis and of first commercial production of a substance are provided; for those which do not occur naturally, this information may allow a reasonable estimate to be made of the date before which no human exposure could have occurred. The dates of first reported occurrence of an exposure are also provided. In addition, methods of synthesis used in past and present commercial production and different methods of production which may give rise to different impurities are described.

Data on production, foreign trade and uses are obtained for representative regions, which usually include Europe, Japan and the USA. It should not, however, be inferred that those areas or nations are necessarily the sole or major sources or users of the agent being evaluated.

Some identified uses may not be current or major applications, and the coverage is not necessarily comprehensive. In the case of drugs, mention of their therapeutic uses does not necessarily represent current practice nor does it imply judgement as to their clinical efficacy.

Information on the occurrence of an agent or mixture in the environment is obtained from data derived from the monitoring and surveillance of levels in occupational environments, air, water, soil, foods and animal and human tissues. When available, data on the generation, persistence and bioaccumulation of the agent are also included.

Statements concerning regulations and guidelines (e.g., pesticide registrations, maximal levels permitted in foods, occupational exposure limits) are included for some countries as indications of potential exposures, but they may not reflect the most recent situation, since such limits are continuously reviewed and modified. The absence of information on regulatory status for a country should not be taken to imply that that country does not have regulations with regard to the agent.

The purpose of the section on analysis is to give the reader an overview of current methods cited in the literature, with emphasis on those widely used for regulatory purposes. No critical evaluation or recommendation of any of the methods is meant or implied. Methods for monitoring human exposure are also given, when available. The IARC publishes a series of volumes, *Environmental Carcinogens: Selected Methods of Analysis*(11), that describe validated methods for analysing a wide variety of agents.

8. BIOLOGICAL DATA RELEVANT TO THE EVALUATION OF CARCINOGENICITY TO HUMANS

The term 'carcinogen' is used in these monographs to denote an agent that is capable of increasing the incidence of malignant neoplasms; the induction of benign neoplasms may in some circumstances (see p. 21) contribute to the judgement that an agent is carcinogenic. The terms 'neoplasm' and 'tumour' are used interchangeably.

Some epidemiological and experimental studies indicate that different agents may act at different stages in the carcinogenic process, probably by fundamentally different mechanisms. In the present state of knowledge, the aim of the *Monographs* is to evaluate evidence of carcinogenicity at any stage in the carcinogenic process independently of the underlying mechanism involved. There is as yet insufficient information to implement classification according to mechanisms of action(6).

Definitive evidence of carcinogenicity in humans can be provided only by epidemiological studies. Evidence relevant to human carcinogenicity may also be provided by experimental studies of carcinogenicity in animals and by other biological data, particularly those relating to humans.

The available studies are summarized by the working groups, with particular regard to the qualitative aspects discussed below. In general, numerical findings are indicated as they appear in the original report; units are converted when necessary for easier comparison. The Working Group may conduct additional analyses of the published data and use them in their assessment of the evidence and may include them in their summary of a study; the results of such supplementary analyses are given in square brackets. Any comments are also made in square brackets; however, these are kept to a minimum, being restricted to those instances in which it is felt that an important aspect of a study, directly impinging on its interpretation, should be brought to the attention of the reader.

9. EVIDENCE FOR CARCINOGENICITY IN EXPERIMENTAL ANIMALS

For several agents (e.g., 4–aminobiphenyl, bis(chloromethyl)ether, diethylstilboestrol, melphalan, 8–methoxypsoralen (methoxsalen) plus UVR, mustard gas and vinyl chloride), evidence of carcinogenicity in experimental animals preceded evidence obtained from epidemiological studies or case reports. Information compiled from the first 41 volumes of the *IARC Monographs*(12) shows that, of the 44 agents for which there is *sufficient* or *limited evidence* of carcinogenicity to humans (see pp. 27–28), all 37 that have been tested adequately experimentally produce cancer in at least one animal species. Although this association cannot establish that all agents that cause cancer in experimental animals also cause cancer in humans, nevertheless, *in the absence of adequate data on humans, it is biologically plausible and prudent to regard agents for which there is sufficient evidence (see p. 28) of carcinogenicity in experimental animals as if they presented a carcinogenic risk to humans*.

The monographs are not intended to summarize all published studies. Those that are inadequate (e.g., too short a duration, too few animals, poor survival; see below) or are judged irrelevant to the evaluation are generally omitted. They may be mentioned briefly, particularly when the information is considered to be a useful supplement to that of other reports or when they provide the only data available. Their inclusion does not, however, imply acceptance of the adequacy of the experimental design or of the analysis and interpretation of their results. Guidelines for adequate long–term carcinogenicity experiments have been outlined (e.g., 13).

The nature and extent of impurities or contaminants present in the agent being evaluated are given when available. Mention is made of all routes of exposure by which the agent has been adequately studied and of all species in which relevant experiments have been performed. Animal strain, sex, numbers per group, age at start of treatment and survival are reported.

Experiments in which the agent was administered in conjunction with known carcinogens or factors that modify carcinogenic effects are also reported. Experiments on the carcinogenicity of known metabolites and derivatives may be included.

(a) Qualitative aspects

The overall assessment of the carcinogenicity of an agent involves several considerations of qualitative importance, including (i) the experimental conditions under which the test was performed, including route and schedule of exposure, species, strain, sex, age, duration of follow–up; (ii) the consistency with which the agent has been shown to be carcinogenic, e.g., in how many species and at which target organ(s); (iii) the spectrum of neoplastic response, from benign tumours to malignant neoplasms; and (iv) the possible role of modifying factors.

Considerations of importance to the Working Group in the interpretation and evaluation of a particular study include: (i) how clearly the agent was defined; (ii) whether the dose was adequately monitored, particularly in inhalation experiments; (iii) whether the doses used were appropriate and whether the survival of treated animals was similar to that of controls; (iv) whether there were adequate numbers of animals per group; (v) whether animals of both sexes were used; (vi) whether animals were allocated randomly to groups; (vii) whether the duration of observation was adequate; and (viii) whether the data were adequately reported. If available, recent data on the incidence of specific tumours in historical controls, as well as in concurrent controls, should be taken into account in the evaluation of tumour response.

When benign tumours occur together with and originate from the same cell type in an organ or tissue as malignant tumours in a particular study and appear to represent a stage in the progression to malignancy, it may be valid to combine them in assessing tumour incidence. The occurrence of lesions presumed to be preneoplastic may in certain instances aid in assessing the biological plausibility of any neoplastic response observed.

Among the many agents that have been studied extensively, there are few instances in which the only neoplasms induced were benign. Benign tumours in experimental animals frequently represent a stage in the evolution of a malignant neoplasm, but they may be 'endpoints' that do not readily undergo transition to malignancy. However, if an agent is found to induce only benign neoplasms, it should be suspected of being a carcinogen and it requires further investigation.

(b) Quantitative aspects

The probability that tumours will occur may depend on the species and strain, the dose of the carcinogen and the route and period of exposure. Evidence of an increased incidence of neoplasms with increased exposure strengthens the inference of a causal association between the exposure and the development of neoplasms.

The form of the dose–response relationship can vary widely, depending on the particular agent under study and the target organ. Since many chemicals require metabolic activation before being converted into their reactive intermediates, both metabolic and pharmacokinetic aspects are important in determining the dose–response pattern. Saturation of steps such as absorption, activation, inactivation and elimination of the carcinogen may produce nonlinearity in the dose–response relationship, as could saturation of processes such as DNA repair(14,15).

(c) Statistical analysis of long–term experiments in animals

Factors considered by the Working Group include the adequacy of the information given for each treatment group: (i) the number of animals studied and the number examined histologically, (ii) the number of animals with a given tumour type and (iii) length of survival. The statistical methods used should be clearly stated and should be the generally accepted techniques refined for this purpose(15,16). When there is no difference in survival between control and treatment groups, the Working Group usually compares the proportions of animals developing each tumour type in each of the groups. Otherwise, consideration is given as to whether or not appropriate adjustments have been made for differences in survival. These adjustments can include: comparisons of the proportions of tumour–bearing animals among the 'effective number' of animals alive at the time the first tumour is discovered, in the case where most differences in survival occur before tumours appear; life–table methods, when tumours are visible or when they may be considered 'fatal' because mortality rapidly follows tumour development; and the Mantel–Haenszel test or logistic regression, when occult tumours do not affect the animals' risk of dying but are 'incidental' findings at autopsy.

In practice, classifying tumours as fatal or incidental may be difficult. Several survival-adjusted methods have been developed that do not require this distinction(15), although they have not been fully evaluated.

10. OTHER RELEVANT DATA IN EXPERIMENTAL SYSTEMS AND HUMANS

(a) Structure–activity considerations

This section describes structure–activity correlations that are relevant to an evaluation of the carcinogenicity of an agent.

(b) Absorption, distribution, excretion and metabolism

Concise information is given on absorption, distribution (including placental transfer) and excretion. Kinetic factors that may affect the dose–reponse relationship, such as saturation of uptake, protein binding, metabolic activation, detoxification and DNA–repair processes, are mentioned. Studies that indicate the metabolic fate of the agent in experimental animals and humans are summarized briefly, and comparisons of data from animals and humans are made when possible. Comparative information on the relationship between exposure and the dose that reaches the target site may be of particular importance for extrapolation between species.

(c) Toxicity

Data are given on acute and chronic toxic effects (other than cancer), such as organ toxicity, immunotoxicity, endocrine effects and preneoplastic lesions. Effects on reproduction, teratogenicity, feto- and embryotoxicity are also summarized briefly.

(d) Genetic and related effects

Tests of genetic and related effects may indicate possible carcinogenic activity. They can also be used in detecting active metabolites of known carcinogens in human or animal body fluids, in detecting active components in complex mixtures and in the elucidation of possible mechanisms of carcinogenesis.

The available data are interpreted critically by phylogenetic group according to the end-points detected, which may include DNA damage, gene mutation, sister chromatid exchange, micronuclei, chromosomal aberrations, aneuploidy and cell transformation. The concentrations (doses) employed are given and mention is made of whether an exogenous metabolic system was required. When appropriate, these data may be represented by bar graphs (activity profiles), with corresponding summary tables and listings of test systems, data and references. Detailed information on the preparation of these profiles is given in an appendix to those volumes in which they are used.

Positive results in tests using prokaryotes, lower eukaryotes, plants, insects and cultured mammalian cells suggest that genetic and related effects (and therefore possibly carcinogenic effects) could occur in mammals. Results from such tests may also give information about the types of genetic effect produced by an agent and about the involvement of metabolic activation. Some endpoints described are clearly genetic in nature (e.g., gene mutations and chromosomal aberrations), others are to a greater or lesser degree associated with genetic effects (e.g., unscheduled DNA synthesis). In-vitro tests for tumour–promoting activity and for cell transformation may detect changes that are not necessarily the result of genetic alterations but that may have specific relevance to the process of carcinogenesis. A critical appraisal of these tests has been published(13).

Genetic or other activity detected in the systems mentioned above is not always manifest in whole mammals. Positive indications of genetic effects in experimental mammals and in humans are regarded as being of greater relevance than those in other organisms. The demonstration that an agent can induce gene and chromosomal mutations in whole mammals indicates that it may have the potential for carcinogenic activity, although this activity may not be detectably expressed in any or all species tested. The relative potency of agents in tests for mutagenicity and related effects is not a reliable indicator of carcinogenic potency. Negative results in tests for mutagenicity in selected tissues from animals treated *in vivo* provide less weight, partly because they do not exclude the possibility of an effect in tissues other than those examined. Moreover, negative results in short–term tests with genetic endpoints cannot be considered to provide evidence to rule out carcinogenicity of agents that act through other mechanisms. Factors may arise in many tests that could give misleading results; these have been discussed in detail elsewhere(13).

The adequacy of epidemiological studies of reproductive outcomes and genetic and related effects in humans is evaluated by the same criteria as are applied to epidemiological studies of cancer.

11. EVIDENCE FOR CARCINOGENICITY IN HUMANS

(a) *Types of studies considered*

Three types of epidemiological studies of cancer contribute data to the assessment of carcinogenicity in humans – cohort studies, case–control studies and correlation studies. Rarely, results from randomized trials may be available. Case reports of cancer in humans exposed to particular agents are also reviewed.

Cohort and case–control studies relate individual exposures to the agent under study to the occurrence of cancer in individuals, and provide an estimate of relative risk (ratio of incidence in those exposed to incidence in those not exposed) as the main measure of association.

In correlation studies, the units of investigation are usually whole populations (e.g., in particular geographical areas or at particular times), and cancer incidence is related to a summary measure of the exposure of the population to the agent under study. Because individual exposure is not documented, however, a causal relationship is less easy to infer from correlation studies than from cohort and case–control studies.

Case reports generally arise from a suspicion, based on clinical experience, that the concurrence of two events – that is, exposure to a particular agent and occurrence of a cancer –

has happened rather more frequently than would be expected by chance. Case reports usually lack complete ascertainment of cases in any population, definition or enumeration of the population at risk and estimation of the expected number of cases in the absence of exposure.

The uncertainties surrounding interpretation of case reports and correlation studies make them inadequate, except in rare instances, to form the sole basis for inferring a causal relationship. When taken together with case–control and cohort studies, however, relevant case reports or correlation studies may add materially to the judgement that a causal relationship is present.

Epidemiological studies of benign neoplasms and presumed preneoplastic lesions are also reviewed by working groups. They may, in some instances, strengthen inferences drawn from studies of cancer itself.

(b) Quality of studies considered

It is necessary to take into account the possible roles of bias, confounding and chance in the interpretation of epidemiological studies. By 'bias' is meant the operation of factors in study design or execution that lead erroneously to a stronger or weaker association between an agent and disease than in fact exists. By 'confounding' is meant a situation in which the relationship between an agent and a disease is made to appear stronger or to appear weaker than it truly is as a result of an association between the agent and another agent that is associated with either an increase or decrease in the incidence of the disease. In evaluating the extent to which these factors have been minimized in an individual study, working groups consider a number of aspects of design and analysis as described in the report of the study. Most of these considerations apply equally to case–control, cohort and correlation studies. Lack of clarity of any of these aspects in the reporting of a study can decrease its credibility and its consequent weighting in the final evaluation of the exposure.

Firstly, the study population, disease (or diseases) and exposure should have been well defined by the authors. Cases in the study population should have been identified in a way that was independent of the exposure of interest, and exposure should have been assessed in a way that was not related to disease status.

Secondly, the authors should have taken account in the study design and analysis of other variables that can influence the risk of disease and may have been related to the exposure of interest. Potential confounding by such variables should have been dealt with either in the design of the study, such as by matching, or in the analysis, by statistical adjustment. In cohort studies, comparisons with local rates of disease may be more appropriate than those with national rates. Internal comparisons of disease frequency among individuals at different levels of exposure should also have been made in the study.

Thirdly, the authors should have reported the basic data on which the conclusions are founded, even if sophisticated statistical analyses were employed. At the very least, they should have given the numbers of exposed and unexposed cases and controls in a case–control study and the numbers of cases observed and expected in a cohort study. Further tabulations by time since exposure began and other temporal factors are also important. In a cohort study, data on all cancer sites and all causes of death should have been given, to avoid the possibility of reporting bias. In a case–control study, the effects of investigated factors other than the agent of interest should have been reported.

Finally, the statistical methods used to obtain estimates of relative risk, absolute cancer rates, confidence intervals and significance tests, and to adjust for confounding should have been clearly stated by the authors. The methods used should preferably have been the generally accepted techniques that have been refined since the mid–1970s. These methods have been reviewed for case–control studies(17) and for cohort studies(18).

(c) Quantitative considerations

Detailed analyses of both relative and absolute risks in relation to age at first exposure and to temporal variables, such as time since first exposure, duration of exposure and time since exposure ceased, are reviewed and summarized when available. The analysis of temporal relationships can provide a useful guide in formulating models of carcinogenesis. In particular, such analyses may suggest whether a carcinogen acts early or late in the process of carcinogenesis(6), although such speculative inferences cannot be used to draw firm conclusions concerning the mechanism of action of the agent and hence the shape (linear or otherwise) of the dose–response relationship below the range of observation.

(d) Criteria for causality

After the quality of individual epidemiological studies has been summarized and assessed, a judgement is made concerning the strength of evidence that the agent in question is carcinogenic for humans. In making their judgement, the Working Group considers several criteria for causality. A strong association (i.e., a large relative risk) is more likely to indicate causality than a weak association, although it is recognized that relative risks of small magnitude do not imply lack of causality and may be important if the disease is common. Associations that are replicated in several studies of the same design or using different epidemiological approaches or under different circumstances of exposure are more likely to represent a causal relationship than isolated observations from single studies. If there are inconsistent results among investigations, possible reasons are sought (such as differences in amount of exposure), and results of studies judged to be of high quality are given more weight than those from studies judged to be methodologically less sound. When suspicion of carcinogenicity arises largely from a single study, these data are not combined with those from later studies in any subsequent reassessment of the strength of the evidence.

If the risk of the disease in question increases with the amount of exposure, this is considered to be a strong indication of causality, although absence of a graded response is not necessarily evidence against a causal relationship. Demonstration of a decline in risk after cessation of or reduction in exposure in individuals or in whole populations also supports a causal interpretation of the findings.

Although the same carcinogenic agent may act upon more than one target, the specificity of an association (i.e., an increased occurrence of cancer at one anatomical site or of one morphological type) adds plausibility to a causal relationship, particularly when excess cancer occurrence is limited to one morphological type within the same organ.

Although rarely available, results from randomized trials showing different rates among exposed and unexposed individuals provide particularly strong evidence for causality.

When several epidemiological studies show little or no indication of an association between an exposure and cancer, the judgement may be made that, in the aggregate, they show evidence of lack of carcinogenicity. Such a judgement requires first of all that the studies giving rise to it meet, to a sufficient degree, the standards of design and analysis described above. Specifically, the possibility that bias, confounding or misclassification of exposure or outcome could explain the observed results should be considered and excluded with reasonable certainty. In addition, all studies that are judged to be methodologically sound should be consistent with a relative risk of unity for any observed level of exposure to the agent and, when considered together, should provide a pooled estimate of relative risk which is at or near unity and has a narrow confidence interval, due to sufficient population size. Moreover, no individual study nor the pooled results of all the studies should show any consistent tendency for relative risk of cancer to increase with increasing level of exposure to the agent. It is important to note that evidence of lack of carcinogenicity obtained in this way from several epidemiological studies can apply only to the type(s) of cancer studied and to dose levels of the agent and intervals between first exposure to it and observation of disease that are the same as or less than those observed in all the studies. Experience with human cancer indicates that, for some agents, the period from first exposure to the development of clinical cancer is seldom less than 20 years; latent periods substantially shorter than 30 years cannot provide evidence for lack of carcinogenicity.

12. SUMMARY OF DATA REPORTED

In this section, the relevant experimental and epidemiological data are summarized. Only reports, other than in abstract form, that meet the criteria outlined on p. 17 are considered for evaluating carcinogenicity. Inadequate studies are generally not summarized: such studies are usually identified by a square–bracketed comment in the text.

(a) Exposures

Human exposure is summarized on the basis of elements such as production, use, occurrence in the environment and determinations in human tissues and body fluids. Quantitative data are given when available.

(b) Experimental carcinogenicity data

Data relevant to the evaluation of carcinogenicity of the agent in animals are summarized. For each animal species and route of administration, it is stated whether an increased incidence of neoplasms was observed, and the tumour sites are indicated. If the agent produced tumours after prenatal exposure or in single–dose experiments, this is also indicated. Dose–response and other quantitative data may be given when available. Negative findings are also summarized.

(c) Human carcinogenicity data

Results of epidemiological studies that are considered to be pertinent to an assessment of human carcinogenicity are summarized. When relevant, case reports and correlation studies are also considered.

(d) Other relevant data

Structure–activity correlations are mentioned when relevant.

Toxicological information and data on kinetics and metabolism in experimental animals are given when considered relevant. The results of tests for genetic and related effects are summarized for whole mammals, cultured mammalian cells and nonmammalian systems.

Data on other biological effects in humans of particular relevance are summarized. These may include kinetic and metabolic considerations and evidence of DNA binding, persistence of DNA lesions or genetic damage in humans exposed to the agent.

When available, comparisons of such data for humans and for animals, and particularly animals that have developed cancer, are described.

13. EVALUATION

Evaluations of the strength of the evidence for carcinogenicity arising from human and experimental animal data are made, using standard terms.

It is recognized that the criteria for these evaluations, described below, cannot encompass all of the factors that may be relevant to an evaluation of the carcinogenicity of an agent. In considering all of the relevant data, the Working Group may assign the agent to a higher or lower category than a strict interpretation of these criteria would indicate.

(a) Degrees of evidence for carcinogenicity in humans and in experimental animals and supporting evidence

It should be noted that these categories refer only to the strength of the evidence that these agents are carcinogenic and not to the extent of their carcinogenic activity (potency) nor to the mechanism involved. The classification of some agents may change as new information becomes available.

(i) *Human carcinogenicity data*

The evidence relevant to carcinogenicity from studies in humans is classified into one of the following categories:

Sufficient evidence of carcinogenicity: The Working Group considers that a causal relationship has been established between exposure to the agent and human cancer. That is, a positive relationship has been observed between the exposure to the agent and cancer in studies in which chance, bias and confounding could be ruled out with reasonable confidence.

Limited evidence of carcinogenicity: A positive association has been observed between exposure to the agent and cancer for which a causal interpretation is considered by the Working Group to be credible, but chance, bias or confounding could not be ruled out with reasonable confidence.

Inadequate evidence of carcinogenicity: The available studies are of insufficient quality, consistency or statistical power to permit a conclusion regarding the presence or absence of a causal association.

Evidence suggesting lack of carcinogenicity: There are several adequate studies covering the full range of doses to which human beings are known to be exposed, which are mutually consistent in not showing a positive association between exposure to the agent and any studied cancer at any observed level of exposure. A conclusion of 'evidence suggesting lack of carcinogenicity' is inevitably limited to the cancer sites, circumstances and doses of exposure and length of observation covered by the available studies. In addition, the possibility of a very small risk at the levels of exposure studied can never be excluded.

In some instances, the above categories may be used to classify the degree of evidence for carcinogenicity of the agent for specific organs or tissues.

(ii) *Experimental carcinogenicity data*

The evidence relevant to carcinogenicity in experimental animals is classified into one of the following categories:

Sufficient evidence of carcinogenicity: The Working Group considers that a causal relationship has been established between the agent and an increased incidence of malignant neoplasms or of an appropriate combination of benign and malignant neoplasms (as described on p. 21) in (a) two or more species of animals or (b) in two or more independent studies in one species carried out at different times or in different laboratories or under different protocols.

Exceptionally, a single study in one species might be considered to provide sufficient evidence of carcinogenicity when malignant neoplasms occur to an unusual degree with regard to incidence, site, type of tumour or age at onset.

In the absence of adequate data on humans, it is biologically plausible and prudent to regard agents for which there is *sufficient evidence* of carcinogenicity in experimental animals as if they presented a carcinogenic risk to humans.

Limited evidence of carcinogenicity: The data suggest a carcinogenic effect but are limited for making a definitive evaluation because, e.g., (a) the evidence of carcinogenicity is restricted to a single experiment; or (b) there are unresolved questions regarding the adequacy of the design, conduct or interpretation of the study; or (c) the agent increases the incidence only of benign neoplasms or lesions of uncertain neoplastic potential, or of certain neoplasms which may occur spontaneously in high incidences in certain strains.

Inadequate evidence of carcinogenicity: The studies cannot be interpreted as showing either the presence or absence of a carcinogenic effect because of major qualitative or quantitative limitations.

Evidence suggesting lack of carcinogenicity: Adequate studies involving at least two species are available which show that, within the limits of the tests used, the agent is not carcinogenic. A conclusion of evidence suggesting lack of carcinogenicity is inevitably limited to the species, tumour sites and doses of exposure studied.

(iii) *Supporting evidence of carcinogenicity*

The other relevant data judged to be of sufficient importance as to affect the making of the overall evaluation are indicated.

(*b*) *Overall evaluation*

Finally, the total body of evidence is taken into accout; the agent is described according to the wording of one of the following categories, and the designated group is given. The categorization of an agent is a matter of scientific judgement, reflecting the strength of the evidence derived from studies in human and in experimental animals and from other relevant data.

Group 1 – The agent is carcinogenic to humans.

This category is used only when there is *sufficient evidence* of carcinogenicity in humans.

Group 2

This category includes agents for which, at one extreme, the degree of evidence of carcinogenicity in humans is almost sufficient, as well as those for which, at the other extreme, there are no human data but for which there is experimental evidence of carcinogenicity. Agents are assigned to either 2A (probably carcinogenic) or 2B (possibly carcinogenic) on the basis of epidemiological, experimental and other relevant data.

Group 2A – The agent is probably carcinogenic to humans.

This category is used when there is *limited evidence* of carcinogenicity in humans and *sufficient evidence* of carcinogenicity in experimental animals. Exceptionally, an agent may be classified into this category solely on the basis of *limited evidence* of carcinogenicity in humans or of *sufficient evidence* of carcinogenicity in experimental animals strengthened by supporting evidence from other relevant data.

Group 2B – The agent is possibly carcinogenic to humans.

This category is generally used for agents for which there is *limited evidence* in humans in the absence of *sufficient evidence* in experimental animals. It may also be used when there is *inadequate evidence* of carcinogenicity in humans or when human data are nonexistent but there is *sufficient evidence* of carcinogenicity in experimental animals. In some instances, an agent for which there is *inadequate evidence* or no data in humans but *limited evidence* of carcinogenicity in experimental animals together with supporting evidence from other relevant data may be placed in this group.

Group 3 – The agent is not classifiable as to its carcinogenicity to humans.

Agents are placed in this category when they do not fall into any other group.

Group 4 – The agent is probably not carcinogenic to humans.

This category is used for agents for which there is *evidence suggesting lack of carcinogenicity* in humans together with *evidence suggesting lack of carcinogenicity* in experimental animals. In some circumstances, agents for which there is *inadequate evidence* of or no data on carcinogenicity in humans but *evidence suggesting lack of carcinogenicity* in experimental animals, consistently and strongly supported by a broad range of other relevant data, may be classified in this group.

References

1. IARC (1987) *IARC Monographs on the Evaluation of Carcinogenic Risks to Humans*, Supplement 6, *Genetic and Related Effects: An Updating of Selected* IARC Monographs *from Volumes 1 to 42*, Lyon

2. IARC (1977) *IARC Monographs Programme on the Evaluation of the Carcinogenic Risk of Chemicals to Humans. Preamble (IARC intern. tech. Rep. No. 77/002)*, Lyon

3. IARC (1978) *Chemicals with* Sufficient Evidence *of Carcinogenicity in Experimental Animals* – IARC Monographs *Volumes 1–17 (IARC intern. tech. Rep. No. 78/003)*, Lyon

4. IARC (1979) *Criteria to Select Chemicals for* IARC Monographs *(IARC intern. tech. Rep. No. 79/003)*, Lyon

5. IARC (1982) *IARC Monographs on the Evaluation of the Carcinogenic Risk of Chemicals to Humans*, Supplement 4, *Chemicals, Industrial Processes and Industries Associated with Cancer in Humans (IARC Monographs, Volumes 1 to 29)*, Lyon

6. IARC (1983) *Approaches to Classifying Chemical Carcinogens According to Mechanism of Action (IARC intern. tech. Rep. No. 83/001)*, Lyon

7. IARC (1987) *IARC Monographs on the Evaluation of Carcinogenic Risks to Humans*, Supplement 7, *Overall Evaluations of Carcinogenicity: An Updating of* IARC Monographs *Volumes 1 to 42*, Lyon

8. IARC (1973–1988) *Information Bulletin on the Survey of Chemicals Being Tested for Carcinogenicity*, Numbers 1–13, Lyon

   Number 1 (1973)   52 pages
   Number 2 (1973)   77 pages
   Number 3 (1974)   67 pages
   Number 4 (1974)   97 pages
   Number 5 (1975)   88 pages
   Number 6 (1976)   360 pages
   Number 7 (1978)   460 pages
   Number 8 (1979)   604 pages
   Number 9 (1981)   294 pages
   Number 10 (1983) 326 pages
   Number 11 (1984) 370 pages
   Number 12 (1986) 385 pages
   Number 13 (1988) 404 pages

9. Muir, C. & Wagner, G., eds (1977–88) *Directory of On–going Studies in Cancer Epidemiology 1977–88 (IARC Scientific Publications)*, Lyon, IARC

10. IARC (1984) *Chemicals and Exposures to Complex Mixtures Recommended for Evaluation in* IARC Monographs *and Chemicals and Complex Mixtures Recommended for Long–term Carcinogenicity Testing (IARC intern. tech. Rep. No. 84/002)*, Lyon

11. *Environmental Carcinogens. Selected Methods of Analysis:*

     Vol. 1. *Analysis of Volatile Nitrosamines in Food (IARC Scientific Publications No. 18)*. Edited by R. Preussmann, M. Castegnaro, E.A. Walker & A.E. Wasserman (1978)

     Vol. 2. *Methods for the Measurement of Vinyl Chloride in Poly(vinyl chloride), Air, Water and Foodstuffs (IARC Scientific Publications No. 22)*. Edited by D.C.M. Squirrell & W. Thain (1978)

     Vol. 3. *Analysis of Polycyclic Aromatic Hydrocarbons in Environmental Samples (IARC Scientific Publications No. 29)*. Edited by M. Castegnaro, P. Bogovski, H. Kunte & E.A. Walker (1979)

     Vol. 4. *Some Aromatic Amines and Azo Dyes in the General and Industrial Environment (IARC Scientific Publications No. 40)*. Edited by L. Fishbein, M. Castegnaro, I.K. O'Neill & H. Bartsch (1981)

     Vol. 5. *Some Mycotoxins (IARC Scientific Publications No. 44)*. Edited by L. Stoloff, M. Castegnaro, P. Scott, I.K. O'Neill & H. Bartsch (1983)

     Vol. 6. N–*Nitroso Compounds (IARC Scientific Publications No. 45)*. Edited by R. Preussmann, I.K. O'Neill, G. Eisenbrand, B. Spiegelhalder & H. Bartsch (1983)

     Vol. 7. *Some Volatile Halogenated Hydrocarbons (IARC Scientific Publications No. 68)*. Edited by L. Fishbein & I.K. O'Neill (1985)

     Vol. 8. *Some Metals: As, Be, Cd, Cr, Ni, Pb, Se, Zn (IARC Scientific Publications No. 71)*. Edited by I.K. O'Neill, P. Schuller & L. Fishbein (1986)

     Vol. 9. *Passive Smoking (IARC Scientific Publications No. 81)*. Edited by I.K. O'Neill, K.D. Brunnemann, B. Dodet & D. Hoffmann (1987)

     Vol. 10. *Benzene and Alkylated Benzenes (IARC Scientific Publications No. 85)*. Edited by L. Fishbein & I.K. O'Neill (1988)

12. Wilbourn, J., Haroun, L., Heseltine, E., Kaldor, J., Partensky, C. & Vainio, H. (1986) Response of experimental animals to human carcinogens: an analysis based upon the IARC Monographs Programme. *Carcinogenesis, 7*, 1853–1863

13. Montesano, R., Bartsch, H., Vainio, H., Wilbourn, J. & Yamasaki, H., eds (1986) *Long-term and Short-term Assays for Carcinogenesis – A Critical Appraisal (IARC Scientific Publications No. 83)*, Lyon, IARC

14. Hoel, D.G., Kaplan, N.L. & Anderson, M.W. (1983) Implication of nonlinear kinetics on risk estimation in carcinogenesis. *Science, 219*, 1032–1037

15. Gart, J.J., Krewski, D., Lee, P.N., Tarone, R.E. & Wahrendorf, J. (1986) *Statistical Methods in Cancer Research*, Vol. 3, *The Design and Analysis of Long-term Animal Experiments (IARC Scientific Publications No. 79)*, Lyon, IARC

16. Peto, R., Pike, M.C., Day, N.E., Gray, R.G., Lee, P.N., Parish, S., Peto, J., Richards, S. & Wahrendorf, J. (1980) *Guidelines for simple, sensitive significance tests for carcinogenic effects in long-term animal experiments*. In: *IARC Monographs on the Evaluation of the Carcinogenic Risk of Chemicals to Humans*, Supplement 2, *Long-term and Short-term Screening Assays for Carcinogens: A Critical Appraisal*, Lyon, IARC, pp. 311–426

17. Breslow, N.E. & Day, N.E. (1980) *Statistical Methods in Cancer Research*, Vol. 1, *The Analysis of Case-control Studies (IARC Scientific Publications No. 32)*, Lyon, IARC

18. Breslow, N.E. & Day, N.E. (1987) *Statistical Methods in Cancer Research*, Vol. 2, *The Design and Analysis of Cohort Studies (IARC Scientific Publications No. 82)*, Lyon, IARC

GENERAL REMARKS

This forty–seventh volume of the *IARC Monographs* comprises six monographs on organic solvents, one on a solvent stabilizer, three on resin monomers and modifiers, two on pigments and one on occupational exposures in paint manufacture and painting.

Of the solvents, petroleum solvents, toluene and xylene are used primarily in paints. Cyclohexanone is used primarily as an intermediate in the production of nylon, although it may be used in some paint products. Dimethylformamide is a polymer and resin solvent, and morpholine is used mainly as an intermediate in the production of rubber chemicals and as a corrosion inhibitor. 1,2–Epoxybutane is used as a solvent stabilizer, particularly in trichloroethylene. Of the resin monomers, bis(2,3–epoxycyclopentyl)ether was used as a resin modifier. Glycidyl ethers have been used similarly; bisphenol A diglycidyl ether is a basic component of many epoxy paints, which polymerizes when the uncured resin is mixed with a curing agent. Phenol was a common resin monomer used in paints and lacquers when phenolformaldehyde resins were in wide use; such resins are still used to glue plywood and other wood products, but other resins have replaced them for the most part in modern paint products. Antimony trioxide is used mainly in fire retardants; both antimony trioxide and antimony trisulfide are used as paint pigments. Titanium dioxide is the most common white pigment used in paints.

The monograph on occupational exposures in paint manufacture and painting covers four broad categories of industry: manufacture of paints and related products; construction painting; painting and related operations in the furniture industry; and painting in the metal industry, including painting of cars and other vehicles. 'Painting' includes lacquering, varnishing and paint removal; preparation is also involved in painting activities. The few studies on radium dial painting were excluded because this activity is not usually considered as one of the painting trades. The carcinogenic risks of paint manufacture and of painting have not been evaluated previously; however, chemical agents to which employees working in painting trades may be exposed have been, and the evaluations of these compounds are presented in Table 1.

Only limited information is available about the extent of exposure to the agents covered by the present volume. The numbers of potentially exposed workers cited in the individual monographs are estimations based on two field studies conducted by the National Institute for Occupational Safety and Health (1974, 1983) in the USA. These estimates depend on the methods used in the two surveys and cannot be extrapolated to provide worldwide figures.

The monograph on petroleum solvents covers marketable hydrocarbon mixtures derived from petroleum that are used mainly as solvents. It supplements three previous

Table 1. Agents encountered in the painting trades that have been evaluated for carcinogenicity in *IARC Monographs* Volumes 1–47

Agent[a]	Degree of evidence for carcinogenicity[b]		Overall evaluation[b]	IARC Vol. (year)
	Human	Animal		
Acrylonitrile	L	S	2A	19 (1979)
Antimony trioxide	I	S	2B	47 (1989)
Antimony trisulfide	I	L	3	47 (1989)
Arsenic and arsenic compounds	S	L	1[c]	23 (1980)
Asbestos	S	S	1	14 (1977)
Attapulgite	I	L	3	42 (1987)
Benzene	S	S	1	29 (1982)
Benzoyl peroxide	I	I	3	36 (1985)
Bisphenol A diglycidyl ether	ND	L	3	47 (1989)
Cadmium and cadmium compounds	L	S	2A	11 (1976)
Carbon blacks	I	I	3	33 (1984)
Carbon–black extracts		S	2B	
Carbon tetrachloride	I	S	2B	20 (1979)
Chlorophenols	L		2B	41 (1981)
Pentachlorophenol		I		20 (1979)
Chromium, hexavalent compounds	S	S	1[c]	23 (1980)
Coal–tars	S	S	1	35 (1983)
Cyclohexanone	ND	I	3	47 (1989)
ortho–Dichlorobenzene	I	I	3	29 (1982)
1,2–Dichloroethane	ND	S	2B	20 (1979)
Dichloromethane	I	S	2B	41 (1981)
Di(2–ethylhexyl)phthalate	ND	S	2B	29 (1982)
Dimethylformamide	L	I	2B	47 (1989)
Epichlorohydrin	I	S	2A	11 (1976)
1,2–Epoxybutane	ND	L	3	47 (1987)
Ethyl acrylate	ND	S	2B	39 (1986)
Formaldehyde	L	S	2A	29 (1982)
Gasoline	I		2B	45 (1989)
Unleaded automotive gasoline		L		
Kerosene, straight–run and hydrotreated		L		45 (1989)
Lead and lead compounds (inorganic)	I	S	2B	23 (1980)
Melamine	ND	I	3	39 (1986)
4,4'–Methylenedianiline	ND	S	2B	39 (1986)
Methyl methacrylate	ND	I	3	19 (1979)
Nickel and nickel compounds	S	S	1[c]	11 (1976)
2–Nitropropane	ND	S	2B	29 (1982)
Petroleum solvents	I		3	47 (1989)
High–boilding aromatic solvents		I		
Special boiling–range solvents		ND		
White spirits		ND		

Table 1 (contd)

Agent[a]	Degree of evidence for carcinogenicity[b]		Overall evaluation[b]	IARC Vol. (year)
	Human	Animal		
meta-Phenylenediamine	ND	I	3	16 (1978)
Polychlorinated biphenyls	L	S	2A	18 (1978)
Polycyclic aromatic hydrocarbons	ND	I/L/S	3/2B/2A	32 (1983)
Polyvinyl acetate	ND	I	3	19 (1979)
Silica, crystalline	L	S	2A	42 (1987)
Styrene	I	L	2B	19 (1979)
Styrene–butadiene copolymers	ND	ND	3	19 (1979)
Styrene oxide	ND	S	2A	19 (1979)
Talc				42 (1987)
Not containing asbestiform fibres	I	I	3	
Containing asbestiform fibres	S	I	1	
Titanium dioxide	I	L	3	47 (1989)
Toluene	I	I	3	47 (1989)
Toluene diisocyanates	ND	S	2B	39 (1986)
1,1,1-Trichloroethane	ND	I	3	20 (1979)
Trichloroethylene	I	L	3	20 (1979)
Vinyl acetate	ND	I	3	39 (1986)
Xylene	I	I	3	47 (1989)

[a]Based on information provided in the monograph on paint manufacture and painting

[b]Based on *IARC Monographs* Suppl. 7 (IARC, 1987) and Volumes 45–47 (IARC, 1989a,b,c); ND, no adequate data; I, inadequate evidence; L, limited evidence; S, sufficient evidence. For definitions of the degrees of evidence and overall evaluations, see Preamble, pp. 27–30.

[c]The evaluation applies to the group of chemicals as a whole and not necessarily to all individual chemicals within the group.

IARC Monographs on petroleum products, which covered petroleum refining and petroleum fuels (IARC, 1989a), bitumens (IARC, 1985a) and mineral oils (IARC, 1984). The monograph on petroleum solvents excludes substances (e.g., toluene, xylenes, benzene and *n*-hexane) that are not generally classified as hydrocarbon mixtures, and it does not include hydrocarbon mixtures such as gasoline, jet fuel and kerosene, which were evaluated in 1988 (IARC, 1989a) and are used primarily as fuels and not as solvents.

The nomenclature and classification of petroleum solvents are not well defined and are not standardized throughout the world. The grouping adopted in this volume is common in Europe and is based on their boiling ranges and solvent strengths. However, the same product may be referred to by many names, and the same name may be used to refer to two very different products. Section 1 of the monograph on petroleum solvents lists commonly used names and synonyms and provides basic physicochemical data that may be used in interpreting the results of studies in other sections. In the sections on experimental and human studies of petroleum solvents, the name of the solvent given by the authors has been used.

Glycidyl ethers are used mainly as components and reactive modifiers of epoxy resins. Those considered in this volume were chosen on the basis that there is at least one study on carcinogenicity available. Diglycidyl resorcinol ether was evaluated previously (IARC, 1985b). Glycidyl ethers that are produced in high or moderate quantities and for which data on genetic or related effects were available were also considered. Bisphenol A diglycidyl ether is an epoxy compound that can be polymerized to a thermosetting resin and can therefore be considered an uncured epoxy resin. Concentrations of bisphenol A diglycidyl ether in epoxy products vary widely.

There was, in general, a paucity of studies of carcinogenicity in experimental animals for some of the most commonly used solvents, which are some of the most widely used chemicals worldwide. This problem is highlighted in the case of petroleum solvents: not a single publication on white spirits was available. A further difficulty was encountered in that the materials used in many investigations may have been commercial substances of variable composition.

Interpretation of epidemiological studies of carcinogenicity and of other toxic and genetic effects for solvents is complicated by the fact that many occupational groups (e.g., painters and lacquerers) are exposed not only to many solvents but also to other agents, as in many other occupations. When the exposure of a population was estimated quantitatively for a specific solvent, or when the agent was stated to be the main or one of the major solvents to which the study population was exposed, the study was included in the monograph on that agent. For sections on epidemiological studies of carcinogenicity in humans, studies were also allocated to different monographs solely on the basis of whether the substance was named as an exposure in the paper, regardless of the method of identifying the exposure or analysing the data. Some studies (e.g., on painters) were used in more than one monograph, and the same study often appears in many sections, giving the relevant outcomes of the same individuals but associated with different exposures. These are identified in the text. The studies now summarized in the monographs on some petroleum solvents, toluene and xylene are those in which the specific agents being considered are mentioned. They represent, however, selected studies in which details are given that suggest that exposure to more than one solvent is usual and that it would be difficult to separate out the effects of any single agent. Exposures that are specified in epidemiological studies of cancer have generally been estimated by an industrial hygienist on the basis of presumed exposures given in work histories or as mentioned by the subjects; the latter gives rise to serious problems of recall bias.

The studies of cancers and other effects in the children of exposed parents present particular problems, since risk may be due to exposures of the mother or father occurring before, during or after the pregnancy. Such timing is usually poorly identified and varies between studies, making comparison difficult.

The majority of the epidemiological studies included in the last monograph are on 'painters' described as such. Details of the length and extent of exposure and of the specific type of work done were rarely given or analysed separately. Data from cross-sectional studies of large numbers of deaths were used because painting is well identified as a trade; however, it should be noted that the occupation as stated on a death certificate is not necessarily comparable to the census data that are used to derive expected numbers. In many case–control studies of specific cancer sites, painting was one of a list of occupational exposures considered; all studies in which findings in painters were mentioned have been included here, but there is no way of knowing how many other studies enquired about painting but did not present the results. It should also be noted that reports of 'exposure to paint' may include nonprofessional use of paints.

References

IARC (1984) *IARC Monographs on the Evaluation of the Carcinogenic Risk of Chemicals to Humans*, Vol. 33, *Polynuclear Aromatic Compounds, Part 2, Carbon Blacks, Mineral Oils and Some Nitroarene Compounds*, Lyon, pp. 87–168

IARC (1985a) *IARC Monographs on the Evaluation of the Carcinogenic Risk of Chemicals to Humans*, Vol. 35, *Polynuclear Aromatic Compounds, Part 4, Bitumens, Coal–tar and Derived Products, Shale–oils and Soots*, Lyon, pp. 39–81

IARC (1985b) *IARC Monographs on the Evaluation of the Carcinogenic Risk of Chemicals to Humans*, Vol. 36, *Allyl Compounds, Aldehydes, Epoxides and Peroxides*, Lyon, pp. 181–188

IARC (1987) *IARC Monographs on the Evaluation of Carcinogenic Risks to Humans*, Suppl. 7, *Overall Evaluations of Carcinogenicity: An Updating of* IARC Monographs *Volumes 1 to 42*, Lyon

IARC (1989a) *IARC Monographs on the Evaluation of Carcinogenic Risks to Humans*, Vol. 45, *Occupational Exposures in Petroleum Refining; Crude Oil and Major Petroleum Fuels*, Lyon

IARC (1989b) *IARC Monographs on the Evaluation of Carcinogenic Risks to Humans*, Vol. 46, *Diesel and Gasoline Engine Exhausts and Some Nitroarenes*, Lyon

IARC (1989c) *IARC Monographs on the Evaluation of Carcinogenic Risks to Humans*, Vol. 47, *Some Organic Solvents, Resin Monomers and Related Compounds, Pigments and Occupational Exposures in Paint Manufacture and Painting*, Lyon

National Institute for Occupational Safety and Health (1974) *National Occupational Hazard Survey 1972–1974*, Cincinnati, OH

National Institute for Occupational Safety and Health (1983) *National Occupational Exposure Survey 1981–1983*, Cincinnati, OH

ORGANIC SOLVENTS

THE MONOGRAPHS

SOME PETROLEUM SOLVENTS[1]

1. Chemical and Physical Data[2]

1.1 Synonyms

The petroleum solvents covered in this monograph are all complex mixtures of hydro-carbons produced from petroleum distillation fractions which have been further refined by one or more process. Unlike petroleum fuels (IARC, 1989), these mixtures are not produced by blending various refinery process streams but rather by further refining and distilling one or two streams to make a product significantly different in chemical composition and much narrower in boiling range (typically 15–30°C) than its source streams.

Petroleum solvents are generally defined and differentiated by certain physical proper-ties and by chemical composition. Among the properties of importance, depending on the intended end–use of the solvent, are boiling range, flash–point, solvent strength (solvency), colour, odour, aromatic content and sulfur content. Some of these properties are interre-lated; for example, increasing the aromatic content often increases solvency, but also in-creases odour. In addition, the same property may be measured in different ways in different industries or in different geographic regions; for example, kauri–butanol value (the amount of petroleum solvent in millilitres required to cause cloudiness in a solution of kauri gum and n–butanol) and aniline point (the minimal temperature for complete mixing of aniline with a petroleum solvent) are both indicators of solvency but kauri–butanol value is more common-ly used in Europe and aniline point more commonly in the USA.

It is not surprising, then, that the same names may be used for solvents that are not identical, and that different classification systems or names may be more common in one industry or geographic region than another. 'White spirits', for example, is a common term in Europe but is seldom used in the USA. The following paragraphs provide a brief overview of

---

[1]Saleable, petroleum–derived hydrocarbon mixtures used principally as solvents
[2]Much of the material presented in this section and in sections 2.1 and 2.3 is taken from reports prepared by CONCAWE (the oil companies' European organization for environmental and health protection), in collabora-tion with the European Chemical Industry Ecology and Toxicology Centre, and the American Petroleum Institute; in these cases, no reference is given.

the classification system adopted for this monograph. The vagaries in the names used for petroleum solvents, however, make it important that some physical/chemical properties be included in any discussion of their toxic effects.

The petroleum solvents covered in this monograph are grouped in three classes generally based on volatility and aromatic content (which is related to solvency). These classes (special boiling range solvents, white spirits and high–boiling aromatic solvents) are widely adopted industrial categories. Figure 1 shows the general relationship among these three broad categories of petroleum solvents, with respect to carbon number, boiling range and solvency.

Common names for various products

Special boiling-range solvents: Benzine; canadol; essence; high–boiling petroleum ether; lacquer diluent; light ligroin; ligroin; naphtha; naphtha 76; petroleum benzin; petroleum ether; refined solvent naphtha; rubber solvent; SBP; special boiling–point solvents; special naphtholite; spezialbenzine; varnish makers' and painters' naphtha; VM & P naphtha [CAS No. 8030–30–6 (National Institute for Occupational Safety and Health); 8032–32–4 (American Conference of Govermental Industrial Hygienists)]

White spirits: DAWS; dearomatized white spirits; 140 flash solvent; HAWS; high aromatic white spirits; kristalloel; lacquer petrol; LAWS; light petrol; low aromatic white spirits; mineral solvent; mineral spirits; mineral turpentine; odourless mineral spirit; petroleum spirits; solvent naphtha; Stoddard solvent [CAS No. 8052–41–3]; terpentina; turpentine substitute

High–boiling aromatic solvents: Naphtha

1.2   Chemical and physical properties

Calculation of conversion factors for converting air concentrations expressed in parts per million to milligrams per cubic metre requires knowledge of the molecular weight of a chemical. Since the molecular weights of complex and variable mixtures such as petroleum solvents cannot be specified, conversion factors have not been included in this monograph. It is noted, however, that, for practical purposes, others have used average molecular weights to generate approximate conversion factors for certain well–defined petroleum solvents (American Conference of Governmental Industrial Hygienists, 1988).

Undated references in parentheses used in sections 1.2 and 1.3 are to national and international specifications as follows: ASTM, American Society for Testing and Materials; BS, British Standards; DIN, Deutsche Industrie–Norm (German Industrial Standard); ISO; International Standards Organization; NF, National Formulary (USA).

(a)   Special boiling-range solvents

Three main types of special boiling-range solvents may be distinguished with regard to production processes:

Type 1 – hydrodesulfurized special boiling-range solvents, which can contain up to 20 wt% aromatic compounds;

Fig. 1. Relationships among the three categories of petroleum solvents[a]

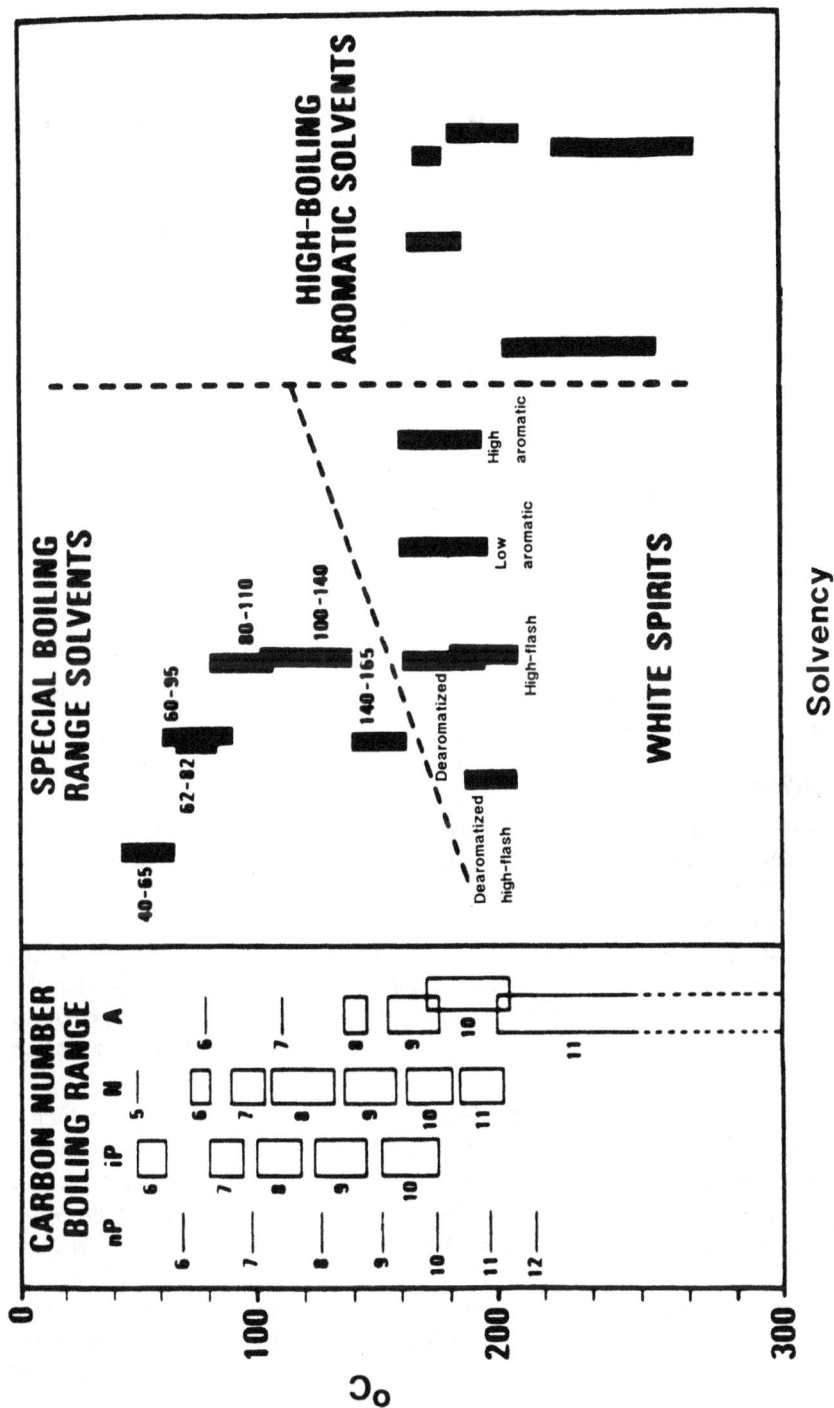

[a]nP, normal paraffins; iP; isoparaffins; N, naphthenic compounds; A, alkyl benzenes

Type 2 – hydrogenated special boiling–range solvents, in which the aromatic compounds have been converted to alicyclic hydrocarbons and the aromatic content is less than 0.02 wt%; and

Type 3 – hydrogenated special boiling–range solvents, to which aromatic compounds (often toluene) are added and which may have an aromatic content of up to 50 wt%.

- (i)   *Description*: Clear, colourless hydrocarbon solvent
- (ii) *Boiling range*: 30–160°C (ASTM D1078, ASTM D850, ASTM D86–82, DIN 51751, NF MO7–002)
- (iii) *Density*: 0.670–0.760 at 15°C (ASTM D1298, ASTM D3505, ASTM D4052–81, ASTM D891, ASTM D1250, DIN 51757, NF T60–101)
- (iv) *Refractive index*: 1.37–1.42 at 20°C (ASTM D1218–82, DIN 53169)
- (v)   *Solubility*: Less than 0.1 wt% in water
- (vi) *Viscosity*: 0.3–0.75 cps at 25°C (ASTM D445–83, DIN 51562, DIN 53015)
- (vii) *Volatility*: 19–0.6; relative evaporation rate, *n*–butyl acetate = 1 (ASTM D3539, DIN 53170)
- (viii) *Reactivity*: Reacts with strong oxidizing agents
- (ix) *Kauri–butanol value*: 30–36 (ASTM D1133)
- (x)   *Aniline point*: 60–64°C (ASTM D611–82, DIN 51775, NF MO7–021)
- (xi) *Flash–point*: <0–32°C (ASTM D56, ASTM D93–80, DIN 51755)
- (xii) *Colour (with a Saybolt colorimeter)*: +30 (ASTM D156–82, DIN 51411)
- (xiii) *Carbon number range*: 4–11

(b)   *White spirits*

The nomenclature of hydrocarbon solvents produced as distillates in the normal refining of petroleum and used, for example, in the coatings, paints and dry–cleaning industries, is vague and confusing. Commercial names may not therefore indicate composition. In Europe, the name encountered most often is 'white spirits', an alternative synonym being 'mineral solvent'. In the USA, the name used most commonly for these solvents is 'mineral spirits'. 'Mineral spirits: type 1 – regular', as specified by ASTM D235, is 'Stoddard solvent' which is similar in composition to the product generally recognized in Europe as white spirits. The terms 'solvent naphtha' and 'turpentine substitute' are used to describe the same materials.

The solvent generally recognized in Europe as white spirits, and that to which the general public is most likely to be exposed in its applications as a paint thinner and brush cleaner, is a petroleum distillate that boils typically in the range 150–205°C with a typical aromatic content of 15–25%. However, products of this composition in the UK are deemed to be 'low aromatics', whereas in France a 'low aromatic white spirit' is one with an aromatic content of 5% or less. In the USA, the latter would be 'type 3 – odourless mineral spirits' (ASTM D235), which has a maximal kauri–butanol value of 29 rather than a limit on aromatic content. Some white spirits of even lower aromatic content (typically less than 1%) are produced by treatment with hydrogen (hydrogenation) in the presence of a catalyst. 'High aromatic white spirit', with a typical aromatic content of 45%, is both specified (ISO 1250, BS 245) and commercially available in Europe.

Three main types of white spirits may be distinguished with regard to production processes:

Type 1 –   hydrodesulfurized white spirits, which generally contain 15–25 wt% of aromatic compounds but may contain up to 45 wt%; in the USA, the aromatic content of such solvents rarely exceeds 15–16% and is often much lower;

Type 2 –   solvent–extracted hydrodesulfurized white spirits containing 3–5 wt% aromatic compounds; and

Type 3 –   hydrogenated white spirits, in which the aromatic hydrocarbons have been converted to alicyclic hydrocarbons and which have an aromatic content of less than 1 wt%.

A further regulated group of products in the USA consists of a blend of the three types of white spirits to meet a maximum of not more than 8 wt% aromatic compounds.

(i)   *Description*: Clear, colourless hydrocarbon solvent

(ii)   *Boiling range*: 130–220°C (ASTM D1078, ASTM D850, ASTM D86, DIN 51751, NF M07–002)

(iii)   *Density*: 0.750–0.797 at 15°C (ASTM D1298, ASTM D3505, ASTM D4052, ASTM D891, ASTM D1250, DIN 51757, NF T60–101)

(iv)   *Refractive index*: 1.41–1.44 at 20°C (ASTM D1218–82, DIN 53169)

(v)   *Solubility*: Less than 0.1 wt% in water

(vi)   *Viscosity*: 0.74–1.65 cps at 25°C (ASTM D445–83, DIN 51562, DIN 53015)

(vii)   *Volatility*: 0.5–<0.01; relative evaporation rate, *n*–butyl acetate = 1 (ASTM D3539, DIN 53170)

(viii) *Reactivity*: Reacts with strong oxidizing agents

(ix)   *Kauri–butanol value*: 29–33 (ASTM D113)

(x)   *Aniline point*: 60–75°C (ASTM D611–82, DIN 51775, NF MO7–021)

(xi)   *Flash–point*: 25–80°C (ASTM D56, ASTM D93–80, DIN 51755)

(xii)   *Colour (with a Saybolt colorimeter)*: +30 (ASTM D156–82, DIN 51411)

(xiii) *Carbon number range*: 7–12

(c)   *High–boiling aromatic solvents*

High–boiling aromatic solvents are complex mixtures of aromatic hydrocarbons with carbon numbers from 8 to 16 and boiling typically in the range of 160–300°C, with lower boiling components in some formulations. The total concentration of aromatic compounds is usually greater than 80% but can be as high as 99%, with the remaining constituents comprising a mixture of aliphatics within the same boiling range. Type 1 high–boiling aromatic solvents are catalytically reformed, whereas Type 2 are processed by solvent extraction.

(i)   *Description*: Clear, colourless hydrocarbon solvent

(ii)   *Boiling-range*: 160–300°C (ASTM D1078, ASTM D850, ASTM D86, DIN 51751, NF M07–002)

(iii) *Density*: 0.879–0.999 at 15°C (ASTM D1298, ASTM D3505, ASTM D4052, ASTM D891, ASTM D1250, DIN 51757, NF T60–101)

(iv)   *Refractive index*: 1.5–1.6 at 20°C (ASTM D1218–82, DIN 53169)

(v)   *Solubility*: Less than 0.1 wt% in water

(vi)   *Viscosity* : 0.8–2.6 cps at 25°C (ASTM D445–83, DIN 51562, DIN 53015)

(vii) *Volatility*: 0.21–< 0.01; relative evaporation rate, *n*-butyl acetate = 1 (ASTM D3539, DIN 53170)

(viii) *Reactivity*: Reacts with strong oxidizing agents

(ix)   *Kauri–butanol value*: 85–89 (ASTM D1133)

(x)   *Aniline point*: 12–15°C (ASTM D611–82, DIN 51775, NF MO7–021)

(xi)   *Flash–point*: 45–110°C (ASTM D56, ASTM D93–80, DIN 51755)

(xii) *Colour (with a Saybolt colorimeter)*: + 30 (ASTM D156–82, DIN 51411)

(xiii) *Carbon number range*: 8–16

1.3   Technical products and impurities

There are five major US producers of special boiling–range solvents, white spirits and high–boiling aromatic solvents. Table 1 presents technical data and specifications for representative solvents from these sources. A similar range of products is available from producers in Europe, Japan and elsewhere.

Detailed analyses were reported for several samples of petroleum solvents used in toxicity studies in the USA (Carpenter *et al.*, 1975a,b,c, 1977). The data are summarized in Table 2.

(a)   Special boiling–range solvents

Trade names: Clairsol; Exxsol DSP; Halpasol; Hydrosol; Indusol; Shellsol; Solfina

Some typical technical grades available in Europe are (SBP is special boiling–point and the numbers refer to boiling range): SBP 40/65, SBP 80/100, SBP 100/140, SBP 40/100, SBP 80/110, SBP 100/160, SBP 60/95, SBP 100/120, SBP 145/160. They may be Types 1, 2 or 3 special boiling range solvents, but the majority available today are of Type 2 (hydrogenated). In the past 20–30 years, there has been a pronounced move from nonhydrogenated grades towards hydrogenated (i.e., minimal aromatics) and low *n*-hexane (less than 5 wt%) grades. Low *n*-hexane grades are produced by distillation and/or blending. Type 1 hydrodesulfurized special boiling range solvents typically contain less than 0.5 wt% benzene; hydrogenated grades (Types 2 and 3) contain less than 0.002 wt% benzene.

According to a large Japanese survey of petroleum distillate solvent samples collected in various parts of the country, the average concentrations of *n*-hexane and benzene in low-boiling samples (corresponding approximately to special boiling–range solvents) were 5.1% v/v (range, 0.0–43%) and 0.8% v/v (range, 0.0–3.5%), respectively (Kasahara *et al.*, 1987). Petroleum ether typically has a high concentration of hexanes containing mainly pentanes and hexanes.

Table 1. Technical data and specifications for representative petroleum solvents from US producers[a]

Solvent	Boiling range (°C)	Flash-point (°C)	Aniline point (°C)	Kauri–butanol value	Composition (vol%)			
					Paraffins (aliphatics)	Naphthenes (alicyclics)	Aromatics	Olefins
Special boiling-range solvents								
Petroleum ether	38–60	<–18	61	26	81	17	2	–
Rubber solvent	40–135	–	60	35	66	33	2	–
	42–137	<–18	60	34	82	14	4	–
	42–138	<–18	57	35	65	31	4	0.2
	44–132	<–18	55	32	84	14	2	–
	64–114	<–18	57	34	80	12	9	0.1
VM & P naphtha	118–138	7	52	39	52	37	11	–
	118–149	10	56	56	69	15	15	–
	118–144	10	61	36	79	12	9	–
	119–129	13	54	38	49	42	10	–
	119–130	13	52	39	47	41	13	–
VM & P naphtha	119–137	11	67	32	57	41	2	0.3
Special naphtholite	121–148	12	56	39	43	49	8	–
	122–139	13	56	46	30	64	6	0.1
Low odour	127–142	18	62	35	42	58	<1	–
White spirits/mineral spirits								
Mineral spirits	149–204	41	71	29	56	37	7	–
(low odour)	154–193	39	62	39	48	46	6	–
	156–199	40	62	35	37	60	2	–
	157–196	41	68	31	57	41	2	0.2
	162–193	44	66	33	50	47	4	–
	179–197	53	84	27	86	14	<1	–

Table 1 (contd)

Solvent	Boiling range (°C)	Flash-point (°C)	Aniline point (°C)	Kauri-butanol value	Composition (vol%) Paraffins (aliphatics)	Naphthenes (alicyclics)	Aromatics	Olefins
Stoddard solvent	144–196	35	55	37	42	47	11	0.2
	149–204	38	63	32	67	20	13	–
	153–198	42	56	37	48	37	15	–
	157–196	41	60	36	54	36	10	0.3
	164–195	44	45	43	23	56	21	–
	166–191	46	59	35	46	40	14	–
Stoddard solvent (low odour)	159–197	42	68	33	47	53	<1	–
	159–203	40	66	32	48	45	7	–
140 Flash solvent	182–210	59	73	28	75	19	5	–
	183–199	60	71	31	60	37	3	0.3
	186–208	61	64	33	38	60	3	–
	187–209	60	63	33	55	39	7	–
	190–203	64	72	31	45	55	<1	–
High–boiling aromatic solvents								
A–100	155–173	43	13	92	2	–	98	–
	157–177	43	13	90	1	–	99	–
A–150	182–204	60	16	90	<1	<1	99	–
	183–210	66	16	92	2	–	98	–

aFrom National Paint and Coatings Association (1984)

Table 2. Chemical composition of some petroleum solvents (in vol%)[a]

Compound	Rubber solvent[b]	VM & P naphtha[b]	Stoddard solvent[c]	High-boiling aromatic solvent
Paraffins (aliphatics)	41.5	55.4	47.7	0.3[d]
C_5	0.3	trace	–	
C_6	17.8	0.3	–	
C_7	23.4	4.0	–	
C_8	–	21.7	0.9	
C_9	–	18.8	9.5	
C_{10}	–	9.1	20.6	
C_{11}	–	1.5	13.3	
C_{12}	–	–	3.4	
Naphthenes (alicyclics)	53.6	32.7	37.6	0.8[e]
Monocycloparaffins	53.6	30.3	26.0	0.2
C_6	23.2	0.5	–	
C_7	30.4	7.5	2.4	
C_8	–	13.2	4.3	
C_9	–	6.5	5.0	
C_{10}	–	2.2	8.4	
C_{11}	–	0.4	4.9	
C_{12}	–	–	1.0	
Dicycloparaffins	–	2.4	11.6	
C_9	–	1.3	2.7	
C_{10}	–	1.1	4.7	
C_{11}	–	–	3.2	
C_{12}	–	–	1.0	
Aromatics	4.9	11.9	14.7	98.9
Benzene	1.5	0.1	0.1	
Alkylbenzenes	3.4	11.7	14.1	87.5
C_7	3.3	1.6	0.4	
C_8	0.1	5.9	1.4	0.7
C_9	–	3.7	7.6	11.7
C_{10}	–	0.5	3.7	56.2
C_{11}	–	–	0.9	17.3
C_{12}	–	–	0.1	1.5
C_{13}	–	–	–	0.1
Naphthalenes	–	–	–	0.2
Indans and tetralins	–	0.1	0.5	9.2

[a]From Carpenter *et al.* (1975a,b,c, 1977)
[b]Special boiling range solvents
[c]White spirits
[d]C_8 to C_{13}
[e]0.2% monocyclic, 0.5% di- and tricyclic

(b)   *White spirits*

Trade names: B.A.S.; C.A.S.; Dilutine; Exxsol D; Halpasol; Hydrosol; Indusol; Sangajol; Shellsol D; Solfina; Solnap; Spirdane; Tetrasol; Varsol

Table 3 lists some of the requirements of specifications for white spirits in France, the Federal Republic of Germany, the UK and the USA and of the International Standards Organization.

The US specification ASTM D235–83 lists the physical and chemical properties of four types of 'mineral spirits'; that shown in the table corresponds to what is generally regarded in Europe as 'white spirits'. Requirements for this grade are also detailed in US Federal Specification P–D–860, Type 1. It generally has a lower aromatic content (1–8%) than European white spirits and less than 0.1 wt% benzene (upper limit). Type 2 (solvent extracted) white spirits contain 3–5 wt% aromatic compounds and less than 0.02 wt% benzene (upper limit). Type 3 (hydrogenated) white spirits must contain less than 0.002 wt% benzene (upper limit). The Federal Republic of Germany's specification, DIN 51632, which also covers requirements for aviation gasoline and turbine fuels, does not specify the level of total aromatic compounds but sets limits on specific aromatic compounds such as toluene and ethylbenzene.

According to a large Japanese survey of petroleum distillate solvent samples, the concentrations of *n*-hexane and benzene in solvents boiling in the range of white spirits were both on average 0.01% v/v (range, 0–0.1% v/v) (Kasahara *et al.*, 1987).

Sulfur levels in all three types of white spirits are in general less than 10 mg/kg (ASTM D1266), although levels in Type 2 (solvent extracted) may be up to approximately 200 mg/kg (depending on the degree of hydrodesulfurization).

In general, three principal technical grades of each of the three types of white spirits are available, as follows: 'low flash' grades (flash–point, 25–30°C; boiling range, 130–160°C); 'regular flash' grades (flash–point, 40–50°C; boiling range, 150–190°C); and 'high flash' grades (flash–point, 60–70°C; boiling range, 190–220°C). These grades are available throughout Europe.

(c)   *High–boiling aromatic solvents*

Trade names: A–100; A–150; Caromax; Hydrosol; Indusol; Shellsol; Solvantar; Solvarex; Solvesso

The three main technical grades of high–boiling aromatic solvents that are available on the western European market are fractions boiling at 160–180°C, 180–200°C and 220–290°C. The sulfur content of these grades is generally less than 10 mg/kg, although the sulfur content of Type 2 (solvent extracted) high–boiling aromatic solvents may reach up to several hundred milligrams per kilogram. The benzene content is less than 0.02 wt% (upper limit). The 1,3,5–trimethylbenzene (mesitylene) content of a high–boiling aromatic solvent boiling in the range of 160–180°C can be up to 10 wt%.

Table 3. Specifications for white spirits in selected countries and internationally

Country, product and specification reference	Distillation IBP/FBP[a] (°C)	Flash-point (°C)	Kauri–butanol value (min/max)	Sulfur content (wt%)	Colour (Saybolt)	Aromatic content
Germany, Federal Republic of						
Testbenzine (white spirits) (DIN 51632)	130 min/ 220 max	21 min	–	–	+ 20 max (Hazen colour number)	–
UK						
Mineral solvents (white spirits, type A) (BS 245:1976)	Approx 130/ 220 max	Above 32	–	–	Not darker than standard colour solution	<25% v/v
Mineral solvents (white spirits, type B) (BS 245:1976)	Approx 130/ 220 max	Above 32	–	–	Not darker than standard colour solution	25–50% v/v
USA						
Mineral spirits type 1 – regular (Stoddard solution) (ASTM D235–83)[b]	149 min/ 208 max	38 min	29/45	–[c]	+ 25 min	–
International standard						
Mineral solvents for paint – white spirits, etc. (ISO 1250)	(technically identical to BS 245:1976)					

[a]IBP, initial boiling–point; FBP, final boiling–point
[b]Also includes specifications for high flash–point (60°C min), odourless (Kauri–butanol value, 29 max) and low dry–point (185 max) types of mineral spirits
[c]Bromine number, max 5

2.1 Production and use

(a) Production

Table 4 presents US production of seven solvent products in 1985. Estimated production of all special boiling-range solvents was 2.5 million tonnes, of which 322 000 tonnes were used in paints and coatings. Table 5 presents total sales and the quantities estimated to have been used in the paints and coatings industry for three major products. Approximately 75% of the aliphatic solvents used in paints and coatings are white spirits (SRI International, 1986).

Table 4. US production of seven solvent products in 1985 (thousands of tonnes)[a]

Solvent	Production
Special boiling-range solvents	
Rubber solvent	54.0
VM & P naphtha	151.2
White spirits	
Odourless white spirits	236.0
White spirits (Stoddard solvent)	324.0
140 Flash solvent	362.1
High-boiling aromatic solvents	
A-100	118.4
A-150	171.5

[a]Data provided by Tracor Technology Resources, Inc.

(i) Special boiling-range solvents

In general, special boiling-range solvents are produced in petroleum refineries. Figure 2 is a simplified flow scheme for the production of hydrogenated special boiling-range solvents (Type 2). Naphtha fractions from the atmospheric distillation of crude oil (light and full-range straight-run naphthas) are first subjected to hydrodesulfurization, defined as mild treatment with hydrogen in the presence of a catalyst in order to remove sulfur, followed by hydrogenation, which is treatment with hydrogen in the presence of a different catalyst. Typically, pressures of 19.7–98.7 atm and temperatures of 200–350°C are used in the hydrogenation of these products.

The hydrogenation step converts most of the aromatic hydrocarbons into alicyclic hydrocarbons, leaving less than 0.02 wt% total aromatic content and less than 0.002 wt% benzene. Alternatively, aromatics can be removed by solvent extraction. After hydrogenation, the solvent stream undergoes fractional distillation into narrow and wide boiling ranges, such as SBP 40/65 and SBP 100/160. The sequence of hydrogenation and distillation can be reversed. For the production of hydrodesulfurized special boiling-range solvents (Type 1),

Table 5. US sales and quantities used in the paints and coatings industry (PCI) of three solvents (thousands of tonnes)[a]

Year	Mineral spirits		VM & P naphthas		Lacquer diluents	
	Total	PCI	Total	PCI	PCI	Total
1973	1134	454	211	116	145	91
1974	1061	426	211	111	140	86
1975	1029	370	206	98	125	66
1976	943	372	197	98	126	68
1977	893	372	183	91	118	70
1978	862	374	177	88	116	70
1979	841	363	175	86	113	68
1980	771	331	150	70	91	54
1981	746	320	132	63	86	50
1982	701	302	116	54	68	39
1983	717	315	127	61	79	45
1984	721	320	134	63	79	45
1985	717	324	141	66	75	43

[a]From SRI International (1986)

fractional distillation occurs immediately after hydrodesulfurization, and no hydrogenation step is performed. Type 3 special boiling-range solvents are produced by simple tank blending with an appropriate aromatic product.

(ii) *White spirits*

White spirits are also produced in petroleum refineries. Figure 2 includes a simplified flow scheme for the production of Type 1 (hydrodesulfurized) and Type 3 (hydrogenated) white spirits. The heavy straight-run naphtha and straight-run kerosene fractions from the atmospheric distillation of crude oil are first subjected to hydrodesulfurization, followed by fractional distillation into the appropriate boiling ranges to produce Type 1 white spirits. The sequence of hydrodesulfurization and distillation may be reversed. Type 1 white spirits are hydrogenated to produce Type 3. Hydrogenation reduces the total aromatic hydrocarbon content to less than 1 wt% and benzene to less than 0.002 wt%. The sequence of hydrogenation and distillation may be reversed. In the USA, Type 3 'odourless' white spirits are usually derived from alkylation process streams and are principally isoparaffins.

Type 2 white spirits are produced by extracting kerosene-range feedstock with sulfolane, sulfur dioxide and N-methylpyrollidone, followed by fractional distillation; hydrodesulfurization may also be carried out.

(iii) *High-boiling aromatic solvents*

High-boiling aromatic solvents are also produced in petroleum refineries (Fig. 2). Heavy straight-run naphtha and straight-run kerosene fractions from the distillation of

Fig. 2. Simplified flow scheme of the production of special boiling–range solvents, white spirits and high–boiling aromatic solvents

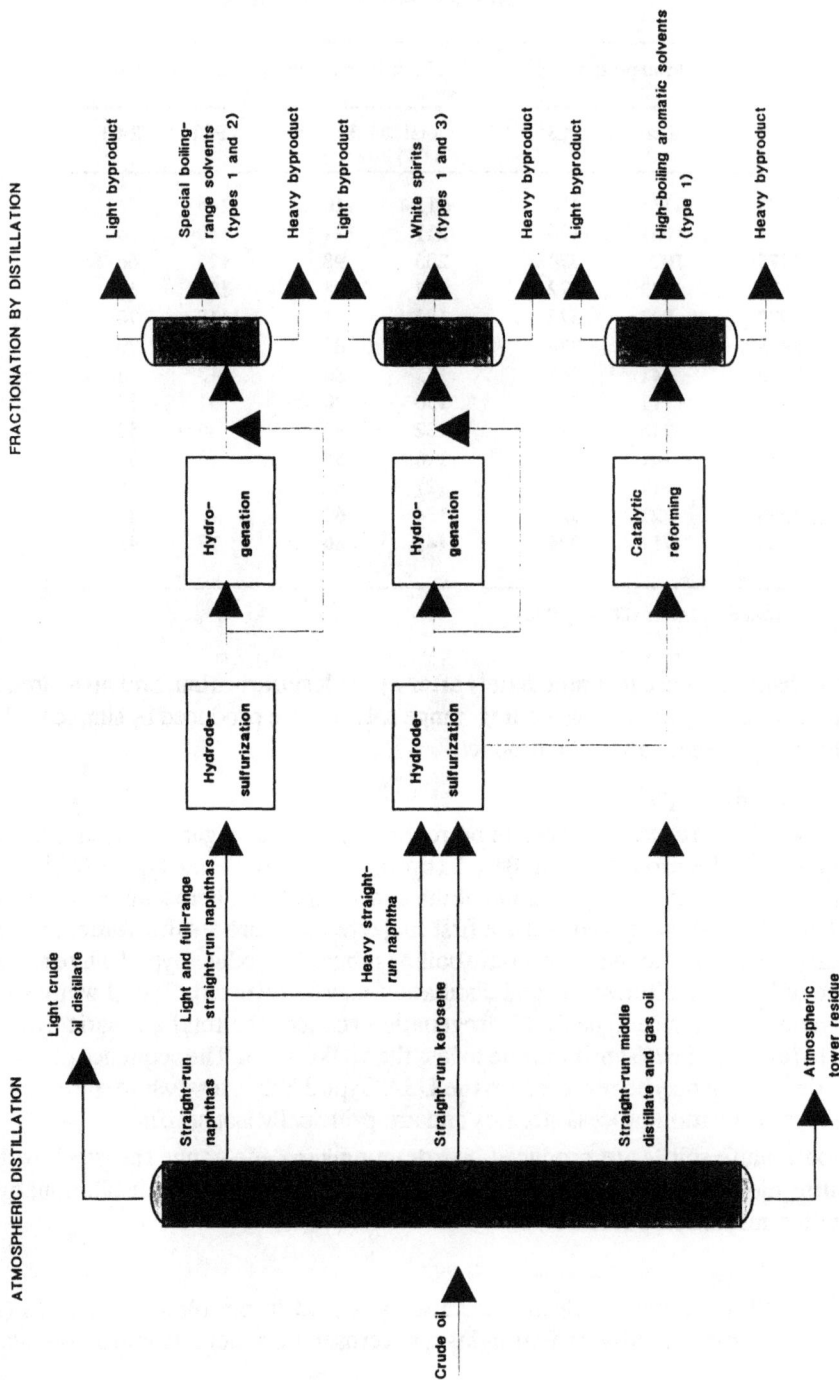

crude oil are first subjected to hydrodesulfurization, then the alicyclic molecules in the naphtha are converted into aromatic compounds in a catalytic reformer by treatment at high temperatures and pressures, liberating hydrogen. Fractional distillation then results in the production of different technical grades of Type 1 high–boiling aromatic solvents.

Type 2 products are produced by extracting middle distillate feedstock with sulfolane, sulfur dioxide and N–methylpyrollidone followed by fractional distillation.

(b) *Use*

(i) *Special boiling range solvents*

Special boiling range solvents are used in a wide range of applications, the principal ones being (approximate percentage of consumption in western Europe): adhesives (34%); paints, lacquers and varnishes (20%); polymerization reaction diluents (10%); rubber industry (11%); and gravure inks and other miscellaneous applications (25%). The aliphatic solvents are used extensively in alkyd resin systems in paints and coatings, as well as for linseed oil and oleoresin varnishes and in nonaqueous dispersion coating systems. These uses are the principal ones worldwide, although the percentages used in the different applications may vary somewhat with geographic region.

Table 6 gives estimated consumption of special boiling range solvents in western Europe in 1972 and 1986. There has been a major shift from special boiling range solvents containing aromatic compounds to those which have been hydrogenated and have a minimal aromatic content, for technical reasons and as a result of concern about human health and the environment. There has also been a move away from products containing high levels of n–hexane to those containing less than 5 wt%, also in relation to health effects (SRI International, 1986).

Table 6. Consumption of petroleum solvents in western Europe (thousands of tonnes)

Solvent	1972	1986
Special boiling–range solvents		
Type 1 and 3 (containing aromatics)	180	40
Type 2 (hydrogenated)	90	210
Total	270	250
White spirits		
Type 1 (hydrodesulfurized)	670	540
Type 2 (solvent extracted)	30	40
Type 3 (hydrogenated)	50	120
Total	750	700
High–boiling aromatic solvents		
Type 1 (catalytically reformed)	160	264
Type 2 (solvent extracted)	40	66
Total	200	330

(ii) *White spirits*

The principal uses for white spirits in western Europe (approximate percentage of consumption) are as follows: paints, lacquers and varnishes (60%); degreasing/industrial cleaning (9%); wood treatment (4%); chemical processes (3%); household, cosmetic and toiletries (3%); and miscellaneous (21%).

Table 6 gives estimated consumption of white spirits in western Europe. There has been a move towards the use of hydrogenated white spirits over the last decade, owing to their reduced odour and for technical reasons. Hydrogenation facilities were first introduced in the mid–1960s, and capacity has been increasing steadily since.

(iii) *High–boiling aromatic solvents*

High–boiling aromatic solvents are used principally in: paints, lacquers and varnishes (58%); agrochemicals (16%); industrial cleaning/degreasing (9%); chemical processes (3%); and miscellaneous (inks, plastisols, lubricant additives, oil field chemicals) (14%). Table 6 also gives estimated western European consumption of these solvents.

(c) *Regulatory status and guidelines*

The US Food and Drug Administration (1988) permits the use of petroleum naphtha in foods if it is a mixture of liquid hydrocarbons, essentially paraffinic and naphthenic in nature, obtained from petroleum and is refined to have a boiling range of 80–150°C and a nonvolatile residue content of 0.002 g/100 ml max and to meet specified ultraviolet absorbance limits.

Examples of occupational exposure limits for petroleum solvents in 12 countries are presented in Table 7.

Table 7. Occupational exposure limits for petroleum solvents[a]

Solvent[b]		Country	Year	Concentration		Interpretation[c]
				ppm	mg/m³	
Petroleum ether		Switzerland	1985	500	2000	TWA
Rubber solvent (naphtha)		Belgium	1984	400	1600	TWA
		Mexico	1985	400	1600	TWA
		Netherlands	1986	400	1600	TWA
		USA (ACGIH)	1988	400	1600	TWA
		Venezuela	1985	400	1600	TWA
VM & P naphtha		Mexico	1985	300	1350	TWA
[CAS No. 8032–32–4]		USA (ACGIH)	1988	300	1350	TWA
White spirits						
Aromaticity	*Boiling range*					
<1%	60–90°C	Finland	1987		350	TWA
					450	STEL
<1%	80–110°C				1200	TWA
					1600	STEL

Table 7 (contd)

Solvent[b]		Country	Year	Concentration		Interpretation[c]
				ppm	mg/m^3	
< 1%	> 110°C	Finland (contd)			900	TWA
					1200	STEL
> 20%	> 110°C				770	TWA
					1020	STEL
100%	> 110°C				240	TWA
					360	STEL
Aromaticity						
< 10%		Norway	1981	200	1050	TWA
> 10%, < 20%				100	575	TWA
> 20%				25	120	TWA
		Sweden	1987	~85	500	TWA
				~110	625	STEL
		UK	1987	100	575	TWA
				125	720	STEL (10–min)
		USSR	1984		300	TWA
Stoddard solvent		Belgium	1984	100	575	TWA
		Chile	1985	80	460	TWA
		Mexico	1985	500	2950	TWA
		Netherlands	1986	100	575	TWA
		Switzerland	1984	100	580	
		USA (OSHA)	1985	500	2950	TWA
		(NIOSH)	1986		350	
					1800	Ceiling (15–min)
[CAS No. 8052–41–3]		(ACGIH)	1988	100	525	TWA
		Venezuela	1985	100	575	TWA
				150	720	Ceiling
Solvent naphtha (as carbon)		USSR	1986		100	TWA

[a]From Direktoratet for Arbeidstilsynet (1981); International Labour Office (1984); Arbeidsinspectie (1986); Institut National de Recherche et de Sécurité (1986); National Institute for Occupational Safety and Health (1986); Cook (1987); Health and Safety Executive (1987); National Swedish Board of Occupational Safety and Health (1987); Työsuojeluhallitus (1987); American Conference of Governmental Hygienists (1988)

[b]OSHA, Occupational Safety and Hygiene Administration; NIOSH, National Insitute for Occupational Safety and Health; ACGIH, American Conference of Governmental Hygienists

[c]TWA, time–weighted average; STEL, short–term exposure limit

2.2 Occurrence

(a) Natural occurrence

Petroleum–derived solvents do not occur naturally in the environment, but only as components of the crude oils from which they are derived (International Programme on Chemical Safety, 1982).

(b)   Occupational exposure

On the basis of one National Occupational Hazard Survey, the US National Institute for Occupational Safety and Health (1974) has estimated that 3500 workers were potentially exposed to petroleum ether (as ligroin [CAS No. 8032-32-4]; a special boiling range solvent) in the USA in 1972-74; 26 400 workers were potentially exposed to rubber solvent. According to a National Occupational Exposure Survey (National Institute for Occupational Safety and Health, 1983), 327 000 workers were potentially exposed to varnish makers' and painters' naphtha (a special boiling range solvent), 160 000 to white spirits and 521 800 to Stoddard solvent in the USA in 1981-83.

Occupational exposure levels have been measured for a wide range of petroleum solvents. The following paragraphs present data available for the types of solvents covered in this monograph.

In a US plant that used a contact adhesive containing rubber solvent and toluene to assemble rubber life rafts, personal time-weighted average (TWA) exposures to rubber solvent ranged from 5 to 12 ppm with an average of 9 ppm (Apol, 1981).

Levels of naphtha have been reported in a variety of US manufacturing plants. In a plant where baseball bats were manufactured, breathing zone levels were 13-190 ppm (mean, 78 ppm; Rivera & Rostand, 1975). During the installation of plastic laminates to kitchen and bathroom counter tops, levels of naphtha were 363-3158 mg/m³ (mean, 1113 mg/m³; Apol, 1980). In the printing industry, air levels of naphtha in personal breathing samples from off-set printers were 94-258.4 mg/m³ (mean, 140.4 mg/m³; Gorman, 1982); levels of varnish makers' and painters' naphtha to which lithographers were exposed were 58-242 mg/m³ (mean, 122.1 mg/m³; Chrostek *et al.*, 1979). In another study, levels of petroleum naphtha in breathing zones and in the general air in the printing industry were 18-106 (average, 53) mg/m³ (Gunter, 1986). Ambient indoor air area levels of naphtha in a large US Federal office building were 8.4-22.7 mg/m³ (Watanabe & Love, 1980). In two shoe manufacturing plants, TWA area concentrations of naphtha were 10-162 mg/m³ (mean, 59 mg/m³) and those of aliphatic naphtha, 19-522 mg/m³ (mean, 305 mg/m³; Tharr *et al.*, 1982). In a plant that used petroleum naphtha as a release agent in the manufacture of refrigerators, levels were 15-147 mg/m³ (Markel & Shama, 1974), and in a plant where window assemblies for cars were manufactured, TWA levels of naphtha were 32-191 mg/m³ (mean, 79 mg/m³; Lucas, 1981).

In a US factory where white spirits were used to clean silk screens, air levels were 137-385 mg/m³ in personal samples and 149-336 mg/m³ in area samples (Geissert, 1975). In a US factory manufacturing naval catapults, exposure to white spirits ranged from < 1 to 2615 mg/m³, with an average of 208 mg/m³ (Gilles, 1975). In a US plant in which white spirits were used to clean automobile starters and generators, the TWA in personal air samples was 43-594 mg/m³, with an average of 275 mg/m³; at the 'Kleen Abrader' (a vibrating machine containing white spirits and abrasive stone pellets), a value of 3967 mg/m³ was found (Levy, 1975). At a US plant manufacturing vinyl floor coverings in which white spirits were used in the solvent/wax solution, personal TWA levels in air were 0.3-8 ppm, with an average of 4.1 ppm (Belanger & Elesh, 1979). Air concentrations of white spirits in the breathing zone of telephone wire splicers, who used it to dissolve petroleum jelly for filling telephone cables, were 79–244 mg/m³, with an average of 144 mg/m³ (Gunter, 1980). At a US tool manufacturing plant, the 8-h TWA exposures of tool repair technicians in 1986 were 20–95 mg/m³ in personal samples, with a mean of 55 mg/m³, and 58–266 mg/m³ (mean, 116 mg/m³) in process air samples (Blade, 1987).

Personal exposures to Stoddard solvent in the press room of two US daily newspapers were 13–319 mg/m³ (average, 83 mg/m³; Hollett *et al.*, 1976) and 5–9 mg/m³ (average, 7 mg/m³; Kronoveter 1977); those in a US printing facility were 0.1–16 mg/m³ (Gunter, 1982). Breathing zone samples taken at a commercial airline maintenance hanger showed concentrations of Stoddard solvent ranging from 363 to 8860 mg/m³, with an average of 3000 mg/m³ (Gunter, 1975a). A US manufacturer of ski boots reported levels of 345–451 mg/m³ Stoddard solvent in the boot finishing and polishing department (Gunter, 1975b). In the dry-cleaning industry, 8-h TWA concentrations of Stoddard solvent were 15–35 ppm (Oberg, 1968). During automobile washing in Finland, TWA concentrations of Stoddard solvent were 5–465 mg/m³ in personal samples and 15–390 mg/m³ in area samples. For heavy vehicles, the corresponding values were 45–805 and 40–685 mg/m³, respectively (Niemelä *et al.*, 1987).

Exposure to petroleum solvents during the manufacture and application of paint products is described in the monograph on occupational exposures in paint manufacture and painting (see p. 329).

(c) Air

Because of their low solubility in water and their relatively high volatility, petroleum solvents are found principally in the air after their release. They may undergo photodegradation in the atmosphere.

2.3 Analysis

For petroleum solvents with final boiling–points up to about 100°C, all of the hydrocarbon components present can be identified and quantified using high–resolution capillary column gas chromatography. For products with a higher final boiling–point, it is difficult to identify all of the components, and analysis is usually for total aliphatic, alicyclic and aromatic hydrocarbons.

A multicolumn gas chromatograph has been designed to separate aromatic, alicyclic and aliphatic components in hydrocarbon solvents that boil at up to 200°C. It consists of three columns – a polar, a nonpolar and a molecular sieve. The first is used to separate aromatic compounds from saturated compounds, the second is used to separate aromatic compounds by boiling–point, and the third to separate alicyclic comounds from aliphatic compounds by carbon number.

Ultraviolet spectrometry can be used to estimate the total aromatic content (ASTM D1019). Sulfur levels are determined by microcoulometry (ASTM D1266).

The total hydrocarbon concentration of various petroleum solvents can be determined in air using a method based on charcoal tube sampling and gas chromatography with flame ionization detection. The working range is 100–2000 mg/m³ for a 5-l air sample (Eller, 1984).

3. Biological Data Relevant to the Evaluation of Carcinogenic Risk to Humans

3.1 Carcinogenicity studies in animals

Inhalation

Rat: Groups of 50 male and 50 female Wistar rats, eight to nine weeks of age, were exposed by inhalation to 0 (controls), 470 ± 29, 970 ± 70 or 1830 ± 130 mg/m³ of a high–boiling aromatic solvent (high–flash aromatic naphtha; ASTM-3734; a 50:50 blend of Shellsol A and Solvesso) containing 13 major components (predominantly C_9 aromatic compounds)[1]. Exposure was for 6 h per day on five days per week for up to 12 months. Ten males and ten females from each group were killed after six months, 25 males and 25 females from each group after 12 months and the remaining animals four months after the end of exposure. A single mammary adenocarcinoma was found in a female in the high–dose group killed at six months. In rats killed at 12 months, there was a glioblastoma in the brain in a low–dose male, a splenic lymphoma in a high–dose male and a leiomyoma of the uterus in a high–dose female (Clark & Bird, 1989). [The Working Group noted the short duration of the study, and the absence of pathological examination of the surviving rats from each group killed at 16 months.]

3.2 Other relevant data

(a) Experimental systems

Absorption, distribution, excretion and metabolism

No data were available to the Working Group.

Toxic effects

The toxicology of petroleum solvents has been reviewed (National Institute for Occupational Safety and Health, 1977; International Programme on Chemical Safety, 1982). A summary of the acute toxicities of mixtures of aromatic and saturated hydrocarbons representing the types of commercial solvents is available (Hine & Zuidema, 1970).

---

[1]*ortho*–Xylene, 2.27%; *n*–propylbenzene, 4.05%; 1–methyl–3–ethylbenzene, 7.14%; 1–methyl–4–ethylbenzene, 16.60%; 1,3,5–trimethylbenzene, 9.35%; 1–methyl–2–ethylbenzene, 7.22%; 1,2,4–trimethylbenzene, 32.70%; 1,2,3–trimethylbenzene, 2.76%; 1–methyl–3–*n*–propylbenzene and 1,2–diethylbenzene, 6.54%; 1–ethyl 3,5–dimethylbenzene, 1.77%; no benzene detected; nonaromatic compounds, 0.46%

Acute oral and inhalation toxicities in rats and percutaneous toxicity in rabbits are shown in Table 8. Toxicity was more pronounced with samples containing higher levels of aromatic compounds than with virtually aromatic free products. The irritating potential of the mixtures for skin and eye (evaluated according to the method of Draize in albino rabbits) ranged from minimally to moderately irritating (Hine & Zuidema, 1970).

(i) *Special boiling range solvents*

A hexane-containing and a heptane-containing petroleum solvent fraction had low toxicity (Table 8). Irritation of the skin was slight and eye irritation minimal, and there was practically no percutaneous toxicity in rabbits (Hine & Zuidema, 1970).

Rats were exposed to either 200 or 500 ppm n-hexane or to petroleum benzine vapour [assumed boiling range, 30–150°C] containing 200 or 500 ppm n-hexane for 12 h per day for 24 weeks. Peripheral nerve function was clearly impaired by 500 ppm n-hexane, slightly impaired by 200 ppm n-hexane and by petroleum benzine vapour containing 500 ppm n-hexane, and impaired even less by petroleum benzine containing 200 ppm n-hexane. The authors concluded that other components of petroleum benzine might inhibit the neurotoxicity of n-hexane (Ono *et al.*, 1982).

The inhalation toxicity of a rubber solvent (boiling range, 76–111°C; composition (in vol%): paraffins, 41; monocycloparaffins (naphthenes), 54; benzene, 1.5; alkyl benzenes, 3.4; monoolefins, 0.1) was studied in mice, rats, cats and beagle dogs. Animals were exposed to concentrations of 480–63 000 ppm. Acute exposure to high concentrations resulted in loss of motor coordination and central nervous system depression. Eye irritation was described in rats (at 9600 ppm for 4 h), toxic stress in dogs (3200–6200 ppm, 4 h) and respiratory rate depression in mice (63 000 ppm). A 4-h LC_{50} of 15 000 ppm was estimated for rats. No significant toxic effect was noted in rats or in beagle dogs exposed subacutely to up to 2000 ppm rubber solvent for 6 h per day on five days per week for 13 weeks (Carpenter *et al.*, 1975c).

The inhalation toxicity of a varnish makers' and painters' naphtha (boiling range, 118–150°C; composition (in vol%): paraffins, 55; monocycloparaffins, 30; dicycloparaffins, 2.5; alkylbenzenes, 12; benzene, 0.1) was also studied in mice, rats, cats and beagle dogs. Animals were exposed to concentrations of 280–15 000 ppm. Acute exposure to high concentrations resulted in loss of motor coordination and central nervous system depression. Eye irritation was described in rats (3400 ppm, 4 h) and dogs (3400 ppm, 2 h) and respiration rate was decreased in mice (> 2600 ppm, 1 min). A 4-h LC_{50} of 3400 ppm was estimated for rats. No significant toxic effect was noted in rats or in beagle dogs exposed subacutely to up to 1200 ppm for 6 h per day on five days per week for 13 weeks (Carpenter *et al.*, 1975a).

(ii) *White spirits*

The effects of a paint thinner (boiling range, 140–190°C) were reported in rats, guinea-pigs, rabbits, beagle dogs and squirrel monkeys. The samples were a complex mixture of 80–86% saturated hydrocarbons, 1% olefins and 13–19% aromatic compounds. Animals were exposed for 8 h a day on five days per week for 30 exposures or continuously (23.5 h per day on seven days per week) for 90 days. In the continuous 90-day experiment (exposure concentrations, 114–1270 mg/m³), increased mortality occurred only among guinea-pigs exposed to 363 mg/m³ or more. Gross examination revealed irritation and congestion of the

Table 8. Acute toxicities of mixtures of aromatic and saturated hydrocarbons representing the three types of commercial solvents[a]

Solvent type (boiling range; °C)	Principal components	Oral LD$_{50}$ (ml/kg bw; rats)[b]	Inhalation 4-h LC$_{50}$ (ppm; rats)	Aspiration mortality (rats)	Percutaneous 4-h LD$_{50}$ (mg/kg bw; rabbits)
Special boiling range solvents					
65–75	C$_6$ normal and isoparaffins (hexanes) and naphthenes (cyclohexane, methyl-cyclopentane)	>25	73 680 (66 310–79 940)		>5.0
91–104	C$_7$ normal and isoparaffins (heptanes) and naphthenes (methylcyclohexane, dimethylcyclopentane)	>25	14 000–16 000		>5.0
White spirits					
156–180	C$_9$ and C$_{10}$ normal and isoparaffins and naphthenes	>25	2000–2600		>5.0
187–212	C$_{11}$ and C$_{12}$ normal and isoparaffins and naphthenes	>25	>710[c]		~5.0
174–216	C$_{12}$ isoparaffins	>25	>792[c]		>5.0
195–260	C$_{13}$ to C$_{16}$ normal and isoparaffins and naphthenes	>25	>263[c]	5/10	>5.0
High–boiling aromatic solvents					
138–141	C$_8$ aromatics (ortho-, meta- and para-xylene; ethyl benzene)	10.0 (7.5–13.3)	6350 (4670–8640)		~5.0
163–203	C$_9$, C$_{10}$ and C$_{11}$ aromatics	4.5 (3.0–6.8)	>2450[c]		~5.0
188–209	C$_{10}$ and C$_{11}$ aromatics	13.3 (7.5–23.7)	>580[c]		~5.0
196–264	C$_{11}$ to C$_{14}$ aromatics	12.3 (8.1–18.7)	>553[c]	5/10	~5.0

[a] From Hine & Zuidema (1970)

[b] Doses above 25 ml/kg bw not practical for testing

[c] Maximum concentration attainable at 25°C

lungs in all species, and the severity of lung irritation and the number of animals involved appeared to be dose related. In the 30–day experiment, the only signs of toxicity were seen in guinea–pigs exposed to the highest concentration (1353 mg/m^3), which had lung irritation, congestion and emphysema on histopathological examination (Rector et al., 1966).

The inhalation toxicity of a Stoddard solvent (boiling range, 153–194°C; composition (in vol%): paraffins, 48; mono– and dicycloparaffins, 26 and 12; alkylbenzenes, 14; benzene, 0.1) was studied in mice, rats, cats and beagle dogs. Animals were exposed to 84–1700 ppm of the solvent for up to 13 weeks. Marked tubular regeneration of the kidneys, which may be indicative of solvent–induced kidney damage, was detected at sacrifice in male rats exposed to 190 or 330 ppm for 13 weeks. Eye irritation and blood around the nostrils was described in rats exposed to 1400 ppm for 8 h. Central nervous system depression was seen in all species at higher exposure concentrations. Depressed respiration rate was observed in mice exposed to 1700 ppm (Carpenter et al., 1975b).

Rats were exposed for 4 h per day for four consecutive days to 214 mg/m^3 (~45 ppm) of white spirits containing mainly C_9–C_{12} hydrocarbons and consisting of approximately 61% paraffins, 20% naphthenes and 19% aromatics (boiling range, 150–195°C). Irritation of the upper respiratory tract (loss of cilia, mucous and basal–cell hyperplasia, squamous metaplasia) was observed (Riley et al., 1984).

The subchronic inhalation toxicity of dearomatized white spirits (boiling range, 155–193°C; containing 58% paraffins, 42% naphthenes and <0.5% aromatics) and of a C_{10}–C_{11} isoparaffin (boiling range, 156–176°C; 100% isoparaffins) was investigated in rats. Animals were exposed for 6 h per day on five days per week for 12 weeks to concentrations of about 5.6 (900 ppm) and 1.9 g/m^3 (300 ppm) of the two solvents, respectively. The primary effect of both solvents after four weeks of exposure was kidney toxicity in male rats (described as mild tubular toxicity), the incidence and severity of which appeared to increase with increasing exposure concentration and duration. No other significant toxic effect was observed at the levels tested (Phillips & Egan, 1984).

In rats exposed by inhalation to industrial white spirits composed of a 99% C_{10}–C_{12} mixture of branched–chain aliphatic hydrocarbons at to 6500 mg/m^3 for 8 h per day on five days per week for 12 months, increased urinary activity of lactic dehydrogenase (a marker of distal tubular damage) and reduced kidney function were seen (Viau et al., 1984).

(iii) High–boiling aromatic solvents

The inhalation toxicity of a high–boiling aromatic solvent (boiling range, 184–206°C; composition (in vol%): alkylbenzenes, 87.5; naphthalenes, 2; indanes, 9; naphthenes, 0.8; paraffins, 0.3) was studied in rats, cats and beagle dogs. No significant toxicity was observed in any species after exposure to 380 mg/m^3 (66 ppm) for 6 h per day on five days per week for 13 weeks. An 8–h exposure of rats to 8700 mg/m^3 of an aerosol of the solvent resulted in early nasal and occular irritation followed by progressive loss of coordination and the death of 2/10 animals (Carpenter et al., 1977).

The inhalation toxicity of the high–boiling aromatic naphtha tested for carcinogenicity (see section 3.1) was investigated in rats. The animals were exposed to 0, 450, 900 or 1800 mg/m^3 of the solvent for 6 h per day on five days per week for up to 12 months. Reductions in body weight gain were observed in males exposed to the highest dose over the first four weeks and in female rats exposed to the two higher concentrations over the first four or 12 weeks. In males exposed to the highest dose, liver and kidney weights were increased at six and 12 months. Various statistically significant haematological changes were observed transiently in male rats up to six months, but were not considered to be of biological relevance by the authors. A preliminary 13–week inhalation study with Shellsol A resulted in increases in liver and kidney weights in rats exposed to 7400 and 3700 mg/m³, and in a low–grade anaemia in rats at 7400, 3700 and 1800 mg/m³ (Clark & Bird, 1989).

Effects on reproduction and prenatal toxicity

As reported in an abstract, female CD rats [number of animals unspecified] were exposed by inhalation for 6 h per day to either commercial white spirits (100 or 300 ppm) or C_8–C_9 isoparaffinic hydrocarbon solvents (400 or 1200 ppm) on days 6–15 of gestation. No maternal effect was reported. Fetuses were examined at term for visceral and skeletal development. Male fetuses exposed to 100 ppm white spirits were heavier than control fetuses, and the incidence of skeletal variations was significantly increased in the group exposed to 1200 ppm isoparaffin (Phillips & Egan, 1981). [The Working Group noted the lack of experimental details available.]

As reported in an abstract, Wistar rats [number of animals unspecified] were exposed by inhalation for 6 h per day to 237, 482 or 953 ppm white spirits on days 6–15 of gestation, or to 950 ppm on days 3–20 of gestation. Maternal toxicity was reported in the high–exposure groups. Fetuses were examined at term for visceral and skeletal development. No skeletal or visceral anomaly was observed, but significant reductions in fetal weight, delays in skeletal development and extra ribs were reported in the group treated on days 3–20 of gestation (Jakobsen et al., 1986). [The Working Group noted the lack of experimental details available.]

Genetic and related effects[1]

A rubber solvent induced chromosomal aberrations but not sister chromatid exchange in human whole–blood cultures (Altenburg et al., 1979).

As reported in an abstract, C_8–C_9 isoparaffinic hydrocarbon solvents (at 400 and 1200 ppm) did not induce dominant lethal mutations in CD rats exposed for 6 h per day on five days per week for eight consecutive weeks prior to mating (Phillips & Egan, 1981).

Special boiling range spirit 100/140 (a mixture of paraffins and cycloparaffins in the C_5–C_{10} range) was not mutagenic to *Salmonella typhimurium* TA1535, TA1537, TA1538, TA98 or TA100 or to *Escherichia coli* WP2 either in the presence or absence of an exogenous metabolic system from Aroclor–induced rat liver. It did not induce mitotic gene conversion

---

[1]Subsequent to the meeting, the Secretariat became aware of a study to investigate the effects of high flash aromatic naphtha in a *Salmonella*/mammalian microsome mutagenicity assay, a hypoxanthine–guanine phosphoribosyl transferase forward mutation assay in Chinese hamster CHO cells, *in–vitro* chromosomal aberration and sister chromatid exchange assays in Chinese hamster CHO cells, and an *in–vivo* chromosomal aberration assay in rat bone marrow (Schreiner et al., 1989).

in *Saccharomyces cerevisiae* strain JD1 or chromosomal aberrations in Chinese hamster ovary cells (Brooks *et al.*, 1988).

A sample of white spirits (boiling range, 160–161°C; 85% aliphatic compounds, 15% aromatic compounds) was not mutagenic to *S. typhimurium* TA1530, TA1535, TA1537, TA1538, TA98 or TA100 either in the presence or absence of an exogenous metabolic system from Aroclor–induced rat liver. It did not induce sister chromatid exchange in human lymphocytes *in vitro* or chromosomal aberrations *in vivo* in mouse bone–marrow cells (Gochet *et al.*, 1984). As reported in an abstract, white spirits (100 and 300 ppm) did not induce dominant lethal mutations in CE rats exposed for 6 h per day on five days per week for eight consecutive weeks prior to mating (Phillips & Egan, 1981).

As reported in an abstract, Stoddard solvent did not induce mutation in *S. typhimurium* [strain unspecified] or in mouse lymphoma L5178Y TK +/– cells. Neither did it induce chromosomal aberrations in rat bone marrow *in vivo* [details not given] (Lebowitz *et al.*, 1979).

(*b*)   *Humans*

Absorption, distribution, excretion and metabolism

White spirits are absorbed through the lungs, and the uptake is enhanced by exercise-induced increase in pulmonary ventilation. Fifteen healthy male volunteers were exposed to 1250 or 2500 mg/m^3 white spirits vapour (boiling–point, 150–200°C; containing 83% aliphatic and 17% aromatic compounds) either at rest for 30 min (one to four times) or with exercise up to 150 W. Blood levels of *n*–decane and 1,2,4–trimethylbenzene (taken to represent aliphatic and aromatic compounds, respectively) were detected after exposure. The exercise-induced increase was more evident for aromatic compounds than for aliphatic compounds, possibly due to different solubilities in blood (Åstrand *et al.*, 1975). In other studies (Milling Pedersen *et al.*, 1984, 1987) in which volunteers were exposed to 10 ppm (600 mg/m^3) white spirits [99% alkanes [38.7% C_{11}; 44.4% C_{12}] and 1% cycloalkanes [mostly C_{10}]), solvent levels were determined in subcutaneous adipose tissue obtained by biopsy. After 46–48 h, about 50% of the initial concentration of the solvent had disappeared.

Toxic effects

[The Working Group noted that some of the information on the adverse effects of petroleum solvents originates from studies of occupational exposures, which generally involve a complex mixture of various organic solvents and other compounds. Reasons for the variable outcomes include differences in exposure, i.e., type of chemical, intensity and duration. Also, selection bias may have occurred, and the examination methods varied, some possibly being influenced by recent, rather than chronic exposures. Finally, the control groups may not have been appropriate, so that the effect of confounders cannot be ruled out. The reader is referred also to the monograph on occupational exposures in paint manufacture and painting.]

Exposure of volunteers to 2700 mg/m^3 of a Stoddard solvent caused eye irritation (Carpenter *et al.*, 1975b). Opacities in the lens of the eye have been recorded in car painters exposed to a mixture of solvents, including white spirits (Elofsson *et al.*, 1980). Stoddard solvent irritates the skin and may cause contact dermatitis (Nethercott *et al.*, 1980), and petroleum ether applied to the skin may also induce severe irritation (Spruit *et al.*, 1970). Solvents with a high aromatic content are more irritant than those of paraffinic origin in the same boiling range (Klauder & Brill, 1947).

Volunteer painters exposed to 100 ppm white spirits containing 17% aromatic compounds complained of irritation in the upper airways (Cohr *et al.*, 1980). Aspiration of petroleum distillates by fire eaters may induce chemical pneumonitis (Agrawal & Srivastava, 1986).

Petroleum solvents may affect the central nervous system. After acute, intense exposure of 'sniffers', there is first an excitatory phase and later a depressive phase (Prockop, 1979). Volunteers exposed to white spirits at 100–400 ppm for 7 h reported headache, fatigue and incoordination, with dose-associated effects on equilibrium, reaction time, visuo–motoric coordination and memory (Cohr *et al.*, 1980). Exposure to 4000 mg/m^3 white spirits for 0.5 h affected reaction time and short–term memory (Gamberale *et al.*, 1975). Nausea, a sense of intoxication and dizziness have been reported in car painters exposed to white spirits and other solvents (mainly toluene; Husman, 1980).

Subjective symptoms originating from the central nervous system, such as headache, fatigue, poor concentration, emotional instability, impaired memory and other intellectual functions, and impaired psychomotor performance have been reported in a series of cross-sectional studies of paint industry workers, house painters, car painters, shipyard painters and floorlayers, all of whom had been exposed to a mixture of solvents, including petroleum solvents (Hänninen *et al.*, 1976; Hane *et al.*, 1977; Elofsson *et al.*, 1980; Husman, 1980; Anshelm Olson, 1982; Lindström & Wickström, 1983; Cherry *et al.*, 1985; Ekberg *et al.*, 1986). Some of these are short- or mid–term effects, others are potentially persistent. In some studies, dose–response relationships were observed between symptoms and lifetime exposure (duration and intensity) to solvents.

Effects on the peripheral nervous system have been seen in car painters, house painters and shipyard painters who were exposed to a mixture of various organic solvents, including white spirits (Seppäläinen *et al.*, 1978; Elofsson *et al.*, 1980; Husman & Karli, 1980; Cherry *et al.*, 1985; Askergren *et al.*, 1988). However, in house painters, no such effect was observed after an average exposure to 40 cm^3/m^3 white spirits (Seppäläinen & Lindström, 1982).

In workers exposed to a glue solvent (rubber solvent; C_5–C_7 aliphatic and alicyclic hydrocarbons), indications of slight renal tubular effects were reported (Franchini *et al.*, 1983). In another study of car painters exposed to low levels of white spirits and toluene, no such effect was observed (Lauwerys *et al.*, 1985).

Conflicting reports of adverse effects on blood and blood–forming organs were reported in house and car painters exposed to white spirits and other solvents (Hane *et al.*, 1977; Elofsson *et al.*, 1980; Angerer & Wulf, 1985). One case of aplastic anaemia was reported following exposure to Stoddard solvent (Prager & Peters, 1970).

Some indication of effects on muscles (raised serum creatine kinase) has been reported after short–term exposure of volunteers to white spirits (Milling Pedersen & Cohr, 1984).

Effects on fertility and pregnancy outcome

Holmberg (1979) reported a case–control study of 120 children with congenital central nervous system defects and an equal number of matched controls. The cases had been registered with the Finnish Register of Congenital Malformations between June 1976 and May 1978. Four mothers of the children with central nervous system defects, but no control mother, reported having worked with white spirits during the first trimester of pregnancy; two of them had also been exposed to other solvents, such as toluene and xylene. One control mother had been exposed to 'mixed aromatic/aliphatic' solvents.

Holmberg *et al.* (1982) reported a case–control study of 378 children with oral clefts and an equal number of controls. The cases had been registered with the Finnish Register of Congenital Malformations between December 1977 and May 1980. Ten mothers of children with oral clefts and two control mothers reported having worked with lacquer petrol (i.e., white spirits; 85% aliphatics, 15% aromatics) during the first trimester of pregnancy. Four of these case mothers and one control had also been exposed to toluene, xylene, acetates, alcohols, acetone, ethylacetate, dichloromethane, turpentine and fluorotrichloromethane.

[The Working Group noted that in these two studies data on exposure were collected retrospectively, after the malformation had been registered.]

Genetic and related effects

No data were available to the Working Group.

3.3 Epidemiological studies of carcinogenicity in humans

The Working Group chose to address only those papers that specifically reported exposure to 'petroleum solvents', e.g., white spirits, and not those that reported only exposure to 'organic solvents' or to unspecified solvents. Studies in which exposures to toluene, xylene or phenol are mentioned are described in the monographs on those compounds.

Relationships between cancer excesses and exposure to solvents in the rubber industry have been investigated in a number of studies, some of which were evaluated previously (IARC, 1982). Wilcosky *et al.* (1984) performed a case–control study on cases of cancer at five sites in a cohort of 6678 male, hourly-paid, active and retired rubber workers identified in 1964 in the USA (McMichael *et al.*, 1975). Detailed records of potential exposure to 25 specific solvents (data are given for only 20 solvents; Wilcosky *et al.*, 1984) were used to classify jobs, and exposure was evaluated using a set of 70 occupational titles that covered all jobs entailing the same process and materials. These titles were then grouped into heavy, medium and light exposure to solvents, according to assessments by industrial hygienists (McMichael *et al.*, 1975). In white males, exposure to 'specialty naphthas' was associated with an increased relative risk (RR) for lymphosarcoma (1.4; 6 cases) and for lymphatic leukaemia (2.8; 8 cases), and exposure to 'VM & P naphtha' with increased RRs for prostatic cancer (1.6; 4 cases) and for lymphatic leukaemia (2.9; 3 cases). [The Working Group noted that the number of cases in each category is small, multiple exposures were evaluated independently of other exposures, and none of the associations is significant.]

Olsson and Brandt (1980) interviewed 25 men aged 20–65 years admitted consecutively to the Department of Oncology, University Hospital, Lund, Sweden, with a recent diagnosis of Hodgkin's disease. Two controls for each case were selected from the computerized Swedish population register and matched for sex and residence. The cases and controls were interviewed using a questionnaire, and daily exposure to solvents for at least one year more than ten years before diagnosis was recorded for all persons. Twelve of the 25 patients with Hodgkin's disease had been exposed occupationally to organic solvents and six of the 50 controls, giving a RR of 6.6 (95% confidence interval (CI), 1.8–23.8). Two of the cases and none of the controls had been exposed to white spirits.

Hardell *et al.* (1984) performed a case–control study on 102 deceased cases of primary liver cancer in men aged 25–80 years in Sweden; 83 were hepatocellular carcinoma, 15 cholangiocarcinoma, three haemangiosarcomas and one an unspecified sarcoma. For each case, two controls were drawn from the national population register and were matched for sex, age, year of death and municipality. Information about various exposures was obtained by written questionnaires from a close relative of each case and control. Exposure to organic solvents for more than one month was stated by 22.4% of the cases with primary liver cancer and by 13.5% of the controls (RR, 1.8; 95% CI, 0.99–3.4). The RR for cases with hepatocellular carcinoma was 2.1 (95% CI, 1.1–4.0). Most of the cases and controls had been exposed to various types of organic solvents, including thinners, turpentine and white spirits. [The Working Group noted that no stratified analyses were presented for alcohol consumption and solvent exposure, although alcohol consumption was associated with high risk.]

A case–control study of cancer at many sites was performed in Montréal, Canada, to generate hypotheses on potential occupational carcinogens (Siemiatycki *et al.*, 1987a,b). About 20 types of cancer were included, and, for each cancer site analysed, controls were selected from among cases of cancer at other sites. Job histories and information on possible confounders were obtained by interview from 3726 men aged 35–70 years with cancer diagnosed at one of 19 participating hospitals between 1979 and 1985. The response rate was 82%. Each job was translated into a series of potential exposures by a team of chemists and hygienists using a check-list of 300 of the most common occupational exposures in Montréal. Cumulative indices of exposure were estimated for a number of occupational exposures: exposure below the median was considered to be 'nonsubstantial' and that above the median to be 'substantial'. Risks associated with exposure to petroleum–derived liquids were analysed separately. A total of 739 men were classified as having been potentially exposed to 'mineral spirits'. The term 'mineral spirits' included white spirits, Stoddard solvent, VM & P naphtha, rubber solvent, benzine and ligroin (30–90% aliphatics, 1–20% aromatics). Those with long (> 20 years), substantial potential exposure were found to have a RR for squamous–cell cancer of the lung of 1.7 (90% CI, 1.2–2.3), based on 44 cases, and a RR for prostatic cancer of 1.8 (90% CI, 1.3–2.6), based on 43 cases. Men with 'substantial' exposure also had a RR for Hodgkin's lymphoma of 2.0 (90% CI, 1.0–4.1), based on 12 cases. There was no increased risk for cancers of the bladder (1.0; 0.8–1.2; 91 cases) or kidney (1.1; 0.8–1.4; 39 cases) or for non-Hodgkin's lymphoma (0.8; 0.6–1.1; 35 cases). The risks were adjusted for age, socioeconomic status, ethnic group, cigarette smoking and blue-/white-collar job history, and for all potential confounders on which information was available. Of the 739 exposed men, 21% had been employed in the construction trade, mostly as painters.

4. Summary of Data Reported and Evaluation

4.1 Exposures

Petroleum solvents are hydrocarbon mixtures which can be grouped into three broad categories on the basis of their boiling ranges and solvent strengths, as follows: special boiling range solvents, boiling range, 30–160°C; white spirits, 130–220°C; and high–boiling aromatic solvents, 160–300°C. Within these broad solvent categories, individual solvents (typically boiling within narrower ranges of 15–30°C) are composed of aliphatic, alicyclic and aromatic hydrocarbons in varying amounts, depending on refining process and end use. Although the content of benzene in petroleum solvents is now generally less than 1% in nonhydrogenated special boiling range solvents and less than 0.1% in other solvents, higher amounts were commonly present in the past.

Exposure to petroleum solvents is widespread in many occupations, including painting, printing, use of adhesives, rubber processing and degreasing. High exposure levels have been measured in many of these occupational environments.

4.2 Experimental carcinogenicity data

A single study in rats exposed by inhalation to a high–boiling aromatic solvent was of insufficient duration to allow an evaluation of carcinogenicity.

4.3 Human carcinogenicity data

In a single case–control study of cancer at many sites, potential long, high exposure to 'mineral spirits' was associated with increased risks for squamous–cell lung cancer and prostatic cancer. In two case–control studies, one of primary liver cancer and one of Hodgkin's disease, an association with organic solvents, including white spirits, was seen. The results of these studies could not be evaluated with regard to petroleum solvents themselves.

4.4 Other relevant data

In humans, petroleum solvents cause nonallergic contact dermatitis and adverse effects on the central nervous system.

In experimental animals, samples of petroleum solvents with a high aromatic content had greater acute toxicity and were more irritating than those that were virtually aromatic-free. A special boiling range solvent containing n–hexane induced chronic toxicity in the peripheral nervous system of experimental animals.

In two studies of malformations in the children of women who had been exposed to petroleum solvents during the first trimester of pregnancy, the numbers of cases were small and the mothers had also been exposed to other substances.

A rubber solvent (special boiling range solvent) induced chromosomal aberrations but not sister chromatid exchange in cultured human cells. Another special boiling range solvent did not induce chromosomal aberrations in cultured mammalian cells, gene conversion in yeast or mutation in bacteria. A sample of white spirits did not induce chromosomal aberrations in mice *in vivo*, sister chromatid exchange in human cells or mutation in bacteria. (See Appendix 1.)

4.5 Evaluation[1]

There is *inadequate evidence* for the carcinogenicity of petroleum solvents in humans.

There is *inadequate evidence* for the carcinogenicity of high–boiling aromatic solvents in experimental animals.

No data were available on the carcinogenicity of special boiling range solvents or white spirits in experimental animals.

Overall evaluation

Petroleum solvents *are not classifiable as to their carcinogenicity to humans (Group 3)*.

5. References

Agrawal, P.K. & Srivastava, D.K. (1986) Hydrocarbon pneumonitis – a hazard of fire–eaters (Letter). *J. Assoc. Phys. India, 34,* 752

Altenburg, L.C., Ray, J.H., Smart, C.E. & Moore, F.B. (1979) Rubber solvent: a clastogenic agent that fails to induce sister–chromatid exchanges. *Mutat. Res., 67,* 331–341

American Conference of Governmental Industrial Hygienists (1988) *Threshold Limit Values and Biological Exposure Indices for 1988–1989,* Cincinnati, OH, pp. 32, 33, 38

Angerer, J. & Wulf, H. (1985) Occupational chronic exposure to organic solvents. XI. Alkylbenzene exposure of varnish workers: effects on hematopoetic system. *Int. Arch. occup. environ. Health, 56,* 307–321

Anshelm Olson, B. (1982) Effects of organic solvents on behavioral performance of workers in the paint industry. *Neurobehav. Toxicol. Teratol., 4,* 703–708

Apol, A.G. (1980) *Procraft, Spokane, Washington (Health Hazard Evaluation Determination Report No. 80–43–679),* Cincinnati, OH, National Institute for Occupational Safety and Health

Apol, A.G. (1981) *Beaufort Air–Sea Equipment Company, Seattle, Washington (Health Hazard Evaluation Report No. HETA 81–023–807),* Cincinnati, OH, National Institute for Occupational Safety and Health

Arbeidsinspectie (Labour Inspection) (1986) *De Nationale MAC–Lijst 1986* [National MAC–List 1986], Voorburg, Ministry of Social Affairs and Work Environment, p. 20

Askergren, A., Beving, H., Hagman, M., Kristensson, J., Linroth, K., Vesterberg, O. & Wennberg, A. (1988) Biological effects of exposure to water-thinned and solvent-thinned paints in house painters (Swed.). *Arb. Hälsa, 4*

---

[1]For definitions of the italicized terms, see Preamble, pp. 27–30.

Åstrand, I., Kilbom, Å. A. & Övrum, P. (1975) Exposure to white spirit. I. Concentration in alveolar air and blood during rest and exercise. *Scand. J. Work Environ. Health, 1*, 15–30

Belanger, P.L. & Elesh, E. (1979) *Kentile Floors, Inc., South Plainfield, New Jersey (Health Hazard Evaluation Report No. HE 78–72–618)*, Cincinnati, OH, National Institute for Occupational Safety and Health

Blade, L.M. (1987) *Snap–on Tools Corporation, Harrisburg, Pennsylvania (Health Hazard Evaluation Report No. 86–387–1810)*, Cincinnati, OH, National Institute for Occupational Safety and Health

Brooks, T.M., Meyer, A.L. & Hutson, D.H. (1988) The genetic toxicology of some hydrocarbon and oxygenated solvents. *Mutagenesis, 3*, 227–232

Carpenter, C.P., Kinkead, E.R., Geary, D.L., Jr, Sullivan, L.J. & King, J.M. (1975a) Petroleum hydrocarbon toxicity studies. II. Animal and human response to vapors of varnish makers' and painters' naphtha. *Toxicol. appl. Pharmacol., 32*, 263–281

Carpenter, C.P., Kinkead, E.R., Geary, D.L., Jr, Sullivan, L.J. & King, J.M. (1975b) Petroleum hydrocarbon toxicity studies. III. Animal and human response to vapors of Stoddard solvent. *Toxicol. appl. Pharmacol., 32*, 282–297

Carpenter, C.P., Kinkead, E.R., Geary, D.L., Jr, Sullivan, L.J. & King, J.M. (1975c) Petroleum hydrocarbon toxicity studies. IV. Animal and human response to vapors of rubber solvent. *Toxicol. appl. Pharmacol., 33*, 526–542

Carpenter, C.P., Geary, D.L., Jr, Meyers, R.C., Nachreiner, D.J., Sullivan, L.J. & King, J.M. (1977) Petroleum hydrocarbon toxicity studies. XIV. Animal and human response to vapors of 'high aromatic solvent'. *Toxicol. appl. Pharmacol., 41*, 235–249

Cherry, N., Hutchins, H. & Waldron, H.A. (1985) Neurobehavioral effects of repeated occupational exposure to toluene and paint solvents. *Br. J. ind. Med., 42*, 291–300

Chrostek, W.J., Flesch, J.P. & DiCostanza, M. (1979) *Federal Communications Commission, Washington, DC (Hazard Evaluation and Technical Assistance Report No. TA 79–7)*, Cincinnati, OH, National Institute for Occupational Safety and Health

Clark, D.G. & Bird, M. (1989) Inhalation toxicity of high flash aromatic naphtha. *Toxicol. ind. Health* (in press)

Cohr, K.-H., Stokholm, J. & Bruhn, P. (1980) *Neurologic response to white spirit exposure*. In: Holmstedt, B., Lauwerys, R., Mercier, M. & Roberfroid, M., eds, *Mechanisms of Toxicity and Hazard Evaluation*, Amsterdam, Elsevier/North–Holland Biomedical Press, pp. 95–102

Cook, W.A. (1987) *Occupational Exposure Limits – Worldwide*, Washington DC, American Industrial Hygiene Association, pp. 34, 125, 202, 215

Direktoratet for Arbeidstilsynet (Directorate for Labour Inspection) (1981) *Administrative Normer for Forurensning i Arbeidsatmosfaere 1981* [Administrative Norms for Pollution in Work Atmosphere 1981] *(No. 361)*, Oslo, pp. 21–22

Ekberg, K., Barregård, L., Hagberg, S. & Sällsten, G. (1986) Chronic and acute effects of solvents on central nervous system functions in floorlayers. *Br. J. ind. Med., 43*, 101–106

Eller, P.M. (1984) *NIOSH Manual of Analytical Methods*, 3rd ed., Vol. 1 *(DHHS (NIOSH) Publ. No. 84–100)*, Washington DC, US Government Printing Office, pp. 1550-1–1550-5

Elofsson, S.-A., Gamberale, F., Hindmarsh, T., Iregren, A., Isaksson, A., Johnsson, I., Knave, B., Lydahl, E., Mindus, P., Persson, H.E., Philipson, B., Steby, M., Struwe, G., Söderman, E., Wennberg, A. & Widén, L. (1980) Exposure to organic solvents. A cross–sectional epidemiologic investigation on occupationally–exposed car and industrial spray painters with special reference to the nervous system. *Scand. J. Work Environ. Health, 6*, 239–273

Franchini, I., Cavatorta, A., Falzoi, M., Lucertini, S. & Mutti, A. (1983) Early indicators of renal damage in workers exposed to organic solvents. *Int. Arch. occup. environ. Health*, *52*, 1-9

Gamberale, F., Annwall, G. & Hultengren, M. (1975) Exposure to white spirit. II. Psychological functions. *Scand. J. Work Environ. Health*, *1*, 31-39

Geissert, J.O. (1975) *Russell Corporation, Alexander City, Alabama (Health Hazard Evaluation Determination Report No. 75-90-236)*, Cincinnati, OH, National Institute for Occupational Safety and Health

Gilles, D. (1975) *Babcock and Wilcox Company, Canton, Ohio (Health Hazard Evaluation Determination Report No. 75-26-245)*, Cincinnati, OH, National Institute for Occupational Safety and Health

Gochet, B., de Meester, C., Léonard, A. & Deknudt, G. (1984) Lack of mutagenic activity of white spirit. *Int. Arch. occup. environ. Health*, *53*, 359-364

Gorman, R. (1982) *Arts Consortium, Cincinnati, Ohio (Health Hazard Evaluation Determination Report No. 82-008-1226)*, Cincinnati, OH, National Institute for Occupational Safety and Health

Gunter, B. (1975a) *Frontier Airlines, Denver, Colorado (Health Hazard Evaluation Determination Report No. 74-138-208)*, Cincinnati, OH, National Institute for Occupational Safety and Health

Gunter, B.J. (1975b) *Lange Company, Broomfield, Colorado (Health Hazard Evaluation Determination Report No. 74-148-239)*, Cincinnati, OH, National Institute for Occupational Safety and Health

Gunter, B.J. (1980) *Mountain Bell, Denver, Colorado (Health Hazard Evaluation Determination Report No. 79-142-697)*, Cincinnati, OH, National Institute for Occupational Safety and Health

Gunter, B.J. (1982) *Jeppesen Sanderson, Englewood, Colorado (Health Hazard Evaluation Report No. HETA 81-261-1085)*, Cincinnati, OH, National Institute for Occupational Safety and Health

Gunter, B.J. (1986) *Hirschfield Press, Denver, Colorado (Health Hazard Evaluation Report No. HETA 85-137-1648)*, Cincinnati, OH, National Institute for Occupational Safety and Health

Hane, M., Axelson, O., Blume, J., Hogstedt, C., Sundell, L. & Ydreborg, B. (1977) Psychological function changes among house painters. *Scand. J. Work Environ. Health*, *3*, 91-99

Hänninen, H., Eskelinen, L., Husman, K. & Nurminen, M. (1976) Behavioral effects of long-term exposure to a mixture of organic solvents. *Scand. J. Work Environ. Health*, *4*, 240-255

Hardell, L., Bengtsson, N.O., Jonsson, U., Eriksson, S. & Larsson, L.G. (1984) Aetiological aspects on primary liver cancer with special regard to alcohol, organic solvents and acute intermittent porphyria – an epidemiological investigation. *Br. J. Cancer*, *50*, 389-397

Health and Safety Executive (1987) *Occupational Exposure Limits, 1987 (Guidance Note EH 40/87)*, London, Her Majesty's Stationery Office, p. 22

Hine, C.H. & Zuidema, H.H. (1970) The toxicological properties of hydrocarbon solvents. *Ind. Med.*, *39*, 215-220

Hollett, B.A., Thoburn, T. & Lucas, J.B. (1976) *Cincinnati Enquirer, Cincinnati, Ohio (Health Hazard Evaluation Determination Report No. 75-187-329)*, Cincinnati, OH, National Institute for Occupational Safety and Health

Holmberg, P.C. (1979) Central-nervous-system defects in children born to mothers exposed to organic solvents during pregnancy. *Lancet*, *ii*, 177-179

Holmberg, P.C., Hernberg, S., Kurppa, K., Rantala, K. & Riala, R. (1982) Oral clefts and organic solvent exposure during pregnancy. *Int. Arch. occup. environ. Health*, *50*, 371-376

Husman, K. (1980) Symptoms of car painters with long-term exposure to a mixture of organic solvents. *Scand. J. Work Environ. Health*, *6*, 19-32

Husman, K. & Karli, P. (1980) Clinical neurological findings among car painters exposed to a mixture of organic solvents. *Scand. J. Work Environ. Health*, *6*, 33-39

IARC (1982) *IARC Monographs on the Evaluation of the Carcinogenic Risk of Chemicals to Humans*, Vol. 28, *The Rubber Industry*, Lyon, pp. 183-221

IARC (1989) *IARC Monographs on the Evaluation of Carcinogenic Risks to Humans*, Vol. 45, *Occupational Exposures in Petroleum Refining, Crude Oil and Major Petroleum Fuels*, Lyon

Institut National de Recherche et de Sécurité (National Institute for Research and Safety) (1986) *Valeurs Limites pour les Concentrations des Substances Dangereuses Dans l'Air des Lieux de Travail* [Limit Values for the Concentrations of Toxic Substances in the Air of Work Places] (*ND 1609-125-86*), Paris, p. 577

International Labour Office (1984) *Occupational Exposure Limits for Airborne Toxic Substances*, 2nd rev. ed. (*Occupational Safety and Health Series 37*), Geneva, pp. 186, 190-191, 217

International Programme on Chemical Safety (1982) *Selected Petroleum Products* (*Environmental Health Criteria 20*), Geneva, World Health Organization, pp. 34-61

Jakobsen, B.M., Hass, U., Juul, F. & Kjaergaard, S. (1986) Prenatal toxicity of white spirit inhalation in the rat (Abstract). *Teratology, 34*, 415

Kasahara, M., Suzuki, H., Takeuchi, Y., Hara, I. & Ikeda, M. (1987) *n*-Hexane, benzene and other aromatic components in petroleum distillate solvents in Japan. *Ind. Health, 25*, 205-214

Klauder, J.V. & Brill, F.A., Jr (1947) Correlation of boiling ranges of some petroleum solvents with irritant action on skin. *Arch. Dermatol. Syph., 56*, 197-215

Kronoveter, K.J. (1977) *Herald Times, Bloomington, Indiana* (*Health Hazard Evaluation Determination Report No. 76-96-390*), Cincinnati, OH, National Institute for Occupational Safety and Health

Lauwerys, R., Bernard, A., Viau, C. & Buchet, J.-P. (1985) Kidney disorders and hematotoxicity of organic solvent exposure. *Scand. J. Work Environ. Health, 11* (*Suppl. 1*), 83-90

Lebowitz, H., Brusick, D., Matheson, D., Jagannath, D.R., Reed, M., Goode, S. & Roy, G. (1979) Commonly used fuels and solvents evaluated in a battery of short-term bioassays (Abstract No. Eb-8). *Environ. Mutagenesis, 1*, 172-173

Levy, B.S.B. (1975) *Safeguard Automotive Company, Baltimore Maryland* (*Health Hazard Evaluation Determination Report No. 74-144-211*), Cincinnati, OH, National Institute for Occupational Safety and Health

Lindström, K. & Wickström, G. (1983) Psychological function changes among maintenance house painters exposed to low levels of organic solvent mixtures. *Acta psychiatr. scand., 67* (*Suppl. 303*), 81-91

Lucas, C. (1981) *Hayes Albion Corp., Spencerville, Ohio* (*Health Hazard Evaluation Determination Report No. HETA 81-186-924*), Cincinnati, OH, National Institute for Occupational Safety and Health

Markel, H.L., Jr & Shama, S.K. (1974) *Whirlpool, Corp., Fort Smith, Arkansas* (*Health Hazard Evaluation Report No. 72-100-121*), Cincinnati, OH, National Institute for Occupational Safety and Health

McMichael, A.J., Spirtas, R., Kupper, L.L. & Gamble, J.F. (1975) Solvent exposure and leukemia among rubber workers: an epidemiologic study. *J. occup. Med., 17*, 234-239

Milling Pedersen, L. & Cohr, K.-H. (1984) Biochemical pattern in experimental exposure of humans to white spirit. II. The effects of repetitive exposures. *Acta pharmacol., 55*, 325-330

Milling Pedersen, L., Larsen, K. & Cohr, K.-H. (1984) Kinetics of white spirit in human fat and blood during short-term experimental exposure. *Acta pharmacol. toxicol., 55*, 308-316

Milling Pedersen, L., Rasmussen, S. & Cohr, K.-H. (1987) Further evaluation of the kinetics of white spirit in human volunteers. *Pharmacol. Toxicol., 60*, 135-139

National Institute for Occupational Safety and Health (1974) *National Occupational Hazard Survey 1972-1974*, Cincinnati, OH, pp. 92, 132

National Institute for Occupational Safety and Health (1977) *Criteria for a Recommended Standard ... Occupational Exposure to Refined Petroleum Solvents*, Cincinnati, OH, Department of Health, Education and Welfare, pp. 81-101

National Institute for Occupational Safety and Health (1983) *National Occupational Exposure Survey 1981-1983*, Cincinnati, OH

National Institute for Occupational Safety and Health (1986) NIOSH recommendations for occupational safety and health standards. *Morb. Mortal. Wkly Rep., 35 (Suppl.),* 28S

National Paint and Coatings Association (1984) *Raw Material Index*, Washington DC

National Swedish Board of Occupational Safety and Health (1987) *Hygieniska Gränsvärden* [Hygienic Limit Values] *(Ordinance AFS 1987:12)*, Solna, p. 30

Nethercott, J.R., Pierce, J.M., Likwornick, G. & Murray, A.H. (1980) Genital ulceration due to Stoddard solvent. *J. occup. Med., 22*, 549-552

Niemelä, R., Pfäffli, P. & Härkönen, H. (1987) Ventilation and organic solvent exposure during car washing. *Scand. J. Work Environ. Health, 13*, 424-430

Oberg, M. (1968) A survey of the petroleum solvent inhalation exposure in Detroit dry cleaning plants. *Am. ind. Hyg. Assoc. J., 29*, 547-550

Olsson, H. & Brandt, L. (1980) Occupational exposure to organic solvents and Hodgkin's disease in men. A case-referent study. *Scand. J. Work Environ. Health, 6*, 302-305

Ono, Y., Takeuchi, Y., Hisanaga, N., Iwata, M., Kitoh, J. & Sugiura, Y. (1982) Neurotoxicity of petroleum benzine compared with n-hexane. *Int. Arch. occup. environ. Health, 50*, 219-229

Phillips, R.D. & Egan, G.F. (1981) Teratogenic and dominant lethal investigation of two hydrocarbon solvents (Abstract No. 53). *Toxicologist, 1*, 15

Phillips, R.D. & Egan, G.F. (1984) Subchronic inhalation exposure of dearomatized white spirit and C_{10}-C_{11} isoparaffinic hydrocarbon in Sprague-Dawley rats. *Fundam. appl. Toxicol., 4*, 808-818

Prager, D. & Peters, C. (1970) Development of aplastic anemia and the exposure to Stoddard solvent. *Blood, 35*, 286-287

Prockop, L. (1979) Neurotoxic volatile substances. *Neurology, 29*, 862-865

Rector, D.E., Steadman, B.L., Jones, R.A. & Siegel, J. (1966) Effects on experimental animals of long-term inhalation exposure to mineral spirits. *Toxicol. appl. Pharmacol., 9*, 257-268

Riley, A.J., Collings, A.J., Browne, N.A. & Grasso, P. (1984) Response of the upper respiratory tract of the rat to white spirit vapour. *Toxicol. Lett., 22*, 125-131

Rivera, R.O. & Rostand, R. (1975) *Hillerich & Bradsby Co., Jeffersonville, Indiana (Health Hazard Evaluation Determination Report No. 74-121-203)*, Cincinnati, OH, National Institute for Occupational Safety and Health

Schreiner, C.A., Edwards, D.A., McKee, R.H., Swanson, M., Wong, Z.A., Schmitt, S. & Beatty, P. (1989) The mutagenic potential of high flash aromatic naphtha. *Cell Biol. Toxicol., 5*, 169-188

Seppäläinen, A.M. & Lindström, K. (1982) Neurophysiological findings among house painters exposed to solvents. *Scand. J. Work Environ. Health, 8*, 131-135

Seppäläinen, A.M., Husman, K. & Mårtenson, C. (1978) Neurophysiological effects of long-term exposure to a mixture of organic solvents. *Scand. J. Work Environ. Health, 4*, 304-314

Siemiatycki, J., Dewar, R., Nadon, L., Gérin, M., Richardson, L. & Wacholder, S. (1987a) Associations between several sites of cancer and twelve petroleum-derived liquids. Results from a case-referent study in Montreal. *Scand. J. Work Environ. Health, 13*, 493-504

Siemiatycki, J., Wacholder, S., Richardson, L., Dewar, R. & Gérin, M., (1987b) Discovering carcinogens in the occupational environment: methods of data collection and analysis of a large case-referent monitoring system. *Scand. J. Work Environ. Health, 13*, 486–492

Spruit, D., Malten, K.E., Lipmann, E.W.R.M. & Liang, T.P. (1970) Horny layer injury by solvents. II. Can the irritancy of petroleum ether be diminished by pretreatment. *Berufsdermatosen, 18*, 269–280

SRI International (1986) *The US Paint Industry: Technology Trends, Markets, Raw Material (formerly NPCA Data Bank Program)*, Menlo Park, CA, pp. 146–150, 209–210

Tharr, D.G., Murphy, D.C. & Mortimer, V. (1982) *Red Wing Shoe Co., Red Wing, Minnesota (Health Hazard Evaluation Report No. HETA 81–455–1229)*, Cincinnati, OH, National Institute for Occupational Safety and Health

Työsuojeluhallitus (National Finnish Board of Occupational Safety and Health) (1987) *HTP–Arvot 1987* [TLV Values 1987], Helsinki, Valtion Painatuskeskus, p. 19

US Food and Drug Administration (1988) Petroleum naphtha. *US Code fed. Regul., Title 21*, Part 172.250, pp. 33–34

Viau, C., Bernard, A. & Lauwerys, R. (1984) Distal tubular dysfunction in rats chronically exposed to a 'white spirit' solvent. *Toxicol. Lett., 21*, 49–52

Watanabe, A.S. & Love, J.R. (1980) *Social Security Administration, Findlay, Ohio (Technical Assistance Report No. 79–56)*, Cincinnati, OH, National Institute for Occupational Safety and Health

Wilcosky, T., Checkoway, H., Marshall, E.G. & Tyroler, H.A. (1984) Cancer mortality and solvent exposures in the rubber industry. *Am. ind. Hyg. Assoc. J., 45*, 809–811

TOLUENE

1. Chemical and Physical Data

1.1 Synonyms

Chem. Abstr. Services Reg. No.: 108–88–3
Chem. Abstr. Name: Methylbenzene
IUPAC Systematic Name: Toluene
Synonyms: Methylbenzol; NCI–CO7272; phenylmethane; toluol

1.2 Structural and molecular formulae and molecular weight

C₇H₈

C_7H_8

Mol. wt: 92.15

1.3 Chemical and physical properties of the pure substance

(a) *Description*: Clear, colourless, inflammable liquid with benzene–like odour (Sandmeyer, 1981; Windholz, 1983)

(b) *Boiling–point*: 110.6°C (Weast, 1985)

(c) *Melting–point*: –95°C (Weast, 1985)

(d) *Density*: 0.8669 (20°/4°C) (Weast, 1985)

(e) *Spectroscopy data*: Infrared, ultraviolet and nuclear magnetic resonance spectral data have been reported (Sadtler Research Laboratories, 1980; Pouchert, 1981, 1983, 1985).

(f) *Solubility*: Soluble in ethanol, benzene, diethyl ether, acetone, chloroform, glacial acetic acid and carbon disulfide; insoluble in water (Hawley, 1981; Sandmeyer, 1981; Windholz, 1983; Weast, 1985)

(g) *Volatility*: Vapour pressure: 28.4 mm Hg at 25°C (Eller, 1984)

(h) *Flash–point*: 4.4°C (Sandmeyer, 1981)

(i) *Reactivity*: Quite stable in air (Clement Associates, 1977). Reacts photochemically with nitrogen oxides or halogens to form nitrotoluene, nitrobenzene and nitrophenol and halogenated products, respectively (Merian & Zander, 1982; US Environmental Protection Agency, 1983)

(*j*)   *Octanol/water partition coefficient*: log P = 2.11–2.80 (Hansch & Leo, 1979)

(*k*)   *Conversion factor*: mg/m³ = 3.77 x ppm[1]

1.4   Technical products and impurities

Trade Names: Antisal 1a; CP 25; Methacide

Toluene is marketed principally as nitration and industrial grades, its purity being dependent on the specific gravity and boiling range of the product (Hoff, 1983). Reagent–grade toluene is available with a purity of greater than 99% (Aldrich Chemical Co., 1988). Technical grades (90–120°C boiling range) are less pure and may contain up to 25% benzene as well as other hydrocarbons (Clement Associates, 1977; Fishbein, 1985).

2. Production, Use, Occurrence and Analysis

2.1   Production and use

(*a*)   *Production*

Toluene is produced during petroleum refining operations, directly as a by–product of styrene manufacture and indirectly as a by–product of coke–oven operations.

It is produced from petroleum as an aromatic mixture with benzene and xylene primarily by catalytic reforming and pyrolytic cracking. Catalytic reforming processes account for about 87% of the total amount of toluene produced in the USA. This process involves dehydrogenation of selected petroleum fractions containing abundant naphthenic hydrocarbons to yield a mixture of aromatics and paraffins. Reforming processes are used to produce a benzene–toluene–xylene reformate from which the individual aromatics are recovered by distillation, washing with nitric acid and redistillation. Only a small fraction of the reformate is used for isolation of the toluene; the bulk of the unseparated toluene in the reformate is used for gasoline blending.

The second largest source of toluene is from pyrolysis gasoline, formed as a by–product during pyrolytic cracking (steam cracking) of heavier hydrocarbons for the manufacture of olefins. Toluene is isolated from pyrolysis gasoline by distillation, removal of olefins and diolefins and redistillation.

---

[1]Calculated from: mg/m³ = (molecular weight/24.45) x ppm, assuming standard temperature (25°C) and pressure (760 mm Hg)

Toluene is also obtained as a by-product during styrene manufacture when ethylbenzene is dehydrogenated. The toluene isolated from the by-product is used for gasoline blending or as feed for benzene manufacture by the hydrodealkylation process. The production of toluene from coke-oven operations is minimal (Hoff, 1983).

The amounts of isolated toluene (and of total toluene) produced in these different ways in the USA in 1978 were as follows: from catalytic reformate, 3.6 (29.5) million tonnes; from pyrolysis gasoline, 376.8 (708) thousand tonnes; as a styrene by-product, 99.8 (145) thousand tonnes; and derived from coal, 65.8 (79.5) thousand tonnes. These quantities represent a total of 4.2 (30) million tonnes (Fishbein, 1985).

Western Europe and Japan are also major producers of this compound. In 1980, over 85% of the toluene produced in the world was accounted for by the USA, western Europe and Japan. In Japan and western Europe, toluene is produced mainly from pyrolysis gasoline. In all three areas, coke-oven light oil provided less than 10% of the toluene supply in 1980 (Fishbein, 1985).

Data on the production of toluene in the major producing countries are presented in Table 1. World production of toluene in 1980 was estimated at more than 5 million tonnes – approximately one-third of the amount of benzene produced. However, an additional 30 million tonnes of toluene are consumed annually as a constituent of motor fuel (Merian & Zander, 1982).

Table 1. Annual production data for toluene^a (thousands of tonnes)

Country	1983	1984	1985	1986	1987
Canada	411	395	472	393	396
France	41	39	40	38	33
Germany, Federal Republic of	314	371	391	478	402
Italy	299	313	348	233	178
Japan	831	784	803	805	882
Mexico	223	216	220	238	313
USA	2560^b	2390^b	2300^b	2640	3050

^{a}From US International Trade Commission (1984, 1985, 1986); Anon. (1987a, 1988)
^{b}Petroleum-derived, not including tar distillation and coke-oven derived toluene

(b) Use

The largest single use of isolated toluene is in the production of benzene *via* the hydro-demethylation process, in which toluene and hydrogen (a reformate by-product) are reacted under high temperature and pressure to yield benzene and methane (Hoff, 1983). This process has been used to balance the supply and demand for benzene (Mannsville Chemical Products Corp., 1981).

The second largest use of toluene is in solvent applications, especially in the paint and coating industry. Significant amounts are also used in inks, adhesives, the leather industry (IARC, 1981), pharmaceuticals and other formulated products. Solvents accounted for 40%

or more of the nonfuel use of toluene in Japan and western Europe in 1980. In the USA in 1981, the use of toluene as a solvent was second only to its use in benzene production *via* hydrodemethylation and accounted for about 26% of nonfuel consumption (Fishbein, 1985).

Isolated toluene is also used directly in several consumer products, such as sanitizing agents, household aerosols, paints and varnishes, paint thinners and antirust preservatives (Fishbein, 1985).

Most of the toluene in the benzene–toluene–xylene mixtures, which is never isolated and remains in various refinery streams, is used in gasoline blending. Toluene has several advantages as a blending agent in gasoline: a high octane number and low volatility, and it blends easily with other inexpensive materials such as *n*–butane, which is highly volatile (Hoff, 1983). It is anticipated that the use of toluene in unleaded gasoline will continue to increase. In 1985, 73–75% of the gasoline used in the USA was unleaded. By 1990, this percentage is expected to rise to 95–100% (Mannsville Chemical Products Corp., 1981). The toluene concentrations in US gasolines are estimated to range from 5 to 22% (wt%; IARC, 1989).

Toluene is used as an intermediate in the production of toluene diisocyanate for use in polyurethane production, and of benzoic acid for use in the manufacture of benzoate and benzyl esters and salts for food preservatives and cosmetic articles such as soaps, perfumes, flavours, creams and lotions. Catalytic disproportionation of toluene has been used to produce benzene and *para*-xylene, with little or no ethylbenzene or *ortho*- or *meta*-xylene. Vinyl toluene, which is produced by alkylation of toluene with ethylene followed by dehydrogenation of ethyltoluene, is used as a modifier in unsaturated polyester resins. Other important chemical products made from toluene include trinitrotoluene and related explosives, benzaldehyde (an important chemical intermediate) and saccharin (Hoff, 1983; US Environmental Protection Agency, 1983; Fishbein, 1985). Small amounts of toluene are used for the manufacture of *para*-cresol, which is used primarily for the manufacture of butylated hydroxytoluene (US Environmental Protection Agency, 1983).

In western Europe, phenol (see monograph, p. 263) is the most important derivative of toluene, followed by toluene diisocyanate and caprolactam. In Japan, toluene is used primarily for benzene and *para*-cresol production (Fishbein, 1985).

Of the estimated 3.3 million tonnes of toluene produced in the USA in 1980, 44% was used to make benzene, 34% to make gasoline, 10% in solvents, 6% to make toluene diisocyanate, and 6% for miscellaneous use (Mannsville Chemical Corp., 1981).

(c)   *Regulatory status and guidelines*

Occupational exposure limits for toluene in 34 countries or regions are presented in Table 2.

2.2 Occurrence

(a)   *Natural occurrence*

Toluene occurs in nature in crude oil (US Environmental Protection Agency, 1983), natural gas deposits and the volatile emissions from volcanoes and forest fires (National Research Council, 1976).

Table 2. Occupational exposure limits for toluene[a]

Country or region	Year	Concentration[b] (mg/m³)	Interpretation[c]
Australia	1984	380	TWA
Austria	1985	750	TWA
Belgium	1985	375	TWA
Brazil	1985	S 290	TWA
Bulgaria	1984	50	TWA
Commission of the European Communities	1986	375	TWA
		1875	Maximum
Chile	1985	S 300	TWA
China	1985	100	TWA
Czechoslovakia	1985	200	Average
		1000	Maximum
Denmark	1988	S 190	TWA
Egypt	1985	375 (100 ppm given)	TWA
Finland	1987	S 375	TWA
		S 565	STEL (15 min)
France	1986	375	TWA
		550	STEL (15 min)
German Democratic Republic	1985	200	TWA
		600	STEL
Germany, Federal Republic of	1988	380	TWA
Hungary	1985	100	TWA
		500	STEL
India	1985	S 375	TWA
		S 560	STEL
Indonesia	1985	375	TWA
Italy	1985	S 300	TWA
Japan	1988	375	TWA
Korea, Republic of	1985	375	TWA
		560	STEL
Mexico	1985	S 750	TWA
Netherlands	1986	S 375	TWA
Norway	1981	280	TWA
Poland	1985	100	TWA
Romania	1985	300	Average
		400	Maximum
Sweden	1987	200	TWA
		400	STEL
Switzerland	1985	S 380	TWA
Taiwan	1985	S 375	TWA
UK	1987	S 375	TWA
		S 560	STEL (10 min)

Table 2 (contd)

Country or region	Year	Concentration[b] (mg/m³)	Interpretation[c]
USA[d]			
OSHA	1988	430	TWA
NIOSH	1986	375	TWA
		750	Ceiling
ACGIH	1988	375	TWA
		560	STEL (15 min)
USSR	1986	50	Ceiling
Venezuela	1985	S 375	TWA
		S 560	Ceiling
Yugoslavia	1985	200	TWA

[a]From Direktoratet for Arbeidstilsynet (1981); International Labour Office (1984); Arbeidsinspectie (1986); Commission of the European Communities (1986); Institut National de Recherche et de Sécurité (1986); National Institute for Occupational Safety and Health (1986); Cook (1987); Health and Safety Executive (1987); National Swedish Board of Occupational Safety and Health (1987); Työsuojeluhallitus (1987); American Conference of Governmental Industrial Hygienists (1988); Arbejdstilsynet (1988); Deutsche Forschungsgemeinschaft (1988)

[b]S, skin notation

[c]TWA, 8-h time-weighted average; STEL, short-term exposure limit

[d]OSHA, Occupational Safety and Health Administration; NIOSH, National Institute for Occupational Safety and Health; ACGIH, American Conference of Governmental Industrial Hygienists

(b)   *Occupational exposure*

On the basis of a US National Occupational Exposure Survey, the National Institute for Occupational Safety and Health (1983) estimated that 1 278 000 workers were potentially exposed to toluene in the USA in 1981–83.

Levels of toluene measured in the air in work environments are summarized in Table 3. In the majority of these environments, concurrent exposure to other solvents is likely to have taken place.

Biological monitoring measurements have also been made. Exposures in the manufacture of trapezoid belts resulted in urinary hippuric acid concentrations of 2.1 g/l in workers in the belt department and 9 g/l in those in the weighing room (Capellini & Alessio, 1971). Exposures to toluene on automatic spray finishing machines in a leather finishing operation resulted in urinary hippuric acid levels ranging from 1.5 to 3.66 g/l, with an average of 2.38 g/l. All nine samples were taken at the end of the work shift. Exposures in the washing and topping department in the same plant resulted in urinary hippuric acid levels of 2.16–5.85 g/l, with an average of 4.48 g/l. Concentrations of toluene in a rubber coating plant resulted in post-shift urinary hippuric acid levels of 2.75–6.8 g/l (average, 3.66; Pagnotto & Lieberman, 1967). Post-shift results of biological monitoring of 35 toluene-exposed printing workers ranged from 0.09 to 3.13 mg/l toluene in blood (average, 1.55 mg/l), 0.33 to 11.6 g/l hippuric

Table 3. Occupational exposures to toluene

Environment	Sampling[a]	Concentration in air	Reference
Printing plants			
Rotogravure plant (Finland)	8-h TWA personal	7–112 ppm (26.4–422 mg/m³)	Mäki-Paakkanen et al. (1980)
Printing plant (Japan)	8-h TWA personal	27.1–53.7 ppm (102–203 mg/m³)	Tokunaga et al. (1974)
Printing plant (Italy)	7-h TWA personal	37–229 mg/m³	De Rosa et al. (1985)
Heliorotogravure printers (Belgium)	Personal		Veulemans et al. (1979)
1st printer		102–667 mg/m³	
2nd printer		120–706 mg/m³	
Helper		81–680 mg/m³	
Printing plant (FRG)	Area	13–49 ppm (48.9–185 mg/m³)	Angerer (1979a)
Printing plant (FRG)	Area	36–269 ppm (136–1014 mg/m³)	Angerer (1985)
Photogravure printing factories (Japan)	Area	4–240 ppm (15–905 mg/m³, average)	Ikeda & Ohtsuji (1969)
Manufacture of trapezoid belts (Italy)			Capellini & Alessio (1971)
Belt department	Air, personal	125 ppm (471 mg/m³, average)	
Weighing room	Air	250 ppm (942 mg/m³, average)	
Waste incinerator (USA)	8-h TWA personal		Decker et al. (1983)
Incinerator workers		0.19 ppm (0.7 mg/m³)	
Laboratory technicians		0.09 ppm (0.3 mg/m³)	
Waste receivers		0.02 ppm (0.1 mg/m³)	
Unloading tank trucks		0.2 ppm (0.8 mg/m³)	
Tank entry (outside)		15 ppm (57 mg/m³)	
Tank entry (inside)		104 ppm (392 mg/m³)	
Plastic processing factories (FRG)	8-h TWA personal	191–309 ppm (720–1165 mg/m³, mean)	Konietzko et al. (1980)

Table 3 (contd)

Environment	Sampling[a]	Concentration in air	Reference
Rubber tyre vulcanization (USA)	Area	0.75–1.5 ppm (2.83–5.66 mg/m^3)	Rappaport & Fraser (1977)
Leather finishing (USA)	Short-term area		Pagnotto & Lieberman (1967)
Automatic spray finishing		19–85 ppm (71–320 mg/m^3)	
Washing and topping		29–195 ppm (109–735 mg/m^3)	
Laboratories (USA)			
Histology	Short-term area	8.9–12.6 ppm (33.6–47.5 mg/m^3)	Kilburn et al. (1985)
Histopathology	8-h TWA personal	2.0–4.2 ppm (7.5–15.8 mg/m^3)	Roper (1980)
Cytopathology	8-h TWA personal	0.17–3.15 ppm (0.6–11.8 mg/m^3)	Roper (1980)
Lithography (Poland)			Moszczyński & Lisiewicz (1985)
1968		ND–420 mg/m^3	
1969		ND–580 mg/m^3	
1972		ND–30 mg/m^3	
1976		ND–81 mg/m^3	
1978		ND–93 mg/m^3	
Manufacture of photographic albums (USA)	TWA personal	0.9–20.0 mg/m^3	Baker & Fannick (1983)
Manufacture of tarpaulins (Finland)	8-h TWA personal	20–200 ppm (75–750 mg/m^3)	Tähti et al. (1981)
Fibrous glasswool plant (USA)	8-h TWA personal	22–66 mg/m^3	Dement et al. (1973)
Golf club and baseball bat manufacturing plant (USA)	8-h TWA personal	3–8 ppm (11–30 mg/m^3)	Rivera & Rostand (1975)
Laminating kitchen counter and bathroom tops (USA)	8-h TWA personal	36–253 mg/m^3	Apol (1980)
Rubber coating plant (USA)	Short-term area	34–120 ppm (128–452 mg/m^3)	Pagnotto & Lieberman (1967)

Table 3 (contd)

Environment	Sampling[a]	Concentration in air	Reference
Rubber sheet manufacture (UK)	TWA personal	3–280 ppm (11–1050 mg/m^3) mean, 57 ppm (215 mg/m^3)	Campbell et al. (1987)
Parquet floorers (FRG)	8-h TWA personal	mean, 86.7 mg/m^3 (max, 750 mg/m^3)	Denkhaus et al. (1986)
Shoemakers (Japan)		15–200 ppm (57–754 mg/m^3, average)	Matsushita et al. (1975)

[a]TWA, time–weighted average
[b]ND, not detected

acid in urine (average, 5.03 g/l), <0.1 to 10.6 mg/l *ortho*-cresol in urine (average, 3.11 mg/l) and 0.1 to 27.1 mg/l phenol in urine (average, 5.29 mg/l; Angerer, 1985). In workers in a factory in the UK that manufactured rubber sheets used in the printing industry, blood toluene levels were 10–18 μmol (0.9–1.6 mg)/l; pre- and post-shift levels of exhaled toluene ranged from 320 to 542 nmol (30–500 μg)/l and urinary hippuric acid levels were 0.72–1.01 mmol (66–93 mg)/mmol creatinine over four years (Campbell *et al.*, 1987). The average concentration of toluene in the blood of parquet floorers was 99 μg/l (max, 2550 μg/l) (Denkhaus *et al.*, 1986).

Occupational exposure of painters and paint manufacturing workers to toluene is described in the monograph on occupational exposures in paint manufacture and painting (see p. 329). Occupational exposures to toluene in petroleum refining and in the production and use of petroleum fuels are described in Volume 45 of the *Monographs* (IARC, 1989).

(c) Air

Toluene is released into the environment during its production, processing (*via* distillation vents), loading and handling and in transportation and storage operations.

Merian (1982) estimated worldwide atmospheric emissions of toluene to be 6.2 million tonnes. Contributions included losses from refineries (40%), automobile exhausts (32%), solvents (16%), petroleum losses to the sea (8%), losses from the chemical industry (2%) and gasoline evaporation (0.8%).

In the USA, total annual emissions of toluene were estimated to be about 450 thousand tonnes, 99.3% of which was released into the atmosphere and 0.7% into waste waterways (Clement Associates, 1977). The US Environmental Protection Agency (1983) estimated that atmospheric emissions of toluene in the USA during its production in 1979 were 3.0 tonnes/year from catalytic reforming, 0.5 tonnes/year from pyrolytic cracking, 0.1 tonnes/year as a styrene by-product and and 0.2 tonnes/year as a coke oven by-product. In 1979, US emissions were estimated to be about 1 million tonnes, 90% of the loss being due to evaporation of gasoline and automobile exhaust emissions (Fishbein, 1985). In Japan, 250 and 600 thousand tonnes of toluene were lost to the environment in 1976 and 1974, respectively, through its use as a solvent in paint and printing ink industries (Merian & Zander, 1982).

Toluene is transported rapidly from water (where it has low solubility) into the atmosphere. Its half-life in water (1 m deep) is about 5 h; that in the atmosphere is 13 h. It is removed from the atmosphere primarily by reactions with atomic oxygen, aryl- or alkyl-peroxy or hydroxyl radicals, and ozone. Because of its rapid oxidation, toluene would not remain long enough in the atmosphere to be influenced by air-to-surface transfer mechanisms (International Programme on Chemical Safety, 1985).

The tropospheric lifetime of toluene is four days, and average worldwide distribution is approximately 0.00075 mg/m³ air. Average atmospheric concentrations of 0.0005–1.31 mg/m³ have been measured, with the highest level being 5.5 mg/m³, in studies from Europe, Canada and the USA, between 1971 and 1980. In the vicinity of an automobile painting plant, levels of 0.06–0.6 mg/m³ were reported 16.5–1.6 km downwind from the painting facility, compared with 0.006 mg/m³ upwind (International Programme for Chemical Safety, 1985). Concentrations of 42 mg/m³ were recorded in the air in the vicinity of a chemical reclamation plant after residents had complained of odour and illnesses (US Environmental Protection Agency, 1983).

Mean atmospheric concentrations of toluene in urban areas around the world in 1971–80 include (in mg/m^3): 0.04 in Canada, 0.002–0.2 in the Federal Republic of Germany, 0.03 in Finland, 0.02 in Japan, 0.02–0.07 in the Netherlands, 0.03–0.05 in South Africa, 0.005 in Sweden, 0.04–0.06 in Switzerland and 0.02–0.06 in the UK (Merian & Zander, 1982; US Environmental Protection Agency, 1983). De Bortoli *et al.* (1984) reported 0.007–0.156 mg/m^3 toluene in 15 samples collected in outdoor air in northern Italy. In the USA, measurements were recorded between 1967 and 1978 for atmospheric concentrations in both urban and rural sites in five major regions of the country. The highest mean concentration was reported in the eastern region (0.15 mg/m^3 in New York and New Jersey), followed by 0.14 mg/m^3 in Los Angeles and urban Alabama. Values reported for other regions were much lower (0.001 and 0.002 mg/m^3 in urban Oklahoma and rural Alabama, respectively), and none was detected in several midwestern states (US Environmental Protection Agency, 1983). Toluene was also detected in the expired air of individuals from a US urban population (mean, 0.0084 mg/m^3; Krotoszynski *et al.*, 1979) and in the interior of cars before (0.5 mg/m^3) and after driving (1.0 mg/m^3; Merian & Zander, 1982). A range of 0.02–0.412 mg/m^3 was found in 48 samples collected at German traffic intersections (Seifert & Abraham, 1982). Levels of 0.004 mg/m^3 toluene were measured in two rural areas in the USA between 1971 and 1978; < 0.001 mg/m^3 was measured in six others (Holzer *et al.*, 1977; Merian & Zander, 1982; US Environmental Protection Agency, 1983). Levels of 97–891 mg/m^3 were measured in the smoke of forest fires (Merian & Zander, 1982).

De Bortoli *et al.* (1984) measured 0.017–0.378 mg/m^3 toluene in 14 homes and in one office building in northern Italy. Levels of 0.15–0.9 mg/m^3 toluene were found in US homes polluted with tobacco smoke (US Environmental Protection Agency, 1983). Toluene has been detected in tobacco smoke (IARC, 1986). Seifert and Abraham (1982) found an average concentration of 0.061 mg/m^3 (range, 0.017–0.116 mg/m^3) in kitchens and other rooms of 15 homes in West Berlin; just outside the walls of these dwellings, the measured concentrations averaged 0.035 mg/m^3 (range, 0.016–0.06 mg/m^3). Mølhave (1979) reported a peak level of 0.61 mg/m^3, based on measurements in 14 rooms in homes in Denmark; and Mølhave and Møller (1979) reported an average concentration in 39 homes of 0.09 mg/m^3.

(d) Water

Drinking-water in Prague, Czechoslovakia, in 1973 contained < 0.1 µg/l toluene (Merian & Zander, 1982), whereas in Toronto, Canada, in 1980 drinking-water contained an average of 2 µg/l (compared to < 1 µg/l before treatment; Otson *et al.*, 1982). Levels of 42–100 µg/l were reported in well water in the vicinity of landfill sites in the USA (US Environmental Protection Agency, 1983). The concentration of toluene in rain water in the Federal Republic of Germany has been reported to be 0.13–0.70 µg/l (US Environmental Protection Agency, 1983; International Programme on Chemical Safety, 1985).

Toluene has been found at concentrations of 1–5 µg/l in water samples from a number of rivers in eastern and midwestern USA, with concentrations ranging up to 12 µg/l in the Mississippi River near New Orleans. Concentrations of 0.8 µg/l have been reported in the

Rhine River in the Federal Republic of Germany and of 1.9 µg/l in Switzerland (Merian & Zander, 1982).

Concentrations of 0.005–0.376 µg/l (mean, 0.061 µg/l) were reported at several coastal sites along the Gulf of Mexico (US Environmental Protection Agency, 1983).

(e)   Soil

Toluene exists in an adsorbed state in soil. In assiciation with clay minerals, its adsorption is inversely proportional to the pH of the soil. Approximately 40–70% of toluene applied to the surface of sandy soils is volatilized. The biodegradation of toluene by microorganisms in the soil ranged from 63–86% after 20 days (Wilson et al., 1981; US Environmental Protection Agency, 1983; Wilson et al., 1983).

(f)   Food

The US Environmental Protection Agency (1983) reported toluene concentrations of < 1 mg/kg in 56 of 59 samples of fish tested; one fish had a level of 35 mg/kg toluene. [It was not clear to the Working Group whether these concentrations were found in whole fish or only in the edible part.]

Toluene was also detected at low concentrations (0.08–0.11 mg/kg) in a few samples of maple syrup packaged in plastic containers (Hollifield et al., 1980).

2.3 Analysis

Methods for the analysis of toluene and its metabolites have recently been reviewed and compiled (Fishbein & O'Neill, 1988) and are summarized in Table 4. Colorimetric detection systems have been developed for toluene in air (ENMET Corp., undated; Matheson Gas Products, undated; Roxan, Inc., undated; The Foxboro Co., 1983; Sensidyne, 1985; National Draeger, Inc., 1987; SKC Inc., 1988).

3. Biological Data Relevant to the Evaluation of Carcinogenic Risk to Humans

3.1 Carcinogenicity studies in animals[1]

(a) Oral administration

Rat: Groups of 40 male and 40 female Sprague–Dawley rats, seven weeks old, were administered 500 mg/kg bw toluene (purity, 98.34%) in olive oil by stomach tube on four to five days per week for 104 weeks. Groups of 50 males and 50 females received olive oil alone and served as controls. All rats were maintained until death, and the study was terminated at

---

[1]The Working Group was aware of a study in progress by inhalation in mice and rats (IARC, 1988).

Table 4. Analytical methods for determining toluene and its metabolites in various matrices[a]

Sample matrix	Sample collection	Sample preparation	Assay procedure	Detection limits	Reference
Air	Passive sampler with charcoal	Desorb (carbon disulfide); inject aliquot using glass capillary column	GC	0.3 mg/m³ x h	Seifert & Abraham (1983)
	Charcoal tube or passive sampler	Desorb (carbon disulfide); inject aliquot; analyse on packed column	GC–FID	0.01 mg/sample	Eller (1984, 1987)
	Passive sampler with charcoal	Desorb (carbon disulfide); inject aliquot; analyse on packed column	GC–FID	0.2 ppm [0.8 mg/m³]	Otson et al. (1983)
Water		Extract with hexane; inject aliquot	GC–FID	5 µg/l	Otson & Williams (1981)
		Heat in water bath at 25°C for 1 h; inject headspace aliquots	GC–MS	1 µg/l	Otson et al. (1982)
Food (maple syrup)	Bulk	Sparge 10 ml with nitrogen; incubate 2 h at 90°C; inject 2 ml of headspace vapours	Headspace GC; GC/MS (confirm)	10–275 µg/l	Fazio & Sherma (1987)
Soil		Wash with distilled water; acidify, steam distill with hexane; remove hexane layer; dry and reduce to 1 ml (nitrogen)	GC–FT–IR	Not given	Gurka & Betowski (1982)
Automobile exhaust gas	Tenax GC polymer adsorbant	Desorb thermally into liquid nitrogen–cooled capillary trap	GC–MS	Not given	Hampton et al. (1982)
Alveolar air	Charcoal	Desorb thermally; inject into glass column	GC–MS	Not given	Apostoli et al. (1982)
Blood	Heparinize	Add redistilled ethyl benzene dissolved in methanol and water; equilibrate at 60°C for 45 min; remove aliquot headspace and inject	GC–FID	0.05 µg/g	Oliver (1982)
	Heparinize	Purge (nitrogen) at room temperature; trap (Tenax TA); desorb thermally; analyse volatiles on column	GC–MS	< 1 µg/l	Cramer et al. (1988)

Table 4 (contd)

Sample matrix	Sample collection	Sample preparation	Assay procedure	Detection limits	Reference
Urine (hippuric acid)		Acidify and extract hippuric acid with chloroform; separate on TLC (para-dimethylamino-benzaldehyde for colour development); extract azalactones with ethanol; determine UV absorbancy	TLC–UV	6 μg	Bieniek et al. (1982)
Tissue (muscle, liver)	Mince	Add sodium hydroxide and olive oil with internal standard (ethylbenzene); incubate at 35°C for 2.5 h; inject aliquot of headspace vapour	GC–FID	2 μg/g	Miyaura & Isono (1985)

[a]Abbreviations: GC, gas chromatography; FID, flame ionization detection; MS, mass spectrometry; FT–IR, Fourier transform/infrared spectrometry; TLC, thin-layer chromatography; UV, ultraviolet spectrometry

week 141. At week 141, thymomas were reported in 1/37 treated males and 2/40 treated females compared to 0/45 and 0/49 controls. Other haemolymphoreticular tumours were reported in 2/37 treated males and 5/40 treated females compared to 3/45 and 1/49 controls (denominators are numbers of rats alive in each group at 58 weeks, when the first haemolymphoreticular tumour was observed). The authors reported an increase in the total numbers of animals with malignant tumours [types unspecified] at 141 weeks: 18/40 treated males and 21/40 treated females compared to 11/45 and 10/49 controls (denominators are numbers of rats alive in each group at 33 weeks, when the first malignant tumour was observed; Maltoni et al., 1983, 1985). [The Working Group noted the incomplete reporting of tumour pathology in this study and that combining different types of tumours is not usually the most appropriate method for evaluating carcinogenicity (IARC, 1980; Montesano et al., 1986).]

(b) Inhalation

Rat: Groups of 120 male and 120 female Fischer 344 rats, about seven weeks of age, were exposed by inhalation to 0, 30, 100 or 300 ppm (0, 113, 377 or 1131 mg/m^3) toluene (purity, >99.98%) for 6 h per day on five days per week for up to 24 months. Interim kills were made in all groups at six months (five rats), 12 months (five rats) and 18 months (20 rats). All surviving rats were killed at 24 months; these comprised 71 male and 70 female controls, 73 males and 75 females given the low dose, 68 males and 76 females given the mid-dose, and 67 males and 75 females given the high dose. No increase in the incidence of tumours was reported in the treated groups (Gibson & Hardisty, 1983). [The Working Group noted the incomplete reporting of data on pathology and that the level of exposure was low.]

(c)   Skin application

Mouse: Toluene was tested as a vehicle control or in combination with various carcinogens in a number of skin painting studies in mice. No skin tumour attributable to toluene alone was observed (Poel, 1962; Frei & Kingsley, 1968; Lijinsky & Garcia, 1972; Doak *et al.*, 1976; Weiss *et al.*, 1986). [The Working Group noted either the small number of animals used in these experiments and the short duration or incomplete reporting of the studies.]

A group of 50 male C3H/HeJ mice, six to ten weeks of age, received applications of 25 μl [21.7 mg] toluene [purity unspecified] on clipped interscapular skin three times a week until death. Mean survival time was 83 weeks. No skin tumour was reported at termination of the study [unspecified], and complete histological examination revealed no treatment-related tumour at other sites (McKee & Lewis, 1987). In a similar study with 50 C3H/HeJ mice (mean survival time, 77 weeks), one skin papilloma was found (McKee *et al.*, 1986).

Seven groups of vehicle controls used for different experiments, each consisting of 50 male C3H/HeJ mice, six to eight weeks old, received applications of 50 mg toluene on the interscapular skin twice a week for 73–120 weeks. Skin tumours [by gross observation] occurred in 3/350 mice (Blackburn *et al.*, 1986).

3.2   Other relevant data

The toxicology of toluene has been reviewed (National Institute for Occupational Safety and Health, 1973; Cohr & Stokholm, 1979; Benignus, 1981a,b; World Health Organization, 1981; International Programme on Chemical Safety, 1985; Anon., 1987b; Low *et al.*, 1988).

(a)   Experimental systems

(i) *Absorption, distribution, excretion and metabolism*

When dogs were exposed to 0.4–0.6 μg/ml toluene vapour, 91–94% was taken up in the lungs (Egle & Gochberg, 1976). Absorption was complete when toluene was given orally to dogs (Knoop & Gehrke, 1925); the blood level in rats increased more slowly after oral administration than after inhalation (Pyykkö *et al.*, 1977). Absorption through the skin of mice *in vivo* was 4.59 μg/cm² per hour (Tsuruta *et al.*, 1987). Toluene penetrated rat skin excised three days after clipping and depilation with cream at a rate one-tenth that of benzene and ten times that of *ortho*-xylene (Tsuruta, 1982).

When ³H–toluene was given to rats either orally or by inhalation, radioactivity 2 h after administration was highest in the adipose tissue, followed by the liver, kidneys and brain (Pyykkö *et al.*, 1977). Similar results were obtained after intramuscular injection of [ring-labelled ¹⁴C]toluene to mice (Ogata *et al.*, 1974) and after intraperitoneal injection of [methyl-¹⁴C]toluene to mice (Koga, 1978). Levels in the cerebrum, cerebellum and spinal cord were comparable to those in blood of rats after intraperitoneal injection (Savolainen, 1978). Toluene levels in brain and blood were linearly related to toluene levels in inhaled air after rats were exposed to 50, 100, 500 or 1000 ppm (189, 377, 1885 or 3770 mg/m³) toluene for 3 h (Benignus *et al.*, 1984). The toluene concentration was higher in brain than in blood immediately after exposure. The decrease after termination of exposure was almost parallel in the two tissues but slightly faster in brain than in blood (Benignus *et al.*, 1981). Less than 2%

radioactivity was excreted in bile within 24 h after intraperitoneal injection of 50 mg/kg bw [14C]toluene to rats (Abou–El–Makarem *et al.*, 1967).

When rabbits were given a single oral dose of 350 mg/kg bw toluene, 19% was exhaled unchanged within 12 h (Smith *et al.*, 1954). In rats given 3H–toluene orally or by inhalation, only 1% or less of the initial radioactivity was found in various tissues 24 h after dosing, except for white adipose tissue which contained 3.5–5% (Pyykkö *et al.*, 1977). Similar results were obtained in mice (Koga, 1978).

Toluene is excreted into the urine primarily as hippuric acid (after side–chain oxidation followed by glycine conjugation) and, to a minute extent, as conjugated cresols (after aromatic hydroxylation and sulfation/glucuronidation; International Programme on Chemical Safety, 1985). Of an orally administered dose of 0.3 g/kg bw given to rabbits, 74% was excreted in urine as hippuric acid within 24 h (El Masry *et al.*, 1956). In rats, 0.04–0.11% and 0.4–1.0% of an oral dose of 100 mg/kg bw toluene were excreted in urine as *ortho*-cresol and *para*-cresol, respectively (Bakke & Scheline, 1970); the ratio of *ortho*- and *para*-cresol:hippuric acid varied depending on exposure intensity and strain of rats (Inoue *et al.*, 1984).

Intraperitoneal injection of 370 mg/kg bw toluene to rats resulted in decreased hepatic glutathione levels and increased urinary thioether excretion, suggesting the formation of mercapturic acid(s) as a minor metabolite(s) (van Doorn *et al.*, 1980). Activation of toluene to covalently binding metabolites has been reported. When [methyl14C]–toluene was incubated with rat liver microsomes in the presence of an NADPH–generating system, part of the radioactivity remained in microsomal components after extensive extraction with various solvents and trichloroacetic acid. Treatment with ribonuclease and protease indicated that the radioactivity bound preferentially to proteins (Pathiratne *et al.*, 1986).

Pregnant C57Bl mice were exposed by inhalation to 14C–toluene [theoretical concentration, 2000 ppm (7540 mg/m3)] for 10 min on days 11, 14 or 17 of gestation, and distribution of the label was determined 0, 0.5, 1, 4 and 24 h after exposure. The label quickly entered the embryo, but uptake was low relative to that in maternal tissues. All fetal activity was extractable, indicating that no firmly bound metabolite was present (Ghantous & Danielson, 1986).

(ii) *Toxic effects*

The oral LD$_{50}$ of toluene in rats has been reported to be about 5 g/kg bw (range, 2.6–7 g/kg bw) depending on age and strain (Wolf *et al.*, 1956; Kimura *et al.*, 1971; Withey & Hall, 1975). The intraperitoneal LD$_{50}$ was reported to be about 1.6 g/kg bw in different strains of rats (Ikeda & Ohtsuji, 1971; Lundberg *et al.*, 1986) and about 1.2 g/kg bw in mice (Schumacher & Grandjean, 1960). The LC$_{50}$ in rats exposed for 6 h was 50 000 mg/m3 (Cameron *et al.*, 1938), and that in mice exposed for 7 h was 19 950 mg/m3 (Svirbely *et al.*, 1943). The estimated dermal LD$_{50}$ in rabbits was about 12 mg/kg bw (Smyth *et al.*, 1969). Toluene is only slightly to moderately irritating for skin and eyes of rabbits (Wolf *et al.*, 1956; International Programme on Chemical Safety, 1985).

Minor weight loss was noted in rats exposed to 500 ppm (1900 mg/m3) toluene for 7 h per day on five days per week for five weeks and in mice exposed to 4000 ppm (15 000 mg/m3) toluene for 3 h per day for ten weeks (Benignus, 1981a).

Acute inhalation of high concentrations of toluene resulted, depending on species, age and concentration, in more or less pronounced central nervous system depression (Carpenter *et al.*, 1976). Inhalation of concentrations of 2600 ppm (9800 mg/m^3) for several hours led to signs of narcotic effects. Inhalation of 12 000 ppm (45 200 mg/m^3) for 5 min produced marked central nervous system depression in mice and rats (Bruckner & Peterson, 1981a,b).

In rats, subchronic inhalation of toluene (1000 ppm [3770 mg/m^3], 12 h per day for 16 weeks) resulted in reversible reduction of mixed nerve conduction velocity (Takeuchi *et al.*, 1981). Disturbance of circardian rhythm was seen with 4000 ppm [15 000 mg/m^3] 4 h per day for four weeks (Hisanaga & Takeuchi, 1983), and behavioural effects were seen with 4000 ppm [15 000 mg/m^3], 2 h per day for 60 days (Ikeda & Miyake, 1978). Behavioural effects have been reported at exposures as low as 150 ppm (560 mg/m^3) for 30 min in rats (Geller *et al.*, 1979; Wood *et al.*, 1983) and 4 mg/m^3 for ten days in mice (Horiguchi & Inoue, 1977). Neurological signs have been recorded in cats exposed to 25 500 mg/m^3 for 10 min per day for 40 days (Contreras *et al.*, 1979). Exposure disturbed the turnover of neurotransmitters (dopamine, norepinephrine and 5–hydroxytryptamine) in the central nervous system of rats after exposure to 300–375 mg/m^3 for one or a few days (Fuxe *et al.*, 1982; Rea *et al.*, 1984).

No significant toxicity (as determined by blood parameters, urinary parameters, organ weights and histopathological examinations of major organs) was seen after oral administration of up to 590 mg/kg bw toluene to female rats for periods of up to six months (Wolf *et al.*, 1956) or daily 6–8-h inhalation exposures to concentrations below 400 ppm [1500 mg/m^3] for up to 24 months in rats (Jenkins *et al.*, 1970; Gibson & Hardisty, 1983) or for up to 127 days in dogs (Jenkins *et al.*, 1970; Carpenter *et al.*, 1976) or monkeys (Jenkins *et al.*, 1970).

In rats exposed to 8000 mg/m^3 toluene for 8 h per day on six days per week for seven weeks, lung irritation but no systematic haematological change was noted. Signs of central nervous system intoxication, incoordination and paralysis of the hind legs, and congestive changes in lung, liver, kidney, heart and spleen were seen in two dogs exposed to toluene concentrations of 7500 mg/m^3 then 10 000 mg/m^3 for 8 h per day on six days per week for six months; both animals died after 180 days (Fabre *et al.*, 1955).

An increase in the number of kidney casts was noted in rats exposed by inhalation to 750 mg/m^3 toluene for 7 h per day on five days per week for five weeks (International Programme on Chemical Safety, 1985). Hyperaemic glomeruli and albuminuria were reported in two dogs exposed to 7500 mg/m^3 then 10 000 mg/m^3 for 8 h per day on six days per week for six months (Fabre *et al.*, 1955).

Changes in the activity of drug–metabolizing enzymes in the liver were reported in rats exposed to 500 ppm (1875 mg/m^3) for 6 h per day for three days (Toftgård *et al.*, 1982) and following oral administration of 0.7 ml/kg bw for two days (Mungikar & Pawar, 1976; Pyykkö, 1980). Reduction in body weight gain and increases in liver weight and in cytochrome P450 and cytochrome b_5 concentrations, but no toxicity–related specific ultrastructural change in the liver, were observed in male rats exposed to toluene at 6000 mg/m^3 for 8 h per day for four weeks and in male and female rats exposed to up to 3500 mg/m^3 for 8 h per day for six months (Ungváry *et al.*, 1980).

(iii) *Effects on reproduction and prenatal toxicity*

Toluene (5–100 μmol [0.5–9.2 mg]/egg) was injected into the air sac of white Leghorn SK 12 chick embryos on day 2 or 6 of incubation; control eggs received an injection of the vehicle (olive oil). The LD_{50} was reported to be in excess of 100 μmol/egg, although this dose caused 100% mortality when given on day 6 (Elovaara *et al.*, 1979).

Toluene was injected into the yolk sac of fresh fertile chicken eggs prior to incubation. Hatchability of the eggs was 85%, 25% and 0 with exposures of 4.3, 8.7 and 17.4 mg/egg, respectively (McLaughlin *et al.*, 1964).

As reported in an abstract, CD–1 mice were exposed by gavage to 0.3 (0.27), 0.5 (0.45) or 1.0 ml (0.9 mg)/kg bw toluene in cottonseed oil on days 6–15 of gestation or to 1.0 ml/kg on days 12–15. No maternal effect was observed in the groups exposed on days 6–15, but significant embryo lethality was seen at all dose levels, and fetal weight was reduced at 0.5 and 1.0 ml/kg bw. Cleft palates were seen at the highest exposure level. In the groups exposed on days 12–15, only maternal toxicity was seen (Nawrot & Staples, 1979).

In a teratology screening assay, two groups of 30 ICR/SIM mice received 0 or 1800 mg/ kg bw per day toluene by oral intubation on days 8–12 of gestation. Dams were allowed to deliver, and the offspring were evaluated for growth and viability in the early neonatal period. No effect was observed in either the dams or the offspring (Seidenberg *et al.*, 1986; Seidenberg & Becker, 1987). Using the same basic protocol, CD–1 mice received 0 or 2350 mg/ kg bw per day (50 mice per group) or 0 and 3000 mg/kg bw per day (groups of 46 and 49 mice, respectively) toluene by oral intubation on days 6–13 of gestation. In the first experiment, exposure to toluene was lethal to one dam; no control died, and there was no other effect on dams or offspring. In the second experiment, 3/49 treated dams died; there was no death in the control group, and no other sign of toxicity was observed in dams or their offspring (Hardin *et al.*, 1987).

In four studies, mice were exposed by inhalation to up to 3770 mg/m³ during various periods of gestation. Exposure to 1500 mg/m³ resulted in maternal mortality after continuous (24 h/day) but not after intermittent (7 h/day) exposure. Fetal viability was not affected in any study. Fetal growth retardation was noted in one study at 500 mg/m³ (24 h/day on days 6–13), but not in another at 1500 mg/m³ (7 h/day on days 6–16). An increased incidence of extra ribs was seen at 3770 mg/m³ (6 h/day on days 1–17) but a lower incidence was reported at 1500 mg/m³ (7 h/day on days 6–16). No treatment–related malformation was seen in any study. In the two studies in which offspring were followed postnatally after exposure at 3770 mg/m³ for 6 h per day on days 1–17, and 1500 mg/m³ for 7 h per day on days 6–16, no effect on postnatal growth or viability was observed (Hudák & Ungváry, 1978; Shigeta *et al.*, 1982; Ungváry & Tátrai, 1985; Courtney *et al.*, 1986). [The Working Group noted that, on the basis of a non–dose–related increase in the frequency of enlarged renal pelvis and a decreased variability in rib profile, Courtenay *et al.* (1986) concluded that toluene was teratogenic to mice.]

CFY rats were exposed by inhalation to 1500 mg/m³ toluene (analytical purity) for 24 h per day on days 9–14 of gestation (19 rats), to 1500 mg/m³ on days 1–8 of gestation (nine rats) or to 1000 mg/m³ for 8 h per day on days 1–21 of gestation (ten rats). There were 26 control females for exposure on days 9–14 and ten control females for the exposures starting on day 1 of gestation. Exposure to 1500 mg/m³ caused mortality in 2/19 and 5/9 dams in the groups exposed on days 9–14 and 1–8, respectively; no other maternal effect was reported. Absence of the tail was reported in 2/213 fetuses exposed on days 9–14 as compared to 0/348 fetuses in the control group. On skeletal examination of the group exposed on days 9–14, 7/102 treated fetuses had fused sternebrae and 22/102 had extra ribs; the incidences in the control group were 2/169 and 0/169, respectively. High exposure levels early in development were accompanied by lower fetal body weights at term but no abnormality; the only effect noted following exposure to low levels throughout gestation (days 1–21) was an increased incidence of signs of skeletal retardation (poorly ossified sternebrae, bipartite vertebra centra and shortened 13th rib). No effect on fetal viability was noted with any exposure regimen (Hudák & Ungváry, 1978).

In a subsequent study, groups of 22 or 20 CFY rats were exposed by inhalation to air or to 1000 mg/m³ toluene for 24 h per day on days 7–14 of gestation. Animals were killed on day 21 of pregnancy and the fetuses were examined by routine teratological techniques. No maternal toxicity was observed in the treated group; an increased incidence of supernumerary ribs was the only effect reported ($p < 0.10$) in the fetuses (Tátrai et al., 1980). [The Working Group noted that it is not clear how the latter data were analysed; it appears that the individual fetus was used as the unit of comparison.] In a further study, exposure by inhalation to 0 or 3600 mg/m³ toluene (analytical purity) for 24 h per day on days 10–13 of gestation did not appear to affect fetal development adversely although it did potentiate the maternal and embryonic toxic effects of acetylsalicylic acid (Ungváry et al., 1983).

Groups of 12 female Nya:NYLAR mice were given 0, 16, 80 or 400 mg/l toluene in the drinking-water from mating throughout gestation and lactation, and the offspring continued to receive toluene in the drinking-water from weaning until the end of testing. Offspring were observed for viability, surface righting ability at seven days of age, eye and ear opening and startle response at 13–14 days of age, open field activity at 35 days of age and rotorod performance at 45–55 days of age. No treatment-related effect was observed for fluid consumption, growth, viability or appearance of developmental landmarks. Mice exposed to 400 mg/l displayed decreased habituation in the open field apparatus; a non-dose-related impairment of rotorod performance was also observed (Kostas & Hotchin, 1981).

Rats of an inbred strain (Tokai high avoiders) were exposed to 0, 100 or 500 ppm (0, 377 or 1885 mg/m³) toluene for 7 h per day from day 13 of gestation to postnatal day 48. Developmental endpoints examined included age at pinna detachment, the presence of downy fur, incisor eruption, eye opening, body weight, a righting reflex and responses in a rotorod test. A learning test (Sidman avoidance) was conducted daily for ten days beginning on day 49, 100 or 150. There was no significant difference between the treated groups with respect to acquisition of developmental landmarks, but body growth was greater in the group exposed to 100 ppm. Treated male offspring in both exposure groups were deficient in acquisition of the learning task at initial, but not later, ages; no consistent effect was noted in the learning behaviour of treated female offspring (Shigeta et al., 1986).

Groups of New Zealand white rabbits were exposed to 0 (60 animals), 500 (ten rabbits) or 1000 (eight rabbits) mg/m³ toluene for 24 h per day on days 7–20 of gestation. Fetuses were examined by routine teratological techniques on day 30 of gestation. Females that received the high dose either died, aborted or had no live fetuses at term. One female in the low–dose group aborted, but no significant fetal effect was noted (Ungváry & Tátrai, 1985). [The Working Group noted that this paper is a compendium of data on rats, mice and rabbits from one laboratory and presents little detail on experimental results.]

(iv) *Genetic and related effects*

The genetic and related effects of toluene have been reviewed (Dean, 1978, 1985; Fishbein, 1985).

Toluene induced a permanent loss of initiation of DNA replication in *Bacillus subtilis* cells (Winston & Matsushita, 1975). It did not produce differential killing in DNA repair-proficient compared to repair–deficient strains of *B. subtilis rec*$^{+/-}$ (McCarroll *et al.*, 1981a) or *Escherichia coli* (McCarroll *et al.*, 1981b). Toluene did not induce SOS activity in *Salmonella typhimurium* TA1535/pSK1002 (Nakamura *et al.*, 1987) and was not mutagenic to *S. typhimurium* TA1535, TA1537, TA1538, TA98, TA100, UTH8413 or UTH8414 either in the presence or absence of an exogenous metabolic system from uninduced or Aroclor–induced rat and Syrian hamster livers (Lebowitz *et al.*, 1979 (abstract); Nestmann *et al.*, 1980; Bos *et al.*, 1981; Spanggord *et al.*, 1982; Haworth *et al.*, 1983; Connor *et al.*, 1985).

As reported in an abstract, toluene induced chromosomal anaphase alterations in *Vicia faba* (Gomez–Arroyo & Villalobos–Pietrini, 1981).

Toluene induced mitotic arrest (C–mitosis) in embryos of the grasshopper, *Melanoplus sanguinipes* (Liang *et al.*, 1983). It did not induce sex-linked recessive lethal mutations or translocations, but did induce sex–chromosome loss and nondisjunction in male *Drosophila melanogaster* at a dose of 1–1.5% toluene administered in food (Rodriguez Arnaiz & Villalobos–Pietrini, 1985a,b). As reported in an abstract, toluene did not induce recessive lethal mutations in *D. melanogaster* exposed to 500 and 1000 mg/kg for 24 h by feeding (Donner *et al.*, 1981).

Toluene induced DNA single–strand breaks (as measured by alkaline elution) in primary cultures of rat hepatocytes (Sina *et al.*, 1983), but did not cause DNA damage or repair, as measured by the 'nick–translation' assay, in cultured human fibroblasts (Snyder & Matheson, 1985). As reported in an abstract, toluene did not induce mutations in mouse lymphoma L5178Y TK$^{+/-}$ cells *in vitro* or chromosomal aberrations in rat bone marrow *in vivo* (Lebowitz *et al.*, 1979). It did not induce sister chromatid exchange or chromosomal aberrations in cultured human lymphocytes *in vitro* (Gerner–Smidt & Friedrich, 1978). [The Working Group noted that the human lymphocytes were tested without an exogenous metabolic system.]

Toluene was reported to induce chromosomal aberrations in the bone–marrow cells of male albino rats after chronic inhalation exposure to 5.4 or 50.7 mg/m^3 on 4 h per day, five days a week for four months (Aristov *et al.*, 1981) or after subcutaneous injection of 0.8 g/kg bw (Dobrokhotov, 1972). Chromosomal aberrations in bone–marrow cells were reported following subcutaneous injection of 1 g/kg bw daily for 12 days to male albino rats (Lyapkalo, 1973). Neither micronuclei nor chromosomal aberrations were observed in male and female CD–1 mice administered two doses of 1720 mg/kg bw toluene (99% pure) at a 24-h interval by oral gavage (Gad–El–Karim *et al.*, 1984). Increases in the frequency of micronuclei and of chromosomal aberrations in rat bone–marrow cells were reported after two intraperitoneal injections of 217 mg/kg bw and 435 mg/kg bw (Roh *et al.*, 1987).

Toluene induced micronuclei in bone–marrow polychromatic erythrocytes of male NMRI and B6C3F1 mice after two intraperitoneal doses of 0.12–0.5 ml/kg bw (0.1–0.44 mg/kg) at a 24–h interval (Mohtashamipur *et al.*, 1985). Pretreatment of male NMRI mice with inducers (phenobarbital, Aroclor 1254, 3-methylcholanthrene) or inhibitors (metyrapone, α–naphthoflavone) of cytochrome P450 enhanced the frequency of micronuclei induced by toluene, while simultaneous injections of toluene and inhibitors decreased the observed clastogenic activities (Mohtashamipur *et al.*, 1987).

Toluene and benzene administered concurrently were reported to have an additive effect on induction of chromosomal aberrations (Dobrokhotov, 1972; Dobrokhotov & Enikeev, 1977). Toluene reduced the number of sister chromatid exchanges induced by benzene when both compounds were administered intraperitoneally to DBA/2 mice (Tice *et al.*, 1982) and reduced the clastogenic activity of benzene when the two compounds were simultaneously administered orally to CD–1 mice (Gad–El–Karim *et al.*, 1984), intraperitoneally to Sprague–Dawley rats (Roh *et al.*, 1987) or subcutaneously to NMRI mice (Tunek *et al.*, 1982).

As reported in an abstract, exposure of male rats by inhalation to 300 ppm (1130 mg/m^3) toluene for 6 h per day on five days per week for 15 weeks did not induce chromosomal aberrations in bone–marrow cells (Donner *et al.*, 1981). As reported in an abstract, oral administration of toluene did not induce chromosomal aberrations in bone–marrow cells or dominant lethal mutations in random–bred male SHR mice (Feldt *et al.*, 1985).

As reported in an abstract, toluene did not inhibit intracellular communication (as measured by metabolic cooperation) in Chinese hamster V79 cells (Awogi *et al.*, 1986).

Toluene did not enhance morphological transformation of Syrian hamster embryo cells by the SA7 adenovirus (Casto, 1981).

It did not induce sperm–head abnormalities in mice (Topham, 1980).

(b) *Humans*

(i) *Absorption, distribution, excretion and metabolism*

Inhalation is a major route of human exposure to toluene, although skin absorption may occur in occupational settings. An average lung uptake of 53.3% was obtained during exposure of volunteers to 271–1177 mg/m^3 toluene for 5 h (Srbová & Teisinger, 1952). Similar results were obtained in later studies: e.g., 57–72% (Piotrowski, 1967), 53% (Nomiyama & Nomiyama, 1974) and 30–50% (Carlsson, 1982) lung uptake. When volunteers immersed their hands in liquid toluene, skin penetration took place at a rate of 14–23 mg/cm^2 per hour (Dutkiewicz & Tyras, 1968). [The Working Group noted that the absorbed amount of toluene was calculated as the difference between the applied and the remaining amount of toluene, and therefore, the absorption rate may be overestimated.] Immersion of one hand in liquid toluene for 30 min resulted in a blood level (taken from the unexposed arm) of toluene twice as high as that after inhalation of 100 ppm (377 mg/m^3) for 4 h (Sato & Nakajima, 1978), indicating that both respiratory and percutaneous absorption are important. Toluene was detected in exhaled air after whole-body skin exposure (with no inhalation) to 600 ppm (2260 mg/m³) toluene for 3.5 h (Riihimäki & Pfäffli, 1978).

After exposure of volunteers to 100 ppm (377 mg/m³) toluene for 2 h, the fall in the concentration of toluene in blood paralleled that in exhaled air. The decay curve consisted of three components with half-times of 1.7, 30 and 180 min [calculated by the Working Group] for the initial 5 min, 5–120 min and 180–300 min, respectively (Sato & Fujiwara, 1972). The biological half-time for the excretion of urinary metabolites among toluene workers was about 7.5 h (Tokunaga *et al.*, 1974). A shorter half-time was observed after exposure of volunteers (Baelum *et al.*, 1987). The half-time of toluene in adipose tissue of exposed workers was 0.5–2.7 days (Carlsson & Ljungquist, 1982). Toluene was present in the blood of printers several days after the end of exposure (Nise & Ørbaek, 1988).

Most [e.g., 68% (Ogata *et al.*, 1970)] of the toluene absorbed undergoes side-chain oxidation followed by glycine conjugation and is excreted in the urine as hippuric acid. *ortho-*, *meta-* and *para*-Cresols were also identified as minor metabolites of toluene (Angerer, 1979; Woiwode *et al.*, 1979; Woiwode & Drysch, 1981). *ortho*-Cresol levels are about 1/1000 of hippuric acid levels in the urine of workers exposed to toluene (Hasegawa *et al.*, 1983; Inoue *et al.*, 1986). The toluene level in blood is closely related to the level in alveolar air; the concentration of metabolites in urine is correlated with both, but less closely (Brugnone *et al.*, 1976).

Levels of hippuric acid, and to a lesser extent *ortho*-cresol, in urine have been studied intensively as indicators of exposure to (De Rosa *et al.*, 1987), and their validity has been established (Pagnotto & Lieberman, 1967; Ikeda & Ohtsuji, 1969; Capellini & Alessio, 1971; Pfäffli *et al.*, 1979; Bergert *et al.*, 1980; Alessio *et al.*, 1981; Hasegawa *et al.*, 1983; De Rosa *et al.*, 1985, 1987). Metabolite levels in urine samples collected near the end of a working day shift (Alessio *et al.*, 1981; Hasegawa *et al.*, 1983; De Rosa *et al.*, 1985) correlated best with the time-weighted average exposure to toluene; toluene accumulates in the body towards the end of a working week (Konietzko *et al.*, 1980) as a reflection of its biological half-time (Tokunaga *et al.*, 1974). Levels of toluene in the blood have also been used since these are low among nonexposed subjects (Szadkowski *et al.*, 1973). The biological monitoring of exposure to toluene has been reviewed (Lauwerys, 1983).

No significant change in toluene metabolism is induced by exposure to toluene under usual occupational conditions (Wallén, 1986). Simultaneous exposure to other solvents, such as benzene, is known to suppress toluene metabolism (Inoue *et al.*, 1988). Toluene metabolism may differ among populations (Inoue *et al.*, 1986).

When a large dose of ethanol was taken in combination with exposure to toluene, toluene metabolism was inhibited due to metabolic competition between the two chemicals. Blood toluene levels were lower in workers who drank regularly, indicating induction of toluene metabolism by continued ethanol intake-induced metabolism (Waldron *et al.*, 1983). A more rapid apparent clearance of toluene from the blood was seen in smokers compared to nonsmokers occupationally exposed to toluene (Wallén, 1986).

(ii) *Toxic effects*

Subjects who intentionally abuse toluene and workers exposed occupationally (mainly printers and painters) are generally also exposed to other organic solvents (International Programme on Chemical Safety, 1985). Metabolic and toxic interactions between toluene and other solvents may enhance or reduce any adverse effect (Swedish Criteria Group for Occupational Standards, 1985).

Slight hyposmia has been noted in printers (Baelum *et al.*, 1982). Moderate and transient effects on the eye (conjunctival irritation and corneal damage) occurred in workers splashed with liquid toluene (Grant, 1962). Eye and upper airway irritation occurred after a 6.5-h exposure to an air level of 100 ppm (377 mg/m^3) toluene (Baelum *et al.*, 1985), and lachrymation was seen at 1500 mg/m^3 (International Programme on Chemical Safety, 1985). An obstructive ventilatory pattern has been recorded in inhalation abusers of spray paint containing toluene (Reyes de la Rocha *et al.*, 1987). Prolonged exposure of the skin to toluene may cause contact dermatitis (Matsushita *et al.*, 1975).

Volunteers exposed to 100 ppm (377 mg/m^3) toluene for 6 h per day for four days suffered from subjective complaints of headache, dizziness and a sensation of intoxication (Andersen *et al.*, 1983). In subjects exposed to 750 mg/m^3 for 8 h, fatigue, muscular weakness, confusion, impaired coordination, enlarged pupils and accommodation disturbances were experienced; at about 3000 mg/m^3, severe fatigue, pronounced nausea, mental confusion, considerable incoordination with staggering gait and strongly affected pupillary light reflexes were observed. After exposure at the high level, muscular fatigue, nervousness and insomnia lasted for several days (International Programme for Chemical Safety, 1985). Heavy accidental exposure leads to coma (Longley *et al.*, 1967; Griffiths *et al.*, 1972; Bakinson & Jones, 1985).

Similar effects have been observed in cross-sectional studies of workers, including painters (see also the monograph on occupational exposures in paint manufacture and painting) exposed to corresponding or lower levels of toluene (Wilson, 1943; Matsushita *et al.*, 1975; Elofsson *et al.*, 1980; Husman, 1980; Baelum *et al.*, 1982; Winchester & Madjar, 1986). [The Working Group noted that there is probably confounding by other agents.]

Initial signs and symptoms of central nervous system effects with an excitatory stage, followed by central nervous system depression, ataxia, depressed consciousness and coma have also been observed in glue sniffers, who may be exposed to very high levels of toluene (Streicher *et al.*, 1981). Generally, these signs and symptoms are reversible (Benignus, 1981a); however, prolonged glue sniffing (two years or more) may result in permanent encephalopathy (Malm & Lying-Tunell, 1980; King *et al.*, 1981; Schikler *et al.*, 1982; Fornazzari *et al.*, 1983). In particular, cerebellar signs have been reported (Fornazzari *et al.*, 1983). Effects on the peripheral nervous system have also been observed in 'sniffers' (Korobkin *et al.*, 1975). [The Working Group noted that the relationship with toluene is severely confounded by concomitant exposure to other solvents (including alcohol) and drugs with known neurotoxicity (e.g., sedatives and neuroleptics).]

In volunteers exposed to 300 ppm (1130 mg/m^3) toluene for a few hours, impairment of simple reaction times was observed (Gamberale & Hultengren, 1972; Winneke, 1982), whereas 300–375 mg/m^3 caused no such effect (Andersen *et al.*, 1983; Dick *et al.*, 1984;

Anshelm Olson *et al.*, 1985; Iregren, 1986). Disturbances of psychomotor performance have been noted in cross-sectional studies of car and industrial painters exposed to a mixture of solvents including toluene (Hänninen *et al.*, 1976; Elofsson *et al.*, 1980; Biscaldi *et al.*, 1981; Winchester & Madjar, 1986; see also the monograph on occupational exposures in paint manufacture and painting). Changes in short-term memory, in other intellectual functions and in mood have also been reported in painters exposed to a mixture of solvents containing toluene (Hänninen *et al.*, 1976; Elofsson *et al.*, 1980; Winchester & Madjar, 1986). However, the data are not consistent: in one study of toluene-exposed factory workers, there was no such effect (Cherry *et al.*, 1985). Neurobehavioural effects have also been found in subjects mostly exposed to toluene, such as in the printing industry (Iregren, 1986; Hänninen *et al.*, 1987), at levels of 300 mg/m³ (Iregren, 1986). However, in one study of printers, no such effect was observed (Struwe & Wennberg, 1983). [The Working Group noted that tests have usually been performed within 24 h after the last exposure, so it cannot be determined if the effects are of short duration or may be prolonged].

After three days of intense exposure (in some cases to the point of unconsciousness) to a mixture containing toluene, workers in a factory suffered memory disturbances that continued for months (Stollery & Flindt, 1988). In one study, there were indications of dyschromalopsia in printers exposed to a mixture of solvents, including toluene (Mergler *et al.*, 1988).

Further indications of effects on the central and peripheral nervous systems, have been reported in car and industrial painters (Seppäläinen *et al.*, 1978; Elofsson *et al.*, 1980; Husman & Karli, 1980) and other workers (Triebig *et al.*, 1983) exposed to mixtures of solvents, including toluene (see also the monograph on occupational exposures in paint manufacture and painting). In a study of toluene workers (Cherry *et al.*, 1985) and in printers exposed almost exclusively to toluene (Struwe & Wennberg, 1983; Antti-Poika *et al.*, 1985), no effect on the peripheral nervous system was observed.

'Sniffers' (Bennett & Forman, 1980; Kroeger *et al.*, 1980; Moss *et al.*, 1980; Voigts & Kaufman, 1983; Batlle *et al.*, 1988) and workers exposed accidentally to toluene (Reisin *et al.*, 1975) have been reported to develop both renal tubular damage (e.g., acidosis) and signs of glomerular damage, with haematuria, pyuria and proteinuria (Voigts & Kaufman, 1983). However, severe toluene poisoning has been reported without kidney disease (Brugnone & Perbellini, 1985), and, in a study of industrial spray painters exposed to a mixture of solvents containing toluene, there was no indication of kidney disease (Greenburg *et al.*, 1942). Later reports of workers (mostly painters but also a series of reports that included photogravure workers) exposed to toluene have indicated slight adverse effects on the kidney (Askergren, 1981; Askergren *et al.*, 1981a,b,c; Franchini *et al.*, 1983), although another study showed no effect (Lauwerys *et al.*, 1985; see also the monographs on some petroleum solvents and on occupational exposures in paint manufacture and painting).

Transient effects on the liver have been reported in a few 'sniffers' (Fornazzari *et al.*, 1983; Suzuki *et al.*, 1983); however, no discernable effect was observed in two subjects in coma following acute toluene intoxication (Brugnone & Perbellini, 1985). Slight effects on the liver have been noted in toluene-exposed workers (Greenburg *et al.*, 1942), including printing workers (Szadkowski *et al.*, 1976) and workers using toluene-based glues (Shiojima *et al.*, 1983), but not among other printers and painters exposed to toluene (Kurppa & Husman, 1982; Waldron *et al.*, 1982; Lundberg & Håkansson, 1985; Boewer *et al.*, 1988).

In one proportionate mortality study (Paganini–Hill *et al.*, 1980) and in one cohort study (Lloyd *et al.*, 1977) of printers, who may be exposed to toluene, there was an excess of liver cirrhosis. [The Working Group noted that the effect cannnot be ascribed with certainty to toluene, since exposure to many other agents had occurred.]

Some early studies (Wilson, 1943; Gattner & May, 1963; Klavis & Wille, 1967) related major myelotoxic effects (leukopenia, anaemia, thrombocytopenia and bone–marrow changes) to exposure to toluene, which were generally associated with benzene contamination of the toluene. Other cross-sectional studies (Bänfer, 1961; Tähti *et al.*, 1981) have displayed no such effect. Some recent studies have shown slightly increased haemoglobin levels (Elofsson *et al.*, 1980) and thrombocytopenia (Beving *et al.*, 1983, 1984) in car painters and workers in paint manufacture exposed to toluene, among other chemicals. Minor changes in white blood cells have also been reported following exposure to toluene (Friborská, 1973; Matsushita *et al.*, 1975).

In human volunteers exposed to 200 ppm (750 mg/m^3) toluene for 6 h per day for two days, the heart rate was increased significantly (Suzuki, 1973).

(iii) *Effects on fertility and on pregnancy outcome*

In the study of Holmberg (1979), described in the monograph on some petroleum solvents, three mothers of children with central nervous system defects, but no control mother, reported having worked with toluene during the first trimester of pregnancy. Two of the mothers of cases had also been exposed to other solvents. In the study of Holmberg *et al.* (1982), described in the monograph on some petroleum solvents, three mothers of children with oral clefts but no control mother reported having worked with toluene during the first trimester of pregnancy. All three had also been exposed to other solvents.

Ericson *et al.* (1984) linked records of female laboratory workers from the 1975 Swedish census to maternity records for 1976. Among the 1161 birth records identified in the laboratory workers, 44 (3.5%) of the children were either born dead or had a significant malformation. Among the 98 354 deliveries in Sweden during the same year, 2504 (2.6%) had a similar outcome. A case–control study of 26 of the children who had died within seven days or who had severe malformations and of 50 controls chosen from among children of laboratory workers was then performed. Exposure to toluene was similar in mothers of cases (8%) and of controls (8%); they had also been exposed to many solvents and other substances.

Axelsson *et al.* (1984) studied the outcome of pregnancy for 745 women born in 1935 and later, who had been engaged in laboratory work at the University of Gothenberg, Sweden, between 1968 and 1979. Data on outcome of pregnancy was obtained by postal survey and from the Medical Birth Register and the Register of Congenital Malformations in Sweden; data on exposure to specific substances were obtained by questionnaire. Toluene exposure during the first trimester of pregnancy was reported by 140 women, 17 of whom (10.2%) had had a spontaneous abortion. This compares with spontaneous abortion rates of 11.5% among women who had not worked in a laboratory during the first trimester and 9.0% among women who had worked in a laboratory but not with solvents during the first trimester. Cases and controls had been exposed to many solvents and other substances.

Taskinen *et al*. (1986) studied the history of spontaneous abortions in women employed in eight Finnish pharmaceutical factories in 1973–80. The identity numbers of the women were linked to the nationwide hospital discharge data for 1973–81; 1795 pregnancies were thus identified, 142 of which were spontaneous abortions. A case–control study was carried out on women with spontaneous abortions who had been employed during the first trimester and three age–matched controls. Toluene exposure during the first trimester was reported by factory physicians for seven of 38 (18%) cases, compared with 14 of 119 (12%) controls. The corresponding overall relative risk (RR) was 1.6 (95% confidence interval [CI], 0.6–4.5); for those exposed less than once a week, the RR was 1.2 (0.2–6.9), and for those exposed more than once a week, the RR was 1.9 (0.6–6.4). Cases and controls had had exposure to many solvents and other substances.

McDonald *et al*. (1987) compared the chemical exposures of 301 women who had given birth to a child with a severe malformation to that of 301 controls, matched by hospital, gra-vidity, educational level, maternal age and date of delivery. Cases and controls were re-stricted to women who had given birth in Montréal in 1982–84 and who had worked for at least 30 h per week during the first 12 weeks of pregnancy. Chemical exposure was assessed by visiting the workplace or by telephone interview with the employer. Overall chemical exposure was assessed to have been more frequent in cases (21%) than in controls (16%). When the solvents were divided into nine chemical categories and the malformation into six anatomical sites, the strongest association was between aromatic solvents and urinary tract abnormalities (nine exposed cases, six of which were hypospadias, *versus* no exposed control). Six of the nine cases were assessed to have been exposed to toluene.

(iv) *Genetic and related effects*

No significant difference in the frequency of chromosomal changes was observed in peripheral blood lymphocytes of 24 workers (aged 29–60) at a rotogravure plant in Italy who were exposed to toluene (mean value, around 200 ppm [750 mg/m³]) for three to 15 years compared with 24 controls matched for age and sex (Forni *et al*., 1971). [The Working Group noted that smoking habits were not considered.]

An excess of chromosomal aberrations (chromatid and isochromatid breaks) was re-ported in the lymphocytes of 14 Swedish workers (aged 23–54) exposed only to toluene for 1.5–26 years (average level, 100–200 ppm [377–750 mg/m³]) in a rotogravure printing factory in comparison with 49 unexposed workers (Funes–Cravioto *et al*., 1977). [The Working Group noted that smoking habits were not considered, and details of controls were not giv-en.]

No increase in the frequency of chromosomal aberrations or sister chromatid exchange was found in the peripheral blood lymphocytes of 32 rotogravure workers in Finland (aged 21–50) exposed to toluene (7–112 ppm) for three to 35 years compared to 15 unexposed sub-jects (Mäki-Paakkanen *et al*., 1980). No increase in the frequency of sister chromatid ex-changes was observed in seven workers in the Swedish paint industry exposed to various sol-vents, including more than 100 mg/m³ toluene (Haglund *et al*., 1980; see also the monograph on occupational exposures in paint manufacture and painting). [The Working Group noted the small number of workers studied.]

Increases in the frequency of sister chromatid exchange, chromatid breaks, chromatid exchanges and gaps were reported in the peripheral lymphocytes of 20 workers (aged 32–60) at a rotogravure plant in the Federal Republic of Germany who had been exposed to toluene (200–300 ppm [750–1130 mg/m³]) for more than 16 years, compared to 24 matched controls (Bauchinger et al., 1982). In an abstract, a synergistic effect of smoking and exposure to toluene on the frequency of sister chromatid exchange was also reported (Bauchinger et al., 1983). In the same plant, a higher incidence of chromatid-type aberrations than in controls was observed up to two years after cessation of exposure to toluene; longer after exposure, the aberration yields reached background level (Schmid et al., 1985).

The frequency of chromosomal aberrations in 20 employees at a rotogravure plant exposed mainly to toluene in various printing inks was no different from that in 23 control workers; an increased frequency was observed in smokers in both groups (Pelclová et al., 1987).

3.3   Epidemiological studies of carcinogenicity in humans

In each of the studies described below, exposures were mixed and overlapping, and these studies are cited in several monographs.

Olsson and Brandt (1980) performed a study on solvent exposure among 25 cases of Hodgkin's disease and 50 controls in Sweden (see the monograph on some petroleum solvents). Exposure to toluene was mentioned by six cases and three controls. All exposed cases and controls were exposed to other solvents.

Austin and Schnatter (1983) performed a case–control study on 21 deceased brain tumour patients and two control groups (80 employees in each) from a cohort of employees at a US petrochemical plant, investigating 37 chemicals. Exposure to 12 of these chemicals was more frequent among cases than controls, but toluene was not among these.

Wilcosky et al. (1984) performed a case–control study of rubber workers in the USA, described in detail in the monograph on some petroleum solvents. Exposure to toluene was associated with an increased risk for prostatic cancer (relative risk (RR), 2.6; three cases) and lymphatic leukaemia (3.0; two cases). Exposure to 'solvent A' (a proprietary mixture containing mostly toluene) was associated with increased RRs for stomach cancer (1.4; 15 cases), lymphosarcoma (2.6; six cases) and lymphatic leukaemia (2.8; seven cases). [The Working Group noted that the number of cases in each category is small, multiple exposures were evaluated independently of other exposures, and none of the associations is significant.]

Carpenter et al. (1988) evaluated the possible association with exposure to 26 chemicals or chemical groups in 89 cases of primary cancers of the central nervous system and 356 matched controls in cohorts of workers at two US nuclear facilities. Toluene, xylene (see monograph, p. 125) and methyl ethyl ketone were evaluated as one chemical group; the matched RR was 2.0 (28 cases; 95% CI, 0.7–5.5) in comparison with nonexposed workers. Almost all cases had had low exposure according to the classification used. The authors reported that the RRs were adjusted for internal and external exposure to radiation. [The

Working Group noted that no separate analysis was reported for the three solvents, nor were exposure levels quantified, and that there were many concurrent exposures.]

4. Summary of Data Reported and Evaluation

4.1 Exposures

Toluene is a major industrial chemical derived mainly from petroleum refining. Major uses of toluene are in the production of benzene and as a solvent in paints, inks and adhesives. Toluene–containing petroleum distillates are extensively and increasingly used in gasoline blending. Toluene is ubiquitous in the environment and is present at high levels in many occupational settings.

4.2 Experimental carcinogenicity data

Toluene was tested for carcinogenicity in one strain of rats by gastric intubation at one dose level and in one strain of rats by inhalation. These studies were inadequate for evaluation. Toluene was used as a vehicle control in a number of skin painting studies. Some of these studies were inadequate for evaluation; in others, repeated application of toluene to the skin of mice did not result in an increased incidence of skin tumours.

4.3 Human carcinogenicity data

Toluene was mentioned as an exposure in four case–control studies involving several anatomical sites of cancer. The results could not be evaluated with regard to toluene itself.

4.4 Other relevant data

In humans, prolonged skin contact with toluene may cause nonallergic contact dermatitis. Exposure to toluene also causes nervous system symptoms and signs. Excessive exposure to toluene may cause adverse effects on the kidney and liver.

Adverse effects on the nervous system have been observed in experimental animals.

In the available studies on spontaneous abortion, perinatal mortality and congenital malformations in humans, the numbers of cases were small and the mothers had also been exposed to other substances.

Embryotoxicity has been seen in some studies in mice and rats but not in rabbits. Embryotoxic effects generally occurred concurrently with maternal toxicity.

Increased frequencies of sister chromatid exchange and chromosomal aberrations in peripheral lymphocytes were observed in one study of workers exposed to toluene but not in two studies of chromosomal aberrations, one of sister chromatid exchange and one in which both effects were investigated. These studies are inconclusive with regard to exposure to toluene.

Toluene induced chromosomal aberrations in rats and micronuclei in mice and rats. Sister chromatid exchange and chromosomal aberrations were not induced in cultured human lymphocytes, in the absence of an exogenous metabolic system. Toluene did not induce morphological transformation in cultured animal cells. Toluene induced DNA damage in cultured animal cells. It did not induce mutation or chromosomal aberrations but induced aneuploidy in *Drosophila*. It did not induce DNA damage or mutation in bacteria. (See Appendix 1.)

4.5 Evaluation[1]

There is *inadequate evidence* for the carcinogenicity of toluene in humans.

There is *inadequate evidence* for the carcinogenicity of toluene in experimental animals.

Overall evaluation

Toluene *is not classifiable as to its carcinogenicity to humans (Group 3)*.

5. References

Abou-El-Makarem, M.M., Millburn, P., Smith, R.L. & Williams, R.T. (1967) Biliary excretion of foreign compounds. Benzene and its derivatives in the rat. *Biochem. J.*, *105*, 1269–1274

Aldrich Chemical Co. (1988) *Aldrich Catalog Handbook of Fine Chemicals*, Milwaukee, WI

Alessio, L., Odone, P., Rivolta, G., Soma, R., Confortini, C. & Colombi, A. (1981) Behaviour of urinary hippuric acid in non-occupationally exposed subjects and in workers with moderate exposure to toluene (Ital.). *Med. Lav.*, *72*, 38–45

American Conference of Governmental Industrial Hygienists (1988) *Threshold Limit Values and Biological Exposure Indices for 1988–1989 for Chemical Substances and Physical Agents in the Work Environment*, Cincinnati, OH, p. 36

Andersen, I., Lundqvist, G.R., Mølhave, L., Pedersen, O.F., Proctor, D.F., Vaeth, M. & Wyon, D.P. (1983) Human response to controlled levels of toluene in six-hour exposures. *Scand. J. Work Environ. Health*, *9*, 405–418

Angerer, J. (1979) Occupational chronic exposure to organic solvents. VII. Metabolism of toluene in man. *Int. Arch. occup. environ. Health*, *43*, 63–67

Angerer, J. (1985) Occupational chronic exposure to organic solvents XII. o–Cresol excretion after toluene exposure. *Int. Arch. occup. environ. Health*, *56*, 323–328

Anon. (1987a) Facts and figures for the chemical industry. *Chem. Eng. News*, *65*, 24

Anon. (1987b) Final report on the safety assessment of toluene. *J. Am. Coll. Toxicol.*, *6*, 77–120

Anon. (1988) Facts and figures for the chemical industry. *Chem. Eng. News*, *66*, 34–82

---

[1]For definitions of the italicized terms, see Preamble, pp. 27–30.

Anshelm Olson, B., Gamberale, F. & Iregren, A. (1985) Coexposure to toluene and p-xylene in man: central nervous functions. Br. J. ind. Med., 42, 117–122

Antti-Poika, M., Juntunen, J., Matikainen, E., Suoranta, H., Hänninen, H., Seppäläinen, A.M. & Liira, J. (1985) Occupational exposure to toluene: neurotoxic effects with special emphasis on drinking habits. Int. Arch. occup. environ. Health, 56, 31–40

Apol, A.G. (1980) Procraft, Spokane, Washington (Health Hazard Evaluation Determination Report No. 80–43–679), Cincinnati, OH, National Institute for Occupational Safety and Health

Apostoli, P., Brugnone, F., Perbellini, L., Cocheo V., Bellomo, M.L. & Silvestri, R. (1982) Biomonitoring of occupational toluene exposure. Int. Arch. occup. environ. Health, 50, 153–168

Arbeidinspectie (Labour Inspection) (1986) De Nationale MAC-Lijst 1986 [National MAC-List 1986] (P145), Voorburg, Ministry of Social Affairs and Work Environment, p. 21

Arbejdstilsynet (Labour Inspection) (1988) Graensevaerdier for Stoffer og Materialer [Limit Values for Substances and Materials] (At-anvisning No. 3.1.0.2), Copenhagen, p. 42

Aristov, V.N., Redkin, Y.V., Bruskin, Z.Z. & Ogleznev, G.A. (1981) Experimental data on the mutagenic effects of toluene, isopropanol and sulfur dioxide (Russ.). Gig. Tr. prof. Zabol., 7, 33–36

Askergren, A. (1981) Studies on kidney function in subjects exposed to organic solvents. III. Excretion of cells in the urine. Acta med. scand., 210, 103–106

Askergren, A., Allgén, A.-G. & Bergstöm, J. (1981a) Studies on kidney function in subjects exposed to organic solvents. II. The effect of desmopressin in a concentration test and the effect of exposure to organic solvents on renal concentration ability. Acta med. scand., 209, 485–488

Askergren, A., Allgén, A.-G., Karlsson, C., Lundberg, I. & Nyberg, E. (1981b) Studies on kidney function in subjects exposed to organic solvents. I. Excretion of albumin and beta-2-microglobulin in the urine. Acta med. scand., 209, 479–483

Askergren, A., Brandt, R., Gullquist, R., Silk, B. & Strandell, T. (1981c) Studies on kidney function in subjects exposed to organic solvents. IV. Effect on 51-Cr-EDTA clearance. Acta med. scand., 210, 373–376

Austin, S.G. & Schnatter, A.R. (1983) A case–control study of chemical exposures and brain tumors in petrochemical workers. J. occup. Med., 25, 313–320

Axelsson, G., Lütz, C. & Rylander, R. (1984) Exposure to solvents and outcome of pregnancy in university laboratory employees. Br. J. ind. Med., 41, 305–312

Awogi, T., Itoh, T. & Tsushimoto, G. (1986) The effect of benzene and its derivatives on metabolic cooperation (Abstract No. 2). Mutat. Res., 164, 263

Baelum, J., Andersen, I. & Mølhane, L. (1982) Acute and subacute symptoms among workers in the printing industry. Br. J. ind. Med., 39, 70–75

Baelum, J., Andersen, I., Lundqvist, G.R., Mølhave, L., Pedersen, O.F., Vaeth, M. & Wyon, D.P. (1985) Response of solvent–exposed printers and unexposed controls to six-hour toluene exposure. Scand. J. Work Environ. Health, 11, 271–280

Baelum, J., Døssing, M., Hansen, S.H., Lundqvist, G.R. & Andersen, N.T. (1987) Toluene metabolism during exposure to varying concentrations combined with exercise. Int. Arch. occup. environ. Health, 59, 281–294

Baker, D. & Fannick, N. (1983) Leather Craftsman, Lynbrook, New York (Health Hazard Evaluation Report No. 81–060–1367), Cincinnati, OH, National Institute for Occupational Safety and Health

Bakinson, M.A. & Jones, R.D. (1985) Gassing due to methylene chloride, xylene, toluene, and styrene reported to Her Majesty's Factory Inspectorate 1961-80. Br. J. ind. Med., 42, 184–190

Bakke, O.M. & Scheline, R.R. (1970) Hydroxylation of aromatic hydrocarbons in the rat. *Toxicol. appl. Pharmacol., 16*, 691–700

Bänfer, W. (1961) Studies on the action of pure toluene on blood composition of pressmen and assistants in photogravure printing (Ger.). *Zbl. Arbeitsmed. Arteitsschutz, 11*, 35–41

Batlle, D.C., Sabatini, S. & Kurtzman, N.A. (1988) On the mechanism of toluene-induced renal tubular acidosis. *Nephron, 49*, 210–218

Bauchinger, M., Schmid, E., Dresp, J., Kolin-Gerresheim, J., Hauf, R. & Suhr, E. (1982) Chromosome changes in lymphocytes after occupational exposure to toluene. *Mutat. Res., 102*, 439–445

Bauchinger, M., Schmid, E., Dresp, J. & Kolin-Gerresheim, J. (1983) Chromosome aberrations and sister-chromatid exchanges in toluene-exposed workers (Abstract No. 17). *Mutat. Res., 113*, 231–232

Benignus, V.A. (1981a) Health effects of toluene: a review. *Neurotoxicology, 2*, 567–588

Benignus, V.A. (1981b) Neurobehavioral effects of toluene. A review. *Neurobehav. Toxicol. Teratol., 3*, 407–415

Benignus, V.A., Muller, K.E., Barton, C.N. & Bittikofer, J.A. (1981) Toluene levels in blood and brain of rats during and after respiratory exposure. *Toxicol. appl. Pharmacol., 61*, 326–334

Benignus, V.A., Muller, K.E., Graham, J.A. & Barton, C.N. (1984) Toluene levels in blood and brain of rats as a function of toluene level in inspired air. *Environ. Res., 33*, 39–46

Bennett, R.H. & Forman, H.R. (1980) Hypokalemic periodic paralysis in chronic toluene exposure. *Arch. Neurol., 37*, 673

Bergert, K.-D., Voigt, H., Neubert, I. & Döhler, R. (1980) Dependence of the excretion of hippuric acid in urine on the concentration of toluene and the respiratory minute volume (Ger.). *Z. ges. inn. Med., 35*, 316–317

Beving, H., Malmgren, R., Olsson, P., Tornling, G. & Unge, G. (1983) Increased uptake of serotonin in platelets from car painters occupationally exposed to mixtures of solvents and organic isocyanates. *Scand. J. Work Environ. Health, 9*, 253–258

Beving, H., Kristensson, J., Malmgren, R., Olsson, P. & Unge, G. (1984) Effect on the uptake kinetics of serotonin (5-hydroxytryptamine) in platelets from workers with long-term exposure to organic solvents. *Scand. J. Work Environ. Health, 10*, 229–234

Bieniek, G., Palys, E. & Wilczok, T. (1982) TLC separation of hippuric, mandelic, and phenylglyoxylic acids from urine after mixed exposure to toluene and styrene. *Br. J. ind. Med., 39*, 187–190

Biscaldi, G.P., Mingardi, M., Pollini, G., Moglia, A. & Bossi, M.C. (1981) Acute toluene poisoning. Electroneurophysiological and vestibular investigations. *Toxicol. Eur. Res., 3*, 271–273

Blackburn, G.R., Deitch, R.A., Schreiner, C.A. & MacKerer, C.R. (1986) Predicting carcinogenicity of petroleum distillation fractions using a modified *Salmonella* mutagenicity assay. *Cell Biol. Toxicol., 2*, 63–84

Boewer, C., Enderlein, G., Wollgast, U., Nawka, S., Palowski, H. & Bleiber, R. (1988) Epidemiological study on the hepatotoxicity of occupational toluene exposure. *Int. Arch. occup. environ. Health, 60*, 181–186

Bos, R.P., Brouns, R.M.E., van Doorn, R., Theuws, J.L.G. & Henderson, P.T. (1981) Non-mutagenicity of toluene, *o-, m-* and *p*-xylene, *o*-methylbenzylalcohol and *o*-methylbenzylsulfate in the Ames assay. *Mutat. Res., 88*, 273–279

Bruckner, J.V. & Peterson, R.G. (1981a) Evaluation of toluene and acetone inhalant abuse. I. Pharmacology and pharmacodynamics. *Toxicol. appl. Pharmacol., 61*, 27–38

Bruckner, J.V. & Peterson, R.G. (1981b) Evaluation of toluene and acetone inhalant abuse. II. Model development and toxicology. *Toxicol. appl. Pharmacol., 61*, 302–312

Brugnone, F. & Perbellini, L. (1985) Toluene coma and liver function. *Scand. J. Work Environ. Health, 11*, 55

Brugnone, F., Perbellini, L., Grigolini, L., Cazzadori, A. & Gaffuri, E. (1976) Alveolar air and blood toluene concentration in rotogravure workers. *Int. Arch. occup. environ. Health, 38*, 45–54

Cameron, G.R., Paterson, J.L.H., de Saram, G.S.W. & Thomas, J.C. (1938) The toxicity of some methyl derivatives of benzene with special reference to pseudocumene and heavy coal–tar naphtha. *J. Pathol. Bacteriol., 46*, 95–107

Campbell, L., Marsh, D.M. & Wilson, H.K. (1987) Towards a biological monitoring strategy for toluene. *Ann. occup. Hyg., 31*, 121–133

Capellini, A. & Alessio, L. (1971) The urinary excretion of hippuric acid in workers exposed to toluene (Ital.). *Med. Lav., 62*, 196–201

Carlsson, A. (1982) Exposure to toluene. Uptake, distribution and elimination in man. *Scand. J. Work Environ. Health, 8*, 43–55

Carlsson, A. & Ljungquist, E. (1982) Exposure to toluene. Concentration in subcutaneous adipose tissue. *Scand. J. Work Environ. Health, 8*, 56–62

Carpenter, A.V., Flanders, W.D., Frome, E.L., Tankersley, W.G. & Fry, S.A. (1988) Chemical exposures and central nervous system cancers: a case–control study among workers at two nuclear facilities. *Am. J. ind. Med., 13*, 351–362

Carpenter, C.P., Geary, D.L., Jr, Myers, R.C., Nachreiner, D.J., Sullivan, L.J. & King, J.M. (1976) Petroleum hydrocarbon toxicity studies. XIII. Animal and human response to vapors of toluene. *Toxicol. appl. Pharmacol., 36*, 473–490

Casto, B. (1981) Detection of chemical carcinogens and mutagens in hamster cells by enhancement of adenovirus transformation. In: Mishra, N., Dunkel, V. & Mehlman, I., eds, *Advances in Modern Environmental Toxicology*, Vol. 1, Princeton, NJ, Senate Press, pp. 241–271

Cherry, N., Hutchins, H. & Waldron, H.A. (1985) Neurobehavioral effects of repeated occupational exposure to toluene and paint solvents. *Br. J. ind. Med., 42*, 291–300

Clement Associates (1977) Toluene. In: *Information Dossiers on Substances Designated by TSCA (Toxic Substances Control Act) Interagency Testing Committee (October, 1977) (Contract NSF–C–Env–77–15417)*, Washington DC

Cohr, K.–H. & Stokholm, J. (1979) Toluene. A toxicological review. *Scand. J. Work Environ. Health, 5*, 71–90

Commission of the European Communities (1986) Limit values for occupational exposure. *Off. J. Eur. Commun., C164*, 6–7

Connor, T.C., Theiss, J.C., Hanna, H.A., Monteith, D.K. & Matney, T.S. (1985) Genotoxicity of organic chemicals frequently found in the air of mobile homes. *Toxicol. Lett., 25*, 33–40

Contreras, C.M., González–Estrada, T., Zarabozo, D. & Fernández–Guardiola, A. (1979) Petit mal and grand mal seizures produced by toluene or benzene intoxication in the cat. *Electroencephalogr. clin. Neurophysiol., 46*, 290–301

Cook, W.A. (1987) *Occupational Exposure Limits – Worldwide*, Washington DC, Americal Industrial Hygiene Association, pp. 35, 126, 155, 220

Courtney, K.D., Andrews, J.E., Springer, J., Ménache, M., Williams, T., Dalley, L. & Graham, J.A. (1986) A perinatal study of toluene in CD–1 mice. *Fundam. appl. Toxicol., 12*, 145–154

Cramer, P.H., Boggess, K.E., Hosenfeld, J.M., Remmers, J.C., Breen, J.J., Robinson, P.E. & Stroup, C. (1988) Determination of organic chemicals of human whole blood. Preliminary method development for volatile organics. *Bull. environ. Contam. Toxicol.*, *40*, 612–618

Dean, B.J. (1978) Genetic toxicology of benzene, toluene, xylenes and phenols. *Mutat. Res.*, *47*, 75–97

Dean, B.J. (1985) Recent findings on the genetic toxicology of benzene, toluene, xylenes and phenols. *Mutat. Res.*, *154*, 153–181

De Bortoli, M., Knöppel, H., Pecchio, E., Peil, A., Rogora, L., Schauenburg, H., Schlitt, H. & Vissers, H. (1984) Integrating 'real life' measurements of organic pollution in indoor and outdoor air of homes in northern Italy. In: Berglund, B., Lindall, T. & Sundell, J., eds, *Indoor Air, Proceedings of the 3rd International Conference on Indoor Air Quality Climatization*, Vol. 4, Stockholm, Swedish Council for Building Research, pp. 21–26

Decker, D.W., Clark, C.S., Elia, V.J., Kominsky, J.R. & Trapp, J.H. (1983) Worker exposure to organic vapors at a liquid chemical waste incinerator. *Am. J. ind. Hyg. J.*, *44*, 296–300

Dement, J., Wallingford, K.M. & Zumwalde, R.D. (1973) *Industrial Hygiene Survey of Owens–Corning Fiberglas, Kansas City, Kansas (Report No. IW 35.16)*, Cincinnati, OH, National Institute for Occupational Safety and Health

Denkhaus, W., von Steldern, D., Botzenhardt, U. & Konietzko, H. (1986) Lymphocyte subpopulations in solvent-exposed workers. *Int. Arch. occup. environ. Health*, *57*, 109–115

De Rosa, E., Brugnone, F., Bartolucci, G.B., Perbellini, L., Bellomo, M.L., Gori, G.P., Sigon, M. & Corona, P.C. (1985) The validity of urinary metabolites as indicators of low exposures to toluene. *Int. Arch. occup. environ. Health*, *56*, 135–145

De Rosa, E., Bartolucci, G.B., Sigon, M., Callegaro, R., Perbellini, L. & Brugnone, F. (1987) Hippuric acid and *ortho*-cresol as biological indicators of occupational exposure to toluene. *Am. J. ind. Med.*, *11*, 529–537

Deutsche Forschungsgemeinschaft (German Research Society) (1988) *Maximale Arbeitsplatzkonzentrationen und Biologische Arbeitsstofftoleranzwerte 1988* [Maximal Concentrations in the Work Place and Biological Tolerance Values for Working Materials 1988] *(Report No. XXIV)*, Weinheim, VCH Verlagsgesellschaft, p. 58

Dick, R.B., Setzer, J.V., Wait, R., Hayden, M.B., Taylor, B.J., Tolos, B. & Putz-Anderson, V. (1984) Effects of acute exposure of toluene and methyl ethyl ketone on psychomotor performance. *Int. Arch. occup. environ. Health*, *54*, 91–109

Direktoratet for Arbeidstilsynet (Directorate for Labour Inspection) (1981) *Administrative Normer for Forurensning i Arbeidsatmosfaere 1981* [Administrative Norms for Pollution in Work Atmosphere 1981] *(No. 361)*, Oslo, p. 2

Doak, S.M.A., Simpson, B.J.E., Hunt, P.F. & Stevenson, D.E. (1976) The carcinogenic response in mice to the topical application of propane sultone to the skin. *Toxicology*, *6*, 139–154

Dobrokhotov, V.B. (1972) The mutagenic effect of benzene and toluene under experimental conditions (Russ.). *Gig. Sanit.*, *37*, 36–39

Dobrokhotov, V.B. & Enikeev, M.I. (1977) The mutagenic action of benzene, toluene and of a mixture of these hydrocarbons in a chronic test (Russ.). *Gig. Sanit.*, *42*, 32–34

Donner, M., Husgafvel-Pursiainen, K., Mäki-Paakkanen, J., Sorsa, M. & Vainio, H. (1981) Genetic effects of in vivo exposure to toluene (Abstract). *Mutat. Res.*, *85*, 293–294

van Doorn, R., Bos, R.P., Brouns, R.M.E., Leijdekkers, C.-M. & Henderson, P.T. (1980) Effect of toluene and xylenes on liver glutathione and their urinary excretion as mercapturic acids in the rat. *Arch. Toxicol.*, *43*, 293–304

Dutkiewicz, T. & Tyras, H. (1968) The quantitative estimation of toluene skin absorption in man. *Int. Arch. Gewerbepathol. Gewerbehyg., 24*, 253–257

Egle, J.L., Jr & Gochberg, B.J. (1976) Respiratory retention of inhaled toluene and benzene in the dog. *J. Toxicol. environ. Health, 1*, 531–538

Eller, P.M. (1984) *NIOSH Manual of Analytical Methods*, 3rd ed., Vol. 1 *(DHHS(NIOSH))* Publ. No. *84–100*), Washington DC, US Government Printing Office, pp. 1500-1-1500-7; 1501-1-1501-7; 8002-1-8002-4

Eller, P.M. (1987) *NIOSH Manual of Analytical Methods*, 3rd ed., 2nd Suppl. *(DHHS(NIOSH) Publ. No. 84–100)*, Washington DC, US Government Printing Office, pp. 4000-1-4000-4

El Masry, A.M., Smith, J.N. & Williams, R.T. (1956) The metabolism of alkyl benzene: *n*-propylbenzene and *n*-butylbenzene with further observations on ethylbenzene. *Biochem. J., 64*, 50–56

Elofsson, S.-A., Gamberale, F., Hindmarsh, T., Iregren, A., Isaksson, A., Johnsson, I., Knave, B., Lydahl, E., Mindus, P., Persson, H.E., Philipson, B., Steby, M., Struwe, G., Söderman, E., Wennberg, A. & Widén, L. (1980) Exposure to organic solvents. A cross-sectional epidemiologic investigation on occupationally exposed car and industrial spray painters with special reference to the nervous system. *Scand. J. Work Environ. Health, 6*, 239–273

Elovaara, E., Hemminki, K. & Vainio, H. (1979) Effects of methylene chloride, trichloroethane, trichloroethylene, tetrachloroethylene and toluene on the development of chick embryos. *Toxicology, 12*, 111–119

ENMET Corp. (undated) *ENMET–Kitagawa Toxic Gas Detector Tubes*, Ann Arbor, MI

Ericson, A., Källén, B., Zetterström, R., Eriksson, M. & Westerholm, P. (1984) Delivery outcome of women working in laboratories during pregnancy. *Arch. environ. Health, 39*, 5–10

Fabre, R., Truhaut, R., Laham, S. & Péron, M. (1955) Toxicological studies on benzene replacement solvents. II. Toluene (Fr.). *Arch. Mal. prof. Méd. Trav. Séc. soc., 16*, 197–215

Fazio, T. & Sherma, J. (1987) *Food Additives Analytical Manual*, Vol. II (*US Food and Drug Administration Publication*), Arlington, VA, Association of Official Analytical Chemists

Feldt, E.G., Zhurkov, V.S. & Sysin, A.N. (1985) Study of the mutagenic effects of benzene and toluene in the mammalian somatic and germ cells (Abstract No. 31). *Mutat. Res., 147*, 294

Fishbein, L. (1985) An overview of environmental and toxicological aspects of aromatic hydrocarbons. II. Toluene. *Sci. total Environ., 42*, 267–288

Fishbein, L. & O'Neill, I.K., eds (1988) *Environmental Carcinogens: Methods of Analysis and Exposure Measurement*, Vol. 10, *Benzene and Alkylated Benzenes (IARC Scientific Publications No. 85)*, Lyon, International Agency for Research on Cancer

Fornazzari, L., Wilkinson, D.A., Kapur, B.M. & Carlen, P.L. (1983) Cerebellar, cortical and functional impairment in toluene abusers. *Acta neurol. scand., 67*, 319–329

Forni, A., Pacifico, E. & Limonta, A. (1971) Chromosome studies in workers exposed to benzene or toluene or both. *Arch. environ. Health, 22*, 373–378

The Foxboro Co. (1983) *Chromatographic Column Selection Guide for Century Organic Vapor Analyzer*, Foxboro, MA

Franchini, I., Cavatorta, A., Falzoi, M., Lucertini, S. & Mutti, A. (1983) Early indicators of renal damage in workers exposed to organic solvents. *Int. Arch. occup. environ. Health, 52*, 1–9

Frei, J.V. & Kingsley, W.F. (1968) Observation on chemically induced regressing tumors of mouse epidermis. *J. natl Cancer Inst., 41*, 1307–1313

Friborská, A. (1973) Some cytochemical findings in the peripheral white blood cells in workers exposed to toluene. *Folia haematol., 99*, 233–237

Funes-Cravioto, F., Kolmodin-Hedman, B., Lindsten, J., Nordenskjöld, M., Zapata-Gayon, C., Lambert, B., Norberg, E., Olin, R. & Swensson, Å. (1977) Chromosome aberrations and sister chromatid exchange in workers in chemical laboratories and a rotoprinting factory and in children of women laboratory workers. *Lancet, ii*, 233–235

Fuxe, K., Andersson, K., Nilsen, O.G., Toftgård. R., Eneroth, P. & Gustafsson, J.-Å. (1982) Toluene and telencephalic dopamine: selective reduction of amine turnover in discrete DA nerve terminal systems of the anterior caudate nucleus by low concentrations of toluene. *Toxicol. Lett., 12*, 115–123

Gad-El-Karim, M.M., Harper, B.L. & Legator, M.S. (1984) Modifications in the myeloclastogenic effect of benzene in mice with toluene, phenobarbital, 3–methylcholanthrene, Aroclor 1254 and SKF–525A. *Mutat. Res., 135*, 225–243

Gamberale, F. & Hultengren, M. (1972) Toluene exposure. II. Psychophysiological functions. *Work Environ. Health, 9*, 131–139

Gattner, H. & May, G. (1963) Panmyelopathy due to toluene action (Ger.). *Zbl. Arbeitsmed. Arbeitsschutz, 13*, 156–157

Geller, I., Hartmann, R.J., Randle, S.R. & Gause, E.M. (1979) Effects of acetone and toluene vapours on multiple schedule performance of rats. *Pharmacol. Biochem. Behav., 11*, 359–399

Gerner-Smidt, P. & Friedrich, U. (1978) The mutagenic effect of benzene, toluene and xylene studied by the SCE technique. *Mutat. Res., 58*, 313–316

Ghantous, H. & Danielson, B.R.G. (1986) Placental transfer and distribution of toluene, xylene and benzene, and their metabolites during gestation in mice. *Biol. Res. Preg., 7*, 98–105

Gibson, J.E. & Hardisty, J.F. (1983) Chronic toxicity and oncogenicity bioassay of inhaled toluene in Fischer–344 rats. *Fundam. appl. Toxicol., 3*, 315–319

Gomez-Arroyo, S. & Villalobos-Pietrini, R. (1981) Chromosomal alterations induced by solvents in *Vicia faba* (Abstract). *Mutat. Res., 85*, 244

Grant, W.M. (1962) *Toxicology of the Eye*, Springfield, IL, Charles C. Thomas, pp. 544–545

Greenburg, L., Mayers, M.R., Heimann, H. & Moskowitz, S. (1942) The effects of exposure to toluene in industry. *J. Am. med. Assoc., 118*, 573–578

Griffiths, W.C., Lipsky, M., Rosner, A. & Martin, H.F. (1972) Rapid identification of and assessment of damage by inhaled volatile substances in the clinical laboratory. *Clin. Biochem., 5*, 222–231

Gurka, D.F. & Betowski, L.D. (1982) Gas chromatographic/Fourier transform infrared spectrometric identification of hazardous waste extract components. *Anal. Chem., 54*, 1819–1824

Haglund, U., Lundberg, I. & Zech, L. (1980) Chromosome aberrations and sister chromatid exchanges in Swedish paint industry workers. *Scand. J. Work Environ. Health, 6*, 291–298

Hampton, C.V., Pierson, W.R., Harvey, T.M., Updegrove, W.S. & Marano, R.S. (1982) Hydrocarbon gases emitted from vehicles on the road. 1. A qualitative gas chromatography/mass spectrometry survey. *Environ. Sci. Technol., 16*, 287–298

Hänninen, H., Eskelinen, L., Husman, K. & Nurminen, M. (1976) Behavioral effects of long-term exposure to a mixture of organic solvents. *Scand. J. Work Environ. Health, 4*, 240–255

Hänninen, H., Antti-Poika, M. & Savolainen, P. (1987) Psychological performance, toluene exposure and alcohol consumption in rotogravure printers. *Int. Arch. environ. Health, 59*, 475–483

Hansch, C. & Leo, A. (1979) *Substituent Constants for Correlation Analysis in Chemistry and Biology*, New York, John Wiley & Sons, p. 218

Hardin, B.D., Schuler, R.L., Burg, J.R., Booth, G.M., Hazelden, K.P., MacKenzie, K.M., Piccirillo, V.J. & Smith, K.N. (1987) Evaluation of 60 chemicals in a preliminary developmental toxicity test. *Teratog. Carcinog. Mutagenesis, 7*, 29–48

Hasegawa, K., Shiojima, S., Koizumi, A. & Ikeda, M. (1983) Hippuric acid and o–cresol in the urine of workers exposed to toluene. *Int. Arch. occup. environ. Health, 52*, 197–208

Hawley, G.G. (1981) *The Condensed Chemical Dictionary*, 10th ed., New York, Van Nostrand Reinhold, p. 1030

Haworth, S., Lawlor, T., Mortelmans, K., Speck, W. & Zeiger, E. (1983) *Salmonella* mutagenicity test results for 250 chemicals. *Environ. Mutagenesis, Suppl. 1*, 3–142

Health and Safety Executive (1987) *Occupational Exposure Limits 1987 (Guidance Note EH 40/87)*, London, Her Majesty's Stationery Ofice, p. 21

Hisanaga, N. & Takeuchi, Y. (1983) Changes in sleep cycle and EEG of rats exposed to 4000 ppm toluene for four weeks. *Ind. Health, 21*, 153–164

Hoff, M.C. (1983) Toluene. In: Mark, H.F., Othmer, D.F., Overberger, C.G., Seaborg, G.T. & Grayson, M., eds, *Kirk–Othmer Encyclopedia of Chemical Technology*, 3rd ed., Vol. 23, New York, John Wiley & Sons, pp. 246–273

Hollifield, H.C., Breder, C.V., Dennison, J.L., Roach, J.A.G. & Adams, W.S. (1980) Container–derived contamination of maple syrup with methyl methacrylate, toluene, and styrene as determined by headspace gas–liquid chromatography. *J. Assoc. off. anal. Chem., 63*, 173–177

Holmberg, P.C. (1979) Central–nervous–system defects in children born to mothers exposed to organic solvents during pregnancy. *Lancet, ii*, 177–179

Holmberg, P.C., Hernberg, S., Kurppa, K., Rantala, K. & Riala, R. (1982) Oral clefts and organic solvent exposure during pregnancy. *Int. Arch. occup. environ. Health, 50*, 371–376

Holzer, G., Shanfield, H., Zlatkis, A., Bertsch, W., Juarez, P., Mayfield, H. & Liebich, H.M. (1977) Collection and analysis of trace organic emissions from natural sources. *J. Chromatogr., 142*, 755–764

Horiguchi, S. & Inoue, K. (1977) Effects of toluene on the wheel–turning activity and peripheral blood findings in mice: an approach to the maximum allowable concentration of toluene. *J. toxicol. Sci., 2*, 363–372

Hudák, A. & Ungváry, G. (1978) Embryotoxic effects of benzene and its methyl derivatives: toluene, xylene. *Toxicology, 11*, 55–63

Husman, K. (1980) Symptoms of car painters with long–term exposure to a mixture of organic solvents. *Scand. J. Work Environ. Health, 6*, 19–32

Husman, K. & Karli, P. (1980) Clinical and neurological findings among car painters exposed to a mixture of organic solvents. *Scand. J. Work Environ. Health, 6*, 33–39

IARC (1980) *IARC Monographs on the Evaluation of the Carcinogenic Risk of Chemicals to Humans*, Suppl. 2, *Long–term and Short–term Screening Assays for Carcinogenesis: A Critical Appraisal*, Lyon

IARC (1981) *IARC Monographs on the Evaluation of the Carcinogenic Risk of Chemicals to Humans*, Vol. 25, *Wood, Leather and Some Associated Industries*, Lyon, p. 222

IARC (1986) *IARC Monographs on the Evaluation of the Carcinogenic Risk of Chemicals to Humans*, Vol. 38, *Tobacco Smoking*, Lyon, p. 97

IARC (1988) *Information Bulletin on the Survey of Chemicals Being Tested for Carcinogenicity*, No. 13, Lyon, p. 265

IARC (1989) *IARC Monographs on the Evaluation of Carcinogenic Risks to Humans*, Vol. 45, *Occupational Exposures in Petroleum Refining; Crude Oil and Major Petroleum Fuels*, Lyon, pp. 159–201

Ikeda, M. & Ohtsuji, H (1969) Significance of urinary hippuric acid determination as an index of toluene exposure. *Br. J. ind. Med., 26*, 244–246

Ikeda, M. & Ohtsuji, H. (1971) Phenobarbital–induced protection against toxicity of toluene and benzene in the rat. *Toxicol. appl. Pharmacol., 20*, 30–43

Ikeda, T. & Miyake, H. (1978) Decreased learning in rats following repeated exposure to toluene: preliminary report. *Toxicol. Lett.*, *1*, 235–239

Inoue, O., Seiji, K., Ishihara, N., Kumai, M. & Ikeda, M. (1984) Increased *o*- and *p*-cresol/hippuric acid ratios in the urine of four strains of rat exposed to toluene at thousands-ppm levels. *Toxicol. Lett.*, *23*, 249–257

Inoue, O., Seiji, K., Watanabe, T., Kasahara, M., Nakatsuka, H., Yin, S., Li, G., Cai, S., Jin, C. & Ikeda, M. (1986) Possible ethnic difference in toluene metabolism: a comparative study among Chinese, Turkish and Japanese solvent workers. *Toxicol. Lett.*, *34*, 167–174

Inoue, O., Seiji, K., Watanabe, T., Kasahara, M., Nakatsuka, H., Yin, S., Li, G., Cai, S., Jin, C. & Ikeda, M. (1988) Mutual metabolic suppression between benzene and toluene in man. *Int. Arch. occup. environ. Health*, *60*, 15–20

Institut National de Recherche et de Sécurité (National Institute for Research and Safety) (1986) *Valeurs Limites pour les Concentrations des Substances Dangereuses Dans l'Air des Lieux de Travail* [Limit Values for Concentrations of Dangerous Substances in the Air of Work Places] (*ND 1609-125-86*), Paris, p. 579

International Labour Office (1984) *Occupational Exposure Limits for Airborne Toxic Substances*, 2nd rev. ed. (*Occupational Safety and Health Series No. 37*), Geneva, pp. 204–205

International Programme on Chemical Safety (1985) *Toluene* (*Environmental Health Criteria 52*), Geneva, World Health Organization

Iregren, A. (1986) Subjective and objective signs of organic solvent toxicity among occupationally exposed workers. An experimental evaluation. *Scand. J. Work Environ. Health*, *12*, 469–475

Jenkins, L.J., Jr, Jones, R.A. & Siegel, J. (1970) Long-term inhalation screening studies of benzene, toluene, *o*-xylene, and cumene on experimental animals. *Toxicol. appl. Pharmacol.*, *16*, 818–823

Kilburn, K.H., Seidman, B.C. & Warshaw, R. (1985) Neurobehavorial and respiratory symptoms of formaldehyde and xylene exposure in histology technicians. *Arch. environ. Health*, *40*, 229–233

Kimura, E.T., Ebert, D.M. & Dodge, P.W. (1971) Acute toxicity and limits of solvent residue for sixteen organic solvents. *Toxicol. appl. Pharmacol.*, *19*, 699–704

King, M.D., Day, R.E., Oliver, J.S., Lush, M. & Watson, J.M. (1981) Solvent encephalopathy. *Br. med. J.*, *283*, 663–665

Klavis, G. & Wille, F. (1967) Difficulties in estimating bone-marrow damage after effects of toluene (Ger.). *Zbl. Arbeitsmed. Arbeitsschutz*, *17*, 174–176

Knoop, F. & Gehrke, M. (1925) Oxidation of acetic acid, acetone and toluene (Ger.). *Hoppe-Seyler's Z. physiol. Chem.*, *146*, 63–71

Koga, K. (1978) Distribution, metabolism and excretion of toluene in mice (Jpn.). *Folia pharmacol. jpn.*, *74*, 687–698

Konietzko, H., Keilbach, J. & Drysch, K. (1980) Cumulative effects of daily toluene exposure. *Int. Arch. occup. environ. Health*, *46*, 53–58

Korobkin, R., Asbury, A.K., Sumner, A.J. & Nielsen, S.L. (1975) Glue-sniffing neuropathy. *Arch. Neurol.*, *32*, 158–162

Kostas, J. & Hotchin, J. (1981) Behavioral effects of low-level perinatal exposure to toluene in mice. *Neurobehav. Toxicol. Teratol.*, *3*, 467–469

Kroeger, R.M., Moore, R.J., Lehman, T.H., Giesy, J.D. & Skeeters, C.E. (1980) Recurrent urinary calculi associated with toluene sniffing. *J. Urol.*, *123*, 89–91

Krotoszynski, B.K., Bruneau, G.M. & O'Neill, H.J. (1979) Measurement of chemical inhalation exposure in urban population in the presence of endogenous effluents. *J. anal. Toxicol.*, *3*, 225–234

Kurppa, K. & Husman, K. (1982) Car painters' exposure to a mixture of organic solvents. Serum activities of liver enzymes. *Scand. J. Work Environ. Health, 8*, 134–140

Lauwerys, R. (1983) Toluene. In: Alessio, L., Berlin, A., Roi, R. & Boni, M., eds, *Human Biological Monitoring of Industrial Chemicals Series*, Luxembourg, Commission of the European Communities, pp. 163–175

Lauwerys, R., Bernard, A., Viau, C. & Buchet, J.-P. (1985) Kidney disorders and hematotoxicity from organic solvent exposure. *Scand. J. Work Environ. Health, 11 (Suppl. 1)*, 83–90

Lebowitz, H., Brusick, D., Matheson, D., Jagannath, D.R., Reed, M., Goode, S. & Roy, G. (1979) Commonly used fuels and solvents evaluated in a battery of short-term bioassays (Abstract No. Eb–8). *Environ. Mutagenesis, 1*, 172–173

Liang, J.C., Hsu, T.C. & Henry, J.E. (1983) Cytogenetic assays for mitotic poisons, the grasshopper embryo system for volatile liquids. *Mutat. Res., 113*, 467–479

Lijinsky, W. & Garcia, H. (1972) Skin carcinogenesis tests of hydrogenated derivatives of anthracene and other polynuclear hydrocarbons. *Z. Krebsforsch., 77*, 226–230

Lloyd, W., Decouflé, P. & Salvin, L. (1977) Unusual mortality experience of printing pressmen. *J. occup. Med., 19*, 543–550

Longley, E.O., Jones, A.T., Welch, R. & Lomaev, O. (1967) Two acute toluene episodes in merchant ships. *Arch. environ. Health, 14*, 481–487

Low, L.K., Meeks, J.R. & Mackerer, C.R. (1988) Health effects of the alkylbenzenes. I. Toluene. *Toxicol. ind. Health, 4*, 49–75

Lundberg, I. & Håkansson, M. (1985) Normal serum activities of liver enzymes in Swedish paint industry workers with heavy exposure to organic solvents. *Br. J. ind. Med., 42*, 596–600

Lundberg, I., Ekdahl, M., Kronevi, T., Lidums, V. & Lundberg, S. (1986) Relative hepatotoxicity of some industrial solvents after intraperitoneal injection or inhalation exposure in rats. *Environ. Res., 40*, 411–420

Lyapkalo, A.A. (1973) Genetic activity of benzene and toluene (Russ.). *Gig. Tr. prof. Zabol., 17*, 24–28

Mäki-Paakkanen, J., Husgafvel-Pursiainen, K., Kalliomäki, P.-L., Tuominen, J. & Sorsa, M. (1980) Toluene-exposed workers and chromosome aberrations. *J. Toxicol. environ. Health, 6*, 775–781

Malm, G. & Lying-Tunell, U. (1980) Cerebellar dysfunction related to toluene sniffing. *Acta neurol. scand., 62*, 188–190

Maltoni, C., Conti, B. & Cotti, G. (1983) Benzene: a multipotential carcinogen. Results of long-term bioassays performed at the Bologna Institute of Oncology. *Am. J. ind. Med., 4*, 589–630

Maltoni, C., Conti, B., Cotti, G. & Belpoggi, F. (1985) Experimental studies on benzene carcinogenicity at the Bologna Institute of Oncology: current results and ongoing research. *Am. J. ind. Med., 7*, 415–446

Mannsville Chemical Products Corp. (1981) *Chemical Products Synopsis: Toluene*, Cortland, NY

Matheson Gas Products (undated) *The Matheson–Kitagawa Toxic Gas Detector System*, East Rutherford, NJ

Matsushita, T., Arimatsu, Y., Ueda, A., Satoh, K. & Nomura, S. (1975) Hematological and neuro-muscular response of workers exposed to low concentration of toluene vapor. *Ind. Health, 13*, 115–121

McCarroll, N.E., Keech, B.H. & Piper, C.E. (1981a) A microsuspension adaptation of the *Bacillus subtilis* 'rec' assay. *Environ. Mutagenesis, 3*, 607–616

McCarroll, N.E., Piper, C.E. & Keech, B.H. (1981b) An *E. coli* microsuspension assay for the detection of DNA damage induced by direct-acting agents and promutagens. *Environ. Mutagenesis, 3*, 429–444

McDonald, J.C., Lavoie, J., Côté, R. & McDonald, A.D. (1987) Chemical exposures at work in early pregnancy and congenital defect: a case-referent study. *Br. J. ind. Med.*, *44*, 527–533

McKee, R.H. & Lewis, S.C. (1987) Evaluation of the dermal carcinogenic potential of liquids produced from the Cold Lake heavy oil deposits of Northeast Alberta. *Can. J. Physiol. Pharmacol.*, *65*, 1793–1797

McKee, R.H., Stubblefield, W.A., Lewis, S.C., Scala, R.A., Simon, G.S. & DePass, L.R. (1986) Evaluation of the dermal carcinogenic potential of tar sands bitumen–derived liquids. *Fundam. appl. Toxicol.*, *7*, 228–235

McLaughlin, J., Jr, Marliac, J.-P., Verrett, J., Mutchler, M.K. & Fitzhugh, O.G. (1964) Toxicity of fourteen volatile chemicals as measured by the chick embryo method. *Am. med. Assoc. ind. Hyg. J.*, *25*, 282–284

Mergler, D., Bélanger, S., De Grosbois, S. & Vachon, N. (1988) Chromal focus of acquired chromatic discrimination loss and solvent exposure among printshop workers. *Toxicology*, *49*, 341–348

Merian, E. (1982) The environmental chemistry of volatile hydrocarbons. *Toxicol. environ. Chem.*, *5*, 167–175

Merian, E. & Zander, M. (1982) Volatile aromatics. In: Hutzinger, O., ed., *Handbook of Environmental Chemistry*, Vol. 3, Part B. *Anthropogenic Compounds*, Berlin (West), Springer, pp. 117–161

Miyaura, S. & Isono, H. (1985) Gas chromatographic determination of toluene in tissue by using a head space method. *Eisei Kagaku*, *31*, 87–94

Mohtashamipur, E., Norpoth, K., Woelke, U. & Huber, P. (1985) Effects of ethylbenzene, toluene and xylene on the induction of micronuclei in bone marrow polychromatic erythrocytes of mice. *Arch. Toxicol.*, *58*, 106–109

Mohtashamipur, E., Sträter, H., Triebel, R. & Norpoth, K. (1987) Effects of pretreatment of male NMRI mice with enzyme inducers or inhibitors on clastogenicity of toluene. *Arch. Toxicol.*, *60*, 460–463

Mølhave, L. (1979) Indoor air pollution due to building materials. In: Fanger, P.O. & Valbjorn, O., eds, *Indoor Climate, Effect on Human Comfort, Performance and Health in Residential, Commercial and Light Industry Buildings, Proceedings of the 1st International Indoor Climate Symposium*, Copenhagen, University of Aarhus, Institute of Hygiene, pp. 89–110

Mølhave, L. & Møller, J. (1979) The atmospheric environment in modern Danish dwellings – measurements in 39 flats. In: Fanger, P.O. & Valbjorn, O., eds, *Indoor Climate, Effect on Human Comfort, Performance and Health in Residential, Commercial and Light Industry Buildings, Proceedings of the 1st International Indoor Climate Symposium*, Copenhagen, University of Aarhus, Institute of Hygiene, pp. 171–186

Montesano, R., Bartsch, H., Vainio, H., Wilbourn, J. & Yamasaki, H., eds (1986) *Long–term and Short–term Assays for Carcinogenesis: A Critical Appraisal (IARC Scientific Publications No. 83)*, Lyon, International Agency for Research on Cancer

Moss, A.H., Gabow, P.A., Kaehny, W.D., Goodman, S.I. & Haut, L.L. (1980) Fanconi's syndrome and distal renal tubular acidosis after glue sniffing. *Ann. intern. Med.*, *92*, 69–70

Moszczyński, P. & Lisiewicz, J. (1985) Occupational exposure to benzene, toluene and xylene and the lymphocyte lysosomal *N*-acetyl-beta-D-glucosaminidase. *Ind. Health*, *23*, 47–51

Mungikar, A.M. & Pawar, S.S. (1976) Hepatic microsomal mixed function oxidase system during toluene treatment and the effect of pretreatment of phenobarbital in adult rats. *Bull. environ. Contam. Toxicol.*, *15*, 198–204

Nakamura, S., Oda, Y., Shimada, T., Oki, I. & Sugimoto, K. (1987) SOS–inducing activity of chemical carcinogens and mutagens in *Salmonella typhimurium* TA1535/pSK 1002: examination with 151 chemicals. *Mutat. Res.*, *192*, 239–246

National Draeger, Inc. (1987) *Detector Tube Products for Gas and Vapor Detection*, Pittsburgh, PA

National Institute for Occupational Safety and Health (1973) *Criteria for a Recommended Standard... Occupational Exposure to Toluene*, Cincinnati, OH

National Institute for Occupational Safety and Health (1983) *National Occupational Hazard Survey 1981–1983*, Cincinnati, OH

National Institute for Occupational Safety and Health (1986) NIOSH recommendations for occupational safety and health standards. *Morbid. Mortal. Wkly Rep. Suppl.*, *35*, 31S

National Research Council (1976) *Vapor–phase Organic Pollutants*, Washington DC, National Academy of Sciences

National Swedish Board of Occupational Safety and Health (1987) *Hygieniska Gränsvärden* [Hygienic Limit Values] *(Ordinance AFS 1987:12)*, Solna, p. 36

Nawrot, P.S. & Staples, R.E. (1979) Embryofetal toxicity and teratogenicity of benzene and toluene in the mouse (Abstract). *Teratology*, *19*, 41A

Nestmann, E.R., Lee, E.G.-H., Matula, T.I., Douglas, G.R. & Mueller, J.C. (1980) Mutagenicity of constituents identified in pulp and paper mill effluents using the *Salmonella*/mammalian–microsome assay. *Mutat. Res.*, *79*, 203–212

Nise, G. & Ørbaek, P. (1988) Toluene in venous blood during and after work in rotogravure printing. *Int. Arch. occup. environ. Health*, *60*, 31–35

Nomiyama, K. & Nomiyama, H. (1974) Respiratory retention, uptake and excretion of organic solvents in man. Benzene, toluene, *n*–hexane, trichloroethylene, acetone, ethyl acetate and ethyl alcohol. *Int. Arch. Arbeitsmed.*, *32*, 75–83

Ogata, M., Tomokuni, K. & Takatsuka, T. (1970) Urinary excretion of hippuric acid and *m*- or *p*–methylhippuric acid in the urine of persons exposed to vapours of toluene and *m*- or *p*–xylene as a test of exposure. *Br. J. ind. Med.*, *27*, 43–50

Ogata, M., Saeki, T., Kira, S., Hasegawa, T. & Watanabe, S. (1974) Distribution of toluene in mouse tissues. *Jpn. J. ind. Health*, *16*, 23–25

Oliver, J.S. (1982) The analytical diagnosis of solvent abuse. *Human Toxicol.*, *1*, 293–297

Olsson, H. & Brandt, L. (1980) Occupational exposure to organic solvents and Hodgkin's disease in men. A case–referent study. *Scand. J. Work Environ. Health*, *6*, 302–305

Otson, R. & Williams, D.T. (1981) Evaluation of a liquid–liquid extraction technique for water pollutants. *J. Chromatogr.*, *212*, 187–197

Otson, R., Williams, D.T. & Bothwell, P.D. (1982) Volatile organic compounds in water in thirty Canadian potable water treatment facilities. *J. Assoc. off. anal. Chem.*, *65*, 1370–1374

Otson, R., Doyle, E.E., Williams, D.T. & Bothwell, P.D. (1983) Survey of selected organics in office air. *Bull. environ. Contam. Toxicol.*, *31*, 222–229

Paganini-Hill, A., Glazer, E., Henderson, B. & Ross, R.K. (1980) Cause-specific mortality among newspaper web pressmen. *J. occup. Med.*, *22*, 542–544

Pagnotto, L.D. & Lieberman, L.M. (1967) Urinary hippuric acid excretion as an index of toluene exposure. *Am. ind. Hyg. Assoc. J.*, *28*, 129–134

Pathiratne, A., Puyear, R.L. & Brammer, J.D. (1986) Activation of ^{14}C-toluene to covalently binding metabolites by rat liver microsomes. *Drug. Metab. Disp.*, *14*, 386–391

Pelclová, D., Rössner, P. & Písková, J. (1987) Cytogenetic analysis of peripheral lymphocytes in workers occupationally exposed to toluene (Czech.). *Prac. Lek.*, *39*, 356–361

Pfäffli, P., Savolainen, H., Kalliomäki, P.-L. & Kalliokoski, P. (1979) Urinary *o*–cresol in toluene exposure. *Scand. J. Work Environ. Health*, *5*, 286–289

Piotrowski, J. (1967) Quantitative evaluation of exposure to toluene in men (Pol.). *Med. Prac.*, *18*, 213–223

Poel, W.E. (1962) Skin as a test site for the bioassay of carcinogens and carcinogen precursors. *Natl Cancer Inst. Monogr.*, *10*, 611–631

Pouchert, C.J., ed. (1981) *The Aldrich Library of Infrared Spectra*, 3rd ed., Milwaukee, WI, Aldrich Chemical Co., p. 1528G

Pouchert, C.J., ed. (1983) *The Aldrich Library of NMR Spectra*, 2nd ed., Vol. 1, Milwaukee, WI, Aldrich Chemical Co., p. 733B

Pouchert, C.J., ed. (1985) *The Aldrich Library of FT–IR Spectra*, Vol. 1, Milwaukee, WI, Aldrich Chemical Co., p. 931B

Pyykkö, K. (1980) Effects of methylbenzenes on microsomal enzymes in rat liver, kidney and lung. *Biochem. biophys. Acta*, *633*, 1–9

Pyykkö, K., Tähti, H. & Vapaatalo, H. (1977) Toluene concentrations in various tissues of rats after inhalation and oral administration. *Arch. Toxicol.*, *38*, 169–176

Rappaport, S.M. & Fraser, D.A. (1977) Air sampling and analysis in a rubber vulcanization area. *Am. ind. Hyg. Assoc. J.*, *38*, 205–210

Rea, T.M., Nash, J.F., Zabik, J.E., Born, G.S. & Kessler, W.V. (1984) Effects of toluene inhalation on brain biogenic amines in the rat. *Toxicology*, *31*, 143–150

Reisin, E., Teicher, A., Jaffe, R. & Eliahou, H.E. (1975) Myoglobinuria and renal failure in toluene poisoning. *Br. J. ind. Med.*, *32*, 163–168

Reyes de la Rocha, S., Brown, M.A. & Fortenberry, J.D. (1987) Pulmonary function abnormalities in intentional spray paint inhalation. *Chest*, *92*, 100–104

Riihimäki, V. & Pfäffli, P. (1978) Percutaneous absorption of solvent vapors in man. *Scand. J. Work Environ. Health*, *4*, 73–85

Rivera, R.O. & Rostand, R. (1975) *Hillerich & Bradsby Co., Jeffersonville, Indiana (Health Hazard Evaluation Determination Report No. 74–121–203)*, Cincinnati, OH, National Institute for Occupational Safety and Health

Rodriguez Arnaiz, R. & Villalobos-Pietrini, R. (1985a) Genetic effects of thinner, benzene and toluene in *Drosophila melanogaster*. 1. Sex chromosome loss and non-disjunction. *Contamin. Amb.*, *1*, 35–43

Rodriguez Arnaiz, R. & Villalobos-Pietrini, R. (1985b) Genetic effects of thinner, benzene and toluene in *Drosophila melanogaster*. 2. Sex linked recessive lethal mutations and translocations II–III. *Contam. Amb.*, *1*, 45–49

Roh, J., Moon, Y.H. & Kim, K.-Y. (1987) The cytogenetic effects of benzene and toluene on bone marrow cells in rats. *Yonsei med. J.*, *28*, 297–309

Roper, P. (1980) *Emory University Pathology Department, Alanta, Georgia (Health Hazard Evaluation Determination Report No. 80–31–693)*, Cincinnati, OH, National Institute for Occupational Safety and Health

Roxan, Inc. (undated) *Precision Gas Detector*, Woodland Hills, CA

Sadtler Research Laboratories (1980) *Standard Spectra Collection, 1980 Cumulative Index*, Philadelphia, PA

Sandmeyer, E. (1981) Aromatic hydrocarbons. In: Clayton, G.D. & Clayton, F.E., eds, *Patty's Industrial Hygiene and Toxicology*, Vol. 2B, *Toxicology*, 3rd rev. ed., New York, John Wiley & Sons, p. 3256

Sato, A. & Fujiwara, Y. (1972) Elimination of inhaled benzene and toluene in man. *Jpn. J. ind. Health*, *14*, 224–225

Sato, A. & Nakajima, T. (1978) Differences following skin or inhalation exposure in the absorption and excretion kinetics of trichloroethylene and toluene. *Br. J. ind. Med.*, *35*, 43–49

Savolainen, H. (1978) Distribution and nervous system binding of intraperitoneally injected toluene. *Acta pharmacol. toxicol.*, *43*, 78–80

Schikler, K.N., Seitz, K., Rice, J.F. & Strader, T. (1982) Solvent abuse associated cortical atrophy. *J. Adolesc. Health Care*, *3*, 37–39

Schmid, E., Bauchinger, M. & Hauf, R. (1985) Chromosome changes with time in lymphocytes after occupational exposure to toluene. *Mutat. Res.*, *142*, 37–39

Schumacher, H. & Grandjean, E. (1960) Comparative studies on narcotic effect and acute toxicity of nine solvents (Ger.). *Arch. Gewerbepathol. Gewerbehyg.*, *18*, 109–119

Seidenberg, J.M. & Becker, R.A. (1987) A summary of the results of 55 chemicals screened for developmental toxicity in mice. *Teratog. Carcinog. Mutagenesis*, *7*, 17–28

Seidenberg, J.M., Anderson, D.G. & Becker, R.A. (1986) Validation of an in vivo developmental toxicity screen in the mouse. *Teratog. Carcinog. Mutagenesis*, *6*, 361–374

Seifert, B. & Abraham, H.-J. (1982) Indoor air concentrations of benzene and some other aromatic hydrocarbons. *Ecotoxicol. environ. Saf.*, *6*, 190–192

Seifert, B. & Abraham, H.J. (1983) Use of passive samplers for the determination of gaseous organic substances in indoor air at low concentration levels. *Int. J. environ. anal. Chem.*, *13*, 237–253

Sensidyne (1985) *The First Truly Simple Precision Gas Detector System*, Largo, FL

Seppäläinen, A.M., Husman, K. & Mårtenson, C. (1978) Neurophysiological effects of long-term exposure to a mixture of organic solvents. *Scand. J. Work Environ. Health*, *4*, 304–314

Shigeta, S., Aikawa, H. & Misawa, T. (1982) Effects of maternal exposure to toluene during pregnancy on mouse embryos and fetuses. *Tokai J. exp. clin. Med.*, *7*, 265–270

Shigeta, S., Misawa, T., Aikawa, H., Momotani, H., Yoshida, T. & Suzuki, K. (1986) Effects of low level toluene exposure during the developing stage of the brain on learning in high avoidance rats. *Jpn. J. ind. Health*, *28*, 445–454

Shiojima, S., Hasegawa, K., Ishihara, N. & Ikeda, M. (1983) Subclinical increases in serum transaminase activities among female workers exposed to toluene at sub–OEL (occupational exposure limit) levels. *Ind. Health*, *21*, 123–126

Sina, J.F., Bean, C.L., Dysart, G.R., Taylor, V.I. & Bradley, M.O. (1983) Evaluation of the alkaline elution/rat hepatocyte assay as a predictor of carcinogenic/mutagenic potential. *Mutat. Res.*, *113*, 357–391

SKC, Inc. (1988) *Comprehensive Catalog and Guide*, Eighty Four, PA

Smith, J.N., Smithies, R.H. & Williams, R.T. (1954) Studies in detoxication. 55. The metabolism of alkylbenzenes. (a) Glucuronic acid excretion following the administration of alkylbenzenes. (b) Elimination of toluene in the expired air of rabbits. *Biochem. J.*, *56*, 317–320

Smyth, H.F., Carpenter, C.P., Weil, C.S., Pozzani, U.C., Striegel, J.A. & Nycum, J.S. (1969) Range-finding toxicity data. List VII. *Am. ind. Hyg. Assoc. J.*, *30*, 470–476

Snyder, R.D. & Matheson, D.W. (1985) Nick translation – a new assay for monitoring DNA damage and repair in cultured human fibroblasts. *Environ. Mutagenesis*, *7*, 267–279

Spanggord, R.J., Mortelmans, K.E., Griffin, A.F. & Simmon, V.F. (1982) Mutagenicity in *Salmonella typhimurium* and structure–activity relationships of wastewater components emanating from the manufacture of trinitrotoluene. *Environ. Mutagenesis, 4,* 163–179

Srbová, J. & Teisinger, J. (1952) Absorption and elimination of toluene in man (Czech.). *Prac. Lek., 4,* 41–47

Stollery, B.T. & Flindt, M.L.H. (1988) Memory sequelae of solvent intoxication. *Scand. J. Work Environ. Health, 14,* 45–48

Streicher, H.Z., Gabow, P.A., Moss, A.H., Kono, D. & Kaehny, W.D. (1981) Syndromes of toluene sniffing in adults. *Ann. intern. Med., 94,* 758–762

Struwe, G. & Wennberg, A. (1983) Psychiatry and neurological symptoms in workers occupationally exposed to organic solvents – results of a differential epidemiological study. *Acta psychiatr. scand., 67,* 68–80

Suzuki, H. (1973) Autonomic nervous responses to experimental toluene exposure in humans. *Jpn. J. ind. Health, 15,* 379–384

Suzuki, T., Kashimura, S. & Umetsu, K. (1983) Thinner abuse and aspermia. *Med. Sci. Law, 23,* 199–202

Svirbely, J.L., Dunn, R.C. & von Oettingen, W.F. (1943) The acute toxicity of vapors of certain solvents containing appreciable amounts of benzene and toluene. *J. ind. Hyg. Toxicol., 25,* 366–373

Swedish Criteria Group for Occupational Standards (1985) Scientific basis for Swedish occupational standards. VI. Mixed solvents, neurotoxic effects. *Arb. Hälsa, 32,* 122–132

Szadkowski, D., Pett, R., Angerer, J., Manz, A. & Lehnert, G. (1973) Occupational chronic exposure to organic solvents. II. Toluene concentrations in blood and excretion rates of metabolites in urine in the supervision of printing workers. *Int. Arch. Arbeitsmed., 31,* 265–276

Szadkowski, D., Pfeiffer, D. & Angerer, J. (1976) Estimation of an occupational toluene exposure with regard to its hepatotoxic relevance (Ger.). *Med. Monatsschr., 30,* 25–28

Tähti, H., Kärkkäinen, S., Pyykkö, K., Rintala, E., Kataja, M. & Vapaatalo, H. (1981) Chronic occupational exposure to toluene. *Int. Arch. occup. environ. Health, 48,* 61–69

Takeuchi, Y., Ono, Y. & Hisanaga, N. (1981) An experimental study on the combined effects of *n*-hexane and toluene on the peripheral nerve of the rat. *Br. J. ind. Med., 38,* 14–19

Taskinen, H., Lindbohm, M.-L. & Hemminki, K. (1986) Spontaneous abortions among women working in the pharmaceutical industry. *Br. J. ind. Med., 43,* 199–205, 432

Tátrai, E., Rodics, K. & Ungváry, G. (1980) Embryotoxic effects of simultaneously applied exposure of benzene and toluene. *Folia morphol., 28,* 286–289

Tice, R.R., Vogt, T.F. & Costa, D.L. (1982) Cytogenetic effects of inhaled benzene in murine bone marrow. In: Tice, R.R., Costa, D.L. & Schaich, K.M., eds, *Genotoxic Effects of Airborne Agents*, New York, Plenum, pp. 257–275

Toftgård, R., Nilsen, O.G. & Gustafsson, J.-Å. (1982) Dose–dependent induction of rat liver microsomal cytochrome P-450 and microsomal enzymatic activities after inhalation of toluene and dichloromethane. *Acta pharmacol. toxicol., 51,* 108–114

Tokunaga, R., Takahata, S., Onoda, M., Ishi–i, T., Sato, K., Hayashi, M. & Ikeda, M. (1974) Evaluation of the exposure to organic solvent mixture. Comparative studies on detection tube and gas–liquid chromotographic methods, personal and stationary sampling and urinary metabolite determination. *Int. Arch. Arbeitsmed., 33,* 257–267

Topham, J.C. (1980) Do induced sperm–head abnormalities in mice specifically identify mammalian mutagens rather than carcinogens? *Mutat. Res., 74,* 379–387

Triebig, G., Bestler, W., Baumeister, P. & Valentin, H. (1983) Investigations on neurotoxicity of chemical substances at the work place. IV. Determination of motor and sensory nerve conduction velocity in persons occupationally exposed to a mixture of organic solvents (Ger.). *Int. Arch. occup. environ. Health*, *52*, 139–150

Tsuruta, H. (1982) Percutaneous absorption of organic solvents. III. On the penetration rates of hydrophobic solvents through the excised rat skin. *Ind. Health*, *20*, 335–345

Tsuruta, H., Iwasaki, K. & Kanno, S. (1987) A method for calculating the skin absorption rate from the amount retained in the whole body of skin-absorbed toluene in mice. *Ind. Health*, *25*, 215–220

Tunek, A., Högstedt, B. & Olofsson, T. (1982) Mechanisms of benzene toxicity. Effects of benzene and benzene metabolites on bone marrow cellularity, number of granulopoietic stem cells and frequency of micronuclei in mice. *Chem.–biol. Interactions*, *39*, 129–138

Työsuojeluhallitus (National Finnish Board of Occupational Safety and Health) (1987) *HTP–Arvot 1987* [TLV Values 1987] (*Safety Bulletin 25*), Helsinki, Valtion Painatuskeskus, p. 26

Ungváry, G. & Tátrai, E. (1985) On the embryotoxic efects of benzene and its alkyl derivatives in mice, rats and rabbits. *Arch. Toxicol.*, *Suppl. 8*, 425–430

Ungváry, G.Y., Mányai, S., Tátrai, E., Szeberény, S.Z., Cseh, R.J., Molnár, J. & Folly, G. (1980) Effect of toluene inhalation on the liver of rats: dependence on sex, dose and exposure time. *J. Hyg. Epidemiol. Microbiol. Immunol.*, *24*, 242–252

Ungváry, G.Y., Tátrai, E., Lőrincz, M. & Barcza, G.Y. (1983) Combined embryotoxic action of toluene, a widely used industrial chemical, and acetylsalicylic acid (aspirin). *Teratology*, *27*, 261–269

US Environmental Protection Agency (1983) *Health Assessment Document for Toluene (Publ. No. PB84-100056)*, Washington DC, US Department of Commerce, National Technical Information Service

US International Trade Commission (1984) *Synthetic Organic Chemicals, US Production and Sales, 1983 (USITC Publ. 1588)* Washington DC, US Government Printing Office

US International Trade Commission (1985) *Synthetic Organic Chemicals, US Production and Sales, 1984 (USITC Publ. 1745)* Washington DC, US Government Printing Office

US International Trade Commission (1986) *Synthetic Organic Chemicals, US Production and Sales, 1985 (USITC Publ. 1892)* Washington DC, US Government Printing Office

Veulemans, H., Van Vlem, E., Janssens, H. & Masschelein, R. (1979) Exposure to toluene and urinary hippuric acid excretion in a group of heliorotagravure printing workers. *Int. Arch. occup. environ. Health*, *44*, 99–107

Voigts, A. & Kaufman, C.E., Jr (1983) Acidosis and other metabolic abnormalities associated with paint sniffing. *South. med. J.*, *76*, 443–447

Waldron, H.A., Cherry, N. & Venables, H. (1982) Solvent exposure and liver function. *Lancet*, *ii*, 1276

Waldron, H.A., Cherry, N. & Johnston, J.D. (1983) The effects of ethanol in blood toluene concentrations. *Int. Arch. occup. environ. Health*, *51*, 365–369

Wallén, M. (1986) Toxicokinetics of toluene in occupationally exposed volunteers. *Scand. J. Work Environ. Health*, *12*, 588–593

Weast, R.C., ed. (1985) *Handbook of Chemistry and Physics*, 66th ed., Cleveland, OH, CRC Press, p. C-518

Weiss, H.S., O'Connell, J.F., Hakaim, A.G. & Jacoby, W.T. (1986) Inhibitory effect of toluene on tumor promotion in mouse skin. *Proc. Soc. exp. Biol. Med.*, *181*, 199–204

Wilcosky, T.C., Checkoway, H., Marshall, E.G. & Tyroler, H.A. (1984) Cancer mortality and solvent exposures in the rubber industry. *Am. ind. Hyg. Assoc. J.*, *45*, 809–811

Wilson, J.T., Enfield, C.G., Dunlap, W.J., Cosby, R.L., Foster, D.A. & Baskin, L.B. (1981) Transport and fate of selected organic pollutants in a sandy soil. *J. environ. Qual.*, *10*, 501–506

Wilson, J.T., McNabb, J.F., Wilson, R.H. & Noonan, M.J. (1983) Biotransformation of selected organic pollutants in ground water. *Dev. ind. Microbiol.*, *24*, 225–233

Wilson, R.H. (1943) Toluene poisoning. *J. Am. med. Assoc.*, *123*, 1106–1108

Winchester, R.V. & Madjar, V.M. (1986) Solvent effects on workers in the paint, adhesive and printing industries. *Ann. occup. Hyg.*, *30*, 307–317

Windholz, M., ed. (1983) *The Merck Index*, 10th ed., Rahway, NJ, Merck & Co., p. 1364

Winneke, G. (1982) Acute behavioral effects of exposure to some organic solvents: psychophysiological aspects. *Acta neurol. scand.*, *66*, 117–129

Winston, S. & Matsushita, T. (1975) Permanent loss of chromosome initiation in toluene–treated *Bacillus subtilis* cells. *J. Bacteriol.*, *123*, 921–927

Withey, R.J. & Hall, J.W. (1975) The joint toxic action of perchloroethylene with benzene or toluene in rats. *Toxicology*, *4*, 5–15

Woiwode, W. & Drysch, K. (1981) Experimental exposure to toluene: further consideration of cresol formation in man. *Br. J. ind. Med.*, *38*, 194–197

Woiwode, W., Wodarz, R., Drysch, K. & Weichardt, H. (1979) Metabolism of toluene in man: gas–chromatographic determination of *o*-, *m*- and *p*-cresol in urine. *Arch. Toxicol.*, *43*, 93–98

Wolf, M.A., Rowe, V.K., McCollister, D.D., Hollingsworth, R.L. & Oyen, F. (1956) Toxicological studies of certain alkylated benzenes and benzene. *Am. med. Assoc. Arch. ind. Health*, *14*, 387–398

Wood, R.W., Rees, D.C. & Laties, V.G. (1983) Behavioral effects of toluene are modulated by stimulus controls. *Toxicol. appl. Pharmacol.*, *68*, 462–472

World Health Organization (1981) *Recommended Health–based Limits in Occupational Exposure to Selected Organic Solvents (Technical Report Series No. 664)*, Geneva

XYLENE

1. Chemical and Physical Data

1.1 Synonyms

Chem. Abstr. Services Reg. Nos: 1330–20–7 (xylene)
95–47–6 (*ortho*-xylene)
108–38–3 (*meta*-xylene)
106–42–3 (*para*-xylene)
Chem. Abstr. Names: 1,2–Dimethylbenzene
1,3–Dimethylbenzene
1,4–Dimethylbenzene

IUPAC Systematic Name: Xylene (*ortho–, meta–, para–*)
Synonyms: *ortho*-Xylene: *ortho*-Dimethylbenzene; *ortho*-methyltoluene; 2-methyltoluene; 1,2-xylene; *ortho*-xylol
meta-Xylene: *meta*-Dimethylbenzene; *meta*-methyltoluene; 3-methyltoluene; 1,3-xylene; *meta*-xylol
para-Xylene: *para*-Dimethylbenzene, *para*-methyltoluene; 4-methyltoluene; 1,4-xylene; *para*-xylol

1.2 Structural and molecular formulae and molecular weight

C_8H_{10}

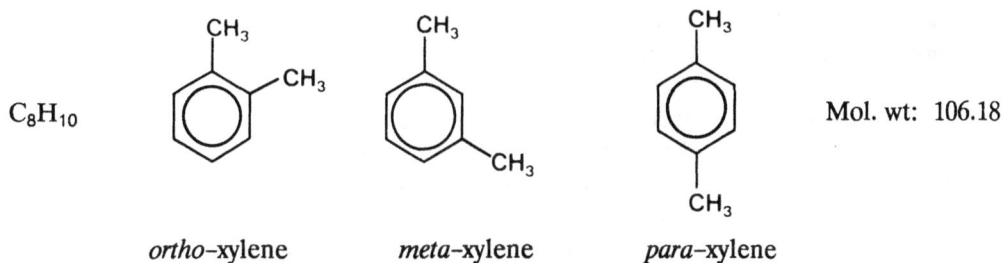

ortho-xylene   *meta*-xylene   *para*-xylene

Mol. wt: 106.18

1.3   Chemical and physical properties of the pure substances

Table 1. Chemical and physical properties of the pure isomers

Property	ortho-Xylene	meta-Xylene	para-Xylene	Reference
Description	Clear, colorless liquid		Crystalline solid	Windholz (1983)
Boiling point (°C)	144.4 32 at 10 mm Hg	139.1 28.1 at 10 mm Hg	138.3 27.2 at 10 mm Hg	Weast (1985)
Melting-point (°C)	−25.2	−47.9	13.3	Weast (1985)
Density	0.8802 at 20°/4°C	0.8642 at 20°/4°C	0.8611 at 20°/4°C	Weast (1985)
Refractive index	1.5055 at 20°C	1.4972 at 20°C	1.4958 at 20°C	Weast (1985)
Spectroscopy data	Infrared, ultraviolet and nuclear magnetic resonance spectral data have been reported			Sadtler Research Laboratories (1980); Pouchert (1981, 1983, 1985)
Solubility	Soluble in ethanol, diethyl ether, acetone, benzene; insoluble in water			Weast (1985)
Volatility (vapour pressure, mm)	6.8 at 25°C	8.3 at 25°C	8.9 at 25°C	Sandmeyer (1981)
Flash-point (°C)	32	29	27	Sandmeyer (1981)
Octanol/water partition coefficient (log P)	2.77–3.12	3.2	3.15	Hansch & Leo (1979)
Conversion factor	$mg/m^3 = 4.34 \times ppm^a$			
Reactivity	Highly inflammable			Hansch & Leo (1979)

[a]Calculated from mg/m^3 = (molecular weight/24.45) $\times$ ppm, assuming standard temperature (25°C) and pressure (760 mm Hg)

1.4   Technical products and impurities

Trade Names: Chromar; Dilan; Scintillar

Xylene is marketed principally as a mixture of *ortho*, *meta* and *para* isomers, generally referred to as 'mixed xylenes'. The individual isomers are also available commercially. Most mixed xylenes contain ethylbenzene, except for a small volume produced by toluene disproportionation (Ransley, 1984). Commercial-grade (mixed) xylene typically is composed of approximately 20% *ortho*-xylene, 40% *meta*-xylene and 20% *para*-xylene, with about 15% ethylbenzene and smaller amounts of toluene, trimethylbenzene (pseudocumene), phenol, thiophene, pyridine and non-aromatic hydrocarbons (National Institute for Occupational Safety and Health, 1975; Clement Associates, 1977). A product of higher purity is reported to contain a minimum of 97% xylene isomer with maximum impurities of 3% ethylbenzene, 0.1% benzene, 0.1% toluene and 0.01% water (Riedel–de Haën, 1984).

Typical *para*-xylene products (99.5% pure) contain 0.3% ethylbenzene, 0.1% *meta*-xylene and 0.1% *ortho*-xylene (Ransley, 1984). All three isomers are available at 99.9% minimal high purity, spectrophotometric grade as well as in 'chemically pure' grades, as follows: *ortho*-xylene, 98% pure; *para*-xylene, 99%; and *meta*-xylene, 99% (Riedel–de Haën, 1984).

2. Production, Use, Occurrence and Analysis

2.1 Production and use

(a) Production

Xylene occurs in petroleum stock, but in very small quantities. It is produced primarily by the catalytic reforming of naphtha streams, which are rich in alicyclic hydrocarbons. The aromatic reformate fractions consist mainly of benzene, toluene and mixed xylenes, xylenes representing the largest fraction. The xylene isomers are separated from the reformate by extraction and distillation on the basis of differences in boiling–point. *ortho*-Xylene, which has the highest boiling point, is separated as the bottom distillate; *para*-xylene is separated by continuous crystallization or adsorption from the mixed xylenes or isomerized from the *meta*-xylene/*para*-xylene distillate; and *meta*-xylene is obtained by selective crystallization or solvent extraction of *meta-para* mixtures (Mannsville Chemical Products Corp., 1981; Ransley, 1984).

Another source of mixed xylenes is pyrolysis gasoline, a by–product that results from cracking of hydrocarbon feeds during olefin manufacture (Fishbein, 1985). The mixed xylene content of pyrolysis gasoline varies, depending upon the feed and the severity of the cracking process. Pyrolysis gasoline is a less efficient source for recovery of mixed xylene than catalytic reformate because it contains large amounts of ethylbenzene.

Mixed xylenes may also be produced from petroleum refining operations by the Toyo Rayon and Atlantic–Richfield processes, in which toluene is transalkylated or disproportionated. Benzene and toluene are the principal products (Fishbein, 1985). Xylenes obtained from this source are 'ethylbenzene free', provided the transalkylation feed stocks are limited to toluene and (polymethyl)benzene (Ransley, 1984).

Less than 1% of the mixed xylenes in the USA are derived from coal. Coal subjected to high–pressure carbonization (coke manufacture) yields crude light oil containing 3–6% mixed xylenes. Every tonne of coal yields 2–3 gallons (7.6–11.4 l) of crude light oil (Ransley, 1984), which may be used as a supplementary source of aromatic compounds in petroleum refining, processed for recovery of light naphtha containing mixed xylenes and styrene, or burned as fuel.

The Mitsubishi Gas Chemical Company (MGCC) process is another commercial method for separating the *meta* isomer from mixed xylenes using a hydrofluoric acid–borofluoride separation technique. It is also a straightforward means of separating the other C_8 aromatic isomers (Ransley, 1984).

The total quantities of mixed xylenes (and the percentages isolated as xylene) produced in the USA in 1978 in the ways described above were as follows: catalytic reformate, 34.9

million tonnes (10%); pyrolysis gasoline, 375 thousand tonnes (52%); toluene disproportion-ation, 90 thousand tonnes (54%); and coal–derived, 15 thousand tonnes (88%). Of the total 35.44 million tonnes produced in 1978, about 11% was isolated (Fishbein, 1985).

Mixed xylenes are also produced in large quantities in Europe and Japan. Data on pro-duction of xylenes in a number of areas are presented in Table 2.

Table 2. Annual production of xylenes (thousands of tonnes)[a]

Country or region	1981	1982	1983	1984	1985	1986	1987
Brazil[b]	86	83	80	84	79	NA[c]	NA
Bulgaria	32	32	29	32	31	NA	NA
Canada	409	431	426	406	415	356	345
China	86	100	104	128	257	NA	NA
Czechoslovakia	99	107	119	119	117	108	NA
France	114	51	90	85	126	113	129
Germany, Federal Republic of	486	459	511	455	495	540	501
Hungary	76	79	93	94	95	94	NA
India	NA	14[b]	18[b]	NA	NA	NA	28
Italy	256	269	365	395	432	405	491
Japan	1202	1225	1264	1401	1523	1570	1767
Korea, Republic of	244	249	300	304	330	490	552
Mexico	104	115	236	268	291	273	381
Portugal	NA	84	110	104	106	NA	NA
Romania	224	254	269	254	225	NA	NA
Spain	40	49	58	53	53	60	NA
Taiwan	241	200	239	291	270	237	NA
Turkey	0.4	0.3	0.2	0.2	0.2	NA	NA
USA	2477	1905	2225	2251	2479	2647	2772
USSR	468	409	556	849	937	962	NA
Yugoslavia	16	NA	11	17	8	NA	NA

[a]From US International Trade Commission (1982, 1983, 1984); Anon. (1985); US International Trade Com-mission (1985, 1986); Anon. (1987); US International Trade Commission (1987); Anon. (1988a,b)
[b]ortho–Xylene
[c]NA, not available

(b)   Use

Mixed xylenes recovered from all sources (petroleum refineries, pyrolysis gasoline, coal–tar) are used in the chemical and solvent industries (Ransley, 1984). Although isolated xylenes are also blended into gasoline to improve octane rating, the reformate without isolation of mixed xylenes or other aromatics is primarily used for gasoline blending. Unleaded premium gasoline has been reported to contain 10–22% xylenes (Korte & Boedefeld, 1978; Ikeda *et al.*, 1984).

Mixed xylenes are also used in the manufacture of perfumes (Sittig, 1985), insecticides, pharmaceuticals and adhesives and in painting, printing, rubber, plastics (Sandmeyer, 1981) and leather industries (IARC, 1981).

In the USA, most of the production of isolated mixed xylenes is separated into the individual isomers for use as chemical intermediates or as solvents (Mannsville Chemical Products Corp., 1981). The approximate distributions of the production of mixed xylenes in the USA are as follows: *para*-xylene, 50–60%; gasoline blending, 10–25%; *ortho*-xylene, 10–15%; solvents, 10%; ethylbenzene, 3%; and *meta*-xylene, 1% (Ransley, 1984). *para*-Xylene is used principally to manufacture terephthalic acid and dimethylterephthalate, used in the production of saturated polyester resins and fibres (Mannsville Chemical Products Corp., 1981). The remaining small amount of *para*-xylene produced is used as a pharmaceutical or pesticide intermediate and in solvents for adhesives and coatings (Hawley, 1981; Anon., 1986). *ortho*-Xylene is used primarily as a feedstock for the manufacture of phthalic anhydride: almost 60% of the *ortho*-xylene produced in the USA in 1978 was used in this way (Fishbein, 1985). It is also used as a chemical intermediate in synthesis of dyes, pharmaceuticals and insecticides (Hawley, 1981; Ransley, 1984). *meta*-Xylene is used in the manufacture of isophthalic acid for polyester resins (Mannsville Chemical Products Corp., 1981) and as a chemical intermediate for dyes and insecticides (Hawley, 1981).

(c)   Regulatory status and guidelines

Occupational exposure limits for xylenes in 32 countries or regions are presented in Table 3.

Table 3. Occupational exposure limits for xylenes (all isomers)[a]

Country or region	Year	Concentration[b] (mg/m^3)	Interpretation[c]
Austria	1985	435	TWA
Belgium	1985	S 435	TWA
Brazil	1985	S 340	TWA
Bulgaria	1985	50	TWA
Commission of the European Communities	1986	435 2175	Average Maximum
Chile	1985	S 348	TWA
China	1985	100	TWA

Table 3 (contd)

Country or region	Year	Concentration[b] (mg/m^3)	Interpretation[c]
Czechoslovakia	1985	200	Average
		1000	Maximum
Denmark	1988	S 217	TWA
Finland	1987	S 435	TWA
		S 655	STEL
France	1986	435	TWA
		650	STEL (15 min)
German Democratic Republic	1985	200	TWA
		600	STEL
Germany, Federal Republic of	1988	440	TWA
Hungary	1985	50	TWA
		100	STEL
India	1985	S 435	TWA
		655	STEL
Indonesia	1985	435	TWA
Italy	1985	S 400	STEL
Japan	1988	435	TWA
Korea, Republic of	1985	435	TWA
		655	STEL
Mexico	1985	S 435	TWA
Netherlands	1986	S 435	TWA
Norway	1981	435	TWA
Poland	1985	100	TWA
Romania	1985	S 300	TWA
		S 400	Maximum
Sweden	1987	S 200	TWA
		S 450	STEL (15 min)
Switzerland	1985	S 435	TWA
Taiwan	1985	435	TWA
UK	1987	S 435	TWA
		S 650	STEL (10 min)
USA[d]			
OSHA	1988	200	TWA
		300	Ceiling
NIOSH	1986	434	TWA
		868	Ceiling (10 min)
ACGIH	1988	435	TWA
		655	STEL (15 min)
USSR	1985	50	Ceiling

Table 3 (contd)

Country or region	Year	Concentration[b] (mg/m^3)	Interpretation[c]
Venezuela	1985	S 435	TWA
		S 655	Ceiling
Yugoslavia	1985	50	TWA

[a]From Direktoratet for Arbeidstilsynet (1981); National Swedish Board of Occupational Safety and Health (1984); Arbeidsinspectie (1986); Commission of the European Communities (1986); Institut National de Recherche et de Sécurité (1986); National Institute for Occupational Safety and Health (1986); Cook (1987); Health and Safety Executive (1987); Työsuojeluhallitus (1987); American Conference of Governmental Industrial Hygienists (1988); Arbejdstilsynet (1988); Deutsche Forschungsgemeinschaft (1988); US Occupational Safety and Health Administration (1988)

[b]S, skin notation

[c]TWA, 8-h time–weighted average; STEL, short–term exposure limit

[d]OSHA, Occupational Safety and Health Administration; NIOSH, National Institute for Occupational Safety and Health; ACGIH, American Conference of Governmental Industrial Hygienists

2.2 Occurrence

(a) Natural occurrence

Mixed xylenes are present in coal–tar, petroleum stocks (Fishbein, 1985) and natural gas (Hillard, 1980) in small quantities.

(b) Occupational exposure

On the basis of a US National Occupational Exposure Survey, the National Institute for Occupational Safety and Health (1983) estimated that 1 106 800 workers were potentially exposed to xylene in the USA in 1981–83.

Levels of xylene to which workers have been exposed are summarized in Table 4. Levels determined during the manufacture and application of paints are described in the monograph on occupational exposures in paint manufacture and painting (see p. 329). Levels of exposure to xylene in petroleum refining and in the manufacture and use of petroleum fuels are reported in Volume 45 of the *Monographs* (IARC, 1989).

Pre– and post–shift concentrations of methyl hippuric acid in the urine of workers in a shipbuilding yard were 0.2–7.1 mg/ml. The workers were using a thinner in spray–painting operations that contained 32.8% *meta*– or *para*-xylene (Ogata *et al.*, 1971). Mean urinary concentrations of methyl hippuric acid in workers in a photograph album manufacturing plant who used a cleaning solvent (complex mixture of 90% C_7–C_9 aliphatic hydrocarbons, 5% toluene and 5% xylene) to remove excess glue were 0.07 g/g creatinine before a shift and 0.48 g/g creatinine afterwards (Baker & Fannick, 1983).

(c) Air

Mixed xylenes are emitted to the ambient air during their production and use from reactor, distillation and crystallization vents. Emissions may also occur during storage, loading and handling. Total emissions of mixed xylenes in the USA in 1978 were estimated to be

4100 tonnes from catalytic reformate, 150 tonnes from pyrolysis gasoline, 18 tonnes from toluene disproportionation and 19 tonnes from coal–derived production. Emissions of total individual isomers were estimated to be 1180 tonnes of *ortho*–xylene, 2900 tonnes of *para*–xylene and 80 tonnes of *meta*–xylene (Fishbein, 1985). Merian (1982) reported that worldwide losses of xylenes into air from refineries, evaporation of gasoline, automobile exhaust and solvent losses are approximately 3 million tonnes.

Table 4. Occupational exposure to xylene

Environment	Sampling[a]	Concentration in air[b]	Reference
Laboratories			
Histology laboratory (USA)	4–h personal	3.2–102 ppm (14–443 mg/m^3)	Kilburn *et al.* (1985)
Histology laboratory [FRG]	8–h TWA personal	(*m*+*p*)–xylene, 56–68 ppm (243–295 mg/m^3) *o*–xylene, 10–13 ppm (43–56 mg/m^3)	Angerer & Lehnert (1979)
Histology laboratory (USA)	8–h TWA personal 8–h TWA area	2.5–72.6 ppm (11–315 mg/m^3) 18.3–28.3 ppm (79–123 mg/m^3)	Roper (1980)
Cytopathology laboratory (USA)	8–h TWA personal 8–h TWA area	1.6–12.8 ppm (7–55 mg/m^3) 15–32 ppm (65–139 mg/m^3	Roper (1980)
Hospital laboratory (USA)	Point	0.6–400 ppm (2.6–1700 mg/m^3)	Klaucke *et al.* (1982)
Chemical plant (Hungary)		Mean, 47–56 mg/m^3	Pap & Varga (1987)
Extraction plant producing xylene from gasoline (USSR)	Air	75–200 mg/m^3 in 35–40% of samples	Sukhanova *et al.* (1969)
Lithography (Poland)			Moszczyński & Lisiewicz (1985)
1968		32–450 mg/m^3; mean, 119 mg/m^3	
1970		110–130 mg/m^3	
1971		ND–360 mg/m^3; mean, 102 mg/m^3	
1974		ND–150 mg/m^3	
1977		15–30 mg/m^3; mean, 17 mg/m^3	
1978		10–506 mg/m^3; mean, 130 mg/m^3	
Manufacture of photograph albums (USA)	Personal TWA	1–56 mg/m^3	Baker & Fannick (1983)
Golf club and baseball bat manufacturing plant (USA)	8–h TWA personal	2–14 ppm (9–61 mg/m^3	Rivera & Rostand (1975)

[a]TWA, time–weighted average

[b]ND, not detected

Mixed xylene are also lost during use, as in the processing of chemicals and solvents, evaporation during transportation, distribution, storage and use of gasoline, in motor vehicle emissions and from agricultural spraying (Fishbein, 1985).

Atmospheric concentrations of total mixed xylenes have been determined at various locations around the world. Mean values and ranges measured between 1961 and 1980 are as follows: (in mg/m^3): France (0.003–0.01), Federal Republic of Germany (rural, 0.001–0.04; urban, 0.15), Japan (0.06–0.39), the Netherlands (urban, 0.07), South Africa (0.02–0.03) and Switzerland (urban, 0.02–0.05). In the USA, mean concentrations of atmospheric xylene at urban sites in California, Texas and New York/New Jersey in 1961–74 were 0.08–0.12, 0.04–0.07 and 0.15 mg/m^3, respectively (Merian & Zander, 1982). Xylene levels of 116–684 mg/m^3 have been reported in smoke from forest fires (Merian & Zander, 1982), and xylene has been detected in cigarette smoke (Holzer *et al.*, 1976). Concentrations of *meta*-xylene in outdoor air in the USA have been reported to range from 0.016 to 0.061 ppm (0.069–0.265 mg/m^3; Fishbein, 1985).

Xylene has been detected in indoor environments as a consequence of cooking, fuel burning and tobacco smoking. The mean concentrations of combined *meta*- and *para*-xylenes in indoor air were 0.029, 0.021 and 0.014 mg/m^3 in kitchens, other rooms and bedrooms, respectively (Seifert & Abraham, 1982; Wallace *et al.*, 1983). Holzer *et al.* (1976) found approximately 50 ppb (0.2 mg/m^3) *meta*- plus *para*-xylene in nonventilated cigarette smoke-filled room air and 18 ppb (0.08 mg/m^3) in the air of a room where no cigarettes had been smoked.

Outdoor air next to dwellings contained 0.009–0.028 mg/m^3 combined *meta*- and *para*-xylenes and that in backyards, 0.0011 mg/m^3 (Seifert & Abraham, 1982; Wallace *et al.*, 1983); 0.0042 mg/m^3 *ortho*-xylene was measured in backyards (Wallace *et al.*, 1983), and 0.1 mg/m^3 *meta*- and *para*-xylenes was measured at traffic intersections (Seifert & Abraham, 1982).

Krotoszynski *et al.* (1979) reported mean levels of 0.001, 0.0003 and 0.0031 mg/m^3 *ortho*-, *meta*- and *para*-xylene, respectively, in expired air of 54 normal, healthy volunteers from an urban population in Chicago, IL, USA. Xylene was also found in breath samples from urban residents of two New Jersey cities in the USA; mean values were 0.0034 mg/m^3 for *ortho*-xylene and 0.009 mg/m^3 for combined *meta*- and *para*-xylene. Levels were higher in persons who pumped their own gasoline or were exposed to auto and truck exhaust (Wallace *et al.*, 1984).

(d)   Water

Xylenes have been identified in surface and drinking-waters, for example in the river Glatt, a tributary of the Rhine. In the USA, levels of 2–8 µg/l were reported in surface water from the Florida Bay and 3–8 µg/l in drinking- and tap-water in New Orleans, LA (Merian & Zander, 1982).

(e)   Animal tissues

Ogata and Miyake (1973) measured mean concentrations of 21.7, 30.1 and 25.0 mg/kg *meta*-, *para*- and *ortho*-xylene in the muscles and 5.2, 26.6 and 6.1 mg/kg of the three isomers, respectively, in the liver of eels (*Aguilla japonica*) exposed to sea water containing 14.1 mg/kg *meta*-xylene and 13.1 mg/kg *ortho*-xylene.

2.3   Analysis

Selected methods for the analysis of xylene in various matrices are listed in Table 5. Methods for the analysis of xylene have recently been reviewed and compiled (Fishbein & O'Neill, 1988).

Colorimetric detection systems have been developed for xylenes in air (The Foxboro Co., 1983; Sensidyne, 1985; National Draeger, Inc., 1987; SKC, 1988; ENMET Corp., undated; Matheson Gas Products, undated; Roxan, Inc., undated).

3. Biological Data Relevant to the Evaluation of Carcinogenic Risk to Humans

3.1   Carcinogenicity studies in animals

Oral administration

Mouse: Groups of 50 male and 50 female B6C3F1 mice, eight weeks of age, received 0, 500 or 1000 mg/kg bw technical-grade xylene (comprising 60.2% *meta*-, 13.6% *para*- and 9.1% *ortho*-xylene with 17% ethylbenzene; purity, 99.7% with 2.8 ppm (0.00028%) benzene as contaminant) in corn oil by stomach tube on five days per week for 103 weeks. The animals were killed in weeks 104–105. No significant difference in mean body weights or survival was observed between control and treated mice. Survival at termination of the experiment was: males – 27 controls, 35 low–dose and 36 high–dose; and females – 36 controls, 35 low–dose and 31 high–dose. No treatment–related increase in the incidence of any tumour was seen in animals of either sex (National Toxicology Program, 1986; Huff *et al.*, 1988).

Rat: Groups of 40 male and 40 female Sprague-Dawley rats, seven weeks of age, were administered 500 mg/kg bw mixed xylenes (*ortho*-, *meta*- and *para*-; purities, >99% [source and percentage composition unspecified]) in olive oil by stomach tube on four to five days per week for 104 weeks. A group of 50 males and 50 females received olive oil only. Rats were maintained until natural death; all rats had died by week 141. At that time, thymomas were reported in 1/34 treated males and 0/36 treated females, compared to 0/45 and 0/49 in the control groups. Other haemolymphoreticular tumours [histology unspecified] were reported in 4/34 treated males and 3/36 treated females, compared to 3/45 and 1/49 controls. (The denominators are numbers of rats alive in each group at 58 weeks when the first haemolymphoreticular tumour was observed.) The authors reported an increase in the total number of animals with malignant tumours [type unspecified] at 141 weeks: in 14/38 treated males and 22/40 treated females compared to 11/45 and 10/49 controls. (The denominators are the number of rats alive in each group at 33 weeks when the first malignant tumour was observed.) (Maltoni *et al.*, 1983, 1985). [The Working Group noted the incomplete reporting of the composition of the test material and of tumour pathology, and that combining different types of tumours is not usually the most appropriate method for evaluating carcinogenicity (IARC, 1980; Montesano *et al.*, 1986).]

Table 5. Analytical methods for the determination of xylene and its metabolites in various matrices

Sample matrix	Sample collection	Sample preparation	Assay procedure	Detection limits	Reference
Air	Passive sampler with charcoal	Desorb (carbon disulfide); inject aliquot; analyse using glass capillary column	GC	0.3 mg/m^3 per h	Seifert & Abraham (1983)
	Charcoal tube	Desorb (carbon disulfide); inject aliquot; analyse on packed column	GC–FID	0.001–0.01 mg/sample	Eller (1984)
Water		Extract with hexane; inject aliquot	GC–FID	5 μg/l	Otson & Williams (1981)
		Heat samples in water bath at 25°C for 1 h; inject headspace aliquots	GC–MS	1 μg/l	Otson et al. (1982)
Automotive exhaust gas	Tenax GC polymer adsorbant cartridge	Desorb thermally into liquid nitrogen-cooled capillary trap	GC–MS	Not given	Hampton et al. (1982)
Breath (air)	Specially designed spirometer containing Tenax–GC cartridge	Dry cartridge over calcium sulfate; desorb thermally in a fused silica capillary column	GC–MS	Not given	Wallace et al. (1983, 1984, 1986)
Blood	Heparinize or antifoam emulsion B	Purge (nitrogen) at room temperature; trap (Tenax TA); desorb thermally; analyse volatiles on column	GC–MS	10 ppt (μg/l)	Cramer et al. (1988)
Tissue (muscle, liver)	Mince tissue	Heat with ethanol and potassium hydroxide; extract with n–hexane; apply extract to silica gel/aluminium trioxide column; elute with n–hexane; concentrate eluate; inject aliquot into GC	GC–FID	Not given	Ogata & Miyake (1973, 1978)
Urine (methylhippuric acid)		After alkaline hydrolysis, extract with diethyl ether at acidic pH; silylate and inject onto GC	GC–FID	Not given	Engström & Riihimäki (1988)

[a]Abbreviations: GC, gas chromatography; FID, flame–ionization detection; MS, mass spectrometry

Groups of 50 male and 50 female Fischer 344/N rats, seven weeks of age, received 0, 250 or 500 mg/kg bw technical-grade xylene (containing 60.2% *meta*-, 13.6% *para*- and 9.1% *ortho*-xylene with 17% ethylbenzene; purity, 99.7% with 2.8 ppm (0.00028%) benzene as contaminant) in corn oil by stomach tube on five days per week for 103 weeks. The animals were killed in weeks 104–105. High-dose males had lower mean body weights from week 59 onwards; body weights of low-dose males and treated females were comparable to those of controls. At termination of the experiment, 36 male controls and 25 males at the low dose and 20 at the high dose were still alive; the differences were due in part to accidental killing of animals. Survival in control and treated females was similar at termination (38 controls, 33 low-dose and 35 high-dose). The incidences of tumours in treated animals of either sex were not significantly higher than that in the control group (National Toxicology Program, 1986; Huff *et al.*, 1988).

3.2   Other relevant data

The toxicology of xylenes has been reviewed (Riihimäki & Engström, 1979; World Health Organization, 1981; Fishbein, 1985; European Chemical Industry Ecology and Toxicology Centre, 1986).

(*a*)   *Experimental systems*

(i)   *Absorption, distribution, excretion and metabolism*

ortho-Xylene was found to penetrate rat skin excised three days after clipping and depilation with cream at a rate that was 1/10 that of toluene and 1/100 that of benzene (Tsuruta, 1982).

In rats exposed to 208 mg/m³ [methyl-¹⁴C]*para*-xylene for 1 h, distribution of radioactivity immediately after termination of the exposure was highest in the kidneys, followed by subcutaneous fat, ischiatic nerve, blood, liver and lungs. Activity was 1/5 to 1/30 of these levels 6 h after the end of exposure (Carlsson, 1981).

Xylenes are metabolized both in the liver and lungs (Carlone & Fouts, 1974; Smith *et al.*, 1982; Toftgård *et al.*, 1986), primarily at a side-chain, to form methylhippuric acid and toluic acid (methylbenzoic acid) glucuronide as major metabolites and methylbenzyl mercapturic acid as a minor metabolite (Carlone & Fouts, 1974; Ogata *et al.*, 1980; van Doorn *et al.*, 1980). They are metabolized to a lesser extent at the aromatic ring to form dimethylphenol (e.g., Toftgård *et al.*, 1986). The ratio among the metabolites varies depending on the isomer (Bray *et al.*, 1949; Bakke & Scheline, 1970) and the species of animal (e.g., Ogata *et al.*, 1980).

Most of the xylene that is absorbed is excreted rapidly into the urine as metabolites. When rabbits were given oral doses of up to 1.8 g each of the three isomers, separately, well over 50% of the radioactivity was recovered in urine within 24 h (Bray *et al.*, 1949).

When 3 mmol/kg *ortho*-, *meta*- or *para*-xylene were given intraperitoneally to rats, urinary excretion of thiocompounds was highest with *ortho*-xylene and much lower with *meta*-xylene and *para*-xylene (van Doorn *et al.*, 1980).

In male rats exposed to *meta*-xylene vapour at concentrations of 200, 1700 or 3200 mg/m³ for 6 h per day on five days per week for two weeks, xylene concentrations in brain and perirenal fat were increased during the second week of exposure (Savolainen & Pfäffli, 1980).

Pregnant mice were exposed by inhalation to ^{14}C-*para*-xylene [theoretical concentration, 2000 ppm (8680 mg/m³)] for 10 min on days 11, 14 or 17 of gestation, and distribution of the label was determined 0, 0.5, 1 and 4 h after exposure. The label quickly entered the embryo, but uptake was low relative to maternal tissues. All fetal activity was extractable, indicating that no firmly bound metabolite was present (Ghantous & Danielsson, 1986).

(ii) Toxic effects

Oral LD$_{50}$ values for *ortho*-xylene, *meta*-xylene, *para*-xylene and the isomer mixture in rats range between 3600 and 5800 mg/kg bw (Wolf *et al.*, 1956; European Chemical Industry Ecology and Toxicology Centre, 1986). The intraperitoneal LD$_{50}$s of the pure isomers in male mice ranged from 1360 to 2100 mg/kg bw (Mohtashampur *et al.*, 1985). An inhalation LC$_{50}$ (4 h) for the isomer mixture in male rats has been determined as 6700 ppm (29 078 mg/m³; Carpenter *et al.*, 1975). The 6 h-inhalation LC$_{50}$s of the pure isomers in female mice were 3900–5300 ppm (17 000–23 000 mg/m³; Bonnet *et al.*, 1979).

A 4-h percutaneous administration of 4400 mg/kg bw of mixed xylenes to three male rabbits resulted in the death of one rabbit on the fifth day after exposure. At dose levels of 1700 mg/kg bw, none of three rabbits died (Hine & Zuidema, 1970).

Ten to 20 applications of undiluted mixed xylenes on the ears or shaved abdomen of rabbits for two or four weeks resulted in moderate to marked erythema and oedema, with superficial necrosis at both sites. After introduction of two drops of mixed xylenes into the rabbit eye, slight conjunctival irritation and transient corneal injury were observed (Wolf *et al.*, 1956). Application of undiluted xylene to the eye caused corneal lesions in cats (Schmid, 1956).

Rats were exposed by inhalation for 4 h to 2500, 5800, 12 000, 26 000 and 43 000 mg/m³ mixed xylenes. All rats at the highest concentration and 4/10 at 26 000 mg/m³ died; xylene-induced pneumonitis was noted in two of the rats that died. Prostration was noted with 43 000 and 12 000 mg/m³ and poor coordination with 5800 mg/m³; no such sign was observed at the lowest exposure concentration. Exposure of four male cats to 41 000 mg/m³ mixed xylene vapour for 2 h resulted in ataxia, spasms and anaesthesia, followed by death (Carpenter *et al.*, 1975).

Exposure of rats by inhalation to *meta*-xylene at concentrations of 200, 1700 or 3200 mg/m³ for 6 h per day on five days per week for two weeks resulted in changes in the activities of brain enzymes (NADPH–diaphorase, azoreductase and superoxide dismutase), which were reversible two weeks after cessation of exposure (Savolainen & Pfäffli, 1980). Changes in open–field behaviour were observed in rats exposed by inhalation to 300 ppm (1300 mg/m³) for 6 h per day for five to 18 weeks (Savolainen *et al.*, 1979).

Intraperitoneal administration of 1 g/kg bw xylene resulted in an increase in serum ornithine carbamyl transferase activity and lipid accumulation in the liver of rabbits and guinea–pigs, indicating liver damage (DiVincenzo & Krasavage, 1974). Similarly, increases in liver enzyme activities in the serum of rats exposed by inhalation to 1500 ppm (6510 mg/m³) *para*-xylene for 4 h (Patel *et al.*, 1979) and to 400 ppm (1730 mg/m³) *meta*-xylene for 6 h per day on five days per week for two weeks (Elovaara, 1982) are indicative of xylene-induced liver damage. Exposure of rats to 600 ppm (2600 mg/m³) xylene during the light period of the day for four weeks (Toftgård et al., 1981) or to 2000 ppm (8680 mg/m³) ortho-, para- or meta-xylene for 6 h per day for three days (Toftgård & Nielsen, 1982) induced microsomal cytochrome P450. Repeated oral administration (1 g/kg per day) of ortho-, meta- or para-xylene to rats for three days (Pyykkö, 1980) or intermittent exposure of rats by inhalation to 300 ppm (1300 mg/m³) xylene on 6 h per day for two weeks (Savolainen et al., 1978) also increased the activities of drug metabolizing enzymes in the liver and kidney. Inhalation exposure of groups of rats to 3000 mg/m³ para-xylene on day 10 or on days 9 and 10 of gestation [daily duration was presumably for 24 h] reduced concentrations of progesterone and 17β-oestradiol in the maternal circulation (Ungváry et al., 1981).

In some rats exposed to 3000 mg/m³ mixed xylenes for 8 h per day on six days per week for 110–130 days, exposure resulted in paralysis of the hind legs, weight loss, a slight decrease in leukocytes, increases in blood urea, urinary blood and albumin, and hyperplasia of the bone marrow. Slight congestion of kidney, liver, heart, adrenal, lung and spleen were observed. Cellular desquamation of glomeruli and necrosis of the convoluted tubules were also reported (Fabre et al., 1960).

Rats, guinea-pigs, monkeys and dogs were exposed either to 780 ppm (3368 mg/m³) ortho-xylene for 8 h per day on five days per week for six weeks or to 78 ppm (337 mg/m³) continuously for 90 days. No significant change in body weight or in haematological parameters and no significant toxicity were observed after histopathological examination of all major organs (Jenkins et al., 1970).

Groups of four male rats and four male dogs were exposed for 6 h per day on five days per week for 13 weeks to 180, 460 or 810 ppm (770, 200 or 3500 mg/m³) mixed xylenes. No significant effect was reported on body weight, haematology, blood chemistry, urine chemistry, organ weight or macroscopic and microscopic pathology at any concentration tested (Carpenter et al., 1975).

Groups of 15 male rats were exposed by inhalation to 3500 ppm (15 200 mg/m³) ortho-xylene for 8 h per day for one or six weeks. Slight decreases in body weight gain and increased liver weight were observed in both groups (Tátrai & Ungváry, 1980).

(iii) *Effects on reproduction and prenatal toxicity*

The teratogenic and developmental effects of xylene have been reviewed (Hood & Otley, 1985).

Groups of 30 Mallard eggs were exposed by immersion for 30 sec in a 1 or 10% aqueous suspension of xylene on day 3 or 8 of incubation; control eggs were immersed in distilled water. No significant effect was observed on the growth, survival or development of embryos examined at day 18 of incubation (Hoffman & Eastin, 1981).

In one study reported in an abstract (Nawrot & Staples, 1980), exposure of CD-1 mice to 0.75 or 1.0 ml/kg bw of any of the three isomers on days 6–15 of gestation was reported to cause maternal toxicity and fetal death; cleft palates were also reported in fetuses exposed to the ortho- and para-isomers. When the experiment was repeated with meta-xylene, a low but statistically significant incidence of cleft palates occurred after repeated exposures to 1.0

ml/kg bw in the absence of overt maternal effects. [The Working Group noted that the doses were incorrectly expressed as mg, rather than ml, in the abstract.] Marks *et al.* (1982) exposed CD-1 mice to 0.52–4.13 g/kg bw mixed xylenes on days 6–15 of gestation. All dams and fetuses at the highest dose died, and dams died at 3.1 g/kg bw. Fetal viability was reduced at this dose, and growth at 2.06 g/kg bw. Cleft palate and wavy ribs were seen with 2.06 g/kg bw and above. [The Working Group noted an error in the paper in converting the dose from volume per kilogram to mass per kilogram.] In ICR/SIM mice given *meta*-xylene at 2000 mg/kg bw on days 8–12 of gestation, no significant effect was seen on maternal toxicity or postnatal growth or on viability of the offspring (Seidenberg *et al.*, 1986).

In one study reported as an abstract, ICR mice were exposed to 0, 500, 1000 and 2000 ppm (2170, 4340 and 8680 mg/m³) xylene on days 6–12 of gestation. It was stated that fetal growth was retarded at the two highest dose levels and that there was a dose–related increase in the frequency of supernumerary ribs and delayed ossification of the sternebrae. At the high dose, growth retardation persisted into the postnatal period (Shigeta *et al.*, 1983). [The Working Group noted that the reporting of the experimental design and results were insufficient to evaluate many of the parameters.] CFLP mice were exposed to 0, 500 or 1000 mg/m³ xylene or to 500 mg/m³ *ortho*-, *meta*- or *para*-xylene for 24 h per day on days 6–15 of gestation. Fetal growth and skeletal retardation were reported at the highest doses (Ungváry & Tátrai, 1985). [The Working Group noted that this paper is a compendium of data on rats, mice and rabbits from one laboratory and presents few details of experimental results.]

In CFY rats, fused sternebrae and extra ribs were observed in fetuses of dams exposed to 1000 mg/m³ xylene for 24 h per day on days 9–14 of pregnancy, in the absence of maternal effects (Hudák & Ungváry, 1978). In another study in CFY rats using levels of 0, 250, 1900 or 3400 mg/m³ xylene given on days 7–15 of gestation, it was stated that maternal effects were moderate and dose–dependent; the highest dose resulted in decreased embryonic viability and fetal growth as well as an increased incidence of extra ribs; skeletal retardation was seen with all three doses (Ungváry & Tátrai, 1985). [The Working Group had the same reservations about this paper as expressed above.] Mirkova *et al.* (1983) reported fetal growth retardation following exposure of Wistar rats to 50 and 500 mg/m³ xylene on days 1–21 of gestation; these effects were not seen with 10 mg/m³. The growth retardation persisted through postnatal day 21. [The Working Group noted that the reporting of the experimental design and results were insufficient to evaluate many of the parameters.]

CFY rats were exposed *via* inhalation to *ortho*-, *meta*- or *para*-xylene (analytical purity) at concentrations of 0, 150, 1500 and 3000 mg/m³ for 24 h per day on days 7–14 of gestation. Food consumption was reduced at the two higher concentrations of *ortho*-xylene and in the groups exposed to the highest level of *meta*- and of *para*-xylene. Exposure to 3000 mg/m³ *meta*-xylene killed 4/30 dams and reduced weight gain in the surviving dams; 2/20 and 7/20 of the females receiving the high doses of *ortho*- and *para*-xylene, respectively, resorbed their entire litters. Increased maternal liver weight:body weight ratios were observed in all groups exposed to *ortho*-xylene. Fetal body weights were reduced by the two highest levels of *ortho*-xylene and by the highest level of *meta*- and of *para*-xylene; fetal viability was affected only by the highest dose level of *para*-xylene. There was no indication that any xylene isomer caused visceral abnormalities in fetuses in a dose-related manner, but skeletal development was retarded by the high concentration of *ortho*-xylene and by all concentrations of *para*-xylene. Extra ribs were seen in significantly more fetuses in the groups exposed to the high doses of *meta*- and *para*-xylene (Tátrai *et al.*, 1979; Hudák *et al.*, 1980; Ungváry *et al.*, 1980). [The Working Group noted that the analysis supporting this observation is based on data on fetuses, rather than on data on litters, the customary unit of comparison.]

Groups of 25 Sprague-Dawley rats were exposed by inhalation to 0, 3500 or 7000 mg/m^3 *para*-xylene (purity, 99%) for 6 h per day on days 7–16 of gestation, and the offspring were evaluated for growth, viability and neurobehavioural development. The high dose level reduced maternal weight gain during the exposure period, but growth, viability, locomotor activity and the acoustic startle response of the offspring were not affected (Rosen *et al.*, 1986).

Groups of New Zealand white rabbits were exposed to 0 (60 animals in a pooled control group), 500 or 1000 mg/m^3 of *ortho*-xylene, *meta*-xylene, *para*-xylene or xylene [composition unspecified] for 24 h per day on days 7–20 of gestation. Fetuses were examined by routine teratological techniques on day 30 of gestation. It was stated that for each solvent the high-dose level produced mild maternal toxicity [no data were presented for the 1000-mg/m^3 *ortho*- and *meta*-xylene group]. Maternal death and abortion were noted with both xylene and *para*-xylene at 1000 mg/m^3. The body weights of female fetuses exposed to 500 mg/m^3 xylene were significantly reduced, but no other effect on fetuses was reported (Ungváry & Tátrai, 1985). [The Working Group noted that this paper is a compendium of data on rats, mice and rabbits from one laboratory and presents few details on experimental results.]

(iv) *Genetic and related effects*

The genetic and related effects of xylene have been reviewed (Dean, 1978, 1985; Fishbein, 1985).

Technical-grade xylene did not produce differential killing in DNA repair-proficient compared to repair-deficient strains of *Bacillus subtilis rec*$^{+/-}$ (McCarroll *et al.*, 1981a) or *Escherichia coli* (McCarroll *et al.*, 1981b). Xylene [grade unspecified] did not induce SOS activity in *Salmonella typhimurium* TA1535/pSK 1002 (Nakamura *et al.*, 1987). *para*-Xylene was not mutagenic to *E. coli* WP2*uvr*A in the presence or absence of an exogenous metabolic system from Aroclor-induced rat liver (Shimizu *et al.*, 1985). *ortho*-, *meta*- and *para*-Xylene, xylene [grade unspecified] and mixed xylenes were not mutagenic to *S. typhimurium* TA1535, TA1537, TA1538, TA98, TA100, UTH8413 or UTH8414 in the presence or absence of an exogenous metabolic system from uninduced or Aroclor-induced rat and Syrian hamster livers (Lebowitz *et al.*, 1979 (abstract); Bos *et al.*, 1981; Haworth *et al.*, 1983; Connor *et al.*, 1985; Shimizu *et al.*, 1985; Zeiger *et al.*, 1987).

As reported in an abstract, exposure to technical-grade xylene (contaminated with 18.3% ethylbenzene), but not exposure to *meta*- or *ortho*-xylene, caused recessive lethal mutations in *Drosophila melanogaster* (Donner *et al.*, 1980).

As reported in an abstract, xylene [grade unspecified] did not induce mutation in mouse lymphoma L5178Y TK$^{+/-}$ cells or chromosomal aberrations in rat bone marrow (Lebowitz *et al.*, 1979). Xylene [grade unspecified] did not induce sister chromatid exchange or chromosomal aberrations in human lymphocytes *in vitro* (Gerner-Smidt & Friedrich, 1978).

[The Working Group noted that the study of human lymphocytes was performed without an exogenous metabolic system.]

None of the three isomers induced micronuclei in the bone marrow of male NMRI mice after two intraperitoneal administrations of 0.12–0.75 ml/kg bw (0.11–0.65 mg/kg bw) at a 24–h interval (Mohtashamipur et al., 1985); however, they enhanced the induction of micronuclei by toluene (Mohtashamipur et al., 1987).

As reported in an abstract, exposure of rats to mixed isomers (300 ppm; 1300 mg/m³) for 6 h per day on five days per week for nine, 14 and 18 weeks did not induce chromosomal aberrations in bone–marrow cells (Donner et al., 1980).

As reported in an abstract, xylene did not inhibit intercellular communication (as measured by metabolic cooperation) in Chinese hamster V79 cells (Awogi et al., 1986).

Xylene [grade unspecified] did not enhance morphological transformation of Syrian hamster embryo cells by the SA7 adenovirus (Casto, 1981).

Rats injected intraperitoneally with 0.5 and 1.5 ml/kg bw (0.44 and 1.32 mg/kg bw) ortho–xylene showed a significant increase in the percentage of abnormal sperm when housed at temperatures of 24–30°C (control: 2.94±1.36; treated: 4.17±1.41) but not at 20–24°C (Washington et al., 1983). The authors interpreted this as a synergistic effect between xylene and temperature.

(b)   Humans

(i) Absorption, distribution, excretion and metabolism

Most of the available information on xylene metabolism in humans deals with meta–xylene.

In volunteers exposed by inhalation, lung retention was practically identical (64%) for the three isomers (Šedivec & Flek, 1976a). In other studies with volunteers, lung retention of meta–xylene was about 60% (Riihimäki et al., 1979) to 75% (Senczuk & Orłowski, 1978). When volunteers immersed their hands in liquid meta–xylene, it was absorbed at 2 μg/cm² per min (Engström et al., 1977). A nine–fold interindividual variation in skin absorption rate was observed among volunteers (Lauwerys et al., 1978). The amount of meta–xylene absorbed after whole–body exposure of volunteers to 600 ppm (2600 mg/m³) vapour, excluding inhalation, for 3.5 h was equivalent to the amount absorbed after inhalation exposure to 20 ppm (87 mg/m³) for the same duration (Riihimäki & Pfäffli, 1978).

More than 70% of meta–xylene absorbed was excreted into the urine as metabolites (Ogata et al., 1970; Engström et al., 1984). A minor portion (~5%, apparently irrespective of the isomer) was exhaled unchanged (Šedivec & Flek, 1976a; Riihimäki et al., 1979; Åstrand et al., 1978).

Elimination of meta–xylene from the body via excretion and inhalation is rapid, with a biological half–time of 1 h for a rapid phase after 6–16 h of exposure and of about 20 h for a slow phase (Riihimäki et al., 1979). About 72% of total urinary metabolites was excreted in the urine within 24 h after termination of exposure to the three isomers (Šedivec & Flek, 1976a). Removal of industrial xylene from subcutaneous adipose tissue, however, is slow (Engström & Bjurström, 1978), with a half–time of 25–128 h for the meta isomer (Engström & Riihimäki, 1979).

Xylenes are primarily metabolized in humans to the corresponding methylhippuric acid (toluric acid); and glycine conjugation is considered to be a rate-limiting step (Riihimäki, 1979). When volunteers were exposed to *ortho-*, *meta-* or *para*-xylene vapour, more than 95% of the absorbed compound was excreted as methylhippuric acid, and only a small portion was excreted as dimethylphenol: 0.86% as 2,3-dimethylphenol and 3,4-dimethylphenol, after exposure to *ortho*-xylene (the ratio between the two dimethylphenols varied depending on individuals), 1.98% as 2,4-dimethylphenol after exposure to *meta*-xylene and 0.05% as 2,5-dimethylphenol after exposure to *para*-xylene (Šedivec & Flek, 1976a). In other experiments in which volunteers were exposed to *meta*-xylene, *meta*-methylhippuric acid in the urine accounted for 72% (Ogata *et al.*, 1970) to 97% (Engström *et al.*, 1984) of the *meta*-xylene absorbed, whereas 2,4-dimethylphenol and 3-methylbenzyl alcohol accounted for 2.5 and 0.05%, respectively (Engström *et al.*, 1984). Similar results were found for *para*-xylene (Ogata *et al.*, 1970). *ortho*-Xylene was metabolized almost exclusively to *ortho*-methylhippuric acid; only trace amounts of *ortho*-toluic acid (*ortho*-methylbenzoic acid) glucuronide were detected in the urine of volunteers exposed to *ortho*-xylene vapour (Ogata *et al.*, 1980).

Methylhippuric acid has therefore been proposed as a marker urinary metabolite for the biological monitoring of factory workers exposed to xylene, and urine collected in the latter half of a shift is recommended for analysis (Lundberg & Sollenberg, 1986; for reviews, see Šedivec & Flek, 1976b; Riihimäki, 1979).

(ii) *Toxic effects*

Some of the information on the adverse effects of xylene on the central and peripheral nervous systems originates from studies of workers exposed occupationally (mainly painters); such workers are generally also exposed to other organic solvents (Seppäläinen *et al.*, 1978; Elofsson *et al.*, 1980; Ekberg *et al.*, 1986). For further information, see the monograph on occupational exposures in paint manufacture and painting.

Most volunteer subjects exposed to 2000 mg/m³ technical xylene for 15 min had eye irritation (Carpenter *et al.*, 1975); workers exposed to a mixture of solvents, including xylene, displayed corneal vacuoles (Schmid, 1956). Similar effects have been described in spray painters exposed to almost pure xylene as a lacquer–diluting agent (Matthäus, 1964).

Exposure of volunteers to technical xylene by inhalation caused irritation of the airways (Carpenter *et al.*, 1975); very high accidental exposure caused pneumonitis (Morley *et al.*, 1970). Ingestion of xylene caused irritation of the gastrointestinal tract (Gosselin *et al.*, 1976).

Skin contact caused a burning sensation and reversible erythema (Lauwerys *et al.*, 1978). Prolonged exposure may cause contact dermatitis (European Chemical Industry Ecology and Toxicology Centre, 1986).

In studies of volunteers exposed to 200 ppm (870 mg/m³) xylene for 8 h, simple reaction time was slowed (Ogata & Nagao, 1970). Heavy accidental exposure may cause narcosis (Bakinson & Jones, 1985) and death (Morley *et al.*, 1970). Goldie (1960) suggested that occupational exposure to xylene in paints provoked epileptic seizures in one case.

In volunteers exposed to 390 mg/m³ or more technical xylene or *meta*-xylene, with or without physical exercise, reaction time, manual coordination, body equilibrium and electroencephalogram were affected (Gamberale *et al.*, 1978; Savolainen & Linnavuo, 1979; Savolainen, 1980; Savolainen *et al.*, 1980a,b; Savolainen & Riihimäki, 1981a,b; Seppäläinen *et al.*, 1981; Savolainen *et al.*, 1984, 1985a,b). In particular, concentration peaks affected performance. Tolerance developed after exposure for a week and disappeared during the weekend.

Transient kidney damage has occasionally been reported in cases of severe, acute xylene poisoning (Morley *et al.*, 1970; Bakinson & Jones, 1985). Furthermore, indications of slight adverse effects on the kidney (Askergren, 1981; Askergren *et al.*, 1981a,b,c; Franchini *et al.*, 1983) have been reported in workers exposed mainly to xylene and toluene (see also the monograph on occupational exposures in paint manufacture and painting).

In cases of severe, acute poisoning, signs of liver damage have been reported (Morley *et al.*, 1970; Bakinson & Jones, 1985).

Aplastic anaemia was reported in one laboratory worker and decreased platelet counts in 12/27 other laboratory workers exposed to technical xylene (containing 0.2% benzene). When exposure to xylene was interrupted, platelet counts returned to normal (Forde, 1973). [The Working Group noted several early reports of effects on blood and blood forming organs, which might have been due to benzene contamination of xylene.]

(iii) *Effects on fertility and on pregnancy outcome*

In their study of female pharmaceutical workers in Finland, Taskinen *et al.* (1986; see the monograph on toluene) also assessed exposure to xylene. Exposure during the first trimester of pregnancy was reported by three of 38 (8%) women who had had a spontaneous abortion compared to four of 199 (3%) control women who had had live births. The corresponding relative risk (RR) was 2.0 (95% confidence interval (CI), 0.4–10.6). Cases and controls had been exposed to many solvents and other substances.

In the study of Swedish female laboratory workers (Axelsson *et al.*, 1984; see the monograph on toluene), 160 women reported having worked in a laboratory with exposure to xylene during the first trimester of pregnancy. The miscarriage rate of 10.3% compares with that of 11.5% among women who had not worked in a laboratory during the first trimester and that of 9.0% among women who had worked in a laboratory but not with solvents during the first trimester. Cases and controls had been exposed to many solvents and other substances.

Ericson *et al.* (1984; see the monograph on toluene) reported that exposure to xylene had been similar for Swedish laboratory workers who had given birth to children who died in early infancy or were malformed (8%) and for women who had had normal births (8%). Cases and controls were exposed to many solvents and other substances.

In the study of Holmberg (1979; described in the monograph on some petroleum solvents), the mother of one child with central nervous system defects and one control mother reported having worked with xylene during the first trimester of pregnancy. Both mothers had also been exposed to other solvents. In the study of Holmberg *et al.* (1982), described in the monograph on some petroleum solvents, three mothers of children with oral clefts but no control mother were reported to have worked with xylenes during the first trimester of pregnancy. The mothers had also been exposed to other solvents.

(iv) *Genetic and related effects*

No increase in the frequency of sister chromatid exchange was observed in ten workers in the Swedish paint industry exposed to various solvents, including more than 100 mg/m³ xylene (Haglund *et al.*, 1980; see also the monograph on occupational exposures in paint manufacture and painting). [The Working Group noted the small number of workers observed.] No increase in sister chromatid exchange was observed in 46 workers at a Hungarian chemical plant exposed to technical xylene (*ortho-*, *meta-* and *para-*xylene, 6–15% ethylbenzene) with an average exposure of nine years to an average of 50 mg/m³, compared with 34 clerical workers from the factory who were used as controls (Pap & Varga, 1987).

3.3 Epidemiological studies of carcinogenicity in humans

In each of the studies described below, exposures were mixed and overlapping, and these studies are cited in several monographs.

Olsson and Brandt (1980) performed a study on exposure to organic solvents among 25 cases of Hodgkin's disease and 50 controls in Sweden (see the monograph on some petroleum solvents). Exposure to xylene was mentioned by four cases but no referent. All exposed cases and referents were exposed to other solvents.

Wilcosky *et al.* (1984) performed a case–control study of rubber workers in the USA (see the monograph on some petroleum solvents). Exposure to xylene was associated with increased risks for prostatic cancer (relative risk (RR), 1.5, eight cases), lymphosarcoma (3.7, four cases) and lymphatic leukaemia (3.3, four cases). [The Working Group noted that the number of cases in each category is small and that multiple exposures were evaluated independently of other exposures. Although the risk for lymphosarcoma in xylene–exposed workers was significantly raised, four significant associations were reported out of the 20 substances, and these associations are based on larger numbers of exposed cases. It was therefore impossible to determine whether a single substance was associated with the risk.]

Carpenter *et al.* (1988) evaluated the possible association with exposure to 26 chemicals or chemical groups in 89 cases of primary cancers of the central nervous system and 356 matched controls in cohorts of workers at two US nuclear facilities. Toluene (see monograph, p. 79), xylene and methyl ethyl ketone were evaluated as one chemical group; the matched RR was 2.0 (28 cases; 95% confidence interval, 0.7–5.5) in comparison with nonexposed workers. Almost all cases had had low exposure according to the classification used. The authors reported that the RRs were adjusted for internal and external exposure to radiation. [The Working Group noted that no separate analysis was performed for the three solvents, nor were exposure levels quantified, and that there were many concurrent exposures.]

4. Summary of Data Reported and Evaluation

4.1 Exposures

Xylene is a major industrial chemical derived mainly from petroleum refining. It occurs in three isomeric forms (*ortho, meta* and *para*) and is produced and used both as 'mixed xylenes' (usually containing 10–15% ethylbenzene) and as the individual isomers. Xylene is used as a solvent in paints, inks, adhesives and insecticides. Xylene-containing petroleum distillates are used extensively and increasingly in gasoline blending.

The individual isomers are used mainly as chemical intermediates in the manufacture of derivatives of phthalic anhydride (from *ortho*-xylene), isophthalic acid (from *meta*-xylene) and terephthalic acid (from *para*-xylene).

Xylene is ubiquitous in the environment. Occupational exposure has been reported in petroleum refining, in the production of xylene and in the use of xylene and its end products.

4.2 Experimental carcinogenicity data

Xylene (technical grade or mixed xylenes) was tested for carcinogenicity in one strain of mice and in two strains of rats by gastric intubation. One study in rats with mixed xylenes was considered inadequate for evaluation. No increase in the incidence of tumours was observed in either mice or rats following the administration of a technical–grade xylene.

No data were available on the individual isomers.

4.3 Human carcinogenicity data

Exposure to xylene has been associated with increased risks for haematopoietic malignancies in two case–control studies, but the number of cases was limited and exposure was to a variety of compounds.

4.4 Other relevant data

In humans, exposure to xylene causes irritant and central nervous system effects. Adverse effects have been observed on the kidney and liver in cases of accidental poisoning. Similar effects have been seen in experimental animals after exposure to xylene at high levels.

In some studies of the reproductive outcome of women exposed to xylene during the first trimester of pregnancy, small excess risks for spontaneous abortion and for congenital malformation were reported. In all of these studies, the numbers of cases were small and the mothers had also been exposed to other substances.

Maternally toxic or near-toxic amounts of xylene have been associated with malformations in mice after oral administration and with embryotoxicity in rabbits, rats and mice after exposure by inhalation.

Sister chromatid exchange was not induced in peripheral lymphocytes of workers in two studies; however, exposure was to a variety of compounds.

None of the three isomers of xylene induced micronuclei in mice *in vivo*. Sister chromatid exchange and chromosomal aberrations were not induced in cultured human lymphocytes, in the absence of an exogenous metabolic system. Xylene of unspecified grade did not induce morphological transformation in cultured animal cells. None of the three isomers or xylene, either alone or in combination, induced mutation in bacteria. Technical-grade xylene did not induce DNA damage in bacteria. (See Appendix 1.)

4.5 Evaluation[1]

There is *inadequate evidence* for the carcinogenicity of xylene in humans.

There is *inadequate evidence* for the carcinogenicity of xylene in experimental animals.

Overall evaluation

Xylene *is not classifiable as to its carcinogenicity to humans (Group 3)*.

5. References

American Conference of Governmental Industrial Hygienists (1988) *Threshold Limit Values and Biological Exposure Indices for 1988–1989*, Cincinnati, OH, p. 38

Angerer, J. & Lehnert, G. (1979) Occupational chronic exposure to organic solvents. VIII. Phenolic compounds – metabolites of alkylbenzenes in man. Simultaneous exposure to ethylbenzene and xylenes. *Int. Arch. occup. environ. Health*, 43, 145–150

Anon. (1985) Facts and figures from the chemical industry. *Chem. Eng. News*, 63, 22–66

Anon. (1986) Chemical profile: para-xylenes. *Chem. Market. Rep., September*, p. 54

Anon. (1987) Facts and figures from the chemical industry. *Chem. Eng. News*, 65, 24–76

Anon. (1988a) Facts and figures from the chemical industry. *Chem. Eng. News*, 66, 34–82

Anon. (1988b) *CHEM–INTELL Database*, Dunstable, UK, Reed Telepublishing Ltd, Chemical Intelligence Service

Arbeidsinspectie (Labour Inspection) (1986) *De Nationale MAC–Lijst 1986* [National MAC-List 1986] (*P145*), Voorburg, Ministry of Social Affairs and Work Environment, p. 21

Arbejdstilsynet (Labour Inspection) (1988) *Graensevaerdier for Stoffer og Materialer* [Limit Values for Substances and Materials] (*At–anvisning No. 3.1.0.2*), Copenhagen, p. 32

Askergren, A. (1981) Studies of kidney function in subjects exposed to organic solvents. III. Excretion of cells in the urine. *Acta med. scand.*, 210, 103–106

Askergren, A., Allgén, L.-G. & Bergström, J. (1981a) Studies of kidney function in subjects exposed to organic solvents. II. The effect of desmopressin in a concentration test and the effect of exposure to organic solvents on renal concentrating ability. *Acta med. scand.*, 209, 485–488

---

[1]For definitions of the italicized terms, see Preamble, pp. 27–30.

Askergren, A., Allgén, L.-G., Karlsson, C., Lundberg, I. & Nyberg, E. (1981b) Studies of kidney function in subjects exposed to organic solvents. I. Excretion of albumin and β-2-microglobulin in the urine. *Acta med. scand.*, *209*, 479–483

Askergren, A., Brandt, R., Gullquist, R., Silk, B. & Strandell, T. (1981c) Studies of kidney function in subjects exposed to organic solvents. IV. Effect on 51-Cr-EDTA clearance. *Acta med. scand.*, *210*, 373–376

Åstrand, I., Engström, J. & Övrum, P. (1978) Exposure to xylene and ethylbenzene. I. Uptake, distribution and elimination in man. *Scand. J. Work Environ. Health*, *4*, 185–194

Awogi, T., Itoh, T. & Tsushimoto, G. (1986) The effect of benzene and its derivatives on metabolic cooperation (Abstract No. 2). *Mutat. Res.*, *164*, 263

Axelsson, G., Lütz, C. & Rylander, R. (1984) Exposure to solvents and outcome of pregnancy in university laboratory employees. *Br. J. ind. Med.*, *41*, 305–312

Baker, D. & Fannick, N. (1983) *Leather Craftsman, Lynbrook, NY (Health Hazard Evaluation Determination Report No. 80–060–1367)*, Cincinnati, OH, National Institute for Occupational Safety and Health

Bakinson, M.A. & Jones, R.D. (1985) Gassings due to methylene chloride, xylene, toluene, and styrene reported to Her Majesty's Factory Inspectorate 1961-80. *Br. J. ind. Med.*, *42*, 184–190

Bakke, O.M. & Scheline, R.R. (1970) Hydroxylation of aromatic hydrocarbons in the rat. *Toxicol. appl. Pharmacol.*, *16*, 691–700

Bonnet, P., Raoult, G. & Gradiski, G. (1979) LC$_{50}$s of major aromatic hydrocarbons (Fr.). *Arch. Mal. prof. Méd. Trav. Séc. soc.*, *40*, 805–810

Bos, R.P., Brouns, R.M.E., van Doorn, R., Theuws, J.L.G. & Henderson, P.T. (1981) Non-mutagenicity of toluene, *o*-, *m*- and *p*-xylene, *o*-methylbenzylalcohol and *o*-methylbenzylsulfate in the Ames assay. *Mutat. Res.*, *88*, 273–279

Bray, H.G., Humphris, B.G. & Thorpe, W.V. (1949) Metabolism of derivatives of toluene. 3. *o*-, *m*- and *p*-Xylenes. *J. Biochem.*, *45*, 241–244

Carlone, M.F. & Fouts, J.R. (1974) In vivo metabolism of *p*-xylene by rabbit lung and liver. *Xenobiotica*, *4*, 705–715

Carlsson, A. (1981) Distribution and elimination of ^{14}C-xylene in rat. *Scand. J. Work Environ. Health*, *7*, 51–55

Carpenter, A.V., Flanders, W.D., Frome, E.L., Tankersley, W.G. & Fry, S.A. (1988) Chemical exposures and central nervous system cancers: a case–control study among workers at two nuclear facilities. *Am. J. ind. Med.*, *13*, 351–362

Carpenter, C.P., Kinkead, E.R., Geary, D.L., Jr, Sullivan, L.J. & King, J.M. (1975) Petroleum hydrocarbon toxicity studies. V. Animal and human response to vapors of mixed xylenes. *Toxicol. appl. Pharmacol.*, *33*, 543–558

Casto, B.C. (1981) Detection of chemical carcinogens and mutagens in hamster cells by enhancement of adenovirus transformation. In: Mishra, N., Dunkel, V. & Mehlman, I., eds, *Advances in Modern Environmental Toxicology*, Vol. 1, Princeton, NJ, Senate Press, pp. 241–271

Clement Associates (1977) Xylene. In: *Information Dossiers on Substances Designated by TSCA (Toxic Substances Control Act) Interagency Testing Committee (October, 1977) (Contract NSF-C-ENV77–15417)*, Washington DC

Commission of the European Communities (1986) Occupational limit values. *Off. J. Eur. Commun.*, *164*, 6–7

Connor, T.C., Theiss, J.C., Hanna, H.A., Monteith, D.K. & Matney, T.S. (1985) Genotoxicity of organic chemicals frequently found in the air of mobile homes. *Toxicol. Lett.*, *25*, 33–40

Cook, W.A. (1987) *Occupational Exposure Limits – Worldwide*, Washington DC, American Industrial Hygiene Association, pp. 37, 126, 157, 224

Cramer, P.H., Boggess, K.E., Hosenfeld, J.M., Remmers, J.C., Breen, J.J., Robinson, P.E. & Stroup, C. (1988) Determination of organic chemicals of human whole blood: preliminary method development for volatile organics. *Bull. environ. Contam. Toxicol.*, *40*, 612–618

Dean, B.J. (1978) Genetic toxicology of benzene, toluene, xylenes and phenols. *Mutat. Res.*, *47*, 75–97

Dean, B.J. (1985) Recent findings on the genetic toxicology of benzene, toluene, xylenes and phenols. *Mutat. Res.*, *154*, 153–181

Deutsche Forschungsgemeinschaft (German Research Society) (1988) *Maximale Arbeitsplatzkonzentrationen und Biologische Arbeitsstofftoleranzwerte 1988* [Maximal Concentrations in the Work Place and Biological Tolerance Values for Working Materials 1988] (*Report No. XXIV*), Weinheim, VCH Verlagsgesellschaft, p. 61

Direktoratet for Arbeidstilsynet (Directorate for Labour Inspection) (1981) *Administrative Normer for Forurensning i Arbeidsatmosfaere 1981* [Administrative Norms for Pollution in Work Atmosphere 1981] (*No. 361*), Olso, p. 22

DiVincenzo, G.D. & Krasavage, W.J. (1974) Serum ornithine carbamyl transferase as a liver response test for exposure to organic solvents. *Am. ind. Hyg. Assoc. J.*, *35*, 21–29

Donner, M., Mäki-Paakkanen, J., Norppa, H., Sorsa, M. & Vainio, H. (1980) Genetic toxicology of xylenes (Abstract No. 9). *Mutat. Res.*, *74*, 171–172

van Doorn, R., Bos, R.P., Brouns, R.M.E., Leijdekkers, C.-M. & Henderson, P.T. (1980) Effect of toluene and xylenes on liver glutathione and their urinary excretion as mercapturic acids in the rat. *Arch. Toxicol.*, *43*, 293–304

Ekberg, K., Barregård, L., Hagberg, S. & Sällsten, G. (1986) Chronic and acute effects of solvents on central nervous system functions in floorlayers. *Br. J. ind. Med.*, *43*, 101–106

Eller, P.M. (1984) *NIOSH Manual of Analytical Methods*, 3rd ed., Vol. 1 (*DHHS (NIOSH) Publ. No. 84–100*), Washington DC, US Government Printing Office, pp. 1500-1–1500-7, 1501-1–1501-7

Elofsson, S.-A., Gamberale, F., Hindmarsh, T., Iregren, A., Isaksson, A., Johnsson, I., Knave, B., Lydahl, E., Mindus, P., Persson, H.E., Philipson, B., Steby, M., Struwe, G., Söderman, E., Wennberg, A. & Widén, L. (1980) Exposure to organic solvents. A cross-sectional epidemiologic investigation on occupationally exposed car and industrial spray painters with special reference to the nervous system. *Scand. J. Work Environ. Health*, *6*, 239–273

Elovaara, E. (1982) Dose-related effects of *m*-xylene inhalation on the xenobiotic metabolism of the rat. *Xenobiotica*, *12*, 345–352

Engström, J. & Bjurström, R. (1978) Exposure to xylene and ethylbenzene. II. Concentration in subcutaneous adipose tissue. *Scand. J. Work Environ. Health*, *4*, 195–203

Engström, J. & Riihimäki, V. (1979) Distribution of *m*-xylene to subcutaneous adipose tissue in short-term experimental human exposure. *Scand. J. Work Environ. Health*, *5*, 126–134

Engström, J. & Riihimäki, V. (1988) Method 11 – determination of methyl hippuric acids in urine by gas chromatography. In: Fishbein, L. & O'Neill, I.K., eds, *Environmental Carcinogens. Methods of Analysis and Exposure Measurement*, Vol. 10, *Benzene and Alkylated Benzenes* (*IARC Scientific Publications No. 85*), Lyon, International Agency for Research on Cancer, pp. 313–318

Engström, J., Husman, K. & Riihimäki, V. (1977) Percutaneous absorption of *m*-xylene in man. *Int. Arch. occup. environ. Health*, *39*, 181–189

Engström, K., Riihimäki, V. & Laine, A. (1984) Urinary disposition of ethylbenzene and *m*-xylene in man following separate and combined exposure. *Int. Arch. occup. environ. Health, 54*, 355–363

ENMET Corp. (undated) *ENMET–Kitagawa Toxic Gas Detector Tubes*, Ann Arbor, MI

Ericson, A., Källén, B., Zetterström, R., Eriksson, M. & Westerholm, P. (1984) Delivery outcome of women working in laboratories during pregnancy. *Arch. environ. Health, 39*, 5–10

European Chemical Industry Ecology and Toxicology Centre (1986) *Joint Assessment of Commodity Chemicals*, No. 6, *Xylenes*, Brussels

Fabre, R., Truhaut, R. & Laham, S. (1960) Toxicological studies on benzene replacement solvents. IV. Xylenes (Fr.). *Arch. Mal. prof., 21*, 301–313

Fishbein, L. (1985) An overview of environmental and toxicological aspects of aromatic hydrocarbons. III. Xylene. *Sci. total Environ., 43*, 165–183

Fishbein, L. & O'Neill, I.K., eds (1988) *Environmental Carcinogens. Methods of Analysis and Exposure Measurement*, Vol. 10, *Benzene and Alkylated Benzenes (IARC Scientific Publications No. 85)*, Lyon, International Agency for Research on Cancer

Forde, J.P. (1973) Xylene affected platelet count. *Occup. Health, November*, 429–433

The Foxboro Co. (1983) *Chromatographic Column Selection Guide for Century Organic Vapor Analyzer*, Foxboro, MA

Franchini, I., Cavatorta, A., Falzoi, M., Lucertini, S. & Mutti, A. (1983) Early indicators of renal damage in workers exposed to organic solvents. *Int. Arch. occup. environ. Health, 52*, 1–9

Gamberale, F., Görel Annwall, B.A. & Hultengren, M. (1978) Exposure to xylene and ethylbenzene. III. Effects on central nervous functions. *Scand. J. Work Environ. Health, 4*, 204–211

Gerner-Smidt, P. & Friedrich, U. (1978) The mutagenic effect of benzene, toluene and xylene studied by the SCE technique. *Mutat. Res., 58*, 313–316

Ghantous, H. & Danielsson, B.R.G. (1986) Placental transfer and distribution of toluene, xylene and benzene, and their metabolites during gestation in mice. *Biol. Res. Preg., 7*, 98–105

Goldie, I. (1960) Can xylene (xylol) provoke convulsive seizures? *Ind. Med. Surg., 29*, 33–35

Gosselin, R.E., Hodge, H.C., Smith, R.P. & Gleason, M.N. (1976) *Clinical Toxicology of Commercial Products. Acute Poisoning*, 4th ed., Baltimore, MD, Williams & Wilkins, pp. 320–323

Haglund, U., Lundberg, I. & Zech, L. (1980) Chromosome aberrations and sister chromatid exchanges in Swedish paint industry workers. *Scand. J. Work Environ. Health, 6*, 291–298

Hampton, C.V., Pierson, W.R., Harvey, T.M., Updegrove, W.S. & Marano, R.S. (1982) Hydrocarbon gases emitted from vehicles on the road. 1. A qualitative gas chromatography/mass spectrometry survey. *Environ. Sci. Technol., 16*, 287–298

Hansch, C. & Leo, A. (1979) *Substituent Constants for Correlation Analysis in Chemistry and Biology*, New York, John Wiley & Sons, p. 232

Hawley, G.G. (1981) *The Condensed Chemical Dictionary*, 10th ed., New York, Van Nostrand Reinhold, pp. 1100–1101

Haworth, S., Lawlor, T., Mortelmans, K., Speck, W. & Zeiger, E. (1983) *Salmonella* mutagenicity test results for 250 chemicals. *Environ. Mutagenesis, Suppl. 1*, 3–142

Health and Safety Executive (1987) *Occupational Exposure Limits 1987 (Guidance Note EH 40/87)*, London, Her Majesty's Stationery Office, p. 22

Hillard, J.H. (1980) Gas, natural. In: Mark, H.F., Othmer, D.F., Overberger, C.G., Seaborg, G.T. & Grayson, M., eds, *Kirk–Othmer Encyclopedia of Chemical Technology*, 3rd ed., Vol. 11, New York, John Wiley & Sons, pp. 630–652

Hine, C.H. & Zuidema, H.H. (1970) The toxicological properties of hydrocarbon solvents. *Ind. Med.*, *39*, 215–220

Hoffman, D.J. & Eastin, W.C., Jr (1981) Effects of industrial effluents, heavy metals, and organic solvents on mallard embryo development. *Toxicol. Lett.*, *9*, 35–40

Holmberg, P.C. (1979) Central-nervous-system defects in children born to mothers exposed to organic solvents during pregnancy. *Lancet*, *ii*, 177–179

Holmberg, P.C., Hernberg, S., Kurppa, K., Rantala, K. & Riala, R. (1982) Oral clefts and organic solvent exposure during pregnancy. *Int. Arch. occup. environ. Health*, *50*, 371–376

Holzer, G., Oró, J. & Bertsch, W. (1976) Gas chromatographic mass-spectrometric evaluation of exhaled tobacco smoke. *J. Chromatogr.*, *126*, 771–785

Hood, R.D. & Ottley, M.S. (1985) Developmental effects associated with exposure to xylene: a review. *Drug. chem. Toxicol.*, *8*, 281–297

Hudák, A. & Ungváry, G. (1978) Embryotoxic effects of benzene and its methyl derivatives: toluene, xylene. *Toxicology*, *11*, 55–63

Hudák, A., Tátrai, E., Lőrincz, M., Barcza, G. & Ungváry, G. (1980) Study of the embryotoxic effect of *ortho*-xylene (Hung.). *Morph. és. Ig. Orv. Szemle*, *20*, 204–209

Huff, J.E., Eastin, W., Roycroft, J., Eustis, S.L. & Haseman, J.K. (1988) Carcinogenesis studies of benzene, methyl benzene, and dimethyl benzenes. *Ann. N.Y. Acad. Sci.*, *534*, 427–440

IARC (1980) *IARC Monographs on the Evaluation of the Carcinogenic Risk of Chemicals to Humans*, Suppl. 2, *Long-term and Short-term Screening Assays for Carcinogenesis: A Critical Appraisal*, Lyon

IARC (1981) *IARC Monographs on the Evaluation of the Carcinogenic Risk of Chemicals to Humans*, Vol. 25, *Wood, Leather and Some Associated Industries*, Lyon

IARC (1989) *IARC Monographs on the Evaluation of Carcinogenic Risks to Humans*, Vol. 45, *Occupational Exposures in Petroleum Refining; Crude Oil and Major Petroleum Fuels*, Lyon

Ikeda, M., Kumai, M., Watanabe, T. & Fujita, H. (1984) Aromatic and other contents in automobile gasoline in Japan. *Ind. Health*, *22*, 235–241

Institut National de Recherche et de Sécurité (National Institute for Research and Safety) (1986) *Valeurs Limites pour les Concentrations des Substances Dangereuses Dans l'Air des Lieux de Travail* [Limit Values for Concentrations of Dangerous Substances in the Air of Work Places] (*ND 1609–125–86*), Paris, p. 581

Jenkins, L.J., Jr, Jones, R.A. & Siegel, S. (1970) Long-term inhalation screening studies of benzene, toluene, *o*-xylene and cumene in experimental animals. *Toxicol. appl. Pharmacol.*, *16*, 818–823

Kilburn, K.H., Seidman, B.C. & Warshaw, R. (1985) Neurobehavioral and respiratory symptoms of formaldehyde and xylene exposure in histology technicians. *Arch. environ. Health*, *40*, 229–233

Klaucke, D., Johansen, M. & Vogt, R.L. (1982) An outbreak of xylene intoxication in a hospital. *Am. J. ind. Med.*, *3*, 173–178

Korte, F. & Boedefeld, E. (1978) Ecotoxicological review of global impact of petroleum industry and its products. *Ecotoxicol. environ. Saf.*, *2*, 55–103

Krotoszynski, B.K., Bruneau, G.M. & O'Neill, H.J. (1979) Measurement of chemical inhalation exposure in urban population in the presence of endogenous effluents. *J. anal. Toxicol.*, *3*, 225–234

Lauwerys, R.R., Dath, T., Lachapelle, J.-M., Buchet, J.-P. & Roels, H. (1978) The influence of two barrier creams on the percutaneous absorption of *m*-xylene in man. *J. occup. Med.*, *20*, 17–20

Lebowitz, H., Brusick, D., Matheson, D., Jagannath, D.R., Reed, M., Goode, S. & Roy, G. (1979) Commonly used fuels and solvents evaluated in a battery of short-term bioassays (Abstract Eb–8). *Environ. Mutagenesis*, *1*, 172–173

Lundberg, I. & Sollenberg, J. (1986) Correlation of xylene exposure and methyl hippuric acid excretion in urine among paint industry workers. *Scand. J. Work Environ. Health*, *12*, 149–153

Maltoni, C., Conti, B. & Cotti, G. (1983) Benzene: a multipotential carcinogen. Results of long-term bioassays performed at the Bologna Institute of Oncology. *Am. J. ind. Med.*, *4*, 589–630

Maltoni, C., Conti, B., Cotti, G. & Belpoggi, F. (1985) Experimental studies on benzene carcinogenicity at the Bologna Institute of Oncology: current results and ongoing research. *Am. J. ind. Med.*, *7*, 415–446

Mannsville Chemical Products Corp. (1981) *Chemical Products Synopsis: Xylenes*, Cortland, NY

Marks, T.A., Ledoux, T.A. & Moore, J.A. (1982) Teratogenicity of a comemrcial xylene mixture in the mouse. *J. Toxicol. environ. Health*, *9*, 97–105

Matheson Gas Products (undated) *The Matheson–Kitagawa Toxic Gas Detector System*, East Rutherford, NJ

Matthäus, W. (1964) Contribution to the corneal lesion of workers involved in surface varnishing in the furniture industry (Ger.). *Klin. Mbl. Augenheilk.*, *144*, 713–717

McCarroll, N.E., Keech, B.H. & Piper, C.E. (1981a) A microsuspension adaptation of the *Bacillus subtilis* 'rec' assay. *Environ. Mutagenesis*, *3*, 607–616

McCarroll, N.E., Piper, C.E. & Keech, B.H. (1981b) An *E. coli* microsuspension assay for the detection of DNA damage induced by direct-acting agents and promutagens. *Environ. Mutagenesis*, *3*, 429–444

Merian, E. (1982) The environmental chemistry of volatile hydrocarbons. *Toxicol. environ. Chem.*, *5*, 167–175

Merian, E. & Zander, M. (1982) Volatile aromatics. In: Hutzinger, O., ed., *Handbook of Environmental Chemistry*, Vol. 3, Part B, *Anthropogenic Compounds*, Berlin (West), Springer, pp. 117–161

Mirkova, E., Zaikov, C., Antov, G., Mikhailova, A., Khinkova, L. & Benchev, I. (1983) Prenatal toxicity of xylene. *J. Hyg. Epidemiol. Microbiol. Immunol.*, *27*, 337–343

Mohtashamipur, E., Norpoth, K., Woelke, U. & Huber, P. (1985) Effects of ethylbenzene, toluene, and xylene on the induction of micronuclei in bone marrow polychromatic erythrocytes of mice. *Arch. Toxicol.*, *58*, 106–109

Mohtashamipur, E., Sträter, H., Triebel, R. & Norpoth, K. (1987) Effects of pretreatment of male NMRI mice with enzyme inducers or inhibitors on clastogenicity of toluene. *Arch. Toxicol.*, *60*, 460–463

Montesano, R., Bartsch, H., Vainio, H., Wilbourn, J. & Yamasaki, H., eds (1986) *Long-term and Short-term Assays for Carcinogens: A Critical Appraisal* (*IARC Scientific Publications No. 83*), Lyon, International Agency for Research on Cancer

Morley, R., Eccleston, D.W., Douglas, C.P., Greville, W.E.J., Scott, D.J. & Anderson, J. (1970) Xylene poisoning: a report on one fatal case and two cases of recovery after prolonged unconsciousness. *Br. med. J.*, *iii*, 442–443

Moszczyński, P. & Lisiewicz, J. (1985) Occupational exposure to benzene, toluene and xylene and the lymphocyte lysosomal *N*-acetyl-beta-D-glucosaminidase. *Ind. Health*, *23*, 47–51

Nakamura, S., Oda, Y., Shimada, T., Oki, I. & Sugimoto, K. (1987) SOS-inducing activity of chemical carcinogens and mutagens in *Salmonella typhimurium* TA1535/pSk1002: examination with 151 chemicals. *Mutat. Res.*, *192*, 239–246

National Draeger, Inc. (1987) *Detector Tube Products for Gas and Vapor Detection*, Pittsburgh, PA

National Institute for Occupational Safety and Health (1975) *Criteria for a Recommended Standard ... Occupational Exposure to Xylene (DHEW (NIOSH) Publ. No. 75–168)*, Washington DC, US Department of Health, Education, and Welfare

National Institute for Occupational Safety and Health (1983) *National Occupational Exposure Survey 1981–83*, Cincinanti, OH

National Institute for Occupational Safety and Health (1986) NIOSH recommendations for occupational safety and health standards. *Morbid. Mortal. Wkly Rep. Suppl.*, *35*, 33S

National Swedish Board of Occupational Safety and Health (1987) *Hygieniska Gränsvärden* [Hygienic Limit Values], (*Ordinance 1987:12*), Solna, p. 38

National Toxicology Program (1986) *Toxicology and Carcinogenesis Studies of Xylenes (Mixed) (60% m-Xylene, 14% p–Xylene, 9% o–Xylene, and 17% Ethylbenzene) (CAS No. 1330–20–7) in F344/N Rats and B6C3F1 Mice (Gavage Studies) (NTP TR 327; NIH Publ. No. 87–2583)*, Research Triangle Park, NC, US Department of Health and Human Services

Nawrot, P.S. & Staples, R.E. (1980) Embryofetal toxicity and teratogenicity of isomers of xylene in the mouse (Abstract No. 65). In: *19th Annual Meeting of the Society of Toxicology, March 9–13, 1980*, Washington DC, Society of Toxicology

Ogata, M. & Miyake, Y. (1973) Identification of substances in petroleum causing objectionable odour in fish. *Water Res.*, *7*, 1493–1504

Ogata, M. & Miyake, Y. (1978) Disappearance of aromatic hydrocarbons and organic sulfur compounds from fish flesh reared in crude oil suspension. *Water Res.*, *12*, 1041–1044

Ogata, M. & Nagao, I. (1970) Urinary m–methyl hippuric acid excretion and physiological changes in persons exposed to 200 ppm m–xylene in an exposure chamber (Jpn.). *Jpn. J. ind. Health*, *10*, 75–79

Ogata, M., Tomokuni, K. & Takatsuka, T. (1970) Urinary excretion of hippuric acid and m– or p–methyl-hippuric acid in the urine of persons exposed to vapours of toluene and m– or p–xylene as a test of exposure. *Br. J. ind. Med.*, *27*, 43–50

Ogata, M., Takatsuka, Y. & Tomokuni, K. (1971) Excretion of hippuric acid and m– or p–methylhippuric acid in the urine of persons exposed to vapours of toluene and m– or p–xylene in an exposure chamber and in workshops with specific reference to repeated exposure. *Br. J. ind. Med.*, *28*, 382–385

Ogata, M., Yamazaki, Y., Sugihara, R., Shimada, Y. & Meguro, T. (1980) Quantitation of urinary o–xy-lene metabolites of rats and human beings by high performance liquid chromatography. *Int. Arch. occup. environ. Health*, *46*, 127–139

Olsson, H. & Brandt, L. (1980) Occupational exposure to organic solvents and Hodgkin's disease in men. A case-referent study. *Scand. J. Work Environ. Health*, *6*, 302–305

Otson, R. & Williams, D.T. (1981) Evaluation of a liquid–liquid extraction technique for water pollutants. *J. Chromatogr.*, *212*, 187–197

Otson, R., Williams, D.T. & Bothwell, P.D. (1982) Volatile organic compounds in water in thirty Canadian potable water treatment facilities. *J. Assoc. off. anal. Chem.*, *65*, 1370–1374

Pap, M. & Varga, C. (1987) Sister–chromatid exchanges in peripheral lymphocytes of workers occupationally exposed to xylenes. *Mutat. Res.*, *187*, 223–225

Patel, J.M., Harper, C., Gupta, B.N. & Drew, R.T. (1979) Changes in serum enzymes after inhalation exposure of p–xylene. *Bull. environ. Contam. Toxicol.*, *21*, 17–24

Pouchert, C.J., ed. (1981) *The Aldrich Library of Infrared Spectra*, 3rd ed., Milwaukee, WI, Aldrich Chemical Co., pp. 564D, 565A, 565E, 566H

Pouchert, C.J., ed. (1983) *The Aldrich Library of NMR Spectra*, 2nd ed., Vol. 1, Milwaukee, WI, Aldrich Chemical Co., pp. 740B, 741A, 742A

Pouchert, C.J., ed. (1985) *The Aldrich Library of FT–IR Spectra*, Vol. 1, Milwaukee, WI, Aldrich Chemical Co., pp. 936D, 938A, 939A, 941B

Pyykkö, K. (1980) Effects of methylbenzenes on microsomal enzymes in rat liver, kidney and lung. *Biochim. biophys. Acta, 633*, 1-9

Ransley, D.L. (1984) Xylenes and ethylbenzene. In: Mark, H.F., Othmer, D.F., Overberger, C.G., Seaborg, G.T. & Grayson, M., eds, *Kirk–Othmer Encyclopedia of Chemical Technology*, 3rd ed., Vol. 24, New York, John Wiley & Sons, pp. 709-744

Riedel–de Haën (1984) *Laboratory Chemicals*, Hanover

Riihimäki, V. (1979) Conjugation and urinary excretion of toluene and m–xylene metabolites in a man. *Scand. J. Work Environ. Health, 5*, 135-142

Riihimäki, V. & Engström, K. (1979) *Xylene* (Swed.) (*Arbete och Hälsa 1997:35*), Nordic Expert Group for Criteria Documents

Riihimäki, V. & Pfäffli, P. (1978) Percutaneous absorption of solvent vapors in man. *Scand. J. Work Environ. Health, 4*, 73-85

Riihimäki, V., Pfäffli, P., Savolainen, K. & Pekari, K. (1979) Kinetics of *m*–xylene in man. General features of absorption, distribution, biotransformation and excretion in repetitive inhalation exposure. *Scand. J. Work Environ. Health, 5*, 217-231

Rivera, R. & Rostand, R. (1975) *Hillerich & Bradsby Co., Jeffersonville, Indiana (Health Hazard Evaluation Determination Report No. 74–121–203)*, Cincinnati, OH, National Institute for Occupational Safety and Health

Roper, P. (1980) *Emory University Pathology Department, Atlanta, Georgia (Health Hazard Evaluation Determination Report No. 80–31–693)*, Cincinnati, OH, National Institute for Occupational Safety and Health

Rosen, M.B., Crofton, K.M. & Chernoff, N. (1986) Postnatal evaluation of prenatal exposure to *p*–xylene in the rat. *Toxicol. Lett., 34*, 223-229

Roxan, Inc. (undated) *Precision Gas Detector*, Woodland Hills, CA

Sadtler Research Laboratories (1980) *Standard Spectra Collection, 1980 Cumulative Index*, Philadelphia, PA

Sandmeyer, E. (1981) Aromatic hydrocarbons. In: Clayton, G.D. & Clayton, F.E., eds, *Patty's Industrial Hygiene and Toxicology*, 3rd rev. ed., Vol. 2B, New York, John Wiley & Sons, p. 3256, 3291-3292

Savolainen, K. (1980) Combined effects of xylene and alcohol on the central nervous system. *Acta pharmacol. toxicol., 46*, 366-372

Savolainen, K. & Linnavuo, M. (1979) Effects of *m*–xylene on human equilibrium measured with a quantitative method. *Acta pharmacol. toxicol., 44*, 315-318

Savolainen, K. & Pfäffli, P. (1980) Dose–dependent neurochemical changes during short–term inhalation exposure to m–xylene. *Arch. Toxicol., 45*, 117-122

Savolainen, K. & Riihimäki, V. (1981a) An early sign of xylene effect on human equilibrium. *Acta pharmacol. toxicol., 48*, 279-283

Savolainen, K. & Riihimäki, V. (1981b) Xylene and alcohol involvement of the human equilibrium system. *Acta pharmacol. toxicol., 49*, 447-451

Savolainen, K., Vainio, H., Helojoki, M. & Elovaara, E. (1978) Biochemical and toxicological effects of short–term, intermittent xylene inhalation exposure and combined ethanol intake. *Arch. Toxicol., 41*, 195-205

Savolainen, H., Pfäffli, P., Helojoki, M. & Tengén, M. (1979) Neurochemical and behavioural effects of long–term intermittent inhalation of xylene vapour and simultaneous ethanol intake. *Acta pharmacol. toxicol., 44*, 200-207

Savolainen, K., Riihimäki, V., Seppäläinen, A.M. & Linnoila, M. (1980a) Effects of short-term *m*-xylene exposure and physical exercise on the central nervous system. *Int. Arch. occup. environ. Health*, *45*, 105-121

Savolainen, K., Riihimäki, V., Vaheri, E. & Linnoila, M. (1980b) Effects of xylene and alcohol on vestibular and visual functions in man. *Scand. J. Work Environ. Health*, *6*, 94-103

Savolainen, K., Kekoni, J., Riihimäki, V. & Laine, A. (1984) Immediate effects of *m*-xylene on the human central nervous system. *Arch. Toxicol., Suppl. 7*, 412-417

Savolainen, K., Riihimäki, V., Luukkonen, R. & Muona, O. (1985a) Changes in the sense of balance correlate with concentrations of *m*-xylene in venous blood. *Br. J. ind. Med.*, *42*, 765-769

Savolainen, K., Riihimäki, V., Muona, O., Kekoni, J., Luukkonen, R. & Laine, A. (1985b) Conversely exposure-related effects between atmospheric *m*-xylene concentrations and human body sense of balance. *Acta pharmacol. toxicol.*, *57*, 67-71

Schmid, E. (1956) Corneal injuries in furniture polishers (Ger.). *Arch. Gewerbepathol. Gewerbehyg.*, *15*, 37-44

Šedivec, V. & Flek, J. (1976a) The absorption, metabolism, and excretion of xylenes in man. *Int. Arch. occup. environ. Health*, *37*, 205-217

Šedivec, V. & Flek, J. (1976b) Exposure test for xylenes. *Int. Arch. occup. environ. Health*, *37*, 219-232

Seidenberg, J.M., Anderson, D.G. & Becker, R.A. (1986) Validation of an in vivo developmental toxicity screen in the mouse. *Teratog. Carcinog. Mutagenesis*, *6*, 361-374

Seifert, B. & Abraham, H.J. (1982) Indoor air concentrations of benzene and some other aromatic hydrocarbons. *Ecotoxicol. environ. Saf.*, *6*, 190-192

Seifert, B. & Abraham, H.J. (1983) Use of passive samplers for the determination of gaseous organic substances in indoor air at low concentration levels. *Int. J. environ. anal. Chem.*, *13*, 237-253

Seńczuk, W. & Orłowski, J. (1978) Absorption of *m*-xylene vapours through the respiratory tract and excretion of *m*-methylhippuric acid in urine. *Br. J. ind. Med.*, *35*, 50-55

Sensidyne (1985) *The First Truly Simple Precision Gas Detector System*, Largo, FL

Seppäläinen, A.M., Husman, K. & Mårtenson, C. (1978) Neurophysiological effects of long-term exposure to a mixture of organic solvents. *Scand. J. Work Environ. Health*, *4*, 304-314

Seppäläinen, A.M., Savolainen, K. & Kovala, T. (1981) Changes induced by xylene and alcohol in human evoked potentials. *Electroenceph. clin. Neurophysiol.*, *51*, 148-155

Shigeta, S., Aikawa, H., Misawa, T. & Suzuki, K. (1983) Fetotoxicity of inhaled xylene in mice (Abstract). *Teratology*, *28*, 22A

Shimizu, H., Suzuki, Y., Takemura, N., Goto, S. & Matsushita, H. (1985) The results of microbial mutation test for forty-three industrial chemicals. *Jpn. J. ind. Health*, *27*, 400-419

Sittig, M. (1985) *Handbook of Toxic and Hazardous Chemicals and Carcinogens*, 2nd ed., Park Ridge, NJ, Noyes, pp. 931-933

SKC (1988) *Comprehensive Catalog and Guide*, Eighty Four, PA

Smith, B.R., Plummer, J.L., Wolf, C.R., Philpot, R.M. & Bend, J.R. (1982) *p*-Xylene metabolism by rabbit lung and liver and its relationship to the selective destruction of pulmonary cytochrome P-450. *J. Pharmacol. exp. Therap.*, *223*, 736-742

Sukhanova, V., Marar'eva, L.M. & Boiko, V.I. (1969) Investigation of functional properties of leukocytes of workers engaged in manufacture of xylene. *Hyg. Sanit.*, *34*, 448-450

Taskinen, H., Lindbohm, M.-L. & Hemminki, K. (1986) Spontaneous abortions among women working in the pharmaceutical industry. *Br. J. ind. Med.*, *43*, 199-205, 432

Tátrai, E. & Ungváry, G. (1980) Changes induced by o-xylene inhalations in the rat liver. *Acta med. acad. sci. hung.*, *37*, 211–216

Tátrai, E., Hudák, A., Barcza, G. & Ungváry, G. (1979) Study of embryotoxic effect of the *meta*-xylene (Hung.). *Egeszsegtudomany*, *23*, 147–151

Toftgård, R. & Nielsen, O.G. (1982) Effects of xylene and xylene isomers on cytochrome P-450 and in vitro enzymatic activities in rat liver, kidney and lung. *Toxicology*, *23*, 192–212

Toftgård, R., Nielsen, O.G. & Gustafsson, J.-Å. (1981) Changes in the rat liver microsomal cytochrome P-450 and enzymatic activities after the inhalation of n-hexane, xylene, methyl ethyl ketone and methylchloroform for four weeks. *Scand. J. Work Environ. Health*, *7*, 31–37

Toftgård, R., Haaparanta, T. & Halpert, J. (1986) Rat lung and liver cytochrome P-450 isozymes involved in the hydroxylation of m-xylene. *Toxicology*, *39*, 225–231

Tsuruta, H. (1982) Percutaneous absorption of organic solvents. III. On the penetration rates of hydrophobic solvents through the excised rat skin. *Ind. Health*, *20*, 335–345

Työsuojeluhallitus (National Finnish Board of Occupational Safety and Health) (1987) *HTP–Arvot 1987* [TLV Values 1987] (*Safety Bulletin 25*), Helsinki, Valtion Painatuskeskus, p. 19

Ungváry, G. & Tátrai, E. (1985) On the embryotoxic effects of benzene and its alkyl derivatives in mice, rats and rabbits. *Arch. Toxicol.*, *Suppl. 8*, 425–430

Ungváry, G., Tátrai, E., Hudák, A., Barcza, G. & Lőrincz, M. (1980) Studies of the embryotoxic effects of *ortho-*, *meta-* and *para*-xylene. *Toxicology*, *18*, 61–74

Ungváry, G., Varga, B., Horváth, E., Tátrai, E. & Folly, G. (1981) Study of the role of maternal sex steroid production and metabolism in the embryotoxicity of *para*-xylene. *Toxicology*, *19*, 262–268

US International Trade Commission (1982) *Synthetic Organic Chemicals, US Production and Sales, 1981* (*USITC Publ. 1292*) Washington DC, US Government Printing Office

US International Trade Commission (1983) *Synthetic Organic Chemicals, US Production and Sales, 1982* (*USITC Publ. 1422*) Washington DC, US Government Printing Office

US International Trade Commission (1984) *Synthetic Organic Chemicals, US Production and Sales, 1983* (*USITC Publ. 1588*) Washington DC, US Government Printing Office

US International Trade Commission (1985) *Synthetic Organic Chemicals, US Production and Sales, 1984* (*USITC Publ. 1745*) Washington DC, US Government Printing Office

US International Trade Commission (1986) *Synthetic Organic Chemicals, US Production and Sales, 1985* (*USITC Publ. 1892*) Washington DC, US Government Printing Office

US International Trade Commission (1987) *Synthetic Organic Chemicals, US Production and Sales, 1986* (*USITC Publ. 2009*) Washington DC, US Government Printing Office

US Occupational Safety and Health Administration (1988) Labor. *US Code Fed. Regul.*, *Title 29*, Part 1910.1000

Wallace, L.A., Pellizzari, E.D., Hartwell, T.D., Sparacino, C. & Zelon, H. (1983) Personal exposure to volatile organics and other compounds indoors and outdoors – the TEAM (total exposure assessment methodology) study. In: *Proceedings of the 76th Annual Meeting of the Air Pollution Control Association, Atlanta, GA*, Atlanta, GA, Air Pollution Control Association, pp. 1–29

Wallace, L.A., Pellizzari, E.D., Hartwell, T.D., Zelon, H., Sparacino, C. & Whitmore, R. (1984) Analysis of exhaled breath of 355 urban residents for volatile organic compounds. In: Bergland, B., Lindvall, T. & Sundell, J., eds, *Indoor Air. Proceedings of the 3rd International Conference on Indoor Air Quality and Climate*, Vol. 4, Stockholm, Swedish Council for Building Research, pp. 15–20

Wallace, L.A., Pellizzari, E.D., Hartwell, T.D., Whitmore, R., Sparacino, C. & Zelon, H. (1986) Total exposure assessment methodology (TEAM) study: personal exposures, indoor-outdoor relationships, and breath levels of volatile organic compounds in New Jersey. *Environ. int.*, *12*, 369–387

Washington, W.J., Murthy, R.C., Doye, A., Eugene, K., Brown, D. & Bradley, I. (1983) Induction of morphologically abnormal sperm in rats exposed to o-xylene. *Arch. Androl.*, *11*, 233–237

Weast, R.C., ed. (1985) *Handbook of Chemistry and Physics*, 66th ed., Cleveland, OH, CRC Press, pp. C-549–C-550

Wilcosky, T.C., Checkoway, H., Marshall, E.G. & Tyroler, H.A. (1984) Cancer mortality and solvent exposures in the rubber industry. *Am. ind. Hyg. Assoc. J.*, *45*, 809–811

Windholz, M., ed. (1983) *The Merck Index*, 10th ed., Rahway, NJ, Merck & Co., pp. 1447–1448

Wolf, M.A., Rowe, V.K., McCollister, D.D., Hollingsworth, R.L. & Oyen, F. (1956) Toxicological studies of certain alkylated benzenes and benzene. *Arch. ind. Health*, *14*, 387–398

World Health Organization (1981) *Recommended Health-based Limits in Occupational Exposure to Selected Organic Solvents. Report of a WHO Study Group (WHO tech. Rep. Ser. 664)*, Geneva, pp. 25–38

Zeiger, E., Anderson, B., Haworth, S., Lawlor, T., Mortelmans, K. & Speck, W. (1987) *Salmonella* mutagenicity tests: III. Results from the testing of 255 chemicals. *Environ. Mutagenesis*, *9* (*Suppl. 9*), 1–110

CYCLOHEXANONE

1. Chemical and Physical Data

1.1 Synonyms

Chem. Abstr. Services Reg. No.: 108–94–1
Chem. Abstr. Name: Cyclohexanone
IUPAC Systematic Name: Cyclohexyl ketone
Synonyms: Ketohexamethylene; pimelic ketone; pimelin ketone

1.2 Structural and molecular formulae and molecular weight

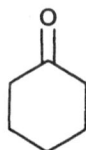

$C_6H_{10}O$ Mol. wt: 98.14

1.3 Chemical and physical properties of the pure substance

(a) *Description*: Colourless liquid with peppermint and acetone odour (Krasavage *et al.*, 1982; Windholz, 1983)

(b) *Boiling–point*: 155.6°C; 47°C at 15 mm Hg (Weast, 1985)

(c) *Melting–point*: –16.4°C (Weast, 1985)

(d) *Density*: 0.948 at 20°C/4°C (Weast, 1985)

(e) *Spectroscopy data*: Nuclear magnetic resonance, infrared and ultraviolet spectral data have been reported (Sadtler Research Laboratories, 1980; Pouchert, 1981, 1983, 1985).

(f) *Solubility*: Miscible with most organic solvents. Soluble in ethanol, diethyl ether, benzene, chloroform and other common organic solvents; soluble in water (150 g/l at 10°C, 50 g/l at 30°C) (Krasavage *et al.*, 1982; Windholz, 1983; Weast, 1985)

(g) *Volatility*: Vapour pressure: 5.2 mm Hg at 25°C (Krasavage *et al.*, 1982)

(h) *Refractive index*: 1.4507 at 20°C (Weast, 1985)

(i) *Flash–point*: 44°C (closed–cup; Krasavage *et al.*, 1982)

(j)   *Conversion factor*: mg/m^3 = 4.0 x ppm[1]

1.4 Technical products and impurities

Trade Names: Anon; Anone; Hexanon; Hytrol O; Nadone; Sextone

Cyclohexanone is available in various grades of purity (98% min, >99.8%). Impurities reported include formic acid (up to 0.05%) and water (up to 0.2%; Fisher & Van Peppen, 1979; Riedel–de Haën, 1984; Eastman Kodak Co., 1985; Aldrich Chemical Co., Inc., 1988).

2. Production, Use, Occurrence and Analysis

2.1 Production and use

(a) Production

Cyclohexanone is produced commercially in several major ways. One widely used process yields cyclohexanol and cyclohexanone by the catalytic oxidation of cyclohexane. The cyclohexanol/cyclohexanone product mixture, also called KA oil, is further reacted to produce adipic acid and hexamethylene diamine, intermediates in the manufacture of nylon 66. Pure cyclohexanone can be produced in high yields by this process either by distillation or by catalytic dehydrogenation of the cyclohexanol (Considine, 1974).

Another important and very efficient process is based on the hydrogenation of phenol. The cyclohexanone produced is further reacted to produce cyclohexanone oxime, an intermediate which then can undergo a Beckmann rearrangement to yield caprolactam (see IARC, 1986; Considine, 1974; Fisher & Van Peppen, 1979), the important intermediate for nylon 6 (see IARC, 1979).

Cyclohexanone production and consumption are determined by the demand for raw materials for nylon. Other uses are minor and have little effect on overall production.

In 1979, approximately 318 000 tonnes of cyclohexanone were produced in the USA (Mannsville Chemical Products Corp., 1979). The US International Trade Commission (1985, 1986, 1987) reported production of approximately 360 000 tonnes each year in 1984 and 1985 and 404 000 tonnes in 1986.

Production of cyclohexanone elsewhere in the world has not been documented.

(b) Use

Cyclohexanone is used predominantly (about 95% of production in the USA) for the synthesis of raw materials used in the production of nylon. The remainder is used as a chemical intermediate in other processes, as an additive or as a high–boiling, slow–drying solvent.

---

[1]Calculated from: mg/m^3 = (molecular weight/24.45) x ppm, assuming standard temperature (25 °C) and pressure (760 mm Hg)

Cyclohexanone is used as a solvent in insecticides, wood stains, paint and varnish removers, spot removers, cellulosics, and natural and synthetic resins and lacquers. Additive uses include detergents, degreasing of metals, mould release agent for paints or varnishes, levelling agent in dyeing and delustering silk, and lube oil additive, especially for aircraft piston–type engines. Cyclohexanone is also used as a monomer in the synthesis of cyclohexanone resins, polyvinyl chloride and its copolymers (see IARC, 1979), and methacrylate ester polymers (International Technical Information Institute, 1979; Hawley, 1981; Windholz, 1983; Sittig, 1985; American Chemical Society, 1987).

(c) Regulatory status and guidelines

Occupational exposure limits for cyclohexanone in 28 countries or regions are presented in Table 1.

Table 1. Occupational exposure limits for cyclohexanone[a]

Country or region	Year	Concentration[b] (mg/m^3)	Interpretation[c]
Australia	1984	200	TWA
Austria	1985	200	TWA
Belgium	1984	200	TWA
Bulgaria	1984	10	TWA
China	1985	50	TWA
Commission of the European Communities	1986	200 1000	TWA Maximum
Czechoslovakia	1985	200 400	Average Maximum
Denmark	1988	100	TWA
Finland	1987	200 250	TWA STEL (15 min)
France	1986	100	TWA
Germany, Federal Republic of	1988	200	TWA
Hungary	1985	20 40	TWA TWA
Indonesia	1985	200	TWA
Italy	1984	200	TWA
Japan	1988	100	TWA
Mexico	1985	200	TWA
Netherlands	1986	200	TWA
Norway	1981	100	TWA
Poland	1984	20	TWA
Rumania	1984	100 200	Average Maximum
Sweden	1984	S 100 S 200	TWA STEL (15 min)
Switzerland	1985	100	TWA
Taiwan	1985	200	TWA
UK	1987	100 400	TWA STEL (10 min)

Table 1 (contd)

Country or region	Year	Concentration[b] (mg/m³)	Interpretation[c]
USA[d]			
OSHA	1983	200	TWA
NIOSH	1983	100	TWA
ACGIH	1988	S 100	TWA
USSR	1986	10	Ceiling
Venezuela	1985	200	TWA
Yugoslavia	1985	200	TWA

[a]From Direktoratet for Arbeidstilsynet (1981); US Occupational Safety and Health Administration (1983); International Labour Office (1984); Arbeidsinspectie (1986); Commission of the European Communities (1986); Institut National de Recherche et de Sécurité (1986); Cook (1987); Health and Safety Executive (1987); National Swedish Board of Occupational Safety and Health (1987); Työsuojeluhallitus (1987); American Conference of Governmental Industrial Hygienists (1988); Arbejdstilsynet (1988); Deutsche Forschungsgemeinschaft (1988)

[b]S, skin notation

[c]TWA, 8-h time-weighted average; STEL, short-term exposure limit

[d]OSHA, Occupational Safety and Health Administration; NIOSH, National Institute for Occupational Safety and Health; ACGIH, American Conference of Governmental Industrial Hygienists

2.2   Occurrence

(a)   *Natural occurrence*

Cyclohexanone is not known to occur as a natural product.

(b)   *Occupational exposure*

On the basis of a US National Occupational Exposure survey, the National Institute for Occupational Safety and Health (1983) estimated that 336 200 workers were potentially exposed to cyclohexanone in the USA in 1981-83.

Mean time-weighted average (TWA) concentrations of 6–28 ppm (24–112 mg/m³; personal samples) and 2.8–23.4 ppm (11–94 mg/m³; area samples) cyclohexanone were detected in a screen–printing plant (Samimi, 1982). Personal 8-h TWA air concentrations of cyclohexanone ranging from 0.4 to 1.1 ppm (1.6–4.4 mg/m³) with a mean of 0.7 ppm (2.8 mg/m³) and area samples containing 0.1–2.0 ppm (0.4–8.0 mg/m³) were reported in a plant that produced paper and vinyl wall coverings (Ordin *et al.*, 1986).

(c)   *Air*

Few data are available on ambient air concentrations of cyclohexanone. Its presence was reported in the air of one house near an offset printing office, but the concentration was not given (Verhoeff *et al.*, 1987).

2.3 Analysis

Cyclohexanone is readily analysed by collecting vapours in air samples by adsorption on chromosorb, desorption with carbon disulfide and determination by gas chromatography with flame–ionization detection. The detection limit was 0.8 ng (equivalent to 0.05 ppm; 0.2 mg/m^3) for a 10–l sample (Elskamp, 1979). Another air sampling method involved extraction of cyclohexanone in distilled water, condensation with furfural in an alkaline medium, acidification with sulfuric acid and colorimetric determination at 550 nm (Domanski, 1977).

An electrometric or colorimetric titration method, which can be used when no other carbonyl compound is present, is based on the reaction of cyclohexanone with hydroxylamine hydrochloride to form the oxime and hydrogen chloride (Fisher & Van Peppen, 1979).

3. Biological Data Relevant to the Evaluation of Carcinogenic Risk to Humans

3.1 Carcinogenicity studies in animals

Oral administration

Mouse: Groups of 52, 52 and 47 male and 52, 50 and 50 female B6C3F$_1$ mice, seven to eight weeks old, were given 0, 6500 or 13 000 mg/l (ppm) cyclohexanone (96% pure) in the drinking–water for 104 weeks. A further group of 41 female mice received 25 000 mg/l (maximum tolerated dose) cyclohexanone over the same period. Survival in the respective groups was 88%, 90% and 70% in males and 86%, 85%, 40% and 15% in females. The incidences of liver–cell adenomas or carcinomas [only combined figures reported] were 16/52, 25/51 and 13/46 in males and 3/52, 6/50, 3/50 and 2/41 in females, respectively. The incidence in low-dose males was statistically significant ($p = 0.041$, adjusted for differences in mortality). In female mice, a statistically significant increase ($p = 0.036$; life–table method) in the incidence of malignant lymphomas and leukaemia was observed in the low–dose group: 8/52 controls, 17/50 at 6500 ppm, 4/50 at 13 000 ppm and 0/41 at 25 000 ppm (Lijinsky & Kovatch, 1986).

Rat: Groups of 52 male and 52 female Fischer 344 rats, seven to eight weeks old, received 0, 3300 or 6500 mg/l (ppm; maximum tolerated dose) cyclohexanone (96% pure) in the drinking–water for 104 weeks. A slight decrease in survival was observed in high–dose females but was not statistically significant. Dose–related reductions in body weight were observed in treated groups. A significant increase ($p = 0.03$) in the incidence of adrenal cortical adenomas was observed in low–dose males (controls, 1/52; low–dose, 7/52; high–dose, 1/51). The incidences of thyroid follicular–cell adenomas–carcinomas [reported in combination] were 1/52, 0/51 and 6/51 ($p = 0.053$) in control, low–dose and high–dose males, respectively. No difference in the incidence of liver tumours was observed between treated and control groups (Lijinsky & Kovatch, 1986).

3.2 Other relevant data

(a) Experimental systems

(i) Absorption, distribution, excretion and metabolism

Cyclohexanone is metabolized in rats, rabbits and dogs to cyclohexanol, which is conjugated with glucuronic acid and excreted mainly in urine; very little cyclohexanone or free cyclohexanol is found in urine (Elliott et al., 1959; Martis et al., 1980; Greener et al., 1982). Cyclohexanone did not accumulate in the body (Martis et al., 1980).

(ii) Toxic effects

The acute oral LD_{50} for cyclohexanone has been reported to be 2.07 and 2.11 g/kg bw in male and female mice, respectively, and 1.80 g/kg bw in male and female rats. The intraperitoneal LD_{50} has been reported to be 1.23 g/kg bw in male mice, 1.13 g/kg bw in male rats, 1.54 g/kg bw in male rabbits and 0.93 g/kg bw in male guinea-pigs. Oral and intraperitoneal administration of cyclohexanone caused narcosis, and death due to central nervous system depression and respiratory arrest. Autopsy revealed peritoneal and intestinal congestion in mice, suggesting an irritant effect (Gupta et al., 1979).

Cyclohexanone did not induce skin allergy in the guinea-pig maximization test (Bruze et al., 1988). Application of 0.2 ml undiluted cyclohexanone to the shaved back of rabbits for 24 h induced marked irritation, which totally disappeared only six days later. Instillation of 99, 80 or 40% cyclohexanone in cottonseed oil caused eye irritation in rabbits. No significant difference was observed in the pentobarbital sleeping-time test in mice that received 120 or 250 mg/kg bw cyclohexanone intraperitoneally on three consecutive days, suggesting no major effect of the compound on hepatic drug metabolizing enzymes (Gupta et al., 1979). These enzymes were not induced by cyclohexanone in beagle dogs (Martis et al., 1980).

Groups of mice received 400–47 000 mg/l (ppm) cyclohexanone in the drinking-water for 13 weeks; one-third of the females and two-thirds of the males in the highest-dose group died during treatment. One male in the group receiving 34 000 mg/l died; the other animals had 15–24% depression of body weight gain, depending on sex. With 47 000 mg/l, focal liver necrosis and hyperplasia in the thymus were observed in some animals. Pathological changes at lower doses were minimal (Lijinsky & Kovatch, 1986).

Exposure of rabbits by inhalation to about 12 000 mg/m³ cyclohexanone for 6 h per day on five days per week for three weeks and to 1200–5560 mg/m³ for ten weeks induced narcosis, loss of coordination and death (2/4 animals) only in the highest exposure group; slight conjunctival irritation was seen at doses of 1200–3000 mg/m³. No toxic effect was observed with exposure to 750 mg/m³ (Treon et al., 1943).

Intravenous administration of about 280 mg/kg bw per day cyclohexanone to beagle dogs for 18–21 days produced a moribund condition and central nervous system effects and liver and kidney toxicity. No significant change in body weight was observed (Koeferl et al., 1981).

Intravenous administration of 50 or 100 mg/kg bw cyclohexanone to rats for 28 consecutive days caused no significant ophthalmological or haematological toxicity or alterations in clinical chemistry, gross pathology or histopathology (Greener et al., 1982).

Guinea–pigs and rabbits were administered 0.5 or 5 mg/kg bw cyclohexanone intravenously or 0.5 ml percutaneously three times a week for three consecutive weeks; lenticular alterations (anterior subcapsular vacuoles) were observed in all groups of guinea–pigs but not in rabbits (Greener & Youkilis, 1984).

Electrophysiological and neuropathological examination of rats receiving intraperitoneal injections of 200 mg/kg bw cyclohexanone twice daily on five days per week for up to 13 weeks revealed no damage to the peripheral nervous system (Perbellini *et al.*, 1981).

(iii) *Effects on reproduction and prenatal toxicity*

Chick embryos were exposed to cyclohexanone vapours [concentration unspecified] either for 3 or 6 h prior to incubation or for 3, 6 or 12 h after 96 h of incubation. Growth retardation was noted in day-13 embryos following exposure for 3 or 6 h prior to or after incubation. In some hatchings exposed after incubation, an abnormal gait was seen (Griggs *et al.*, 1971).

Dietary administration of 1% cyclohexanone to TB or NMRI mice for several generations was reported to affect the viability and growth of first-generation males and females. No such effect was seen in animals of the second generation (Gondry, 1972). [The Working Group noted that the viability of both control and treated animals was low.]

CF1 mice received daily intraperitoneal injections of 50 mg/kg bw cyclohexanone for 28 days; beginning on the tenth day of treatment and throughout the exposure period, females were housed with an untreated male. On the last day of treatment, females were killed and the uterus examined for dead, resorbed and viable fetuses. No adverse effect was noted in the seven exposed litters (Hall *et al.*, 1974).

CD-1 mice were exposed by oral intubation to 0 (25 mice) or 800 (24 mice) mg/kg bw cyclohexanone per day on days 8–12 of gestation and the offspring were evaluated for growth and viability over the first three postnatal days. No treatment–related maternal or developmental effect was observed in this teratology screening assay (Chernoff & Kavlock, 1983), nor were effects detected when the offspring were observed until 250 days of age (Gray & Kavlock, 1984; Gray *et al.*, 1986).

In a similar study, groups of 28 ICR mice were exposed by oral intubation to 0 or 2200 mg/kg bw cyclohexanone per day on days 8–12 of gestation. The treatment was lethal to 6/28 females, but no maternal mortality was observed in the control group. Maternal weight gain during the treatment period was significantly reduced in the treated group, and two females had completely resorbed litters. Pup weight at birth and on postnatal day 3 was significantly reduced in the treated group, but litter size and viability were normal (Seidenberg *et al.*, 1986)

(iv) *Genetic and related effects*

Cyclohexanone was not mutagenic to four strains (TA1535, TA1537, TA98 and TA100) of *Salmonella typhimurium* in the presence or absence of an exogenous metabolic system in a plate incorporation assay (Haworth *et al.*, 1983).

It was reported in an abstract (Aaron *et al.*, 1985) that exposure of Chinese hamster ovary cells to cyclohexanone just as they were entering the S-phase induced sister chromatid exchange and gene mutation in the absence, but not in the presence of an exogenous metabolic system. Under these conditions, no chromosomal aberration was induced in the presence or absence of an exogenous metabolic system.

Cyclohexanone at 10^{-2}, 10^{-3} and 10^{-4} M induced chromosomal aberrations in cultured human leucocytes (Collin, 1971; Lederer et al., 1971). It also produced an increase in the frequency of chromosomal damage in human lymphocytes both in terms of ploidy and structural changes (Dyshlovoi et al., 1981).

Chromosomal abnormalities were induced in bone–marrow cells of male rats (*Rattus norvegicus*) 6, 24 and 48 h after subcutaneous injection of three doses each of 0.1, 0.5 and 1.0 g/kg bw cyclohexanone (maximum tolerated dose). Abnormalities increased with dose and decreased with time, and consisted of chromatid gaps, breaks, centric fusions, centromeric attenuation, chromatid exchanges and polyploidy (de Hondt et al., 1983).

(*b*) *Humans*

(i) *Absorption, distribution, excretion and metabolism*

No data were available to the Working Group.

(ii) *Toxic effects*

Allergic contact dermatitis to a cyclohexanone resin [composition unspecified] was reported on patch testing of five patients with paint–related allergies (Bruze et al., 1988). Irritation of the eyes, nose and throat was described in a review in volunteers exposed to cyclohexanone (Krasavage et al., 1982).

In 100 workers exposed by inhalation (3.7 mg/m³) and *via* skin contact (10^{-4} mg/cm² on the hands) during the production of caprolactam to cyclohexanone, no difference in nervous system function, blood and respiration was reported relative to 49 controls. There was some indication of liver disorders among a subgroup of workers, 30–39 years old, with more than five years' exposure to cyclohexanone (Bereznyak, 1984).

(iii) *Effects on fertility and on pregnancy outcomes*

No data were available to the Working Group.

(iv) *Genetic and related effects*

No data were available to the Working Group.

3.3 Epidemiological studies of carcinogenicity to humans

No data were available to the Working Group.

4. Summary of Data Reported and Evaluation

4.1 Exposures

Cyclohexanone is a synthetic organic liquid used primarily as an intermediate in the production of nylon. Other minor applications are as an intermediate, additive and solvent in a variety of products. Occupational exposure levels have been measured in some industries.

4.2 Experimental carcinogenicity data

Cyclohexanone was tested for carcinogenicity by oral administration in the drinking-water in one strain of mice and one strain of rats. In mice, there was a slight increase in the incidence of tumours that occur commonly in this strain, only in animals given the low dose. In rats, a slight increase in the incidence of adrenal cortical adenomas occurred in males treated with the low dose.

4.3 Human carcinogenicity data

No data were available to the Working Group.

4.4 Other relevant data

No significant systemic toxicity was reported in humans or experimental animals. No significant prenatal toxicity was observed in mice.

Cyclohexanone induced chromosomal aberrations and ploidy changes in cultured human cells and in rats. It did not induce mutation in bacteria. (See Appendix 1.)

4.5 Evaluation[1]

There is *inadequate evidence* for the carcinogenicity of cyclohexanone in experimental animals.

No data were available from studies in humans on the carcinogenicity of cyclohexanone.

Overall evaluation

Cyclohexanone *is not classifiable as to its carcinogenicity to humans (Group 3)*.

5. References

Aaron, C.S., Brewen, J.G., Stetka, D.G., Bleicher, W.T. & Spahn, M.C. (1985) Comparative mutagenesis in mammalian cells (CHO) in culture: multiple genetic endpoint analysis of cyclohexanone *in vitro* (Abstract). *Environ. Mutagenesis, 7 (Suppl. 3)*, 60–61

Aldrich Chemical Co., Inc. (1988) *Aldrich Catalog Handbook of Fine Chemicals*, Milwaukee, WI, p. 425

American Chemical Society (1987) *Chemcyclopedia 87*, Washington DC, p. 62

---

[1]For definitions of the italicized terms, see Preamble, pp. 27–30.

American Conference of Governmental Industrial Hygienists (1988) *Threshold Limit Values and Biological Exposure Indices for 1988–1989*, Cincinnati, OH, p. 16

Arbeidsinspectie (Labour Inspection) (1986) *De Nationale MAC-Lijst 1986* [National MAC-List 1986] (*P145*), Voorburg, Ministry of Social Affairs and Work Environment, p. 10

Arbejdstilsynet (Labour Inspection) (1988) *Graensevaerdier for Stoffer og Materialer* [Limit Values for Substances and Materials] (*At-anvisning No. 3.1.0.2*), Copenhagen, p. 15

Bereznyak, I.V. (1984) Hazards of cyclohexanone penetration through the skin of workers engaged in caprolactam production (Russ.). *Gig. Tr. prof. Zabol.*, 3, 52–54

Bruze, M., Boman, A., Bergqvist-Karlson, A., Björkner, B., Wahlberg, J.E. & Voog, E. (1988) Contact allergy to a cyclohexanone resin in humans and guinea-pigs. *Contact Derm.*, 18, 46–49

Chernoff, N. & Kavlock, R.J. (1983) A teratology test system which utilizes postnatal growth and viability in the mouse. In: Waters, M., Sandhy, S., Lewtas, J., Claxton, L., Chernoff, N. & Nesnow, S., eds, *Short-term Bioassays in the Analysis of Complex Environmental Mixtures*, New York, Plenum, pp. 417–427

Collin, J.-P. (1971) Cytogenetic effect of sodium cyclamate, cyclohexanone and cyclohexanol (Fr.). *Diabète*, 19, 215–221

Commission of the European Communities (1986) List of occupational limit values. *Off. J. Eur. Commun.*, C164, 6–7

Considine, D.M., ed. (1974) *Chemical and Process Technology Encyclopedia*, New York, McGraw Hill, pp. 337–338

Cook, W.A. (1987) *Occupational Exposure Limits – Worldwide*, Washington DC, American Industrial Hygiene Association, pp. 119, 177

Deutsche Forschungsgemeinschaft (German Research Society) (1988) *Maximale Arbeitsplatzkonzentrationen und Biologische Arbeitsstofftoleranzwerte 1988* [Maximal Concentrations in the Workplace and Biological Tolerance Values for Working Materials 1988] (*Report No. XXIV*), Weinheim, VCH Verlagsgesellschaft, p. 27

Direktoratet for Arbeidstilsynet (Directorate for Labour Inspection) (1981) *Administrative Normer for Forurensning i Arbeidsatmosfaere 1981* [Administrative Norms for Pollution in Work Atmosphere 1981] (*No. 361*), Oslo, p. 8

Domanski, W. (1977) Determination of cyclohexanone in air by the furfural method. *Pr. Cent. Inst. Ochr. Pr.*, 27, 181–188

Dyshlovoi, V.D., Boiko, N.L., Shemetun, A.M. & Kharchenko, T.I. (1981) Cytogenetic action of cyclohexanone (Russ.). *Gig. Sanit.*, 5, 76–77

Eastman Kodak Co. (1985) *Kodak Laboratory Chemicals*, Rochester, NY, p. 148

Elliott, T.H., Parke, D.V. & Williams, R.T. (1959) The metabolism of cyclo[^{14}C]hexane and its derivatives. *Biochem. J.*, 72, 193–200

Elskamp, C.J. (1979) Cyclohexanone. In: *OSHA Analytical Methods Manual, 1985*, Salt Lake City, UT, Organic Methods Evaluation Branch, Occupational Safety and Health Administration, p. 01–1

Fisher, W.B. & Van Peppen, J.F. (1979) Cyclohexanone. In: Mark, H.F., Othmer, D.F., Overberger, C.G., Seaborg, G.T. & Grayson, M., eds, *Kirk-Othmer Encyclopedia of Chemical Technology*, 3rd ed., Vol. 7, New York, John Wiley & Sons, pp. 413–416

Gondry, E. (1972) Studies on the toxicity of cyclohexylamine, cyclohexanone and cyclohexanol, metabolites of cyclamate (Fr.). *J. exp. Toxicol.*, 5, 227–238

Gray, L.E., Jr & Kavlock, R.J. (1984) An extended evaluation of an in vivo teratology screen utilizing postnatal growth and viability in the mouse. *Teratog. Carcinog. Mutagenesis*, 4, 403–426

Gray, L.E., Jr, Kavlock, R.J., Ostby, J., Ferrell, J., Rogers, J. & Gray, K. (1986) An evaluation of figure-eight maze activity and general behavioral development following prenatal exposure to forty chemicals: effects of cytosine arabinoside, dinocap, nitrofen and vitamin A. *Neurotoxicology, 7*, 449–462

Greener, Y. & Youkilis, E. (1984) Assessment of the cataractogenic potential of cyclohexanone in guinea pigs and rabbits. *Fundam. appl. Toxicol., 4*, 1055–1066

Greener, Y., Martis, L. & Indacochea–Redmond, N. (1982) Assessment of the toxicity of cyclohexanone administered intravenously to Wistar and Gunn rats. *J. Toxicol. environ. Health, 10*, 385–396

Griggs, J.H., Weller, E.M., Palmisano, P.A. & Niedermeier, W. (1971) The effect of noxious vapors on embryonic chick development. *Am. J. med. Sci., 8*, 342–345

Gupta, P.K., Lawrence, W.H., Turner, J.E. & Autian, J. (1979) Toxicological aspects of cyclohexanone. *Toxicol. appl. Pharmacol., 49*, 525–533

Hall, I.H., Carlson, G.L., Abernethy, G.S. & Piantadosi, C. (1974) Cycloalkanones. 4. Antifertility activity. *J. med. Chem., 17*, 1253–1257

Hawley, G.G. (1981) *The Condensed Chemical Dictionary*, 10th ed., New York, Van Nostrand Reinhold, p. 297

Haworth, S., Lawlor, T., Mortelmans, K., Speck, W. & Zeiger, E. (1983) *Salmonella* mutagenicity test results for 250 chemicals. *Environ. Mutagenesis, Suppl. 1*, 3–142

Health and Safety Executive (1987) *Occupational Exposure Limits 1987 (Guidance Note EH 40/87)*, London, Her Majesty's Stationery Office, p. 12

de Hondt, H.A., Temtamy, S.A. & Abd–Aziz, K.B. (1983) Chromosomal studies on laboratory rats (*Rattus norvegicus*) exposed to an organic solvent (cyclohexanone). *Egypt. J. genet. Cytol., 12*, 31–40

IARC (1979) *IARC Monographs on the Evaluation of the Carcinogenic Risk of Chemicals to Humans*, Vol. 19, *Some Elastomers, Plastics and Synthetic Elastomers, and Acrolein*, Lyon, pp. 120–130

IARC (1986) *IARC Monographs on the Evaluation of the Carcinogenic Risk of Chemicals to Humans*, Vol. 39, *Some Chemicals Used in Plastics and Elastomers*, Lyon, pp. 247–276

Institut National de Recherche et de Sécurité (National Institute for Research and Safety) (1986) *Valeurs Limites pour les Concentrations des Substances Dangereuses Dans l'Air des Lieux de Travail* [Limit Values for Concentrations of Dangerous Substances in the Air of Work Places] (*ND 1609–125–86*), Paris, p. 563

International Labour Office (1984) *Occupational Exposure Limits for Airborne Toxic Substances*, 2nd rev. ed. (*Occupational Safety and Health Series No. 37*), Geneva, pp. 82–83

International Technical Information Institute (1979) *Toxic and Hazardous Industrial Chemicals Safety Manual for Handling and Disposal with Toxicity and Hazard Data*, Tokyo, pp. 144–145

Koeferl, M.T., Miller, T.R., Fisher, J.D., Martis, L., Garvin, P.J. & Dorner, J.L. (1981) Influence of concentration and rate of intravenous administration on the toxicity of cyclohexanone in beagle dogs. *Toxicol. appl. Pharmacol., 59*, 215–229

Krasavage, W.J., O'Donoghue, J.L. & Divincenzo, G.D. (1982) Ketones: cyclohexanone. In: Clayton, G.D. & Clayton, F.E., eds, *Patty's Industrial Hygiene and Toxicology*, 3rd ed., Vol. 2C, New York, John Wiley & Sons, pp. 4722–4723, 4780–4782

Lederer, J., Collin, J.-P., Pottier-Arnould, A.-M. & Gondry, E. (1971) Cytogenetic and teratogenic effect of cyclamate and its metabolites (Fr.). *Thérapeutique, 47*, 357–363

Lijinsky, W. & Kovatch, R.M. (1986) Chronic toxicity study of cyclohexanone in rats and mice. *J. natl Cancer Inst., 77*, 941–949

Mannsville Chemical Products Corp. (1979) *Chemical Products Synopsis: Cyclohexanone*, Cortland, NY

Martis, L., Tolhurst, T., Koeferl, M.T., Miller, T.R. & Darby, T.D. (1980) Disposition kinetics of cyclohexanone in beagle dogs. *Toxicol. appl. Pharmacol.*, *55*, 545–553

National Institute for Occupational Safety and Health (1983) *National Occupational Exposure Survey 1981–83*, Cincinnati, OH

National Swedish Board of Occupational Safety and Health (1987) *Hygienska Gränsvärden* [Hygienic Limit Values] *(Ordinance 1987:12)*, Solna, p. 16

Ordin, D.L., Seixas, N.S. & Liveright, T. (1986) *Laminating Corporation of America, Eatontown, NJ (Health Hazard Evaluation Determination Report No. 83–270–1656)*, Cincinnati, OH, National Institute for Occupational Safety and Health

Perbellini, L., De Grandis, D., Semenzato, F. & Bongiovanni, L.G. (1981) Experimental study on the neurotoxicity of cyclohexanol and cyclohexanone (Ital.). *Med. Lav.*, *2*, 102–107

Pouchert, C.J., ed. (1981) *The Aldrich Library of Infrared Spectra*, 3rd ed., Milwaukee, WI, Aldrich Chemical Co., pp. 256B

Pouchert, C.J., ed. (1983) *The Aldrich Library of NMR Spectra*, 2nd ed., Vol. 1, Milwaukee, WI, Aldrich Chemical Co., pp. 394C

Pouchert, C.J., ed. (1985) *The Aldrich Library of FT–IR Spectra*, Vol. 1, Milwaukee, WI, Aldrich Chemical Co., pp. 432C

Riedel–de Haën (1984) *Riedel–de Haën Laboratory Chemicals*, Hanover, p. 280

Sadtler Research Laboratories (1980) *Standard Spectra Collection, 1980 Cumulative Index*, Philadelphia, PA

Samimi, B. (1982) Exposure to isophorone and other organic solvents in a screen printing plant. *Am. ind. Hyg. Assoc. J.*, *43*, 43–48

Seidenberg, J.M., Anderson, D.G. & Becker, R.A. (1986) Validation of an in vivo developmental toxicity screen in the mouse. *Teratog. Carcinog. Mutagenesis*, *6*, 361–374

Sittig, M. (1985) *Handbook of Toxic and Hazardous Chemicals and Carcinogens*, 2nd ed., Park Ridge, NJ, pp. 280–281

Treon, J.F., Crutchfield, W.E., Jr & Kitzmiller, K.V. (1943) The physiological response of animals to cyclohexane, methylcyclohexane, and certain derivatives of these compounds. *J. ind. Hyg. Toxicol.*, *25*, 323–347

Työsuojeluhallitus (National Finnish Board of Occupational Safety and Health) (1987) *HTP–Arvot 1987* [TLV Values 1987] *(Safety Bulletin No. 25)*, Helsinki, Valtion Painatuskeskus, p. 24

US International Trade Commission (1985) *Synthetic Organic Chemicals, US Production and Sales, 1984 (USITC Publ. 1745)* Washington DC, US Government Printing Office

US International Trade Commission (1986) *Synthetic Organic Chemicals, US Production and Sales, 1985 (USITC Publ. 1892)*, Washington DC, US Government Printing Office

US International Trade Commission (1987) *Synthetic Organic Chemicals, US Production and Sales, 1986 (US ITC Publ. 2009)*, Washington DC, US Government Printing Office

US Occupational Safety and Health Administration (1983) Air contaminants. *US Code fed. Regul.*, *Title 29*, No. 1910.1000, p. 600

Verhoeff, A.P., Wilders, M.M.W., Monster, A.C. & Van Wijnen, J.H. (1987) Organic solvents in the indoor air of two small factories and surrounding houses. *Int. Arch. occup. environ. Health*, *59*, 153–163

Weast, R.C., ed. (1985) *Handbook of Chemistry and Physics*, 66th ed., Cleveland, OH, CRC Press, p. C–228

Windholz, M., ed. (1983) *The Merck Index*, 10th ed., Rahway, NJ, Merck & Co., p. 391

DIMETHYLFORMAMIDE

1. Chemical and Physical Data

1.1 Synonyms

Chem. Abstr. Services Reg. No.: 68–12–2
Chem. Abstr. Name: N,N–Dimethylformamide
Synonyms: N,N–Dimethylmethanamide; DMF; DMFA; DMF (amide); N–formyldimethylamine

1.2 Structural and molecular formulae and molecular weight

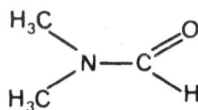

C₃H₇NO   Mol. wt: 73.1

1.3 Chemical and physical properties of the pure substance

From E.I. duPont de Nemours & Co. (1986) unless otherwise specified.

(a) *Description*: Clear, colourless hygroscopic liquid (Eberling, 1980) with slight amine odour (Windholz, 1983)
(b) *Boiling–point*: 153.0 °C
(c) *Freezing–point*: –61.0 °C
(d) *Flash–point*: 67 °C (open–cup); 58 °C (closed–cup)
(e) *Density*: 0.949 g/ml at 20 °C
(f) *Viscosity*: 0.802 cp at 25 °C
(g) *Spectroscopy data*: Nuclear magnetic resonance and infrared spectral data have been reported (Sadtler Research Laboratories, 1980; Pouchert, 1981, 1983, 1985)
(h) *Solubility*: Soluble in water, acetone, alcohols, benzene, chloroform, diethyl ether, esters and chlorinated and aromatic hydrocarbons; limited solubility in aliphatic hydrocarbons

(i)   *Refractive index*: 1.428 at 25°C

(j)   *Volatility*: Vapour pressure, 3.7 mm Hg at 25°C

(k)   *Stability*: Photodegrades when exposed to ultraviolet radiation (or strong sun-light), with formation of dimethylamine and formaldehyde (see IARC, 1987)

(l)   *Reactivity*: Reacts violently when mixed with oxidizing agents, such as perchlo-rates, nitrates, permanganates, chromates, nitric acid, chromic acid, halogens and some cleaning solutions; may cause fire or explosion when reacted with any halo-genated hydrocarbon in the presence of metal; generates carbon monoxide va-pours when heated to decomposition (E.I. DuPont de Nemours & Co., 1988a); can attack copper, brass and other copper alloys (Eberling, 1980)

(m)   *Octanol/water partition coefficient*: log P = −1.01 (Hansch & Leo, 1979)

(n)   *Conversion factor*: mg/m^3 = 2.99 x ppm[1]

1.4 Technical products and impurities

Dimethylformamide is available commercially with the following specifications: puri-ty, approximately 99.9%; water, 0.03–0.05% (max), typically, 0.01%; N–methylformamide, 100 ppm (max); dimethylamine, 15–20 ppm (max), typically, 6 ppm; iron, 0.05 ppm (max), typi-cally 0.01 ppm; methanol, 100 ppm (max); formic acid, 20 ppm (max), typically, 7 ppm (Eberl-ing, 1980; Air Products and Chemicals, Inc., 1985; E.I. duPont de Nemours & Co., 1986).

2. Production, Use, Occurrence and Analysis

2.1 Production and use

(a) Production

Dimethylformamide was first synthesized in 1893. In a one–stage process, a solution of dimethylamine in methanol reacts with carbon monoxide in the presence of sodium methy-late or with metal carbonyls at 110–150°C and pressures of 1.5–2.5 MPa (15–25 atm). In the two–stage process, methyl formate is first produced from carbon monoxide and methanol under high pressure at 60–100°C in the presence of sodium methylate. The methyl formate is distilled and then reacts with dimethylamine at 80–100°C and low pressure. The product is purified by distillation (Eberling, 1980).

Worldwide production capacity was estimated to be about 225 000 tonnes in 1979, ap-proximately half of which was located in Europe (Eberling, 1980). By 1983, estimated world-

---

[1]Calculated from: mg/m^3 = (molecular weight/24.45) x ppm, assuming standard temperature (25°C) and pres-sure (760 mm Hg)

wide capacity had dropped to 181 600 tonnes and worldwide production was only about 100 000 tonnes. These decreases were a result of a decline in consumer demand for 'wet-look' fabrics. In 1983, production capacity in North America was 54 400 tonnes. US consumption was about 18 100 tonnes in 1977–78 and decreased to 13 600 tonnes in 1983. US production of dimethylformamide was estimated to be 23 000–27 000 tonnes in 1987 (E.I. DuPont de Nemours & Co., 1988b).

Mexico, Taiwan, Brazil and the Republic of Korea were estimated to have a combined capacity for dimethylformamide production of 29 500–31 800 tonnes in 1983 (Anon., 1983). Total production capacity for dimethylformamide in Japan was estimated in 1985 to be 41 000 tonnes per year; 60% was used for the production of polyurethane artificial leather, 30% for export to North America and south–east Asia and the rest used as solvents for fabric materials and resins (Anon., 1985).

(b) Use

(i) Polymer and resin solvent

Dimethylformamide is used as a solvent for many vinyl–based polymers in the manufacture of films, fibres and coatings, and as a booster or cosolvent for both high molecular-weight polyvinyl chlorides and vinyl chloride–vinyl acetate copolymers in the manufacture of protective coatings, films, printing inks and adhesive formulations. Since it is a highly polar solvent capable of hydrogen bonding, it is effective as a solvent for polar polymers with strong intermolecular forces. Dimethylformamide is used as a solvent for making polyurethane lacquers for clothing and accessories made of synthetic leather, and its use in leather tanneries has been reported (Levin et al., 1987). Dimethylformamide has been used as a solvent for certain epoxy resin curing agents, such as dicyandiamide and meta-phenylenediamine, and acts as a catalyst in accelerating cure at elevated temperatures. It has been widely used as a solvent in the production of fibres and films based on polyacrylonitrile (E.I. duPont de Nemours & Co., 1986).

(ii) Separations

Dimethylformamide is used commercially as a selective solvent to recover high purity acetylene from hydrocarbon feed streams. It is also used as a scrubbing solvent for the purification of ethylene and propylene, and has become a major solvent for extracting and separating butadiene from hydrocarbon streams (E.I. duPont de Nemours & Co., 1986).

(iii) Selective solvent extractions

Dimethylformamide is used in petroleum processing for the separation of non–paraffinic from paraffinic hydrocarbons and is the preferred solvent for extraction of condensed-ring polycyclic aromatic compounds from wax. Aqueous dimethylformamide has been used as a selective solvent for the separation of polycarboxylic acids, such as isophthalic from terephthalic acid, brassylic from azelaic acid and sebacic from adipic acid and fatty acid oxidation products (E.I. duPont de Nemours & Co., 1986).

(iv) Miscellaneous

Dimethylformamide has been used as a reactant in many organic synthetic preparations, as a component in cold formulation industrial paint strippers and as a solvent for electrolytes, particularly in high–voltage capacitors. Dimethylformamide is also used as a combination quench and solvent cleaner for hot–dipped tinned articles (E.I. duPont de Nemours & Co., 1986).

(c)   *Regulatory status and guidelines*

The US Food and Drug Administration (1988) permits the use of dimethylformamide as a component of adhesives used in articles intended for use in packaging, transporting or holding food.

Occupational exposure limits for dimethylformamide in 29 countries or regions are presented in Table 1.

Table 1. Occupational exposure limits for dimethylformamide[a]

Country	Year	Concentration[b] (mg/m^3)	Interpretation[c]
Australia	1984	S 30	TWA
Austria	1985	S 60	TWA
Belgium	1985	S 30	TWA
Brazil	1985	24	TWA
Bulgaria	1985	S 10	TWA
China	1985	S 10	TWA
Czechoslovakia	1985	30	Average
		60	Maximum
Denmark	1988	S 30	TWA
Finland	1987	S 30	TWA
		S 60	STEL (15 min)
France	1986	S 30	TWA
German Democratic Republic	1985	S 10	TWA
Germany, Federal Republic of	1988	S 60	TWA
Hungary	1985	S 10	TWA
		S 20	STEL
Indonesia	1985	S 30	TWA
Italy	1985	S 30	TWA
Japan	1988	S 30	TWA
Mexico	1985	60	TWA
Netherlands	1986	S 30	TWA
Norway	1981	S 30	TWA
Poland	1985	10	TWA
Rumania	1985	S 20	Average
		S 50	Maximum
Sweden	1987	S 30	TWA
		S 45	STEL (15 min)
Switzerland	1985	S 30	TWA
Taiwan	1985	S 30	TWA
UK	1987	S 30	TWA
		S 60	STEL (10 min)

Table 1 (contd)

Country	Year	Concentration[b] (mg/m^3)	Interpretation[c]
USA[d]			
OSHA	1987	30	TWA
ACGIH	1988	S 30	TWA
USSR	1986	10	Ceiling
Venezuela	1985	S 30	TWA
		S 60	Ceiling
Yugoslavia	1985	S 10	TWA

[a]From Direktoratet for Arbeidstilsynet (1981); International Labour Office (1984); Arbeidsinspectie (1986); Institut National de Recherche et de Sécurité (1986); Cook (1987); Health and Safety Executive (1987); National Swedish Board of Occupational Safety and Health (1987); Työsuojeluhallitus (1987); US Occupational Safety and Health Administration (1987); American Conference of Governmental Industrial Hygienists (1988); Arbejdstilsynet (1988); Deutsche Forschungsgemeinschaft (1988)
[b]S, skin notation
[c]TWA, 8-h time-weighted average; STEL, short-term exposure limit
[d]OSHA, Occupational Safety and Health Administration; ACGIH, American Conference of Governmental Industrial Hygienists

2.2 Occurrence

(a) Natural occurrence

Dimethylformamide is not known to occur as a natural product.

(b) Occupational exposure

On the basis of a US National Occupational Exposure Survey, the National Institute for Occupational Safety and Health (1983) estimated that 94 000 workers were potentially exposed to dimethylformamide in the USA in 1981–83. Levels of exposure to dimethylformamide are given in Table 2.

(c) Environmental occurrence

No data were available to the Working Group on the environmental occurrence of dimethylformamide.

2.3 Analysis

Methods have been reported for the analysis of dimethylformamide in air and water, and as its metabolite, methylformamide in biological media. Dimethylformamide has been determined in air by drawing air samples through charcoal or silica gel adsorption tubes, desorption with an appropriate solvent and analysis by gas chromatography with flame-ionization detection or high-pressure liquid chromatography. Lower limits of detection for these

Table 2. Occupational exposure to dimethylformamide

Environment	Biological monitoring[a] (no. of samples)	Concentration in air[b] and monitoring method	Reference
Solvent extraction in a chemical plant	Analysed as MF in urine < 10 μl/l (288) > 20 μl/l (15) 54 μl/l (1) 62 μl/l (1) 77 μl/l (1)	ND–200 ppm (600 mg/m³); area samples by detector tube	Lyle et al. (1979)
Polyurethane surface-treatment of synthetic leather	0.4–19.6 mg MF in urine/ day	ND–5.13 ppm (15 mg/m³); 8-h TWA personal samples	Yonemoto & Suzuki (1980)
Artificial leather factory	0.4–7.1 mg/m³; mean, 1.5 mg/m³ DMF in alveolar air	1.1–20.9 mg/m³; mean, 5.3 mg/m³; 8-h TWA personal samples	Brugnone et al. (1980)
Polyurethane lacquering for textile substrate	12 mg/l MF in urine	Mean, 28.4 mg/m³; 8-h TWA personal samples	Pozzoli et al. (1981)
Polyurethane production plant		Mean, 1.3–1.8 mg/m³; 8-h TWA personal samples 1.3 mg/m³; area samples	Rimatori & Carelli (1982)
Textile dye manufacturing plant		0.83 mg/m³; 8-h TWA 15.6 mg/m³; area sample	Zey et al. (1987)
Synthetic fibre plant	8.9–13.2 μg/l MF in urine (geometric mean)	3.4–3.6 mg/m³; 8-h TWA (geometric mean)	Dixon et al. (1983)
Acrylic fibre plant	10.3–63 mg MF/g creatine, daily means	Mean, 1.0–46.6 mg/m³; 8-h TWA area samples	Lauwerys et al. (1980)
Amine processing plant		mean, 12.3 mg/m³	Berger et al. (1985)

[a]MF, methylformamide; DMF, dimethylformamide
[b]ND, not detected

methods are in the range of 0.5–1.0 mg/m³ (Lipski, 1982; Rimatori & Carelli, 1982; Eller, 1985; Guenier et al., 1986; Stránsky, 1986). Colorimetric detection systems have been developed for dimethylformamide in air (Matheson Gas Products, undated; Roxan, Inc., undated; The Foxboro Co., 1983; Sensidyne, 1985; National Draeger, Inc., 1987; SKC, 1988).

Gas chromatography with flame–ionization or mass spectrometric detection has been used for the analysis of aqueous solutions of dimethylformamide by direct injection (Kubelka et al., 1976).

A method has been reported for the direct analysis of dimethylformamide in breath samples using a modified portable quadrupole mass spectrometer. The lower limit of detection was 0.5 mg/m³ (Wilson & Ottley, 1981). Personal exposures to dimethylformamide have also been monitored by gas chromatographic analysis of urine for N–methylformamide

(Barnes & Henry, 1974). This compound has been shown, however, to originate mainly from thermal degradation during the analysis of N-hydroxymethyl-N-methylformamide, which is the main metabolite present in urine (Scailteur & Lauwerys, 1984).

3. Biological Data Relevant to the Evaluation of Carcinogenic Risk to Humans

3.1 Carcinogenicity studies in animals[1]

(a) Oral administration

Rat: One group of 15 and one group of five BD rats [sex and age unspecified] were given 75 or 150 mg/kg bw dimethylformamide [purity unspecified] per day in the drinking-water until a total dose of 38 g/kg bw had been given to both groups. The total experimental period was 107 weeks (mean survival time, 76 weeks). No tumour was observed (Druckrey *et al.*, 1967). [The Working Group noted the small number of animals used and the incomplete reporting of the results.]

(b) Subcutaneous administration

Rat: Two groups of 12 BD rats [sex and age unspecified] received weekly subcutaneous injections of 200 or 400 mg/kg bw dimethylformamide [purity unspecified] until total doses of 8 and 20 g/kg bw had been given, which was at 104 weeks for the low-dose group and 109 weeks for the high-dose group. No tumour was observed (Druckrey *et al.*, 1967). [The Working Group noted the small number of animals used and the incomplete reporting of the results.]

(c) Intraperitoneal administration

Rat: Groups of 20 male and 20 female MRC rats, 13–14 weeks of age, received weekly intraperitoneal injections of 0.1 ml dimethylformamide (distilled, gas chromatography grade) for ten weeks (total dose, 1 ml [949 mg]). A group of 15 male and 15 female rats served as untreated controls. Median survival times were 87 weeks for treated males and 96 weeks for treated females, 92 weeks for control males and 100 weeks for control females. The experiment was terminated at 115 weeks. In the treated groups, 9/18 males and 11/19 females had tumours at different sites; in the control groups, 4/14 males and 5/14 females had tumours. A total of 13 tumours (three malignant) occurred in treated males and 17 (nine malignant) in females; untreated males had four (benign) tumours and untreated females, eight

---

[1]The Working Group was aware of a study in progress in mice and rats by inhalation (IARC, 1988)

(two malignant). A few uncommon tumours were reported in treated animals: an embryonal-cell carcinoma of the testis in one male, and two colon adenocarcinomas and a squamous-cell carcinoma of the rectum in females (Kommineni, 1972). [The Working Group noted the small number of animals, the unequal group sizes, the short duration of treatment and the incomplete description of some of the pathological results.]

3.2 Other relevant data

The toxicology of dimethylformamide has been reviewed (Kennedy, 1986; Lauwerys, 1986).

(a) Experimental systems

(i) Absorption, distribution, excretion and metabolism

Dimethylformamide is readily absorbed by mammals following its oral administration, dermal contact or inhalation exposure (Massmann, 1956; Kimmerle & Eben, 1975a; Kennedy, 1986). After rats were exposed for 4 h by inhalation, dimethylformamide and its main metabolite were distributed uniformly throughout the tissues; almost all was removed within two days (Lundberg et al., 1983).

The main metabolic pathway of dimethylformamide in rodents involves hydroxylation of the methyl group to form N-hydroxymethyl-N-methylformamide (Brindley et al., 1983; Scailteur & Lauwerys, 1984; Scailteur et al., 1984). Liver is the main organ in which metabolism occurs (Scailteur et al., 1984). Other metabolites excreted in rodent urine include N-methylformamide (Scailteur & Lauwerys, 1984), monomethylamine and dimethylamine, each of which constituted less than 5% of the administered dose (Kestell et al., 1987). Some unmetabolized dimethylformamide is also excreted, to a greater extent in female rats than males (Scailteur et al., 1984). When ^{14}C-dimethylformamide (labelled in the formyl group) was administered to mice, 83% of the dose was recovered in urine within 24 h. Of this amount, 56% was excreted as N-hydroxymethyl-N-methylformamide and 5% as unmetabolized dimethylformamide; 3% of the dose administered was excreted as N-(hydroxymethyl)formamide or formamide and 18% as unidentified metabolites (Brindley et al., 1983). [The Working Group noted that, until recently, N-methylformamide was considered to be the main metabolite; however, N-hydroxymethyl-N-methylformamide is broken down to N-methylformamide during gas chromatographic analysis.]

Dimethylformamide has been shown to cross the placenta after exposure of rats by inhalation (Sheveleva et al., 1977).

(ii) Toxic effects

The oral LD_{50} for dimethylformamide has been reported to be 3.8–6.8 g/kg bw in mice, 2.0–7.6 g/kg bw in rats, 3.4 g/kg bw in guinea-pigs and 3–4 g/kg bw in gerbils. The intraperitoneal LD_{50} has been reported to be 1.1–6.2 g/kg bw in mice, 1.4–4.8 g/kg bw in rats, 4 g/kg bw in guinea-pigs, 3–4 g/kg bw in gerbils, 1 g/kg bw in rabbits and 0.3–0.5 g/kg bw in cats. The intravenous LD_{50} was 2.5–4.1 g/kg bw in mice, 2–3.0 g/kg bw in rats, 1.0 g/kg bw in guinea-pigs, 1–1.8 g/kg bw in rabbits and 0.5 g/kg bw in dogs. The subcutaneous LD_{50} was 3.5–6.5 g/kg bw in mice, 3.5–5 g/kg bw in rats, 2 g/kg bw in rabbits and 3–4 g/kg bw in gerbils. An intramuscular LD_{50} of 3.8–6.5 g/kg bw has been reported in mice, and dermal LD_{50}s of 11 and 1.5 g/kg bw have been reported for rats and rabbits, respectively (Massmann, 1956; Davis & Jenner, 1959; Thiersch, 1962; Kutzche, 1965; Druckrey et al., 1967; Spinazzola et al., 1969; Kimura et al., 1971; Llewellyn et al., 1974; Bartsch et al., 1976; Stula & Krauss, 1977; Kennedy, 1986). A 2–h inhalational LD_{50} of 9400 mg/m³ was reported in mice and a 4–h inhalation LD_{50} of > 2500 ppm (7500 mg/m³) in rats (Clayton et al., 1963). Dimethylformamide was more toxic in younger than in older rats, with oral LD_{50}s of < 1 g/kg bw in newborn, 1.4 g/kg bw in 14–day–old, 4.0 g/kg bw in young adult and 6.8 g/kg bw in adult animals (Kimura et al., 1971).

Rats survived a single 4–h exposure to saturated vapours of dimethylformamide [dose unspecified] (Smyth & Carpenter, 1948); no mortality was observed when rats were exposed to 2500 ppm saturated vapours of dimethylformamide for 4 h, but deaths occurred when the period was extended to 6 h (Clayton et al., 1963).

Slight skin irritation was observed after skin applications of 2.5 g/kg bw dimethylformamide to mice; no such irritation was found in rabbits similarly treated with 0.5 g/kg (Wiles & Narcisse, 1971). Moderate corneal injury and moderate to severe conjunctivitis were observed after application of 0.01 ml dimethylformamide on the corneal surface or of 50% in the conjunctival sac of rabbits (Massmann, 1956; Williams et al., 1982).

Feeding of dimethylformamide to mice (160, 540, 1850 mg/kg) and to rats (215, 750, 2500 mg/kg) in the diet for more than 100 days resulted in a slight increase in liver weights in both species but no evidence of histopathological damage in the liver or other tissues (Becci et al., 1983). When dimethylformamide was given to gerbils at concentrations of 10 000, 17 000, 34 000 and 66 000 mg/l in the drinking–water, mortality and severe liver toxicity (necrotic foci) were observed in a dose–dependent fashion at the three higher dose levels (Llewellyn et al., 1974).

In several experiments, rats were exposed by inhalation to 100–1200 ppm dimethylformamide, for up to about 120 days. Liver toxicity (as evaluated by clinical chemistry and/or gross pathological and histopathological examination) was seen after prolonged exposure and at higher concentrations (Massman, 1956; Clayton et al., 1963; Tanaka, 1971; Craig et al., 1984). Liver necrosis was also seen in mice given 150–1200 ppm (450–3600 mg/m³) dimethylformamide (Craig et al., 1984); and toxicity was observed in guinea–pigs after several daily intragastric administrations of 10 ml of the undiluted compound (Martelli, 1960). In one study, however, inhalation exposure of rats and cats to 1000 ppm (3000 mg/m³) for 6 h per day for two months induced no toxic effect in liver (Hofmann, 1960), and no macroscopic effect was seen in the liver of rats exposed to 600 ppm (1800 mg/m³) dimethylformamide (Schottek, 1970). After mice, rats, rabbits, guinea–pigs and dogs were exposed to 58 aerosolized doses of 23 ppm (69 mg/m³) dimethylformamide for 5.5 h and 426 ppm (1300 mg/m³) for a further 30 min, no adverse clinical sign was seen in rodents. One of four dogs had decreased systolic blood pressure, and all four had degenerative changes in heart muscle. Liver weights were elevated in all species, except guinea–pigs, and liver fat content was increased in rats. No other toxic change, as evaluated by haematology or tissue histopathology, was detected (Clayton et al., 1963).

Kidney toxicity was seen in gerbils given dimethylformamide in the drinking–water (17 000, 34 000 and 66 000 mg/l) for up to 80 days (Llewellyn et al., 1974) and in guinea–pigs given several daily oral administrations of 10 ml of the undiluted compound (Martelli, 1960). Exposure of rats and cats to 1000 ppm (3000 mg/m³) dimethylformamide by inhalation for 6 h per day for two months did not induce kidney toxicity (Hofmann, 1960).

(iii) *Effects on reproduction and prenatal toxicity*

As reported in an abstract, intraperitoneal administration of 1.24 ml (1.2 g)/kg bw dimethylformamide to NMRI mice on days 6–15 of gestation had no teratogenic effect, although monomethylformamide at a dose of 0.1 ml/kg induced a high incidence of fetal death and malformation (Gleich, 1974).

Groups of 12–30 AB Jena-Halle or C57Bl mice were given intraperitoneal injections of 170–2100 mg/kg bw dimethylformamide on either one or several days of gestation, and the fetuses were examined for growth, morphology and viability. Single injections of 2100 mg/kg bw into Jena-Halle strain mice on day 3, 7 or 9 of gestation were reported to be embryotoxic. [The Working Group noted that no statistical analysis was included in the table of experimental results and that it is not clear what the effect was.] Treatment of AB Jena Halle mice with 600 or 1080 mg/kg bw and of C57Bl mice with 1080 mg/kg bw on days 1–14 of gestation induced a high incidence of malformations in both strains. Defects included deficient ossification of the occipital and parietal bones, and open eyes (Scheufler & Freye, 1975).

Rats were exposed by inhalation to 0, 0.05 or 0.6 mg/m³ dimethylformamide for 4 h per day on days 1–19 of gestation. No maternal effect was observed, but fetal growth was reduced at the lower dose and growth retardation and postimplantation embryonic death were seen at the higher dose. The number of postnatal deaths was increased in the higher dose group (Sheveleva & Osina, 1973).

Groups of 22–23 Long-Evans rats were exposed by inhalation to 0, 18 or 172 ppm (54 or 515 mg/m³) dimethylformamide for 6 h per day on days 6–15 of gestation, and the fetuses were examined by routine teratological techniques. No clinical sign of systemic toxicity was reported in the exposed females, and no effect on fetal viability or morphology was observed. The growth of fetuses in the high–dose group was retarded, but they showed normal skeletal development (Kimmerle & Machemer, 1975).

As reported in an abstract, Sprague–Dawley rats were exposed to 0, 32 or 301 ppm (96 or 900 mg/m³) dimethylformamide vapours for 6 h per day on days 6–15 of gestation. Slight maternal toxicity and fetal growth retardation were reported at the highest dose level (Keller & Lewis, 1981).

Dimethylformamide was one of several acetamides and formamides administered in a teratology study by oral gavage to rabbits on days 6–18 of gestation. Doses were 0 (24 rabbits), 46.4 µl (44 mg)/kg bw (12 rabbits), 68.1 µl/kg (65 mg/kg) (18 rabbits) and 200 µl/kg (190 mg/kg) (11 rabbits). A dose–related increase in the incidence of internal hydrocephalus was noted in fetuses. In the high–dose group, maternal toxicity, abortion, retardation of fetal growth and additional malformations (umbilical hernia, eventratio simplex, exophthalmus, cleft palate and abnormal positioning of limbs) were also observed (Merkle & Zeller, 1980).

Groups of three to nine Sprague-Dawley rats and four to five New Zealand white rabbits received dermal application of dimethylformamide (commercial grade with less that 2% impurities). Rats were treated for several 1–3–day periods during the middle of gestation while rabbits were exposed on days 8–16. The administered dose was 200 mg/kg bw to rabbits and 600–2400 mg/kg bw to rats. It was reported that the test agent caused an increase in the rate of embryonic death in rats at a dose that also resulted in maternal mortality. Subcutaneous haemorrhages were observed in fetuses exposed during days 12 and 13 or 11–13, but the authors did not consider these to be toxicologically significant. No adverse effect was noted in the few rabbits that were studied (Stula & Krauss, 1977).

(iv) *Genetic and related effects*

Dimethylformamide was one of 42 chemicals selected for study in the International Collaborative Program for the Evaluation of Short-term Tests for Carcinogens (de Serres & Ashby, 1981), in which 30 assay systems were included and more than 50 laboratories contributed data. Dimethylformamide gave negative results in five studies for DNA repair in prokaryotes, 16 studies for mutation in bacteria, five studies for mutation or mitotic recombination in yeast, three studies for DNA repair in cultured human cells, three studies for sister chromatid exchange in cultured animal cells, one study for mutation in cultured animal cells, one study for mutation in cultured human cells, two studies for chromosomal aberrations in cultured animal cells, one study for sex–linked recessive lethal mutation in *Drosophila*, one study for sister chromatid exchange in bone marrow and liver of mice, three studies for micronuclei in mice, and one sperm morphology assay. In most of the in–vitro studies, dimethylformamide was tested both in the presence and absence of an exogenous metabolic system. Dimethylformamide gave inconclusive results in one study of lambda induction. It gave positive results in one study of differential toxicity in yeast. It induced mutation in *Salmonella typhimurium* TA1538 and TA98 in one test with metabolic activation. It induced DNA damage in *Saccharomyces cerevisiae* in one study and aneuploidy in *S. cerevisiae* D6 both in the presence and absence of an exogenous metabolic system in a single study. Dimethylformamide gave positive results in one study for mitotic recombination in yeast.

In many other studies, dimethylformamide did not induce mutation in *S. typhimurium* TA1530, TA1531, TA1532, TA1535, TA1537, TA1538, TA98, TA100 or TA1964 either in the presence or absence of an exogenous metabolic system (Green & Savage, 1978 [solvent control]; Purchase *et al.*, 1978; Antoine *et al.*, 1983; Falck *et al.*, 1985; Mortelmans *et al.*, 1986). Negative results were also obtained with *Escherichia coli* WP2*uvrA* in the presence of an exogenous metabolic system (Falck *et al.*, 1985). Dimethylformamide enhanced the mutagenicity of tryptophan–pyrolysate in *S. typhimurium* TA98 in the presence of an exogenous metabolic system (Arimoto *et al.*, 1982).

Dimethylformamide induced a slight increase in unscheduled DNA synthesis in primary rat hepatocyte cultures in one study (Williams, 1977) but not in two others (Williams & Laspia, 1979; Ito, 1982). It gave negative responses in the hepatocyte primary culture/DNA repair assay using mouse or hamster hepatocytes (McQueen *et al.*, 1983; Klaunig *et al.*, 1984).

Dimethylformamide had no effect on the frequency of recessive chlorophyll and embryonic lethal mutations in *Arabidopsis thaliana* (Gichner & Veleminský, 1987). In the same system, dimethylformamide altered the mutagenic activity of known mutagens (Gichner & Veleminský, 1986, 1987). It did not induce sex–linked recessive lethal mutations or somatic mutation in *Drosophila* (Fahmy & Fahmy, 1972, 1983). [The Working Group noted that dimethylformamide was used as a solvent control in these experiments.]

Dimethylformamide was reported to induce a marginal mutagenic response in L5178Y TK$^{+/-}$ mouse lymphoma cells in the absence but not in the presence of an exogenous metabolic system (McGregor *et al.*, 1988); in similar studies, negative results were obtained (Mitchell *et al.*, 1988; Myhr & Caspary, 1988).

In one study, dimethylformamide did not increase the incidence of chromosomal aberrations or of sister chromatid exchange in human peripheral blood lymphocytes *in vitro* (highest no-effect dose, 80 000 µg/ml; Antoine *et al.*, 1983). In another study, chromosomal aberrations were reported in human peripheral lymphocyte cultures treated with dimethylformamide (lowest effective dose, 0.007 µg/ml; Koudela & Spazier, 1979).

In Balb/c mice injected intraperitoneally with 0.2, 20 or 2000 mg/kg bw dimethylformamide, no increase in the frequency of micronuclei in bone-marrow cells was observed (Antoine *et al.*, 1983), and no increase was seen in the frequency of sperm abnormalities after five doses of 0.1–1.5 ml/kg bw (Topham, 1980) or after 0.2–2000 mg/kg bw (Antoine *et al.*, 1983). It induced micronuclei in the bone marrow of Kunming mice after single (1 mg/kg) or multiple (3×1 mg/kg) intraperitoneal injections (Ye, 1987).

As reported in an abstract, no dominant lethal effect was observed in groups of ten Sprague–Dawley rats exposed by inhalation to dimethylformamide for 6 h per day for five consecutive days (Lewis, 1979).

Dimethylformamide did not induce morphological transformation in Syrian hamster embryo cells (Pienta *et al.*, 1977), nor did it induce transformation of hamster embryo cells after transplacental exposure by intraperitoneal injection (Quarles *et al.*, 1979). [The Working Group noted that since dimethylformamide was being used as a solvent control in these experiments, no other control was available and only one dose was tested.]

Dimethylformamide inhibited intercellular communication (as measured by metabolic cooperation) between Chinese hamster V79 *hprt*$^{+/-}$ cells (Chen *et al.*, 1984).

(b)   *Humans*

(i) *Absorption, distribution, excretion and metabolism*

Dimethylformamide in liquid or vapour form is readily absorbed through the skin, by inhalation or after oral exposure (Maxfield *et al.*, 1975). It is rapidly metabolized and excreted in the urine in the form of *N*-hydroxymethyl–*N*-methylformamide and, to a small extent, *N*-methylformamide, *N*-hydroxymethylformamide and unmetabolized dimethylformamide (Scailteur & Lauwerys, 1984, 1987). In volunteers exposed by inhalation, *N*-hydroxymethyl–*N*-methylformamide (measured as *N*-methylformamide) was detected in urine 4 h after onset of exposure; almost complete elimination had taken place by 24 h (Kimmerle & Eben, 1975b). *N*-methylformamide, formed from *N*-hydroxymethyl–*N*-methylformamide during gas chromatographic analysis, has been measured in the urine of exposed workers. Urinary measurements showed a dose-relationship to airborne levels of dimethylformamide after exposure by inhalation (Lauwerys *et al.*, 1980); however, extensive skin contact may markedly influence the dose absorbed. Other methods for assessing exposure can include measurements of dimethylformamide in blood or exhaled air (Lauwerys, 1986).

(ii) *Toxic effects*

Accidental dermal and inhalation exposure has been reported to cause liver injury, with symptoms of abdominal pain, vomiting, hypertension and elevated levels of urinary bilirubin and serum transaminases. Some dermal irritation and hyperaemia were seen. After the disappearance of clinical signs, 11 days after exposure, a liver biopsy revealed septal fibrosis and accumulation of mononuclear cells (Potter, 1973, 1974). In other cases of chronic exposure in work place settings (to 14–60 mg/m³), irritation of the eyes, upper respiratory tract and digestive tract were observed (Tomasini *et al.*, 1983).

High exposures at various work places have been reported to cause nausea, vomiting, colic (Reinl & Urban, 1965), gastrointestinal abnormalities, hepatopathy (Aldyreva *et al.*, 1980; Paoletti *et al.*, 1982; Redlich *et al.*, 1987), cardiovascular abnormalities and nervous system disorders (Aldyreva *et al.*, 1980). Of five persons exposed occupationally [concentration unspecified], four had increased levels of serum amylase, suggesting pancreatitis (Chary, 1974).

Exposure to dimethylformamide through the skin in an acrylic fibre production plant led to five cases of intoxication, with gastritis, gastroesophagitis and hepatic dysfunction. These effects were reversible on removal from exposure (Guirguis, 1981).

In a study of 100 workers exposed to dimethylformamide (determined as 22 mg/m³ by 8–h TWA personal sampling) in two factories producing artificial polyurethane leather (mean period of exposure, five years), headache, dyspepsia and hepatic–type digestive impairment could be specifically associated with chronic exposure. Increased levels of γ–glutamyl transpeptidase demonstrated minimal hepatocellular damage (Cirla *et al.*, 1984). No sign of liver function change was reported in other studies of persons exposed to up to 60 ppm (180 mg/m³) dimethylformamide (Kennedy, 1986).

Polyacrylonitrile fibre production workers exposed to 30–60 ppm dimethylformamide for three to five years complained of fatigue, weakness, numbness of the extremities and eye and throat irritation (Kennedy, 1986). Skin sensitivity, allergic dermatitis, eczema and vitiligo have also been reported (Bainova, 1975; Kennedy, 1986).

Occupational exposure to dimethylformamide followed by consumption of alcohol has resulted in alcohol intolerance, dermal flushing (especially of the face), severe headache and dizziness (Reinl & Urban, 1965; Lyle *et al.*, 1979; Tomasini *et al.*, 1983).

(iii) *Effects on fertility and pregnancy outcome*

No data were available to the Working Group.

(iv) *Genetic and related effects*

In a study of 20 workers exposed to mono–, di– and trimethylamines as well as dimethylformamide in the German Democratic Republic, the mean workplace concentrations during one year before blood sampling were: 12.3 mg/m³ (range, 5.6–26.4) dimethylformamide, 5.3 mg/m³ (range, 1.2–10.1) monomethylformamide and 0.63 mg/m³ (range, 0.01–3.3) dimethylamine, which were within the maximal admissible range in the country. Eighteen unexposed employees from the same factory were used as controls. Increases in the frequency of chromosomal gaps and breaks were observed in 1.4% of the exposed group compared to 0.4% of controls (Berger *et al.*, 1985). [The Working Group noted the low number of chromosomal breaks observed in the controls, and that the possible effect of smoking was not accounted for.]

Chromosomal aberrations in peripheral lymphocytes were also reported in another study of workers who had been exposed occupationally to dimethylformamide with trace quantities of methylethylketone, butyl acetate, toluene, cyclohexanone and xylene. Sampling at two four–month intervals, when exposure was to an average of 180 and 150 mg/m³ dimethylformamide, respectively, showed an increase in the frequency of chromosomal aberrations; but subsequent sampling at three six–month intervals, when average exposures were to 50, 40 and 35 mg/m³, showed no increase (Koudela & Spazier, 1981).

It was reported in an abstract that there was no evidence for an increased frequency of chromosomal aberrations in peripheral lymphocytes of a group of workers exposed to dimethylformamide [details not given] (Šrám et al., 1985).

3.3   Epidemiological studies of carcinogenicity to humans

Ducatman et al. (1986) reported three cases of testicular germ–cell tumour in 1981–83 among 153 white men who repaired the exterior surfaces and electrical components of F4 Phantom jet aircraft in the USA. This finding led to surveys of two other repair shops at different geographical locations, in one of which the same type of aircraft was repaired and in another at which different types of aircraft were repaired. Four among 680 white male workers in the same type of repair shop had a history of testicular germ–cell cancers (0.95 expected) occurring in 1970–83. No case of testicular germ–cell cancer was found among the 446 white men employed at the facility where different types of aircraft were repaired. Of the seven cases, five were seminomas and two were embryonal–cell carcinomas. All seven men had long work histories in aircraft repair. There were many common exposures to solvents in the three facilities, but the only exposure identified as unique to the F4 Phantom jet aircraft repair facilities where the cases occurred was to a solvent mixture containing 80% dimethylformamide [20% unspecified]. Three of the cases had been exposed to this mixture with certainty and three cases had probably been exposed. Other cases of cancer were not searched for, and cases were found through foremen and from filed death certificates. The authors suggested that underreporting was possible.

Levin et al. (1987), in a letter to the Editor, described three cases of embryonal–cell carcinoma of the testis in workers at one leather tannery in the USA, all of whom had worked as swabbers on the spray lines in leather finishing. According to the authors, all the tanneries they had surveyed used dimethylformamide, as well as a wide range of dyes and solvents. [The Working Group noted that the number of workers from which these three cases arose was not given and that other cancers were not looked for.]

Chen et al. (1988a) studied cancer incidence among 2530 actively employed workers with potential exposure to dimethylformamide in 1950–70 and 1329 employees with exposure to dimethylformamide and acrylonitrile at an acrylic fibre manufacturing plant in South Carolina, USA (O'Berg et al., 1985). Cancer incidence rates for the company (1956–84) and US national rates (1973–77) were used to calculate expected numbers of cases. For all workers exposed to dimethylformamide (alone or with acrylonitrile), the standardized incidence ratio

(SIR) based on company rates for all cancers combined was 110 ([95% confidence interval (CI), 88–136]; 88 cases); the SIR based on national rates was 92. The SIR for cancer of the buccal cavity and pharynx was 344 ([172–615]; 11 cases) based on company rates and 167 based on US rates. More cancer cases than expected from company rates (34 cases: SIR, 134; [98–195]) were found among wage employees exposed to dimethylformamide alone, due mainly to eight carcinomas of the buccal cavity or pharynx *versus* 1.0 expected (SIR, 800; [345–1580]). An additional case occurred in salaried employees exposed to dimethylformamide alone (SIR, 167); four of these tumours were cancers of the lip. No such excess was found among the workers exposed to both dimethylformamide and acrylonitrile (two observed; SIR, 125, based on company rates). The authors reported no association with intensity or duration of exposure: low and moderate exposure, SIR, 420 (five cases); high exposure, SIR, 300 (six cases). 'Low' exposure was defined as no direct contact with liquids containing any dimethylformamide, even with protective equipment, and workplace levels consistently below 10 ppm (30 mg/m^3) in air (no odour of dimethylformamide evident). 'Moderate' exposure was defined as intermittent contact with liquids containing >5% dimethylformamide, and workplace levels sometimes >10 ppm (more than once per week); dimethylformamide-laden materials handled but fumes contained the levels described above. 'High' exposure was defined as frequent contact with liquids containing >5% dimethylformamide, and workplace levels often >10 ppm, use of breathing protection often required for 15 min to 1 h; dimethylformamide vapour frequently >10 ppm when handling pure dimethylformamide or dimethylformamide–containing materials. [The Working Group noted that the exposure categories do not seem to be mutually exclusive.] One case of testicular cancer was found among the 3859 workers exposed to dimethylformamide (alone or with acrylonitrile), with 1.7 expected based on company rates; no case of liver cancer was seen. [The Working Group noted that the company rates may be more relevant for comparison, as there were only actively employed persons among the exposed and because the US rates are based on a limited time period, 1973–77. No data on tobacco use, alcohol consumption or other occupational exposures were given.]

Chen *et al.* (1988b) analysed mortality in 1950–82 in the same cohort among both active and pensioned employees. Expected numbers (adjusted for age and time period) were based on company rates. For all workers exposed to dimethylformamide (alone or with acrylonitrile), the standardized mortality ratio (SMR) for lung cancer was 124 (33 cases; [95% CI, 85–174]). An increased risk for lung cancer was found in the cohort exposed only to dimethylformamide (19 cases; SMR, 141; [84–219]) but not in that exposed to dimethylformamide and acrylonitrile. There were three deaths from cancer of the buccal cavity and pharynx (SIR, 188) in all persons exposed to dimethylformamide (alone or with acrylonitrile). No other excess cancer risk was reported. [The Working Group noted that no information on loss to follow–up or on death certificates is given in this report or whether these deaths were included in the incidence study reported above.]

4. Summary of Data Reported and Evaluation

4.1 Exposures

Dimethylformamide is a synthetic organic liquid used mainly as an industrial solvent in the manufacture of films, fibres, coatings and adhesives, in the purification of hydrocarbons in petroleum refining and in other chemical processes. Exposure to dimethylformamide may occur through inhalation and dermal absorption. Occupational exposure has been reported during manufacturing processes and during use of products in which dimethylformamide is a solvent.

4.2 Experimental carcinogenicity data

Dimethylformamide was tested for carcinogenicity by oral administration and subcutaneous injection in one strain of rats. In a study in which dimethylformamide was administered by intraperitoneal injection in another strain of rats, a small number of uncommon tumours was observed in treated rats. All of these studies were inadequate for evaluation.

4.3 Human carcinogenicity data

An excess risk for testicular germ–cell tumours was identified among workers involved in aircraft repair who had been exposed to a solvent mixture containing 80% dimethylformamide. An excess risk for cancer of the buccal cavity or pharynx and a nonsignificant excess of lung cancer, but no excess risk for testicular cancer, were observed in workers exposed to dimethylformamide at a plant manufacturing acrylic fibres. No adjustment was made for possible confounding variables in either study.

4.4 Other relevant data

Liver toxicity and dermatitis have been observed in persons occupationally exposed to dimethylformamide. Dimethylformamide also induces liver toxicity in experimental animals.

Dimethylformamide induced malformations in mice following intraperitoneal administration and in rabbits following oral (but not dermal) exposure. Fetal growth retardation but no malformation was seen following exposure of rats by inhalation.

An increased frequency of chromosomal aberrations was observed in peripheral lymphocytes of industrial workers exposed to dimethylformamide in one study. Another study showed an increased frequency but was inconclusive because the workers were also exposed to other industrial chemicals.

Dimethylformamide did not induce sister chromatid exchange or micronuclei in mice. It did not induce DNA damage, mutation or sister chromatid exchange in cultured human cells but gave equivocal results for chromosomal aberrations. It did not induce chromosomal aberrations, sister chromatid exchange, mutation or DNA damage in cultured animal cells. It inhibited intercellular communication in cultured animal cells. It did not induce mutation in *Drosophila*, plants or yeast nor mitotic recombination in yeast. It induced DNA damage and aneuploidy in yeast. Dimethylformamide did not induce mutation or DNA damage in bacteria. (See Appendix 1.)

4.5   Evaluation[1]

There is *limited evidence* for the carcinogenicity of dimethylformamide in humans.

There is *inadequate evidence* for the carcinogenicity of dimethylformamide in experimental animals.

Overall evaluation

Dimethylformamide *is possibly carcinogenic to humans (Group 2B)*.

5. References

Agrelo, C. & Amos, H. (1981) Nuclear enlargement in HeLa cells and fibroblasts. In: de Serres, F.J. & Ashby, J., eds, *Evaluation of Short-term Tests for Carcinogens. Report of the International Collaborative Programme (Progress in Mutation Research, Vol. I)*, Amsterdam, Elsevier, pp. 245-248

Air Products and Chemicals, Inc. (1985) *Amines from Air Products – More Than Chemicals, Chemistry*, Allentown, PA

Aldyreva, M.V., Bortsevich, S.V., Palugushina, A.I., Sidorova, N.V. & Tarasova, L.A. (1980) Effect of dimethylformamide on the health of workers manufacturing polyurethane synthetic leather (Russ.). *Gig. Tr. prof. Zabol., 6*, 24-28

American Conference of Governmental Industrial Hygienists (1988) *Threshold Limit Values and Biological Exposure Indices for 1988-1989*, Concinnati, OH, p. 19

Anon. (1983) Aliphatic organics. *Chem. Mark. Rep., 224*, 11-12

Anon. (1985) N,N-Dimethylformamide (Jpn.). *Fine Chem., 14*, 37-39

Antoine, J.L., Arany, J., Léonard, A., Henrotte, Jenar-Dubuisson, G. & Decat, G. (1983) Lack of mutagenic activity of dimethylformamide. *Toxicology, 26*, 207-212

Arbeidsinspectie (Labour Inspection) (1986) *De Nationale MAC-Lijst 1986* [National MAC-List 1986] *(P145)*, Voorburg, Ministry of Social Affairs and Work Environment, p. 11

Arbejdstilsynet (Labour Inspection) (1988) *Graensevaerdier for Stoffer og Materialer* [Limit Values for Substances and Materials] *(At-anvisning No. 3.1.0.2)*, Copenhagen, p. 17

---

[1]For definitions of the italicized terms, see Preamble, pp. 27-30.

Arimoto, S., Nakano, N., Ohara, Y., Tanaka, K. & Hayatsu, H. (1982) A solvent effect in the mutagenicity of tryptophan–pyrolysate mutagens in the *Salmonella*/mammalian microsome assay. *Mutat. Res.*, *102*, 105–112

Bainova, A. (1975) Assessment of skin lesions in the production of bulana polyacrylonitrile fibres (Russ.). *Dermatol. Venerol.*, *14*, 92–97

Baker, R.S.U. & Bonin, A.M. (1981) Study of 42 coded compounds with the *Salmonella*/mammalian microsome assay (University of Sydney). In: de Serres, F.J. & Ashby, J., eds, *Evaluation of Short-term Tests for Carcinogens. Report of the International Collaborative Program (Progress in Mutation Research, Vol. I)*, Amsterdam, Elsevier, pp. 249–260

Barnes, J.R. & Henry, N.W., III (1974) The determination of N–methylformamide and N–methylacetamide in urine. *Am. ind. Hyg. Assoc. J.*, *35*, 84–87

Bartsch, W., Sponer, G., Dietmann, K. & Fuchs, G. (1976) Acute toxicity of various solvents in the mouse and rat. LD_{50} of ethanol, diethylacetamide, dimethylformamide, dimethylsulfoxide, glycerine, N–methylpyrrolidone, polyethylene glycol 400, 1,2–propane diol and Tween 20. *Arzneimittel.-Forsch.*, *26*, 1581–1583

Becci, P.J., Voss, K.A., Johnson, W.D., Gallo, M.A. & Babish, J.G. (1983) Subchronic feeding study of N,N–dimethylformamide in rats and mice. *J. Am. Coll. Toxicol.*, *2*, 371–378

Berger, H., Haber, I., Wüncher, G. & Bittersohl, G. (1985) Epidemiological studies on exposure to dimethylformamide (Ger.). *Z. ges. Hyg.*, *31*, 366–368

Brindley, C., Gescher, A. & Ross, D. (1983) Studies on the metabolism of dimethylformamide in mice. *Chem.-biol. Interactions*, *45*, 387–392

Brooks, T.M. & Dean, B.J. (1981) Mutagenic activity of 42 coded comounds in the *Salmonella*/microsome assay with preincubation (Pollards Wood Research Station). In: de Serres, F.J. & Ashby, J., eds, *Evaluation of Short-term Tests for Carcinogens. Report of the International Collaborative Program (Progress in Mutation Research, Vol. I)*, Amsterdam, Elsevier, pp. 261–270

Brugnone, F., Perbellini, L. & Gaffuri, E. (1980) N,N–Dimethylformamide concentration in environmental and aveolar air in an artificial leather factory. *Br. J. ind. Med.*, *37*, 185–188

Chary, S. (1974) Dimethylformamide: a cause of acute pancreatitis? (Lett.) *Lancet*, *ii*, 356

Chen, J.L., Fayerweather, W.E. & Pell, S. (1988a) Cancer incidence of workers exposed to dimethylformamide and/or acrylonitrile. *J. occup. Med.*, *30*, 813–818

Chen, J.L., Fayerweather, W.E. & Pell, S. (1988b) Mortality study of workers exposed to dimethylformamide and/or acrylonitrile. *J. occup. Med.*, *30*, 819–321

Chen, T.-H., Kavanagh, T.J., Chang, C.C. & Trosko, J.E. (1984) Inhibition of metabolic cooperation in Chinese hamster V79 cells by various organic solvents and simple compounds. *Cell Biol. Toxicol.*, *1*, 155–171

Cirla, A.M., Pisati, G., Invernizzi, E. & Torricelli, P. (1984) Epidemiological study of workers exposed to low dimethylformamide concentrations. *Med. Lav.*, *6*, 149–156

Clayton, J.W., Jr, Barnes, J.R., Hood, D.B. & Schepers, G.W.H. (1963) The inhalation toxicity of dimethylformamide (DMF). *Am. ind. Hyg. Assoc. J.*, *24*, 144–154

Cook, W.A. (1987) *Occupational Exposure Limits – Worldwide*, Washington DC, American Industrial Hygiene Association, pp. 24, 120, 137, 183

Craig, D.K., Weir, R.J., Wagner, W. & Groth, D. (1984) Subchronic inhalation toxicity of dimethylformamide in rats and mice. *Drug chem. Toxicol.*, *7*, 551–571

Dambly, C., Toman, Z. & Radman, M. (1981) Zorotest. In: de Serres, F.J. & Ashby, J., eds, *Evaluation of Short–term Tests for Carcinogens. Report of the International Collaborative Program (Progress in Mutation Research, Vol. I)*, Amsterdam, Elsevier, pp. 219–223

Davis, K.J. & Jenner, P.M. (1959) Toxicity of three drug solvents. *Toxicol. appl. Pharmacol., 1*, 576–578

Dean, B.J. (1981) Activity of 27 coded compounds in the RL₁ chromosome assay. In: de Serres, F.J. & Ashby, J., eds, *Evaluation of Short–term Tests for Carcinogens. Report of the International Collaborative Program (Progress in Mutation Research, Vol. I)*, Amsterdam, Elsevier, pp. 570–579

Deutsche Forschungsgemeinschaft (German Research Society) (1988) *Maximale Arbeitsplatzkonzentrationene und Biologische Arbeitsstofftoleranzwerte 1988* [Maximal Concentrations in the Workplace and Biological Tolerance Values for Working Materials 1988] *(Report No. XXIV)*, Weinheim, VCH Verlagsgesellschaft, p. 33

Direktoratet for Arbeidstilsynet (Directorate for Labour Inspections) (1981) *Administrative Normer for Forurensning i Arbeidsatmosfaere 1981* [Administrative Norms for Pollution in Work Atmosphere 1981] *(No. 361)*, Oslo, p. 10

Dixon, S.W., Graepel, G.J. & Looney, W.C. (1983) Seasonal effects on concentrations of monomethylformamide in urine samples. *Am. ind. Hyg. Assoc. J., 44*, 273–275

Druckrey, H., Preussmann, R., Ivankovic, S. & Schmähl, D. (1967) Organotropic carcinogenic action of 65 different N–nitroso compounds in BD rats (Ger.). *Z. Krebsforsch., 69*, 103–201

Ducatman, A.M., Conwill, D.E. & Crawl, J. (1986) Germ cell tumors of the testicle among aircraft repairmen. *J. Urol., 136*, 834–836

Eberling, C.L. (1980) Dimethylformamide. In: Mark, H.F., Othmer, D.F., Overberger, C.G., Seaborg, G.T. & Grayson, M., eds, *Kirk–Othmer Encyclopedia of Chemical Technology*, 3rd ed., Vol. 11, New York, John Wiley & Sons, pp. 263–268

E.I. duPont de Nemours & Co. (1986) *Dimethylformamide (DMF) – Properties, Uses, Storage and Handling*, Wilmington, DE

E.I. duPont de Nemours & Co. (1988a) *Material Safety Data Sheet, Dimethylformamide*, Wilmington, DE

E.I. duPont de Nemours & Co. (1988b) *Dimethyl Formamide Meeting, 20 March 1988*, Wilmington, DE

Eller, P.M. (1985) *NIOSH Manual of Analytical Methods*, 3rd ed., Suppl. 1 *(DHHS (NIOSH) Publ. No. 84–100)*, Washington DC, US Government Printing Office, pp. 2204–1 – 2204–4

Evans, E.L. & Mitchell, A.D. (1981) Effects of 20 coded chemicals on sister chromatid exchange frequencies in cultured Chinese hamster cells. In: de Serres, F.J. & Ashby, J., eds, *Evaluation of Short–term Tests for Carcinogens. Report of the International Collaborative Program (Progress in Mutation Research, Vol. I)*, Amsterdam, Elsevier, pp. 538–550

Fahmy, O.G. & Fahmy, M.J. (1972) Mutagenic selectivity of the RNA–forming genes in relation to the carcinogenicity of alkylating agents and polycyclic aromatics. *Cancer Res., 32*, 550–557

Fahmy, M.J. & Fahmy, O.G. (1983) Differential induction of altered gene expression by carcinogens at mutant alleles of a *Drosophila* locus with a transposable element. *Cancer Res., 43*, 801–807

Falck, K., Partanen, P., Sorsa, M., Suovaniemi, O. & Vainio, H. (1985) Mutascreen, an automated bacterial mutagenicity assay. *Mutat. Res., 150*, 119–125

The Foxboro Co. (1983) *Chromatographic Column Selection Guide for Century Organic Vapor Analyzer*, Foxboro, MA

Gatehouse, D. (1981) Mutagenic activity of 42 coded compounds in the 'microtiter' fluctuation test. In: de Serres, F.J. & Ashby, J., eds, *Evaluation of Short–term Tests for Carcinogens. Report of the International Collaborative Program (Progress in Mutation Research, Vol. I)*, Amsterdam, Elsevier, pp. 376–386

Gichner, T. & Veleminský, J. (1986) Organic solvents inhibit the mutagenicity of promutagens dimethyl-nitrosamine and methylbutyl nitrosamine in a higher plant *Arabidopsis thaliana*. *Mutagenesis, 1*, 107–109

Gichner, T. & Veleminský, J. (1987) The organic solvents acetone, ethanol and dimethylformamide potentiate the mutagenic activity of *N*-methyl-*N'*-nitro-*N*-nitrosoguanidine, but have no effect on the mutagenic potential of *N*-methyl-*N*-nitrosourea. *Mutat. Res, 192*, 31–35

Gleich, J. (1974) The influence of simple acid amides on fetal development of mice (Abstract). *Naunyn-Schmiedeberg's Arch. Pharmacol., 282 (Suppl.)*, R25

Green, M.L.H. (1981) A differential killing test using an improved repair-deficient strain of *Escherichia coli*. In: de Serres, F.J. & Ashby, J., eds, *Evaluation of Short-term Tests for Carcinogens. Report of the International Collaborative Program (Progress in Mutation Research, Vol. I)*, Amsterdam, Elsevier, pp. 183–194

Green, N.R. & Savage, J.R. (1978) Screening of safrole, eugenol, their ninhydrin positive metabolites and selected secondary amines for potential mutagenicity. *Mutat. Res., 57*, 115–121

Guenier, J.P., Lhuillier, F. & Muller, J. (1986) Sampling of gaseous pollutants on silica gel with 1400 mg tubes. *Ann. occup. Hyg., 30*, 103–114

Guirguis, S. (1981) Dimethylformamide intoxication in acrylic fibre production. *Med. Lav., 3*, 137–140

Gupta, R.S. & Goldstein, S. (1981) Mutagen testing in the human fibroblast diphtheria toxin resistance (HF Dip^r) system. In: de Serres, F.J. & Ashby, J., eds, *Evaluation of Short-term Tests for Carcinogens. Report of the International Collaborative Program. (Progress in Mutation Research, Vol. I)*, Amsterdam, Elsevier, pp. 614–625

Hansch, C. & Leo, A. (1979) *Substituent Constants for Correlation Analysis in Chemistry and Biology*, New York, John Wiley & Sons, p. 180

Health and Safety Executive (1987) *Occupational Exposure Limits 1987 (Guidance Note 40/87)*, London, Her Majesty's Stationery Office, p. 13

Hofmann, H.T. (1960) On the question of hepatotoxic activity of dimethylformamide (Ger.). *Naunyn-Schmiedebergs Arch. exp. Pathol., 240*, 38–39

Hubbard, S.A., Green, M.H.L., Bridges, B.A., Wain, A.J. & Bridges, J.W. (1981) Fluctuation test with S9 and hepatocyte activation (University of Sussex). In: de Serres, F.J. & Ashby, J., eds, *Evaluation of Short-term Tests for Carcinogens. Report of the International Collaborative Program (Progress in Mutation Research, Vol. I)*, Amsterdam, Elsevier, pp. 360–370

IARC (1987) *IARC Monographs on the Evaluation of Carcinogenic Risks to Humans*, Suppl. 7, *Overall Evaluations of Carcinogenicity: An Updating of* IARC Monographs *Volumes 1 to 42*, Lyon, pp. 211–216

IARC (1988) *Information Bulletin on the Survey of Chemicals Being Tested for Carcinogenicity*, No. 13, Lyon, p. 236

Ichinotsubo, D., Mower, H. & Mandel, M. (1981a) Testing of a series of paired compounds (carcinogen and noncarcinogenic structural analog) by DNA repair-deficient *E. coli* strains. In: de Serres, F.J. & Ashby, J., eds, *Evaluation of Short-term Tests for Carcinogens. Report of the International Collaborative Program (Progress in Mutation Research, Vol. I)*, Amsterdam, Elsevier, pp. 195–198

Ichinotsubo, D., Mower, H. & Mandel, M. (1981b) Mutagen testing of a series of paired compounds with the Ames *Salmonella* testing system (University of Hawaii at Manoa). In: de Serres, F.J. & Ashby, J., eds, *Evaluation of Short-term Tests for Carcinogens. Report of the International Collaborative Program (Progress in Mutation Research, Vol. I)*, Amsterdam, Elsevier, pp. 298–301

Institut National de Recherche et de Sécurité (National Institute for Research and Safety) (1986) *Valeurs Limites pour les Concentrations des Substances Dangereuses Dans l'Air des Lieux de Travail* [Limit Values for Concentrations of Dangerous Substances in the Air of Work Places] (*ND 1609-125-86*), Paris, p. 565

International Labour Office (1984) *Occupational Exposure Limits for Airborne Toxic Substances*, 2nd rev. ed. (*Occupational Safety and Health Series No. 37*), Geneva, p. 102

Ito, N. (1982) Unscheduled DNA synthesis induced by chemical carcinogens in primary cultures of adult rat hepatocytes. *Mie med. J., 32*, 53–60

Jagannath, D.R., Vultaggio, D.M. & Brusick, D.J. (1981) Genetic activity of 42 coded compounds in the mitotic gene conversion assay using *Saccharomyces cerevisiae* strain D4. In: de Serres, F.J. & Ashby, J., eds, *Evaluation of Short–term Tests for Carcinogens. Report of the International Collaborative Program (Progress in Mutation Research, Vol. I)*, Amsterdam, Elsevier, pp. 456–467

Jotz, M.M. & Mitchell, A.D. (1981) Effects of 20 coded chemicals on the forward mutation frequency at the thymidine kinase locus in L5178Y mouse lymphoma cells. In: de Serres, F.J. & Ashby, J., eds, *Evaluation of Short–term Tests for Carcinogens. Report of the International Collaborative Program (Progress in Mutation Research, Vol. I)*, Amsterdam, Elsevier, pp. 580–593

Kada, T. (1981) The DNA–damaging activity of 42 coded compounds in the rec–assay. In: de Serres, F.J. & Ashby, J., eds, *Evaluation of Short–term Tests for Carcinogens. Report of the International Collaborative Program (Progress in Mutation Research, Vol. I)*, Amsterdam, Elsevier, pp. 175–182

Kassinova, G.B., Kovaltsova, S.V., Margin, S.V. & Zakharov, I.A. (1981) Activity of 40 coded compounds in differential inhibition and mitotic crossing–over assays in yeast. In: de Serres, F.J. & Ashby, J., eds, *Evaluation of Short–term Tests for Carcinogens. Report of the International Collaborative Program (Progress in Mutation Research, Vol. I)*, Amsterdam, Elsevier, pp. 434–455

Keller, C.A. & Lewis, S.C. (1981) Inhalation teratology study of N,N–dimethylformamide (DMF) (Abstract). *Teratology, 23*, 45A

Kennedy, G.L., Jr (1986) Biological effects of acetamide, formamide, and their monomethyl and dimethyl derivatives. *CRC crit. Rev. Toxicol., 17*, 129–182

Kestell, P., Threadgill, M.D., Gescher, A., Gledhill, A.P., Shaw, A.J. & Farmer, P.B. (1987) An investigation of the relationship between the hepatotoxicity and the metabolism of *N*–alkylformamides. *J. Pharmacol. exp. Ther., 240*, 265–270

Kimmerle, G. & Eben, A. (1975a) Metabolism studies of *N,N*–dimethylformamide. I. Studies in rats and dogs. *Int. Arch. Arbeitsmed., 34*, 109–126

Kimmerle, G. & Eben, A. (1975b) Metabolism studies of *N,N*–dimethylformamide. II. Studies in persons. *Int. Arch. Arbeitsmed., 34*, 127–136

Kimmerle, G. & Machemer, L. (1975) Studies with *N,N*–dimethylformamide for embryotoxic and teratogenic effects on rats after dynamic inhalation. *Int. Arch. Arbeitsmed., 34*, 167–175

Kimura, E.T., Ebert, D.M. & Dodge, P.W. (1971) Acute toxicity and limits of solvent residue for sixteen organic solvents. *Toxicol. appl. Pharmacol., 19*, 699–704

Kirkhart, B. (1981) Micronucleus test on 21 compounds. In: de Serres, F.J. & Ashby, J., eds, *Evaluation of Short–term Tests for Carcinogens. Report of the International Collaborative Program (Progress in Mutation Research, Vol. I)*, Amsterdam, Elsevier, pp. 698–704

Klaunig, J.E., Goldblatt, P.J., Hinton, D.E., Lipsky, M.M. & Trump, B.F. (1984) Carcinogen induced unscheduled DNA synthesis in mouse hepatocytes. *Toxicol. Pathol., 12*, 119–125

Kommineni, C. (1972) *Pathological Studies of Aflatoxin Fractions and Dimethylformamide in MRC Rats*, PhD Thesis, Omaha, Nebraska University

Koudela, K. & Spazier, K. (1979) Effect of dimethylformamide on human peripheral lymphocytes. *Cesk. Hyg.*, *24*, 432–436

Koudela, K. & Spazier, K. (1981) Results of cytogenetic examination of persons working in an environment of increased concentration of dimethylformamide vapours in the atmosphere (Czech.). *Prak. Lek*, *33*, 121–123

Kubelka, V., Mitera, J., Rábl, V. & Mostecký, J. (1976) Chromatographic determination of DMA and DMF in aqueous medium. *Water Res.*, *10*, 137–138

Kutzsche, A. (1965) The toxicology of dimethylformamide (Ger.). *Arzneimittel.–Forsch.*, *15*, 618–624

Lauwerys, R.R. (1986) Dimethylformamide. In: Alessio, L., Berlin, A., Boni, M. & Roi, R., eds, *Biological Indicators for the Assessment of Human Exposure to Industrial Chemicals*, Luxembourg, Commission of the European Communities, pp. 17–27

Lauwerys, R.R., Kivits, A., Lhoir, M., Rigolet, P., Houbeau, D., Buchet, J.P. & Roels, H.A. (1980) Biological surveillance of workers exposed to dimethylformamide and the influence of skin protection on its percutaneous absorption. *Int. Arch. occup. environ. Health*, *45*, 189–203

Levin, S.M., Baker, D.B., Landrigan, P.J., Monoghan, S.V., Frumin, E., Braithwaite, M. & Towne, W. (1987) Testicular cancer in leather tanners exposed to dimethylformamide. *Lancet*, *ii*, 1153

Lewis, S.C. (1979) Dominant lethal mutagenic bioassay of dimethyl formamide (DMF) (Abstract No. Ea–7). *Environ. Mutagenesis*, *1*, 166

Lipski, K. (1982) Liquid chromatographic determination of dimethyl formamide, methylene bisphenyl isocyanate and methylene bisphenyl amine in air samples. *Ann. occup. Hyg.*, *25*, 1–4

Llewellyn, G.C., Hastings, W.S., Kimbrough, T.D., Rea, F.W. & O'Rear, C.E. (1974) The effects of dimethylformamide on female mongolian gerbils, *Meriones unguiculatus*. *Bull. environ. Contam. Toxicol.*, *11*, 467–473

Loprieno, N. (1981) Screening of coded carcinogen–noncarcinogenic chemicals by a forward–mutation system with the yeast *Schizosaccharomyces pombe*. In: de Serres, F.J. & Ashby, J., eds, *Evaluation of Short–term Tests for Carcinogens. Report of the International Collaborative Program (Progress in Mutation Research, Vol. I)*, Amsterdam, Elsevier, pp. 424–433

Lundberg, I., Pehrsson, A., Lundberg, S., Kronevi, T. & Lidums, V. (1983) Delayed dimethylformamide biotransformation after high exposures in rats. *Toxicol. Lett.*, *17*, 29–34

Lyle, W.H., Spence, T.W.M., McKinneley, W.M. & Duckers, K. (1979) Dimethylformamide and alcohol intolerance. *Br. J. ind. Med.*, *36*, 63–66

MacDonald, D.J. (1981) *Salmonella*/microsome tests on 42 coded chemicals (University of Edinburgh). In: de Serres, F.J. & Ashby, J., eds, *Evaluation of Short–term Tests for Carcinogens. Report of the International Collaborative Program (Progress in Mutation Research, Vol. I)*, Amsterdam, Elsevier, pp. 285–297

Martelli, D. (1960) Toxicology of dimethylformamide (Ital.). *Med. Lav.*, *51*, 123–128

Martin, C.N. & McDermid, A.C. (1981) Testing of 42 coded compounds for their ability to induce unscheduled DNA repair synthesis in HeLa cells. In: de Serres, F.J. & Ashby, J., eds, *Evaluation of Short–term Tests for Carcinogens. Report of the International Collaborative Program (Progress in Mutation Research, Vol. I)*, Amsterdam, Elsevier, pp. 533–537

Massmann, W. (1956) Toxicological investigations on dimethylformamide. *Br. J. ind. Med.*, *13*, 51–54

Matheson Gas Products (undated) *The Matheson–Kitagawa Toxic Gas Detector System*, East Rutherford, NJ

Matsushima, T., Takamoto, Y., Shirai, A., Sawamura, M. & Sugimura, T. (1981) Reverse mutation test on 42 coded comounds in the *E. coli* WP2 system. In: de Serres, F.J. & Ashby, J., eds, *Evaluation of Short–term Tests for Carcinogens. Report of the International Collaborative Program (Progress in Mutation Research, Vol. I)*, Amsterdam, Elsevier, pp. 387–395

Maxfield, M.E., Barnes, J.R., Azar, A. & Trochimowicz, H.T. (1975) Urinary excretion of metabolite following experimental human exposures to dimethylformamide or to dimethylacetamide. *J. occup. Med., 17*, 506–511

McGregor, D.B., Brown, A., Cattanach, P., Edwards, I., McBride, D. & Caspary, W.J. (1988) Responses of the L5178Ytk[+]/tk[−] mouse lymphoma cell forward mutation assay. II. 18 coded chemicals. *Environ. mol. Mutagenesis, 11*, 91–118

McQueen, C.A., Kreiser, D.M. & Williams, G.M. (1983) The hepatocyte primary culture/DNA repair assay using mouse or hamster hepatocytes. *Environ. Mutagenesis, 5*, 1–8

Mehta, R.D. & von Borstel, R.C. (1981) Mutagenic activity of 42 encoded compounds in the haploid yeast reversion assay, strain XV185–14C. In: de Serres, F.J. & Ashby, J., eds, *Evaluation of Short–term Tests for Carcinogens. Report of the International Collaborative Program (Progress in Mutation Research, Vol. I)*, Amsterdam, Elsevier, pp. 414–423

Merkle, J. & Zeller, H. (1980) Studies on acetamides and formamides for embryotoxic and teratogenic activities in the rabbit (Ger.). *Arzneimittel.–Forsch./Drug Res., 30*, 1557–1562

Mitchell, A.D., Rudd, C.J. & Caspary, W.J. (1988) Evaluation of the L5178Y mouse lymphoma cell mutagenesis assay: intralaboratory results for sixty–three coded chemicals tested at SRI International. *Environ. mol. Mutagenesis, 12 (Suppl. 13)*, 37–101

Mohn, G.R., Vogels–Bouter, S. & van der Horst–van der Zon, J. (1981) Studies on the mutagenic activity of 20 coded compounds in liquid tests using the multipurpose strain *Escherichia coli* K–12/343/113 and derivatives. In: de Serres, F.J. & Ashby, J., eds, *Evaluation of Short–term Tests for Carcinogens. Report of the International Collaborative Program (Progress in Mutation Research, Vol. I)*, Amsterdam, Elsevier, pp. 396–413

Mortelmans, K., Haworth, S., Lawlor, T., Speck, W., Tainer, B. & Zeiger, E. (1986) *Salmonella* mutagenicity tests: II. Results from the testing of 270 chemicals. *Environ. Mutagenesis, 8 (Suppl. 7)*, 1–119

Myhr, B.C. & Caspary, W.J. (1988) Evaluation of the L5718Y mouse lymphoma cell mutagenesis assay: intralaboratory results for sixty–three coded chemicals tested at Litton Bionetics, Inc. *Environ. mol. Mutagenesis, 12(Suppl. 13)*, 103–194

Nagao, M. & Takahashi, Y. (1981) Mutagenic activity of 42 coded compounds in the *Salmonella*/microsome assay (National Cancer Center Research Institute). In: de Serres, F.J. & Ashby, J., eds, *Evaluation of Short–term Tests for Carcinogens. Report of the International Collaborative Program (Progress in Mutation Research, Vol. I)*, Amsterdam, Elsevier, pp. 302–313

Natarajan, A.T. & van–Kesteren–van Leeuwen, A.C. (1981) Mutagenic activity of 20 coded compounds in chromosome aberrations/sister chromatid exchange assay using Chinese hamster ovary (CHO) cells. In: de Serres, F.J. & Ashby, J., eds, *Progress in Mutation Research*, Vol. I, *Evaluation of Short–term Tests for Carcinogens. Report of the International Collaborative Program (Progress in Mutation Research, Vol. I)*, Amsterdam, Elsevier, pp. 551–559

National Draeger, Inc. (1987) *Detector Tube Products for Gas and Vapor Detection*, Pittsburgh, PA

National Institute for Occupational Safety and Health (1983) *US National Occupational Exposure Survey 1981–83*, Cincinnati, OH

National Swedish Board of Occupational Safety and Health (1987) *Hygienska Gränsvärden* [Hygienic Limit Values] *(Ordinance 1987:12)*, Solna, p. 18

O'Berg, M.T., Chen, J.L., Burke, C.A., Walrath, J. & Pell, S. (1985) Epidemiologic study of workers exposed to acrylonitrile: an update. *J. occup. Med.*, *27*, 835–840

Paika, I.J., Beauchesne, M.T., Randall, M., Schreck, R.R. & Latt, S.A. (1981) In vivo SCE analysis of 20 coded compounds. In: de Serres, F.J. & Ashby, J., eds, *Evaluation of Short-term Tests for Carcinogens. Report of the International Collaborative Program (Progress in Mutation Research, Vol. I)*, Amsterdam, Elsevier, pp. 672–681

Paoletti, A., Fabri, G. & Marini Bettolo, P. (1982) An unusual case of abdominal pain due to dimethylformamide intoxication (Ital.). *Minerva med.*, *73*, 3407–3410

Parry, J.M. & Sharp, D. (1981) Induction of mitotic aneuploidy in the yeast strain *D6* by 42 coded compounds. In: de Serres, F.J. & Ashby, J., eds, *Evaluation of Short-term Tests for Carcinogens. Report of the International Collaborative Program (Progress in Mutation Research, Vol. I)*, Amsterdam, Elsevier, pp. 468–480

Perry, P.E. & Thomson, E.J. (1981) Evaluation of the sister chromatid exchange method in mammalian cells as a screening system for carcinogens. In: de Serres, F.J. & Ashby, J., eds, *Evaluation of Short-term Tests for Carcinogens. Report of the International Collaborative Program (Progress in Mutation Research, Vol. I)*, Amsterdam, Elsevier, pp. 560–569

Pienta, R.J., Poiley, J.A. & Lebherz, W.B., III (1977) Morphological transformation of early passage golden Syrian hamster embryo cells derived from cryopreserved primary cell cultures as a reliable in vitro bioassay for identifying diverse carcinogens. *Int. J. Cancer*, *19*, 642–655

Potter, H.P. (1973) Dimethylformamide-induced abdominal pain and liver injury. *Arch. environ. Health*, *27*, 340–341

Potter, H.P. (1974) Toxicity of dimethylformamide (Lett.). *Lancet*, *ii*, 1084

Pouchert, C.J., ed. (1981) *The Aldrich Library of Infrared Spectra*, 3rd ed., Milwaukee, WI, Aldrich Chemical Co., p. 443D

Pouchert, C.J., ed. (1983) *The Aldrich Library of NMR Spectra*, 2nd ed., Vol. 1, Milwaukee, WI, Aldrich Chemical Co., p. 639A

Pouchert, C.J., ed. (1985) *The Aldrich Library of FT–IR Spectra*, Vol. 1, Milwaukee, WI, Aldrich Chemical Co., p. 758D

Pozzoli, L., Cottica, D., Ghittori, S. & Catenacci, G. (1981) Monitoring of occupational exposure to dimethylformamide by passive personal samplers and monomethylformamide urinary excretion. *Med. Lav.*, *3*, 161–163

Purchase, I.F.H., Longstaff, E., Ashby, J., Styles, J.A., Anderson, D., Lefevre, P.A. & Westwood, F.R. (1978) An evaluation of 6 short-term tests for detecting organic chemical carcinogens. *Br. J. Cancer*, *37*, 873–903

Quarles, J.M., Sega, M.W., Schenley, C.K. & Lijinsky, W. (1979) Transformation of hamster fetal cells by nitrosated pesticides in a transplacental assay. *Cancer Res.*, *39*, 4525–4533

Redlich, C., Sparer, J., Cowan, D., Beckett, W., Miller, H., Cherniack, M. & Cullen, M. (1987) Outbreak of occupational hepatitis – Connecticut. *Morb. Mortal. Wkly Rep.*, *36*, 101–102

Reinl, W. & Urban, H.J. (1965) Illness after dimethylformamide (Ger.). *Int. Arch. Gewerbepathol. Gewerbehyg.*, *21*, 333–346

Richold, M. & Jones, E. (1981) Mutagenic activity of 42 coded compounds in the *Salmonella*/microsome assay (Huntington Research Centre). In: de Serres, F.J. & Ashby, J., eds, *Evaluation of Short-term Tests for Carcinogens. Report of the International Collaborative Program (Progress in Mutation Research, Vol. I)*, Amsterdam, Elsevier, pp. 314–322

Rimatori, V. & Carelli, G. (1982) Charcoal sampling and gas chromatographic determination on *N,N*–dimethylformamide in air samples from a polyurethane plant. *Scand. J. Work Environ. Health, 8,* 20–23

Robinson, D.E. & Mitchell, A.D. (1981) Unscheduled DNA synthesis response of human fibroblasts, WI–38 cells, to 20 coded chemicals. In: de Serres, F.J. & Ashby, J., eds, *Evaluation of Short–term Tests for Carcinogens. Report of the International Collaborative Program (Progress in Mutation Research, Vol. I),* Amsterdam, Elsevier, pp. 517–527

Rosenkranz, H.S., Hyman, J. & Leifer, Z. (1981) DNA polymerase deficient assay. In: de Serres, F.J. & Ashby, J., eds, *Progress in Mutation Research,* Vol. I, *Evaluation of Short–term Tests for Carcinogens. Report of the International Collaborative Program (Progress in Mutation Research, Vol. I),* Amsterdam, Elsevier, pp. 210–218

Rowland, I. & Severn, B. (1981) Mutagenicity of carcinogens and noncarcinogens in the *Salmonella*/microsome test (British Industrial Biological Research Association). In: de Serres, F.J. & Ashby, J., eds, *Evaluation of Short–term Tests for Carcinogens. Report of the International Collaborative Program (Progress in Mutation Research, Vol. I),* Amsterdam, Elsevier, pp. 323–332

Roxan, Inc. (undated) *Precision Gas Detector,* Woodland Hills, CA

Sadtler Research Laboratories (1980) *Standard Spectra Collection, 1980 Cumulative Index,* Philadelphia, PA

Salamone, M.F., Heddle, J.A. & Katz, M. (1981) Mutagenic activity of 41 compounds in the in vivo micronucleus assay. In: de Serres, F.J. & Ashby, J., eds, *Evaluation of Short–term Tests for Carcinogens. Report of the International Collaborative Program (Progress in Mutation Research, Vol. I),* Amsterdam, Elsevier, pp. 686–697

Scailteur, V. & Lauwerys, R.R. (1984) *In vivo* metabolism of dimethylformamide and relationship to toxicity in the male rat. *Arch. Toxicol., 56,* 87–91

Scailteur, V. & Lauwerys, R.R. (1987) Dimethylformamide (DMF) hepatotoxicity. *Toxicology, 43,* 231–238

Scailteur, V., de Hoffmann, E., Buchet, J.P. & Lauwerys, R. (1984) Study on in vivo and in vitro metabolism of dimethylformamide in male and female rats. *Toxicology, 29,* 221–234

Scheufler, H. & Freye, H.–A. (1975) The embryotoxic and teratogenic effects of dimethylformamide (Ger.). *Dtsch. Gesundh. Wes., 30,* 455–459

Schottek, W. (1970) Studies in experimental animals on the toxicity of repeated administrations of dimethylformamide (Ger.). *Acta biol. med. ger., 25,* 359–361

Sensidyne (1985) *The First Truly Simple Precision Gas Detector System,* Largo, FL

de Serres, F.J. & Ashby, J., eds (1981) *Evaluation of Short–term Tests for Carcinogens. Report of the International Collaborative Program (Progress in Mutation Research, Vol. 1),* Amsterdam, Elsevier

Sharp, D.C. & Parry, J.M. (1981a) Induction of mitotic gene conversion by 41 coded compounds using the yeast culture *JD1.* In: de Serres, F.J. & Ashby, J., eds, *Evaluation of Short–term Tests for Carcinogens. Report of the International Collaborative Program (Progress in Mutation Research, Vol. 1),* Amsterdam, Elsevier, pp. 491–501

Sharp, D.C. & Parry, J.M. (1981b) Use of repair–deficient strains of yeast to assay the activity of 40 coded compounds. In: de Serres, F.J. & Ashby, J., eds, *Evaluation of Short–term Tests for Carcinogens. Report of the International Collaborative Program (Progress in Mutation Research, Vol. 1),* Amsterdam, Elsevier, pp. 502–516

Sheveleva, G.A. & Osina, S.A. (1973) Experimental investigation of the embryotropic action of dimethylformamide (Russ.). *Toksikol. Nov. Prom. Khim. Veshchestv., 13,* 75–82

Sheveleva, G.A., Sivochalova, O.V., Osina, S.A. & Salnikova, L.S. (1977) Permeability of the placenta for dimethylformamide (Russ.). *Akush. Ginekol.*, *5*, 44–45

Simmon, V.F. & Shepherd, G.F. (1981) Mutagenic activity of 42 coded compounds in the *Salmonella*/microsome assay (SRI International). In: de Serres, F.J. & Ashby, J., eds, *Evaluation of Short-term Tests for Carcinogens. Report of the International Collaborative Program (Progress in Mutation Research, Vol. 1)*, Amsterdam, Elsevier, pp. 333–342

SKC (1988) *Comprehensive Catalog and Guide*, Eighty Four, PA

Skopek, T.R., Andon, B.M., Kaden, D.A. & Thilly, W.G. (1981) Mutagenic activity of 42 coded compounds using 8-azaguanine resistance as a genetic marker in *Salmonella typhimurium*. In: de Serres, F.J. & Ashby, J., eds, *Evaluation of Short-term Tests for Carcinogens. Report of the International Collaborative Program (Progress in Mutation Research, Vol. 1)*, Amsterdam, Elsevier, pp. 371–375

Smyth, H.F., Jr & Carpenter, C.P. (1948) Further experience with the range finding test in the industrial toxicology laboratory. *J. ind. Hyg. Toxicol.*, *30*, 63–68

Spinazzola, A., Devoto, G., Zedda, S., Carta, G. & Satta, G. (1969) Experimental dimethylformamide intoxication. I. Behavior of peripheral and medullary blood cells in rabbits with acute DMFA intoxication (Ital.). *Folia med.*, *52*, 739–746

Šrám, R.J., Landa, K., Holá, N. & Roznícková, I. (1985) The use of the cytogenetic analysis of peripheral lymphocytes as a method for checking the level of MAC in Czechoslovakia (Abstract No. 87). *Mutat. Res.*, *147*, 322

Stranský, V. (1986) The determination of *N,N*-dimethylformamide in working atmosphere by the method of gas chromatography after sampling on activated charcoal (Czech.). *Prac. Lek.*, *38*, 15–19

Stula, E.F. & Krauss, W.C. (1977) Embryotoxicity in rats and rabbits from cutaneous application of amide-type solvents and substituted ureas. *Toxicol. appl. Pharmacol.*, *41*, 35–55

Tanaka, K.-I. (1971) Toxicity of dimethylformamide (DMF) to the young female rat. *Int. Arch. Arbeitsmed.*, *28*, 95–105

Thiersch, J.B. (1962) Effects of acetamides and formamides on the rat litter *in utero*. *J. Reprod. Fertil.*, *4*, 219–220

Tomasini, M., Todaro, A., Piazzoni, M. & Peruzzo, G.F. (1983) Pathology of dimethylformamide: observations on 14 cases (Ital.). *Med. Lav.*, *74*, 217–220

Topham, J.C. (1980) Do induced sperm-head abnormalities in mice specifically identify mammalian mutagens rather than carcinogens? *Mutat. Res.*, *74*, 379–387

Topham, J.C. (1981) Evaluation of some chemicals by the sperm morphology assay. In: de Serres, F.J. & Ashby, J., eds, *Evaluation of Short-term Tests for Carcinogens. Report of the International Collaborative Program (Progress in Mutation Research, Vol. 1)*, Amsterdam, Elsevier, pp. 718–722

Trueman, R.W. (1981) Activity of 42 coded compounds in the *Salmonella* reverse mutation test (Imperial Chemical Industries, Ltd). In: de Serres, F.J. & Ashby, J., eds, *Evaluation of Short-term Tests for Carcinogens. Report of the International Collaborative Program (Progress in Mutation Research, Vol. 1)*, Amsterdam, Elsevier, pp. 343–350

Tsuchimoto, T. & Matter, B.E. (1981) Activity of coded comopunds in the micronucleus test. In: de Serres, F.J. & Ashby, J., eds, *Evaluation of Short-term Tests for Carcinogens. Report of the International Collaborative Program (Progress in Mutation Research, Vol. 1)*, Amsterdam, Elsevier, pp. 705–711

Tweats, D.J. (1981) Activity of 42 coded compounds in a differential killing test using *Escherichia coli* strains WP2, WP67 (*uvrA polA*), and CM871 (*uvrA lexA recA*). In: de Serres, F.J. & Ashby, J., eds, *Evaluation of Short-term Tests for Carcinogens. Report of the International Collaborative Program* (*Progress in Mutation Research, Vol. 1*), Amsterdam, Elsevier, pp. 199–209

Työsuojeluhallitus (National Finnish Board of Occupational Safety and Health) (1987) *HTP-Arvot 1987* [TLV Values 1987] (*Safety Bulletin 25*), Helsinki, Valtion Painatuskeskus, p. 13

US Food and Drug Administration (1988) Adhesives. *US Code fed. Regul., Title 21*, Part 175.105, pp. 126–141

US Occupational Safety and Health Administration (1987) Labor. *US Code fed. Regul., Title 29*, Part 1910.1000, p. 678

Venitt, S. & Crofton-Sleigh, C. (1981) Mutagenicity of 42 coded compounds in a bacterial assay using *Escherichia coli* and *Salmonella typhimurium* (Chester Beatty Research Institute). In: de Serres, F.J. & Ashby, J., eds, *Evaluation of Short-term Tests for Carcinogens. Report of the International Collaborative Program* (*Progress in Mutation Research, Vol. 1*), Amsterdam, Elsevier, pp. 351–360

Wiles, J.S. & Narcisse, J.K., Jr (1971) The acute toxicity of dimethylformamides in several animal species. *Am. ind. Hyg. Assoc. J., 32*, 539–545

Williams, G. (1977) Detection of chemical carcinogens by unscheduled DNA synthesis in rat liver primary cell cultures. *Cancer Res., 37*, 1845–1851

Williams, G.A. & Laspia, M.F. (1979) The detection of various nitrosamines in the hepatocyte primary culture/DNA repair test. *Cancer Lett., 6*, 199–206

Williams, S.J., Graepel, G.J. & Kennedy, G.L. (1982) Evaluation of ocular irritancy potential: intralaboratory variability and effect of dosage volume. *Toxicol. Lett., 12*, 235–241

Wilson, H.K. & Ottley, T.W. (1981) The use of a transportable mass spectrometer for the direct measurement of industrial solvents in breath. *Biomed. Mass Spectr., 8*, 606–610

Windholz, M., ed. (1983) *The Merck Index*, 10th ed., Rahway, NJ, Merck & Co., p. 473

Würgler, F.E. & Graf, U. (1981) Mutagenic activity of 10 coded compounds in the *Drosophila* sex-linked recessive lethal assay. In: de Serres, F.J. & Ashby, J., eds, *Evaluation of Short-term Tests for Carcinogens. Report of the International Collaborative Program* (*Progress in Mutation Research, Vol. 1*), Amsterdam, Elsevier, pp. 666–672

Ye, G. (1987) The effect of *N,N*-dimethylformamide on the frequency of micronuclei in bone marrow polychromatic erythrocytes of mice (Chin.). *Zool. Res., 8*, 27–32

Yonemoto, J. & Suzuki, S. (1980) Relation of exposure to dimethylformamide vapor and the metabolite, methylformamide, in urine of workers. *Int. Arch. occup. environ. Health, 46*, 159–165

Zey, J.N., Singal, M., Smith, A.B. & Caplan, P.E. (1987) *Blackman Uhler Chemical Company (Synalloy Corporation), Spartanburg, South Carolina (Health Hazard Evaluation Report No. 85–159–1827*), Cincinnati, OH, National Institute for Occupational Safety and Health

Zimmermann, F.K. & Scheel, I. (1981) Induction of mitotic gene conversion in strain *D7* of *Saccharomyces cerevisiae* by 42 coded compounds. In: de Serres, F.J. & Ashby, J., eds, *Evaluation of Short-term Tests for Carcinogens. Report of the International Collaborative Program* (*Progress in Mutation Research, Vol. 1*), Amsterdam, Elsevier, pp. 481–490

MORPHOLINE

1. Chemical and Physical Data

1.1 Synonyms

Chem. Abstr. Services Reg. No.: 110–91–8
Chem. Abstr. Name: Morpholine
Synonyms: Diethylene imidoxide; diethylene oximide; diethylenimide oxide; 1–oxa–4–azacyclohexane; tetrahydro–*para*–isoxazine; tetrahydro–1,4–isoxazine; tetrahydro–1,4–oxazine; tetrahydro–(2*H*)–1,4–oxazine; tetrahydro–(4*H*)–1,4–oxazine; tetrahydro–*para*–oxazine

1.2 Structural and molecular formulae and molecular weight

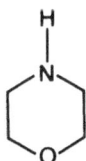

C₄H₉NO

Mol. wt: 87.12

1.3 Chemical and physical properties of the pure substance

(a) *Description*: Clear, colourless, hygroscopic liquid with ammonia–like odour (Weast, 1985; Texaco Chemical Co., 1986)

(b) *Boiling–point*: 128.3°C, 24.8°C at 10 mm Hg (Weast, 1985)

(c) *Freezing–point*: –4.7°C (Weast, 1985); –4.9°C (Texaco Chemical Co., 1986)

(d) *Viscosity*: 2.23 cp at 20°C (Texaco Chemical Co., 1986)

(e) *Flash–point*: 35°C (closed cup) (Mjos, 1978; Texaco Chemical Co., 1986)

(f) *Spectroscopy data*: Nuclear magnetic resonance, infrared and mass spectral data have been reported (Sadtler Research Laboratories, 1980; Pouchert, 1981, 1983; Ohnishi, 1984; Hunt *et al.*, 1985; Pouchert, 1985).

(g) *Solubility*: Soluble in water, acetone, benzene, carbon tetrachloride, diethyl ether, ethanol, methanol, *n*–heptane, propylene glycol methyl ether and ethylene glycol (Weast, 1985; Dow Chemical Co., 1985; Texaco Chemical Co., 1986)

(h)   *Refractive index*: 1.4548 at 20°C (Weast, 1985); 1.4545 at 20°C (Texaco Chemical Co., 1986)

(i)   *Volatility*: Vapour pressure, 7 mm Hg at 20°C (Texaco Chemical Co., 1986)

(j)   *Reactivity*: As a base (pK_B, 5.64), reacts readily with most acids to form corresponding salts; reacts with carbon dioxide under anhydrous conditions to form carbamates and with carbon disulfide to form dithiocarbamates (Mjos, 1978; Dow Chemical Co., 1985; Texaco Chemical Co., 1986). Nitrosation is known to occur, leading to the formation of *N*–nitrosomorpholine (see IARC, 1978).

(k)   *Specific gravity*: 1.0005 at 20°/4°C (Weast, 1985); 1.002 at 20°/20°C (Texaco Chemical Co., 1986)

(l)   *Octanol/water partition coefficient*: log P, –1.08 (Hansch & Leo, 1979)

(m)   *Conversion factor*: mg/m³ = 3.56 × ppm[1]

1.4   Technical products and impurities

Trade names: BASF 238; Drewamine

Morpholine is generally marketed as a high–purity product with the following specifications: assay (min), 99.0%; distillation range, 126.0–130.0°C; specific gravity (20°/20°C), 1.001–1.004; water (max), 0.2–0.5% (Air Products and Chemicals, Inc., 1985; Dow Chemical Co., 1985; Texaco Chemical Co., 1986). It is also available in 40% and 88% solutions with water.

2. Production, Use, Occurrence and Analysis

2.1   Production and use

(a)   Production

The major process for the production of morpholine until the early 1970s was the acid–catalysed dehydration of diethanolamine. This process has largely been replaced by a process based on the reaction of diethylene glycol with ammonia at high temperatures and pressures, with or without a catalyst (Mannsville Chemical Products Corp., 1981).

Typically, diethylene glycol and ammonia are combined in the presence of hydrogen and a catalyst at a temperature between 150–400°C and a pressure between 3–40 MPa (30–400 atm). The hydrogenation catalyst may be any one of a number of metals. Excess ammonia is stripped from the crude reaction mixture, and morpholine is obtained by fractional distillation (Mjos, 1978).

---

[1]Calculated from: mg/m³ = (molecular weight/24.45) × ppm, assuming standard temperature (25°C) and pressure (760 mmg Hg)

Production of morpholine in the USA in 1981 was estimated to be in the range of 13 000 tonnes – about 40% higher than output ten years earlier. Imports to the USA may have been about 1100–1600 tonnes per year in the late 1970s; in 1981, they were reported to be in the range of 700 tonnes. About 1400 tonnes per year are exported from the USA (Mannsville Chemical Products Corp., 1981).

Data on production of morpholine elsewhere in the world were not available to the Working Group.

(b) Use

Morpholine is typically used as follows: rubber chemicals, 40%; corrosion inhibitors, 30%; waxes and polishes, 5%; optical brighteners, 5%; and miscellaneous, 20% (Mannsville Chemical Products Corp., 1981).

(i) Rubber chemicals

Morpholine is an important intermediate for rubber–processing chemicals, especially in the production of delayed–action rubber accelerators, which are added during the vulcanization process to reduce the possibility of prevulcanization during the mixing stages of fabrication. Morpholine–derived sulfur compounds give a high rate of cure during vulcanization, with a reduced tendency to overcure. These include morpholinium N,N–oxydiethylenedithiocarbamate and 2–(N–morpholinothio)benzothiazole. Morpholine derivatives, such as N,N'–dithiodimorpholine and N,N'–tetrathiodimorpholine, are also used to stabilize halogenated butyl rubber against heat–ageing effects (Mjos, 1978; Dow Chemical Co., 1985).

(ii) Corrosion control

Morpholine is used to control corrosion in steam condensate systems. It neutralizes carbon dioxide and other corrosive acid components in steam and condensate, aids in maintaining the proper pH throughout the system, has a suitable vapour pressure and aqueous solubility, and is stable at temperatures up to 288°C. Morpholine and its salts with fatty acids from animal and vegetable oils may be used as corrosion inhibitors for steel or tin plate to be used in contact with food. Morpholine has been used as a corrosion inhibitor in the natural gas and pipeline industry (Dow Chemical Co., 1985).

(iii) Waxes and polishes

Morpholine reacts readily with fatty acids, forming emulsifying agents, which are used in the formulation of water–resistant waxes and polishes for automobiles, floors, leather and furniture. As the film of polish emulsion gradually dries, morpholine evaporates to form a film highly resistant to waterspotting and deterioration. An example of this type of polish is a carnauba wax formulation with the following composition (in parts by weight): carnauba wax, 11.2; oleic acid, 2.4; morpholine, 2.2; water, 67.0. A typical silicone automobile polish–cleaner is composed as follows (in parts by weight): silicone fluid, 4.0; oleic acid, 2.5; morpholine, 1.5; Stoddard solvent (see the monograph on some petroleum solvents, p.), 19; kerosene (deodourized), 2; water, 57; abrasive, 14. The following formulation is representative of silicone–containing furniture polishes (in parts by weight): silicone fluid, 5.0; VM & P naphtha (see the monograph on some petroleum solvents, p. 43) (high flash), 30.0; oleic acid, 2.5; morpholine, 1.5; water, 60.6; water–soluble resin, 0.4 (Dow Chemical Co., 1985).

(iv) *Optical brighteners*

Morpholine is an important intermediate in the manufacture of optical brighteners. The diaminostilbene triazine type brightener with morpholine as a substituent on one of the triazine rings is used in home laundry detergents since it is stable to chlorine bleaches (Texaco Chemical Co., 1986).

(v) *Miscellaneous*

Morpholine and its salts have been reported to be used as components of protective coatings applied on fruits and vegetables (Ohnishi, 1984; US Food and Drug Administration, 1988).

The compound has been used in the preparation of pharmaceuticals, including such diverse products as analgesics, local anaesthetics, respiratory and circulatory stimulants, antispasmodics and soluble sulfanilamides (Dow Chemical Co., 1985).

Morpholine derivatives have been used widely in the textile industry, as softening agents for cellulosic fibres, ingredients of rayon spinning baths, sizing emulsifiers, textile lubricants, whitening agents and dyes (Dow Chemical Co., 1985).

Cosmetic products that may incorporate morpholine–based compounds include hair conditioners, deodorants, shampoos, mouthwashes and cosmetic creams. Stabilizers and antioxidants for lubricating oils, soluble oils for cutting tools and carbon remover compounds have been prepared from morpholine. This compound has a high selectivity and solvency for aromatics and can be used to extract benzene (see IARC, 1982), toluene (see monograph, p. 79) and xylene (see monograph, p. 125) economically from petroleum feedstock. Other products synthesized from morpholine include ion–exchange resins, dyes for electrolytic recording inks and photographic chemicals (Dow Chemical Co., 1985).

(c) *Regulatory status and guidelines*

Morpholine can be used as a component of food, provided that: (i) it is used as the salt(s) of one or more of the fatty acids meeting certain requirements, as a component of protective coatings applied to fresh fruits and vegetables; and (ii) it is used at a level not in excess of that reasonably required to produce its intended effect (US Food and Drug Administration, 1988).

Morpholine is allowed as a boiler–water additive either alone or in combination with other substances at up to 10 ppm (36 mg/m³) concentration in steam, when the steam may contact food, excluding milk or milk products. Morpholine and morpholine fatty acid salts derived from animal or vegetable oils are allowed for use as corrosion inhibitors for steel and tin plate. Additional regulations cover its use as a component of adhesives, as a defoaming agent component for use in the manufacture of paper and paperboard and as a component of animal glue (US Food and Drug Administration, 1987).

Occupational exposure limits for morpholine in 18 countries are presented in Table 1.

Table 1. Occupational exposure limits for morpholine[a]

Country	Year	Concentration[b] (mg/m³)	Interpretation[c]
Australia	1984	S 70	TWA
Austria	1985	70	TWA
Belgium	1985	S 70	TWA
Denmark	1988	S 70	TWA
Finland	1987	S 70	TWA
		S 105	STEL (15 min)
France	1986	70	TWA
		105	STEL (15 min)
Germany, Federal Republic of	1988	S 70	TWA
Indonesia	1985	S 70	TWA
Netherlands	1986	S 70	TWA
Norway	1981	S 70	TWA
Romania	1985	S 40	Average
		S 60	Maximum
Sweden	1987	70	TWA
		110	STEL (15 min)
Switzerland	1985	S 70	TWA
UK	1987	S 70	TWA
		S 105	STEL (10 min)
USA[d]			
OSHA	1987	70	TWA
ACGIH	1988	S 70	TWA
		S 105	STEL (15 min)
USSR	1986	S 0.5	Ceiling
Venezuela	1985	S 70	TWA
		S 105	STEL
Yugoslavia	1984	S 70	TWA

[a]From Direktoratet for Arbeidstilsynet (1981); International Labour Office (1984); Arbeidsinspectie (1986); Institut National de Recherche et de Sécurité (1986); Cook (1987); Health and Safety Executive (1987); National Swedish Board of Occupational Safety and Health (1987); Työsuojelahallitus (1987); US Occupational Safety and Health Administration (1987); American Conference of Governmental Industrial Hygienists (1988); Arbejdstilsynet (1988); Deutsche Forschungsgemeinschaft (1988)

[b]S, skin notation

[c]TWA, 8–h time-weighted average; STEL, short–term exposure limit

[d]OSHA, Occupational Safety and Health Administration; ACGIH, American Conference of Governmental Hygienists

2.2 Occurrence

(a) Natural occurrence

Morpholine is not known to occur as a natural product.

(b)   *Occupational exposure*

On the basis of a US National Occupational Exposure Survey, the National Institute for Occupational Safety and Health (1983) estimated that 117 000 workers were potentially exposed to morpholine in the USA in 1981–83.

Eight–hour time–weighted average personal exposures of up to 0.4 ppm (1.4 mg/m³) morpholine were reported in a plant that manufactured a rubber accelerator, 4–morpholinyl–2–benzothiazole disulfide (Taft & Stroman, 1979).

(c)   *Food and beverages*

Morpholine has been detected in many samples of foodstuffs. In six fish samples, the following concentrations were determined (μmol/100 g) [mg/kg]: tinned tuna, ≤0.7 [≤0.6]; frozen ocean perch, 10 [9]; frozen cod, < 0.3 [< 0.3]; spotted trout, 7 [6]; small mouth bass, ≤0.8 [≤0.7]; salmon, 1.2 [1.0]. Two meat samples also contained morpholine: baked ham, 0.6 μmol/100 g [0.5 mg/kg];; sausages, 0.5 μmol/100 g [0.4 mg/kg]. Six beverage samples had the following concentrations (μmol/100 g) [mg/kg]: evaporated milk, 0.2 [0.2]; coffee, 1 [1]; tea, < 0.1 [< 0.1]; tinned beer, 0.5 [0.4]; bottled beer, ≤0.3 [≤0.2]; wine, ≤0.8 [≤0.7] (Singer & Lijinsky, 1976).

Morpholine was also detected at an average concentration of 0.2 mg/kg in five samples of baked ham. None was detected (limit of detection, 0.01 mg/kg) in fish sausage, cod roe, spinach or fermented soya bean (Hamano *et al.*, 1981). Morpholine was tentatively identified by gas chromatographic retention time in an extract of cooked bacon (Rounbehler & Fine, 1982).

The morpholine concentrations in paper and paperboard food packages ranged from 0.10 to 0.84 mg/kg, with a mean of 0.38 mg/kg. Those materials that contained *N*–nitrosomorpholine (see IARC, 1978) usually contained higher concentrations of morpholine. Morpholine was detected at a concentration of 0.018 mg/kg in flour packaged in a paper sack that contained morpholine (Hotchkiss & Vecchio, 1983).

Residual levels of morpholine in the rinds of commercial citrus fruits ranged from undetectable (< 0.02 mg/kg) to 71.1 mg/kg. Levels in the flesh of citrus fruits and in marmalade were much lower, from undetectable to 0.7 mg/kg (Ohnishi, 1984).

2.3   Analysis

Singer and Lijinsky (1976) analysed foodstuffs for naturally occurring secondary amines by gas–liquid chromatography–mass spectrometry of *para*–toluenesulfonamide (tosylamide) derivatives. Tosylamides were prepared by steam distillation of the homogenized sample, extraction and reaction with *para*–toluenesulfonyl chloride under alkaline conditions. The lower quantifiable limit for morpholine in foods and beverages by this method, using dual flame–ionization/Coulson detectors, was approximately 0.1–0.3 mg/kg.

Secondary amines in foods have also been quantified by gas chromatography (GC) of benzene sulfonamide derivatives. Derivatives were prepared by extraction of the homogenized sample with acidified methanol, followed by a series of solvent exchanges and reaction with benzenesulfonyl chloride under alkaline conditions. Using this method, the detection limit for morpholine, analysed by capillary GC with a flame photometric detector, was 0.01 mg/kg (Hamano *et al.*, 1981).

Morpholine has been detected (lower limit of detection, 0.02 mg/kg) in citrus fruits by steam distillation, column chromatography and GC analysis as the free amine with a flame-ionization detector (FID). Identity of the GC peak was confirmed by mass spectrometry (Ohnishi, 1984).

GC–FID, after a multi-step extraction and solvent exchange procedure, has been used to quantify morpholine in blood plasma, urine and biological tissues (Tombropoulos, 1979).

The sensitivity and selectivity of the thermal energy analyser (TEA) in the GC analysis of nitrosamines has been used to advantage in the determination of *N*-nitrosomorpholine and morpholine in foods and food packaging materials. Morpholine itself has been determined by nitrosation of extracts or distillates prior to GC–TEA analysis (lower limit of detection, approximately 0.003 mg/kg). A modified TEA, which converts any organic nitrogen to the nitrosyl radical, has been used to analyse directly for morpholine in foods and beverages (Rounbehler *et al.*, 1980; Rounbehler & Fine, 1982).

3. Biological Data Relevant to the Evaluation of Carcinogenic Risk to Humans

3.1 Carcinogenicity studies in animals[1,2]

(a) Oral administration

Mouse: As part of a toxicity study to investigate the nitrosation of morpholine, groups of 20 male and 20 female random-bred Swiss mice, six to 11 weeks of age, were fed a diet containing 6.33 g/kg morpholine (purified) for 28 weeks and observed for a further 12 weeks. A group of 80 males and 80 females received no treatment. Surviving mice were killed in week 40. Lung adenomas were counted grossly, and the numbers of adenoma-bearing mice and the total numbers of lung adenomas per mouse were comparable in the treated and untreated groups: 5/38 and 20/144 mice had five and 26 adenomas (both sexes pooled), respectively. There was a significant increase in the incidence of extrapulmonary tumours in treated mice (Greenblatt *et al.*, 1971). [The Working Group noted the short duration of the study.]

Groups of 50 male and 50 female B6C3F1 mice, six weeks old, received 0, 0.25 or 1.0% morpholine oleic acid salt [purity unspecified] in the drinking-water for 96 weeks followed by

---

[1]Studies on the carcinogenicity of morpholine in combination with nitrite and of *N*-nitrosomorpholine were evaluated previously, in Volume 17 of the *IARC Monographs* (1978); *N*-nitrosomorpholine is carcinogenic to experimental animals.

[2]The Working Group was aware of a study in progress in rats by oral administration of morpholine (IARC, 1988).

eight weeks on tap water. Survival was not affected, but body weight was significantly reduced in high–dose males and in females of both treatment groups. No significant increase in the incidence of any tumour occurred in treated animals. The incidence of squamous epithelial hyperplasia of the forestomach was increased in high–dose males (Shibata *et al.*, 1987a).

Rat: In a long–term study to investigate the nitrosation of morpholine, groups of pregnant Sprague–Dawley rats and their offspring, treated from conception, were fed reagent-grade morpholine at 0 or 1000 mg/kg of diet and an F_2 generation was treated similarly. Median survival was 117 weeks for treated and 109 weeks for control rats. In the group of 104 rats treated with morpholine (F_1 and F_2 generations combined), three developed liver–cell carcinomas, two lung angiosarcomas and two malignant brain gliomas. A group of 156 untreated controls (F_1 and F_2 combined) did not develop these tumours. The diet was stated to contain no detectable nitrite or *N*-nitrosomorpholine, and the rats were given only distilled drinking-water. The authors noted that the liver and lung tumours might be the result of the interaction of morpholine with nitrite of unknown origin to form *N*-nitrosomorpholine (Newberne & Shank, 1973; Shank & Newberne, 1976).

Hamster: In a long–term study to investigate the nitrosation of morpholine, groups of pregnant random–bred Syrian golden hamsters were treated from conception with reagent-grade morpholine at 0 or 1000 mg/kg of diet. Animals were killed at 110 weeks of age. Median survival time was 72 weeks for controls and 68 weeks for animals treated with morpholine. One liver–cell tumour and four angiosarcomas were observed among 23 control animals, but none of these tumours were seen in 22 treated animals of both sexes combined (Shank & Newberne, 1976). [The Working Group noted the small number of animals used.]

(b)   Inhalation

Rat: Groups of 70 male and 70 female Sprague–Dawley rats, approximately nine weeks of age, were exposed by inhalation to 0, 10, 50 or 150 ppm (0, 36, 178 or 534 mg/m³) morpholine (purity, 99.16%) for 6 h per day on five days per week for up to 104 weeks. Levels of nitrates and nitrites in the drinking-water were reported to be < 0.1 mg/l and 0.01 mg/l, respectively. An interim kill of ten males and ten females per group at was done 53 weeks. The experiment was terminated in week 105. Survival at termination in control, low-, mid- and high–dose groups was 40, 44, 33 and 32 males and 35, 27, 32 and 35 females. Exposure to morpholine was associated with dose–related increases in inflammation of the cornea, inflammation and squamous metaplasia of the turbinate epithelium, and necrosis of the turbinate bones in the nasal cavity in rats of both sexes. No significant increase in the incidence of tumours was seen in rats of either sex (Harbison *et al.*, 1989). [The Working Group noted that only tissues from the respiratory tract and eyes were examined histologically for mid- and low–dose groups.]

3.2 Other relevant data

(a) Experimental systems

(i) Absorption, distribution, excretion and metabolism

Morpholine is absorbed after oral, dermal and inhalation administration. In rats, it was distributed to all organs and was eliminated rapidly (Tanaka et al., 1978). In mice, rats, hamsters, guinea-pigs and rabbits, almost all ingested or intravenously or intraperitoneally injected morpholine was excreted unchanged in the urine (Tanaka et al., 1978; Van Stee et al., 1981; Sohn et al., 1982). In the urine of guinea-pigs, 20% of an administered dose was identified as N-methylmorpholine-N-oxide (Sohn et al., 1982). In mice, rats and hamsters, N-nitrosomorpholine is formed following concomitant administration of morpholine and nitrite or nitrous oxide (e.g., IARC, 1978; Van Stee et al., 1983). N-Nitrosomorpholine was formed when morpholine was added to human saliva (Wishnok & Tannenbaum, 1977).

(ii) Toxic effects

The oral LD_{50} for morpholine was reported to be 1.05 g/kg bw in rats (Smyth et al., 1954); that for 1:4 aqueous dilutions was 1.6 g/kg bw in rats and 0.9 g/kg bw in guinea-pigs (Shea, 1939). The skin penetration LD_{50} for morpholine was reported to be 0.5 ml(g)/kg in rabbits (Smyth et al., 1954). The maximal exposure time without mortality to concentrated morpholine vapour by inhalation was estimated to be about 1 h, and 6/6 rats survived at least 14 days after an 8-h exposure to 8000 ppm (28 500 mg/m³; Smyth et al., 1954). As reported in an abstract, acute inhalation studies have given LC_{50} values of 2250 ppm (8010 mg/m³) in male and 2150 ppm (7650 mg/m³) in female rats and of 1450 ppm (5160 mg/m³) in male and 1900 ppm (6760 mg/m³) in female mice (Lam & Van Stee, 1978).

Morpholine is a strong skin and eye irritant in rabbits (Smyth et al., 1954) and guinea-pigs; it caused skin necrosis after dermal application of 0.9 g/kg bw diluted or undiluted, unneutralized compound to rabbits and of 0.9 g/kg bw undiluted, unneutralized compound to guinea-pigs (Shea, 1939).

Morpholine (undiluted, unneutralized) caused necrosis of the liver and tubular necrosis of the kidney in guinea-pigs after dermal application of 0.9 g/kg bw. Similar effects were observed in rabbits following dermal application of 0.9 g/kg bw dilute morpholine. A single oral administration (0.1–10 g/kg bw) of undiluted, unneutralized morpholine caused haemorrhages in the stomach and small intestine in guinea-pigs and rats; treatment of guinea-pigs for up to 30 days (0.5 g/kg bw) caused necrosis of the liver and renal tubules. Liver and kidney necrosis also occurred after exposure of rats by inhalation to 12 000 or 18 000 ppm (42 720 or 64 000 mg/m³) for up to 42 h (Shea, 1939).

Morpholine given to rats at a level of 1 g/kg in the diet caused fatty degeneration of the liver after 270 days (Sander & Bürkle, 1969). In mice given 0.15–2.5% morpholine oleic acid salt in the drinking-water for 13 weeks, a significant increase in urine specific gravity and plasma urea nitrogen and an increase in the relative weights of the kidneys were observed in males and females in the two highest dose groups. Slight cloudy swelling of the proximal tubules of the kidneys was seen in animals of each sex given the highest dose (Shibata et al., 1987b).

When rats were exposed by inhalation to 250 ppm (890 mg/m³) morpholine for 6 h per day on five days per week, there were signs of an irritant effect after one week. In some of the animals killed after seven weeks, and in almost all killed after 13 weeks, there were focal erosions and squamous metaplasia in the maxilloturbinates. Similar effects were noted in a few animals exposed to 100 ppm (356 mg/m³) for 13 weeks (Conaway et al., 1984a). It was reported in an abstract that rats exposed by inhalation to 450 ppm (1600 mg/m³) morpholine for 6 h per day on five days per week for eight weeks developed changes in sensory areas (eyes, nose), had decreased body weight gains and had increased organ:body weight ratios for the lungs and kidneys (Lam & Van Stee, 1978).

After 33 exposures to 250 ppm (890 mg/m³) in air, rabbits displayed increased enzyme activities in alveolar macrophages, indicating airway damage (Tombropoulos et al., 1983).

(iii) *Effects on reproduction and prenatal toxicity*

No data were available to the Working Group.

(iv) *Genetic and related effects*

Morpholine was not mutagenic to *Salmonella typhimurium* TA1535, TA1537, TA98 or TA100 either in the presence or in the absence of an exogenous metabolic activation system from Aroclor-induced rats (Haworth et al., 1983). A morpholine fatty acid salt was not mutagenic to *S. typhimurium* TA1535, TA1537, TA92, TA94, TA98 or TA100 either in the presence or in the absence of an exogenous metabolic activation system (Ishidate et al., 1984). It did not induce mutation in *S. typhimurium* TA1590 when used as indicator strain in a host-mediated assay in NMRI mice (Braun et al., 1977), or in *S. typhimurium* TA1530 in CD-1 mice (Edwards et al., 1979).

Morpholine did not induce DNA repair in primary cultures of rat hepatocytes (Conaway et al., 1984b).

Morpholine did not cause chromosomal aberrations, micronuclei, morphological transformation, or 8-azaguanine- or ouabain-resistant mutants in hamster embryo cells following administration by stomach tube to mothers on day 11 or 12 of pregnancy (Inui et al., 1979). A morpholine fatty acid salt did not induce chromosomal aberrations in Chinese hamster fibroblasts (Ishidate et al., 1984).

In an abstract, morpholine was reported to have a weak positive effect on the induction of mutations in L5178 TK$^{+/-}$ mouse lymphoma cells without metabolic activation. In the same abstract, an increase in the frequency of morphological transformation of Balb/3T3 cells was also reported (Conaway et al., 1982).

(b) *Humans*

(i) *Absorption, distribution, excretion and metabolism*

No data were available to the Working Group.

(ii) *Toxic effects*

Lachrymation, rhinitis and lower airway irritation have been reported in humans exposed to 'high levels' of morpholine in an industrial setting. Corneal oedema with 'hazy vision' and halo phenomena around lights have also been reported (National Research Council, 1983; Grant, 1986).

(iii) *Effects on fertility and on pregnancy outcome*

No data were available to the Working Group.

(iv) *Genetic and related effects*

No data were available to the Working Group.

3.3 Epidemiological studies of carcinogenicity to humans

No data were available to the Working Group.

4. Summary of Data Reported and Evaluation

4.1 Exposures

Morpholine is a synthetic organic liquid used mainly as an intermediate in the production of rubber chemicals and optical brighteners, as a corrosion inhibitor in steam condensate systems, as an ingredient in waxes and polishes and as a component of protective coatings on fresh fruits and vegetables. Occupational exposure may occur during the production of morpholine and in its various uses, but data on exposure levels are sparse. It has been detected in samples of foodstuffs and beverages.

4.2 Experimental carcinogenicity data

Morpholine was tested for carcinogenicity by oral administration in two strains of mice, one strain of rats and one strain of hamsters. The studies in one of the strains of mice and in hamsters were considered inadequate for evaluation. In the other strain of mice, no significant increase in the incidence of tumours was seen in treated animals. In the study in rats, a few tumours of the liver and lung occurred in treated animals. Morpholine was also tested by inhalation exposure in rats; it did not increase the incidence of tumours over that in controls.

4.3 Human carcinogenicity data

No data were available to the Working Group.

4.4 Other relevant data

Morpholine is an irritant in humans and experimental animals. It caused kidney damage in experimental animals.

Morpholine did not induce micronuclei, chromosomal aberrations or mutation in hamsters. It did not induce morphological transformation, chromosomal aberrations or DNA damage in cultured animal cells. It did not induce mutations in bacteria. (See Appendix 1.)

4.5 Evaluation[1]

There is *inadequate evidence* for the carcinogenicity of morpholine in experimental animals.

No data were available from studies in humans on the carcinogenicity of morpholine.

Overall evaluation

Morpholine *is not classifiable as to its carcinogenicity to humans (Group 3)*.

5. References

Air Products and Chemicals, Inc. (1985) *Amines from Air Products – More than Chemicals, Chemistry*, Allentown, PA

American Conference of Governmental Industrial Hygienists (1988) *Threshold Limit Values and Biological Exposure Indices for 1988–1989*, Concinnati, OH, p. 28

Arbeidsinspectie (Labour Inspection) (1986) *De Nationale MAC–Lijst 1986* [National MAC-List 1986] (*P145*), Voorburg, Ministry of Social Affairs and Work Environment, p. 18

Arbejdstilsynet (Labour Inspection) (1988) *Graensevaerdier for Stoffer og Materialer* [Limit Values for Substances and Materials] (*At–ansvisning No. 3.1.0.2*), Copenhagen, p. 25

Braun, R., Schöneich, J. & Ziebarth, D. (1977) In vitro formation of *N*–nitroso compounds and detection of their mutagenic activity in the host–mediated assay. *Cancer Res., 37*, 4572–4579

Conaway, C.C., Myhr, B.C., Rundell, J.O. & Brusick, D.J. (1982) Evaluation of morpholine, piperazine and analogues in the L5178Y mouse lymphoma assay and BALB/3T3 transformation assay (Abstract No. Dg–7). *Environ. Mutagenesis, 4*, 390

Conaway, C.C., Coate, W.B. & Voelker, R.W. (1984a) Subchronic inhalation toxicity of morpholine in rats. *Fundam. appl. Toxicol., 4*, 465–472

Conaway, C.C., Tong, C. & Williams, G.M. (1984b) Evaluation of morpholine, 3–morpholine, and *N*–substituted morpholines in the rat hepatocyte primary culture/DNA repair test. *Mutat. Res., 136*, 153–157

Cook, W.A. (1987) *Occupational Exposure Limits – Worldwide*, Washington DC, American Industrial Hygiene Association, pp. 30, 147, 202

Deutsche Forschungsgemeinschaft (German Research Society) (1988) *Maximale Arbeitsplatzkonzentrationen und Biologische Arbeitsstofftoleranzwerte 1988* [Maximal Concentrations at the Workplaces and Biological Tolerance Values for Working Materials 1988] (*Report No. XXIV*), Weinheim, VCH Verlagsgesellschaft, p. 47

Direktoratet for Arbeidstilsynet (Directorate for Labour Inspection) (1981) *Administrative Normer for Forurensning i Arbeidsatmosfaere 1981* [Administrative Norms for Pollution in the Work Atmosphere 1981] (*No. 361*), Olso, p. 17

---

[1]For definitions of the italicized terms, see Preamble, pp. 27–30.

Dow Chemical Co. (1985) *Morpholine*, Midland, MI

Edwards, G., Whong, W.-Z. & Speciner, N. (1979) Intrahepatic mutagenesis assay: a sensitive method for detecting *N*-nitrosomorpholine and in vivo nitrosation of morpholine. *Mutat. Res., 64*, 415–423

Grant, W.M. (1986) *Toxicology of the Eye*, 3rd ed., Springfield, IL, C.C. Thomas, pp. 75–76, 642

Greenblatt, M., Mirvish, S. & So, B.T. (1971) Nitrosamine studies: induction of lung adenomas and concurrent administration of sodium nitrite and secondary amines in Swiss mice. *J. natl Cancer Inst., 46*, 1029–1034

Hamano, T., Mitsuhashi, Y. & Matsuki, Y. (1981) Glass capillary gas chromatography of secondary amines in foods with flame photometric detection after derivatization with benzenesulfonyl chloride. *Agric. Biol. Chem., 45*, 2237–2243

Hansch, C. & Leo, A. (1979) *Substituent Constants for Correlation Analysis in Chemistry and Biology*, New York, John Wiley & Sons, p. 186

Harbison, R.D., Marino, D.J., Conaway, C.C., Rubin, L.F. & Gandy, J. (1989) Chronic morpholine exposure of rats. *Fundam. appl. Toxicol., 12*, 491–507

Haworth, S., Lawlor, T., Mortelmans, K., Speck, W. & Zeiger, A. (1983) *Salmonella* mutagenicity test results for 250 chemicals. *Environ. Mutat., Supplement 1*, 3–142

Health and Safety Executive (1987) *Occupational Exposure Limits 1987 (Guidance Note EH 40/87)*, London, Her Majesty's Stationery Office, p. 18

Hotchkiss, J.H. & Vecchio, A.J. (1983) Analysis of direct contact paper and paperboard food packaging for *N*-nitrosomorpholine and morpholine. *J. Food Sci., 48*, 240–242

Hunt, D.F., Shabanowitz, J., Harvey, T.M. & Coates, M. (1985) Scheme for the direct analysis of organics in the environment by tandem mass spectrometry. *Anal. Chem., 57*, 525–537

IARC (1978) *IARC Monographs on the Evaluation of the Carcinogenic Risk of Chemicals to Humans*, Vol. 17, *Some N-Nitroso Compounds*, Lyon, pp. 263–280

IARC (1982) *IARC Monographs on the Evaluation of the Carcinogenic Risk of Chemicals to Humans*, Vol. 29, *Some Industrial Chemicals and Dyestuffs*, Lyon, pp. 93–148, 395–398

IARC (1988) *Information Bulletin on the Survey of Chemicals Being Tested for Carcinogenicity*, No. 13, Lyon, p. 101

Institut National de Recherche et de Sécurité (National Institute for Research and Safety) (1986) *Valeurs Limites pour les Concentrations des Substances Dangereuses Dans l'Air des Lieux de Travail* [Limit Values for Concentrations of Dangerous Substances in the Air of Work Places] (*ND 1609–125–86*), Paris, p. 572

International Labour Office (1984) *Occupational Exposure Limits for Airborne Toxic Substances*, 2nd rev. ed. (*Occupational Safety and Health Series No. 37*), Geneva, p. 152

Inui, N., Nishi, Y., Taketomi, M., Mori, M., Yamamoto, M., Yamada, T. & Tanimura, A. (1979) Transplacental mutagenesis of products formed in the stomach of golden hamsters given sodium nitrite and morpholine. *Int. J. Cancer, 24*, 365–372

Ishidate, M., Jr, Sofuni, T., Yoshikawa, K., Hayashi, M., Nohmi, T., Sawada, M. & Matsuoka, A. (1984) Primary mutagenicity screening of food additives currently used in Japan. *Food chem. Toxicol., 22*, 623–636

Lam, H.F. & Van Stee, E.W. (1978) A re-evaluation of the toxicity of morpholine (Abstract No. 2459). *Fed. Proc., 37*, 679

Mannsville Chemical Products Corp. (1981) *Chemical Products Synopsis: Morpholine*, Cortland, NY

Mjos, K. (1978) Cyclic amines. In: Mark, H.F., Othmer, D.F., Overberger, C.G., Seaborg, G.T. & Grayson, M., eds, *Kirk–Othmer Encyclopedia of Chemical Technology*, 3rd ed., Vol. 2, New York, John Wiley & Sons, pp. 295-308

National Institute for Occupational Safety and Health (1983) *US National Occupational Exposure Survey 1981–83*, Cincinnati, OH

National Research Council (1983) *An Assessment of the Health Risks of Morpholine and Diethylaminoethanol (PB 85–122661)*, Washington DC, National Academy Press

National Swedish Board of Occupational Safety and Health (1987) *Hygienska Gränsvärder* [Hygienic Limit Values] *(Ordinance 1987:12)*, Solna, p. 32

Newberne, P.M. & Shank, R.C. (1973) Induction of liver and lung tumours in rats by the simultaneous administration of sodium nitrite and morpholine. *Food Cosmet. Toxicol.*, *11*, 819-825

Ohnishi, T. (1984) Studies on mutagenicity of the food additive morpholine (fatty acid salt) (Jpn.). *Jpn. J. Hyg.*, *39*, 729-748

Pouchert, C.J., ed. (1981) *The Aldrich Library of Infrared Spectra*, 3rd ed., Milwaukee, WI, Aldrich Chemical Co., p. 225H

Pouchert, C.J., ed. (1983) *The Aldrich Library of NMR Spectra*, 2nd ed., Vol. 1, Milwaukee, WI, Aldrich Chemical Co., p. 335C

Pouchert, C.J., ed. (1985) *The Aldrich Library of FT–IR Spectra*, Vol. 1, Milwaukee, WI, Aldrich Chemical Co., p. 225H

Rounbehler, D.P. & Fine, D.H. (1982) Specific detection of amines and other nitrogen–containing compounds with a modified TEA analyser. In: Bartsch, H., O'Neill, I.K., Castegnaro, M. & Okada, M., eds, *N–Nitroso Compounds: Occurrence and Biological Effects (IARC Scientific Publications No. 41)*, Lyon, International Agency for Research on Cancer, pp. 209-219

Rounbehler, D.P., Reisch, J. & Fine, D.H. (1980) Some recent advances in the analysis of volatile *N*–nitrosamines. In: Walker, E.A., Castegnaro, M., Griciute, L. & Börzsönyi, M., eds, *N–Nitroso Compounds: Analysis, Formation and Occurrence (IARC Scientific Publications No. 31)*, Lyon, International Agency for Research on Cancer, pp. 403-417

Sadtler Research Laboratories (1980) *Standard Spectra Collection, 1980 Cumulative Index*, Philadelphia, PA

Sander, J. & Bürkle, G. (1969) Induction of malignant tumours in rats by simultaneous feeding of nitrite and secondary amines (Ger.). *Z. Krebsforsch.*, *73*, 54-66

Shank, R.C. & Newberne, P.M. (1976) Dose–response study of the carcinogenicity of dietary sodium nitrite and morpholine in rats and hamsters. *Food Cosmet. Toxicol.*, *14*, 1-8

Shea, T.E., Jr (1939) The acute and sub–acute toxicity of morpholine. *J. ind. Hyg. Toxicol.*, *2*, 236-245

Shibata, M.-A., Kurata, Y., Ogiso, T., Tamano, S., Fukushima, S. & Ito, N. (1987a) Combined chronic toxicity and carcinogenicity studies of morpholine oleic acid salt in B6C3F1 mice. *Food chem. Toxicol.*, *25*, 569-574

Shibata, M.-A., Kurata, Y., Tamano, S., Ogiso, T., Fukushima, S. & Ito, N. (1987b) 13–Week subchronic toxicity study with morpholine oleic acid salt administered to B6C3F1 mice. *J. Toxicol. environ. Health*, *22*, 187-194

Singer, G.M. & Lijinsky, W. (1976) Naturally occurring nitrosatable compounds. I. Secondary amines in foodstuffs. *J. agric. Food Chem.*, *24*, 550-553

Smyth, H.F., Jr, Carpenter, C.P., Weil, C.S. & Pozzani, U.C. (1954) Range–finding toxicity data. List V. *Arch. ind. Hyg. occup. Med.*, *10*, 61-68

Sohn, O.S., Fiala, E.S., Conaway, C.C. & Weisburger, J.H. (1982) Metabolism and disposition of morpholine in the rat, hamster, and guinea pig. *Toxicol. appl. Pharmacol.*, *64*, 486–491

Taft, R.M. & Stroman, R.E. (1979) *Goodyear Tire and Rubber Co., Niagara Falls, New York (Health Hazard Evaluation Determination Report No. 78–131–586)*, Cincinnati, OH, National Institute for Occupational Safety and Health

Tanaka, A., Tokieda, T., Nambaru, S., Osawa, M. & Yamaha, T. (1978) Excretion and distribution of morpholine salts in rats. *J. Food Hyg. Soc.*, *19*, 329–334

Texaco Chemical Co. (1986) *Morpholine*, Austin, TX

Tombropoulos, E.G. (1979) Micromethod for the gas chromatographic determination of morpholine in biological tissues and fluids. *J. Chromatogr.*, *164*, 95–99

Tombropoulos, E.G., Koo, J.O., Gibson, W. & Hook, G.E.R. (1983) Induction by morpholine of lysosomal α–mannosidase and acid phosphatase in rabbit alveolar macrophages *in vivo* and *in vitro*. *Toxicol. appl. Pharmacol.*, *70*, 1–6

Työsuojeluhallitus (National Finnish Board of Occupational Safety and Health) (1987) *HTP–Arvot 1987* [TLV Values 1987] (*Safety Bulletin 25*), Helsinki, Valtion Painatuskeskus, p. 21

US Food and Drug Administration (1987) Subpart D. Specific usage additives. *US Code fed. Regul.*, *Title 21*, Parts 173.310, 175.105, 176.210, 178.3120, 178.3300, pp. 117–119, 126–140, 196–197, 321–322, 326

US Food and Drug Administration (1988) Morpholine. *US Code fed. Regul.*, *Title 21*, Part 172.250, p. 33

US Occupational Safety and Health Administration (1987) Labor. *US Code fed. Regul.*, *Title 29*, Part 1910.1000, p. 680

Van Stee, E.W., Wynns, P.C. & Moorman, M.P. (1981) Distribution and disposition of morpholine in the rabbit. *Toxicology*, *20*, 53–60

Van Stee, E.W., Sloane, R.A., Simmons, J.E. & Brunnemann, K.D. (1983) In vivo formation of *N*-nitrosomorpholine in CD-1 mice exposed by inhalation to nitrogen dioxide and by gavage to morpholine. *J. natl Cancer Inst.*, *70*, 375–379

Weast, R.C., ed. (1985) *CRC Handbook of Chemistry and Physics*, 66th ed., Boca Raton, FL, CRC Press, p. C-355

Wishnok, J.S. & Tannenbaum, S.R. (1977) An unknown salivary morpholine metabolite. Identification of the metabolite leads to the discovery of a new biochemical reaction of secondary amines. *Anal. Chem.*, *49*, 715A–718A

SOLVENT STABILIZER

1,2-EPOXYBUTANE

1. Chemical and Physical Data

1.1 Synonyms

Chem. Abstr. Services Reg. No.: 106–88–7
Chem. Abstr. Name: Ethyloxirane
IUPAC Systematic Name: 1,2–Butylene oxide
Synonyms: 1–Butene oxide; 1,2–butene oxide; 1,2–butylene epoxide; α–butylene oxide; 1–butylene oxide; epoxybutane; ethyl ethylene oxide; 2–ethyloxirane

1.2 Structural and molecular formulae and molecular weight

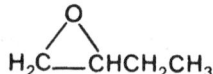

C_4H_8O Mol. wt: 72.12

1.3 Chemical and physical properties of the pure substance:

(*a*) *Description*: Clear, colourless liquid with pungent odour (Dow Chemical Co., 1988)

(*b*) *Boiling–point*: 63.3°C (Weast, 1985)

(*c*) *Melting–point*: < –60°C (Parmeggiani, 1983)

(*d*) *Density*: 0.8312 (20°/20°C) (Hawley, 1981); 0.837 (17°C/4°C) (Weast, 1985)

(*e*) *Spectroscopy data*: Nuclear magnetic resonance and infrared spectral data have been reported (Sadtler Research Laboratories, 1980; Pouchert, 1981, 1983, 1985).

(*f*) *Solubility*: Miscible with most organic solvents (National Toxicology Program, 1988); very soluble in ethanol and diethyl ether (Weast, 1985); soluble in water (Hawley, 1981)

(*g*) *Volatility*: Vapour pressure, 140 mm Hg at 20°C (Dow Chemical Co., 1988)

(*h*) *Refractive index*: 1.3851 at 20°C (Weast, 1985)

(i)   *Flash–point*: –22°C (closed–cup) (Dow Chemical Co., 1988)

(j)   *Reactivity*: Extremely inflammable; reacts with water and other source of labile hydrogen, especially in the presence of acids, bases or other oxidizing substances. Reactive monomer which can polymerize exothermically. Undergoes atmospheric hydrolysis; atmospheric half–life for oxidation estimated to be six days (Hine *et al.*, 1981; National Toxicology Program, 1988; Dow Chemical Co., 1988)

(k)   *Conversion factor*: $mg/m^3 = 2.95 \times ppm$[1]

1.4   Technical products and impurities

1,2–Epoxybutane is available at a purity of 98%, 99% or >99%. 1,2–Butanediol has been identified as an impurity (<0.2%) (Riedel–de Haën, 1984; Aldrich Chemical Co., Inc., 1988; National Toxicology Program, 1988).

2. Production, Use, Occurrence and Analysis

2.1   Production and Use

(a)   *Production*

1,2–Epoxybutane has been prepared by the chlorohydrin process from 1–butene (α–butene) and chlorine in water. The butylene chlorohydrin produced is dehydrochlorinated with calcium hydroxide to yield 1,2–epoxybutane, with calcium chloride as a byproduct (Considine, 1974; Hine *et al.*, 1981). A newer process based on the catalysed reaction of olefins with hydroperoxides produces high yields of epoxides and the corresponding alcohols. 1,2–Epoxybutane is reported to be produced commercially by the epoxidation of 1–butene with peroxyacetic acid (Considine, 1974; Hoff *et al.*, 1978).

It has been reported that 3.6 thousand tonnes of 1,2–epoxybutane were produced in the USA in 1978 (National Toxicology Program, 1988). Data on production elsewhere in the world were not available.

(b)   *Use*

1,2–Epoxybutane is widely used as a stabilizer for chlorinated hydrocarbon solvents. More than 75% of 1,2–epoxybutane produced commercially is added to chlorine–containing materials such as trichloroethylene to act as an acid–scavenger (Hine *et al.*, 1981; National Toxicology Program, 1988).

---

[1]Calculated from: $mg/m^3 = (molecular\,weight/24.45) \times ppm$, assuming standard temperature (25°C) and pressure (760 mm Hg)

This compound is also used as a chemical intermediate for the production of butylene glycols and their derivatives (polybutylene glycols, mixed poly glycols and glycol ethers and esters), butanol–amines, surface–active agents, and other products such as gasoline additives (Hine *et al.*, 1981; Parmeggiani, 1983). It has been reported to be used as a corrosion inhibiting additive (at 0.25–0.5%) during the preparation of vinyl chloride and its copolymer resins (Hoff *et al.*, 1978).

(c) *Regulatory status and guidelines*

Standards for 1,2–epoxybutane have not been established by any country in the form of regulations or guidelines. However, US manufacturers have recommended a voluntary standard of 40 ppm (118 mg/m³) for an 8–h time–weighted average threshold limit value (National Toxicology Program, 1988).

2.2 Occurrence

(a) *Natural occurrence*

1,2–Epoxybutane is not known to occur naturally.

(b) *Occupational exposure*

On the basis of a US National Occupational Exposure Survey, the National Institute for Occupational Safety and Health (1983) estimated that 47 900 workers were potentially exposed to 1,2–epoxybutane in the USA in 1981-83. No data on levels of exposure to 1,2–epoxybutane were available to the Working Group.

2.3 Analysis

In a general method for epoxides, 1,2–epoxybutane was determined spectrophotometrically after adding a colour–forming reagent, 4–(*para*–nitrobenzyl)pyridine. Maximal absorbance is observed at 574 nm (Agarwal *et al.*, 1979).

3. Biological Data Relevant to the Evaluation of Carcinogenic Risk to Humans

3.1 Carcinogenicity studies in animals

(a) *Inhalation*

Mouse: Groups of 50 male and 50 female B6C3F1 mice, seven to nine weeks old, were exposed by inhalation to 0, 50 or 100 ppm (150 or 300 mg/m³) 1,2–epoxybutane (purity, >99%) for 6 h per day on five days a week for 102 weeks. Survival was comparable in all groups of males (41/50, 45/50 and 33/50 mice killed at termination) but was reduced in high-dose females after week 86 (29/50, 25/50 and 9/50 mice killed at termination). A single squamous-cell papilloma was found in the nasal cavity of a male receiving 100 ppm. Treatment-related, non-neoplastic nasal changes included inflammation, erosion, hyperplasia and squamous metaplasia of the nasal epithelium and atrophy of the olfactory sensory epithelium at both dose levels (Dunnick *et al.*, 1988; National Toxicology Program, 1988).

Rat: Groups of 50 male and 50 female Fischer 344/N rats, seven to nine weeks old, were exposed by inhalation to 0, 200 or 400 ppm (600 or 1200 mg/m^3) 1,2-epoxybutane (purity, >99%) for 6 h per day on five days a week for 103 weeks. Survival at termination of the experiment was 30/50, 18/50 and 23/50 for males and 32/50, 21/50 and 22/50 for females in the control, low-dose and high-dose groups, respectively. The incidences of papillary adenomas of the nasal cavity were 7/50 ($p < 0.05$, adjusted for differences in mortality) for high-dose males and 2/50 for high-dose females. No such tumour occurred in controls or in animals receiving the low dose. The historical background incidence of nasal papillary adenomas was 2/1977 (0.1%) in male rats. The incidences of alveolar/bronchiolar carcinomas were 0/50, 1/50 and 4/49 for males and 1/50, 0/49 and 0/50 for females in the control, low-dose and high-dose groups; and those of alveolar/bronchiolar adenomas was 0/50, 1/50 and 1/49 for males and 1/50, 0/49 and 1/50 for females, respectively. Taken together, the incidence of alveolar/bronchiolar adenomas and carcinomas in high-dose males was statistically significantly increased compared with that in controls ($p < 0.05$, adjusted for mortality); moreover, a significant positive trend was found for the incidence of adenomas and carcinomas combined ($p < 0.02$; Cochrane-Armitage test). The historical background incidence of pulmonary alveolar/bronchiolar tumours in male rats of this strain was 0.7% (2% for chamber controls in this laboratory). The neoplasms of the nose and lungs did not cause early death of the animals. Treatment-related, non-neoplastic changes in the nose included inflammation, epithelial hyperplasia and squamous metaplasia of the nasal epithelium and atrophy of the olfactory sensory epithelium at both dose levels (Dunnick *et al.*,1988; National Toxicology Program, 1988).

(*b*) *Skin application*

Mouse: A group of 30 female ICR/Ha mice, eight weeks old, received applications of 10% 1,2-epoxybutane (purity established by boiling-point: 62–64°C) in acetone (reagent grade) on shaved dorsal skin three times per week for 77 weeks. One control group of 40 female mice received applications of 100% acetone three times a week for 85 weeks, and 100 female mice served as untreated controls. Animals in both groups were killed in week 85. No visible skin reaction and no tumour was observed in any of the groups (Van Duuren *et al.*, 1967).

(*c*) *Combined exposure*

Mouse: Groups of 50 male and 50 female Swiss ICR/Ha mice, five weeks of age, received 2400 mg/kg bw (males) or 1800 mg/kg bw (females) corn oil (control), 2400 or 1800 mg/kg bw trichloroethylene or 2400 or 1800 mg/kg bw trichloroethylene containing 0.8% 1,2-epoxybutane (97.8% pure; used as a stabilizer) up to week 35, and then half of this dose, containing 0.4% 1,2-epoxybutane, in corn oil by gavage five times per week from week 40 to week 69. The experiment was terminated at week 106. Survival was reduced in both treated groups, but did not differ between the two groups; at week 106, six male and three female controls, no trichloroethylene–treated animal and one male and one female mouse treated with trichloroethylene plus 1,2-epoxybutane were still alive. Squamous–cell carcinomas of the forestomach occurred in 3/49 males (p = 0.029, age–adjusted) and 1/48 females treated with trichloroethylene plus 1,2-epoxybutane. No such tumour occurred in animals treated with trichloroethylene alone or with corn oil. Two of the tumours in males metastasized to the lung, liver or abdominal cavity. Hyperkeratosis of the forestomach was observed in groups treated with trichloroethylene and with trichloroethylene plus 1,2-epoxybutane [incidence and severity unspecified] (Henschler et al., 1984). [The Working Group noted the two–fold reduction in dose to treated animals due to toxicity.]

3.2 Other relevant data

(a) Experimental systems

(i) Absorption, distribution, excretion and metabolism
When a single dose of 1.9 mmol (137 mg)/kg bw 1,2-epoxybutane was administered by gavage to rabbits or of 2.5 mmol (180 mg)/kg bw to rats, 4% and 11% of the dose, respectively, was excreted in the urine as 2-hydroxybutyl mercapturic acid (James et al., 1968).

(ii) Toxic effects
The oral LD_{50} of 1,2-epoxybutane in rats has been reported to be 1.4 ml/kg bw (1.17 g/kg bw) and the dermal LD_{50}, 2.1 ml/kg bw (1.76 g/kg bw) (Weil et al., 1963). Acute exposure of rats to 4000 ppm (11 800 mg/m^3) 1,2-epoxybutane by inhalation for 4 h resulted in the death of one of six animals; concentrations of 8000 ppm (23 600 mg/m^3) resulted in 100% mortality (Smyth et al., 1962).

When rats received a single 4–h exposure to 500–6550 ppm (1475–19 320 mg/m^3) 1,2-epoxybutane by inhalation, all animals exposed to 6550 ppm died during the exposure period, but no death occurred in the other groups. Ocular discharge and dyspnoea were observed in males and females at 2050 and 6550 ppm (6050 and 19 320 mg/m^3) and eye irritation at 1400 ppm (4130 mg/m^3). Mice similarly exposed to 400–2050 ppm (1200–6050 mg/m^3) 1,2-epoxybutane exhibited the same toxic effects; all mice exposed to 2050 ppm, 4/5 males and 4/5 females exposed to 1420 ppm and 1/5 males exposed to 400 ppm died during the study. The 4 h–LC_{50} for mice was calculated to be about 1000 ppm (2950 mg/m^3) (National Toxicology Program, 1988).

Application of 1,2-epoxybutane to the eye of rabbits resulted in corneal injury (Weil et al., 1963).

Rats and mice were exposed by inhalation to 400, 800 or 1600 ppm (1180, 2360 or 4720 mg/m^3) 1,2-epoxybutane for 6 h per day on five days per week for two weeks. All mice in the high–dose group had died by the third day of exposure; all rats exposed to this concentration survived but had pronounced retardation of growth. Inflammatory and degenerative changes in the nasal mucosa, myeloid hyperplasia in bone marrow and elevated mean white blood–cell counts were found in high–dose male and female rats; focal corneal cloudiness was observed in high- and medium-dose rats. Similar toxic effects were observed in mice, with the exception of haematological changes. In a similar experiment, exposure of mice and rats to 75, 150 or 600 ppm (220, 440 or 1770 mg/m³) 1,2-epoxybutane for 13 weeks did not result in treatment-related mortality; slight growth retardation was observed in high-dose female rats and mice. Inflammatory and degenerative changes in the nasal mucosa were observed in both species and myeloid hyperplasia in bone marrow in male rats only (Miller *et al.*, 1981). Similar toxic effects were observed in rats and mice exposed for 6 h on five days per week to 400–6400 ppm (1180–18 900 mg/m³) 1,2-epoxybutane for two weeks or to 50–800 ppm (148–2360 mg/m³) for 13 weeks (National Toxicology Program, 1988).

1,2-Epoxybutane alkylated deoxyguanosine at the N-7 position (Hemminki *et al.*, 1980).

(iii) *Effects on reproduction and prenatal toxicity*

Groups of 38–45 Wistar rats were exposed by inhalation to 0, 250 or 1000 ppm (738 or 2950 mg/m³) 1,2-epoxybutane (>99% pure) for 7 h per day on five days per week during a three-week pregestational period, or for 7 h per day on days 1–19 of gestation, or were exposed during the combined pregestational and gestational period. Fetuses were examined by routine teratological techniques on day 21 of gestation. Exposure to the high dose prior to and during gestation was lethal to one of 42 females, and maternal body weight gain was depressed in all groups exposed to 1000 ppm. Fetal growth and viability were not affected by exposure to 1,2-epoxybutane, and there was no indication of dose-related malformations in the offspring (Sikov *et al.*, 1981).

Groups of 24–49 New Zealand rabbits were exposed by inhalation to 0, 250 or 1000 ppm 1,2-epoxybutane (>99% pure) for 7 h per day on days 1–24 of gestation. Fetuses were examined by standard teratological techniques on day 30 of gestation. Exposure to 1000 ppm was lethal to 14/24 does and 250 ppm to 6/48 does; no effect on maternal body weight gain was observed in survivors. An indication of decreased pregnancy rate at term was seen in the high-dose group, but this may have been a result of differential toxicity in pregnant and non-pregnant females. Litter size appeared to be reduced, and embryonic mortality was increased in the high-dose group. Of the eight surviving fetuses at this exposure level, one was stunted and had a hypoplastic tail and unilateral renal agenesis (Sikov *et al.*, 1981).

(iv) *Genetic and related effects*

The genetic activity of 1,2-epoxybutane has been reviewed (Ehrenberg & Hussain, 1981). It is a directly acting alkylating agent.

1,2-Epoxybutane has been shown to induce SOS repair activity in *Salmonella typhimurium* TA1525/pSK1002 (Nakamura *et al.*, 1987) and to produce differential killing zones in various *pol* and *rec* proficient and deficient strains of *Escherichia coli* (Rosenkranz & Poirier, 1979; McCarroll *et al.*, 1981). It produced streptomycin-resistant mutants in *Klebsiella pneumoniae* (Voogd *et al.*, 1981; Knaap *et al.*, 1982).

It was shown to be mutagenic to *E. coli* WP2 *uvr*A⁻ in one study (McMahon *et al.*, 1979) but not in others (Dunkel *et al.*, 1984). It gave positive results only in *S. typhimurium* base-pair substituting strains (TA1535 and TA100), although frameshift tester strains were also tested both in the presence and absence of an exogenous metabolic system (Chen *et al.*, 1975;

Rosenkranz & Speck, 1975; McCann *et al.*, 1975; Speck & Rosenkranz, 1976; Henschler *et al.*, 1977; De Flora, 1979; McMahon *et al.*, 1979; Rosenkranz & Poirier, 1979; De Flora, 1981; Weinstein *et al.*, 1981; de Meester *et al.*, 1982, abstract; De Flora *et al.*, 1984; Gervasi *et al.*, 1985; Canter *et al.*, 1986; Hughes *et al.*, 1987, abstract; Rosman *et al.*, 1987; National Toxicology Program, 1988). Negative results have been reported from four laboratories in five strains (Dunkel *et al.*, 1984), and Simmon (1979a) reported negative results in six strains.

1,2–Epoxybutane induced forward mutation in *Schizosaccharomyces pombe* P1 (Migliore *et al.*, 1982) and mitotic recombination in *Saccharomyces cerevisiae* D3 (Simmon, 1979b). It was weakly mutagenic at the adenine locus in *Neurospora crassa* (Kolmark & Giles, 1955).

1,2–Epoxybutane induced sex–linked recessive lethal mutations (Knaap *et al.*, 1982; National Toxicology Program, 1988) and translocations (National Toxicology Program, 1988) in *Drosophila melanogaster* after either feeding or injection.

It produced no effect in the hepatocyte rat primary culture/DNA repair test (Williams *et al.*, 1982) but did produce differential toxicity as measured by relative growth in repair–deficient Chinese hamster ovary cells (Hoy *et al.*, 1984).

It was reported in an abstract that 1,2–epoxybutane induced mutation in L5178Y $TK^{+/-}$ mouse lymphoma cells in the absence of an exogenous metabolic system (Myhr *et al.*, 1981). In the same system, Amacher *et al.* (1980) reported positive results with a dose–reponse in the absence of an exogenous metabolic system; McGregor *et al.* (1987), the National Toxicology Program (1988), Mitchell *et al.* (1988) and Myhr and Caspary (1988) reported positive results both in the presence and absence of an exogenous metabolic system. 1,2–Epoxybutane also gave weakly positive results for the induction of 6–thioguanine–resistant mutations in L5178Y mouse lymphoma cells (Knaap *et al.*, 1982).

1,2–Epoxybutane induced chromosomal aberrations and sister chromatid exchange in Chinese hamster ovary cells with and without an exogenous metabolic system (National Toxicology Program, 1988).

It induced morphological transformation of Syrian hamster embryo cells (Pienta, 1980; Dunkel *et al.*, 1981) and increased transformation in Rauscher murine leukaemia virus–infected Fischer 344 rat embryo cells (Price & Mishra, 1980; Dunkel *et al.*, 1981). It did not induce morphological transformation in Balb/3T3 cells (Dunkel *et al.*, 1981).

(b)   Humans

No data were available to the Working Group.

3.3   Case reports and epidemiological studies of carcinogenicity to humans

No data were available to the Working Group.

4. Summary of Data Reported and Evaluation

4.1 Exposures

1,2–Epoxybutane is a synthetic organic liquid used primarily as a stabilizer for chlorinated hydrocarbon solvents. Other applications include its use as a chemical intermediate and corrosion inhibitor. Measurements of occupational exposure levels have not been reported.

4.2 Experimental carcinogenicity data

1,2–Epoxybutane was tested for carcinogenicity by inhalation exposure in one study in mice and in one study in rats, producing nasal papillary adenomas in rats of both sexes and pulmonary alveolar/bronchiolar tumours in male rats. It did not induce skin tumours when tested by skin application in one study in mice. Oral administration of trichloroethylene containing 1,2–epoxybutane to mice induced squamous–cell carcinomas of the forestomach, whereas administration of trichloroethylene alone did not.

4.3 Human carcinogenicity data

No data were available to the Working Group.

4.4 Other relevant data

1,2–Epoxybutane caused inflammatory and degenerative changes in the nasal mucosa and myeloid hyperplasia in the bone marrow in rats and mice. It did not cause prenatal toxicity in rats or rabbits.

1,2–Epoxybutane induced morphological transformation in cultured animal cells. It induced sister chromatid exchange, chromosomal aberrations and mutation in cultured animal cells, but did not induce DNA damage. It induced sex–linked recessive lethal mutations and translocations in *Drosophila*, mitotic recombination in yeast and mutation in yeast and fungi. 1,2–Epoxybutane induced DNA damage and mutation in bacteria. (See Appendix 1.)

4.5 Evaluation[1]

There is *limited evidence* for the carcinogenicity of 1,2–epoxybutane in experimental animals.

---

[1]For definitions of the italicized terms, see Preamble, pp. 27–30.

No data were available from studies in humans on the carcinogenicity of 1,2–epoxybutane.

Overall evaluation

1,2–Epoxybutane *is not classifiable as to its carcinogenicity to humans (Group 3)*.

5. References

Agarwal, S.C., Van Duuren, B.L. & Kneip, T.J. (1979) Detection of epoxides with 4–(*p*–nitrobenzyl)pyridine. *Bull. environ. Contam. Toxicol.*, *23*, 825–829

Aldrich Chemical Co., Inc. (1988) *Aldrich Catalog Handbook of Fine Chemicals*, Milwaukee, WI, p. 671

Amacher, D.E., Paillet, S.C., Turner, G.N., Ray, V.A. & Salsburg, D.S. (1980) Point mutations at the thymidine kinase locus in L5178Y mouse lymphoma cells. II. Test validation and interpretation. *Mutat. Res.*, *72*, 447–474

Canter, D.A., Zeiger, E., Haworth, S., Lawlor, T., Mortelmans, K. & Speck, W. (1986) Comparative mutagenicity of aliphatic epoxides in *Salmonella. Mutat. Res.*, *172*, 105–138

Chen, C.C., Speck, W.T. & Rosenkranz, H.S. (1975) Mutagenicity testing with *Salmonella typhimurium* strains. II. The effect of unusual phenotypes on the mutagenic response. *Mutat. Res.*, *28*, 31–35

Considine, D.M., ed. (1974) *Chemical and Process Technology Encyclopedia*, New York, McGraw Hill, pp. 443–445, 933

De Flora, S. (1979) Metabolic activation and deactivation of mutagens and carcinogens. *Ital. J. Biochem.*, *28*, 81–103

De Flora, S. (1981) Study of 106 organic and inorganic compounds in the *Salmonella*/microsome test. *Carcinogenesis*, *2*, 283–298

De Flora, S., Zanacchi, P., Camoirano, A., Bennicelli, C. & Badolati, G.S. (1984) Genotoxic activity and potency of 135 compounds in the Ames reversion test and in a bacterial DNA–repair test. *Mutat. Res.*, *133*, 161–198

Dow Chemical Co. (1988) *Material Safety Data Sheet; 1,2–Epoxybutane*, Midland, MI

Dunkel, V.C., Pienta, R.J., Sivak, A. & Traul, K.A. (1981) Comparative neoplastic transformation responses of Balb/3T3 cells, Syrian hamster embryo cells and Rauscher murine leukemia virus–infected Fischer 34 rat embryo cells to chemical carcinogens. *J. natl Cancer Inst.*, *67*, 1303–1315

Dunkel, V.C., Zeiger, E., Brusick, D., McCoy, E., McGregor, D., Mortelmans, K., Rosenkranz, H.S. & Simmon, V.F. (1984) Reproducibiity of microbial mutagenicity assays: I. Tests with *Salmonella typhimurium* and *Escherichia coli* using a standardized protocol. *Environ. Mutagenesis, 6 (Suppl. 2)*, 1–254

Dunnick, J.K., Eustis, S.L., Piegorsch, W.W. & Miller, R.A. (1988) Respiratory tract lesions in F344/N rats and B6C3F1 mice after inhalation exposure to 1,2–epoxybutane. *Toxicology*, *50*, 69–82

Ehrenberg, L. & Hussain, S. (1981) Genetic toxicity of some important epoxides. *Mutat. Res.*, *86*, 1–113

Gervasi, P.G., Citti, L., Del Monte, M., Longo, V. & Benetti, D. (1985) Mutagenicity and chemical reactivity of epoxidic intermediates of the isoprene metabolism and other structurally related compounds. *Mutat. Res.*, *156*, 77–82

Hawley, G.G. (1981) *Condensed Chemical Dictionary*, 10th ed., New York, Van Nostrand Reinhold, p. 165

Hemminki, K., Falck, K. & Vainio, H. (1980) Comparison of alkylation rates and mutagenicity of directly acting industrial and laboratory chemicals. Epoxides, glycidyl ethers, methylating and ethylating agents, halogenated hydrocarbons, hydrazine derivatives, aldehydes, thiuram and dithiocarbamate derivatives. *Arch. Toxicol.*, *46*, 277–285

Henschler, D., Eder, E., Neudecker, T. & Metzler, M. (1977) Carcinogenicity of trichloroethylene: fact or artifact? *Arch. Toxicol.*, *37*, 233–236

Henschler, D., Elsässer, H., Romen, W. & Eder, E. (1984) Carcinogenicity study of trichloroethylene, with and without epoxide stabilizers, in mice. *J. Cancer Res. clin. Oncol.*, *107*, 149–156

Hine, C., Rowe, V.K., White, E.R., Darmer, K.I., Jr & Youngblood, G.T. (1981) Epoxy compounds. In: Clayton, G.D. & Clayton, F.E., eds, *Patty's Industrial Hygiene and Toxicology*, 3rd rev. ed., Vol 2A, New York, John Wiley & Sons, pp. 2162–2165

Hoff, M.C., Im, U.K., Hauschildt, W.F. & Puskas, I. (1978) Butylenes. In: Mark, H.F., Othmer, D.F., Overberger, C.G., Seaborg, G.T. & Grayson, M., eds, *Kirk–Othmer Encyclopedia of Chemical Technology*, 3rd ed., Vol. 4, New York, John Wiley & Sons, p. 366

Hoy, C.A., Salazar, E.P. & Thompson, L.H. (1984) Rapid detection of DNA–damaging agents using repair-deficient CHO cells. *Mutat. Res.*, *130*, 321–332

Hughes, T.J., Simmons, D.S., Monteith, L.G., Myers, L.E. & Claxton, L.D. (1987) Mutagenicity of 31 organic compounds in a modified preincubation Ames assay with *Salmonella typhimurium* strains TA100 and TA102 (Abstract No. 123). *Environ. Mutagenesis*, *9* (*Suppl. 8*), 49

James, S.P., Jeffery, D.A., Waring, R.H. & Wood, P.B. (1968) Some metabolites of 1-bromobutane in the rabbit and the rat. *Biochem. J.*, *109*, 727–736

Knaap, A.G.A.C., Voogd, C.E. & Kramers, P.G.N. (1982) Comparison of the mutagenic potency of 2-chloroethanol, 2-bromoethanol, 1,2-epoxybutane, epichlorohydrin and glycidaldehyde in *Klebsiella pneumoniae*, *Drosophila melanogaster* and L5178Y mouse lymphoma cells. *Mutat. Res.*, *101*, 199–208

Kolmark, G. & Giles, N.H. (1955) Comparative studies of monoepoxides as inducers of reverse mutations in *Neurospora*. *Genetics*, *40*, 890–902

McCann, J., Choi, E., Yamasaki, E. & Ames, B.N. (1975) Detection of carcinogens as mutagens in the *Salmonella*/microsome test. Assay of 300 chemicals. *Proc. natl Acad. Sci. USA*, *72*, 5135–5139

McCarroll, N.E., Piper, C.E. & Keech, B.H. (1981) An *E. coli* microsuspension assay for the detection of DNA damage induced by direct-acting agents and promutagens. *Environ. Mutagenesis*, *3*, 429–444

McGregor, D.B., Martin, R., Cattanach, P., Edwards, I., McBride, D. & Caspary, W.J. (1987) Responses of the L5178Y tK$^+$/tK$^-$ mouse lymphoma cell forward mutation assay to coded chemicals. I: Results for nine compounds. *Environ. Mutagenesis*, *9*, 143–160

McMahon, R.E., Cline, J.C. & Thompson, C.Z. (1979) Assay of 855 test chemicals in ten tester strains using a new modification of the Ames test for bacterial mutagens. *Cancer Res.*, *39*, 682–693

de Meester, C., Mercier, M. & Poncelet, F. (1982) Comparative mutagenic activity of epoxides derived from 1,3-butadiene (Abstract No. 78). *Mutat. Res.*, *97*, 204–205

Migliore, L., Rossi, A.M. & Loprieno, N. (1982) Mutagenic action of structurally related alkene oxides on *Schizosaccharomyces pombe*: the influence, 'in vitro', of mouse-liver metabolizing system. *Mutat. Res.*, *102*, 425–437

Miller, R.R., Quast, J.F., Ayres, J.A. & McKenna, M.J. (1981) Inhalation toxicity of butylene oxide. *Fundam. appl. Toxicol.*, *1*, 319–324

Mitchell, A.D., Rudd, C.J. & Caspary, W.J. (1988) Evaluation of the L5178Y mouse lymphoma cell mutagenesis assay: intralaboratory results for sixty-three coded chemicals tested at SRI International. *Environ. mol. Mutagenesis, 12 (Suppl. 13)*, 37–101

Myhr, B.C. & Caspary, W.J. (1988) Evaluation of the L5178Y mouse lymphoma cell mutagenesis assay: intralaboratory results for sixty-three coded chemicals tested at Litton Bionetics, Inc. *Environ. mol. Mutagenesis, 12(Suppl. 13)*, 103–194

Myhr, B., Tajiri, D., Mitchell, A., Caspary, W. & Lee, Y. (1981) Blind evaluation of chemicals in the L5178Y/TK[+/-] mouse lymphoma forward assay (Abstract No. P55). *Environ. Mutagenesis, 3*, 324

Nakamura, S.-I., Oda, Y., Shimada, T., Oki, I. & Sugimoto, K. (1987) SOS inducing activity of chemical carcinogens and mutagens in *Salmonella typhimurium* TA1535/pSK1002: examination with 151 chemicals. *Mutat. Res., 192*, 239–246

National Institute for Occupational Safety and Health (1983) *National Occupational Exposure Survey 1981–83*, Cincinnati, OH

National Toxicology Program (1988) *Toxicology and Carcinogenesis Studies of 1,2–Epoxybutane (CAS No. 106–88–7) in F344/N Rats and B6C3F1 Mice (Technical Report No. 329; NIH Publication No. 88–2585)*, Research Triangle Park, NC, US Department of Health and Human Services

Parmeggiani, L., ed. (1983) Epoxy compounds. In: *Encyclopaedia of Occupational Health and Safety*, 3rd rev. ed., Vol. 1, Geneva, International Labour Office, pp. 770–773

Pienta, R.J. (1980) Transformation of Syrian hamster embryo cells by diverse chemicals and correlation with their reported carcinogenic and mutagenic activities. In: de Serres, F.J. & Hollaender, A., eds, *Chemical Mutagens. Principles and Methods for Their Detection*, Vol. 6, New York, Plenum, pp. 175–202

Pouchert, C.J., ed. (1981) *The Aldrich Library of Infrared Spectra*, 3rd ed., Milwaukee, WI, Aldrich Chemical Co., pp. 135H

Pouchert, C.J., ed. (1983) *The Aldrich Library of NMR Spectra*, 2nd ed., Vol. 1, Milwaukee, WI, Aldrich Chemical Co., pp. 194A

Pouchert, C.J., ed. (1985) *The Aldrich Library of FT–IR Spectra*, Vol. 1, Milwaukee, WI, Aldrich Chemical Co., pp. 228D

Price, P.J. & Mishra, N.K. (1980) The use of Fischer rat embryo cells as a screen for chemical carcinogens and the role of the nontransforming type 'C' RNA tumor viruses in the assay. *Adv. mod. environ. Toxicol., 1*, 213–239

Riedel-de Haën (1984) *Riedel–de Haën Laboratory Chemicals*, Hanover, p. 397

Rosenkranz, H.S. & Poirier, L.A. (1979) Evaluation of the mutagenicity and DNA–modifying activity of carcinogens and noncarcinogens in microbial systems. *J. natl Cancer Inst., 62*, 873–892

Rosenkranz, H.S. & Speck, W.T. (1975) Mutagenicity of metronidazole: activation by mammalian liver microsomes. *Biochem. biophys. Res. Commun., 66*, 520–525

Rosman, L.B., Gaddamichi, V. & Sinsheimer, J.E. (1987) Mutagenicity of aryl propylene and butylene oxides with Salmonella. *Mutat. Res., 189*, 189–204

Sadtler Research Laboratories (1980) *Standard Spectra Collection, 1980 Cumulative Index*, Philadelphia, PA

Sikov, M.R., Cannon, W.C., Carr, D.B., Miller, R.A., Montgomery, L.F. & Phelps, D.W. (1981) *Teratologic Assessment of Butylene Oxide, Styrene Oxide and Methyl Bromide (NIOSH Technical Report 81–124)*, Cincinnati, OH, National Institute for Occupational Safety and Health

Simmon, V.F. (1979a) In vitro mutagenicity assays of chemical carcinogens and related compounds with *Salmonella typhimurium. J. natl Cancer Inst., 62*, 893–399

Simmon, V.F. (1979b) In vitro assays for recombinogenic activity of chemical carcinogens and related compounds with *Saccharomyces cerevisiae* D3. *J. natl Cancer Inst.*, 62, 901–909

Smyth, H.F., Jr, Carpenter, C.P., Weil, C.S., Pozzani, U.C. & Striegel, J.A. (1962) Range–finding toxicity data: list VI. *Am. ind. Hyg. Assoc. J.*, 23, 95–107

Speck, W.T. & Rosenkranz, H.S. (1976) Mutagenicity of azathioprine. *Cancer Res.*, 36, 108–109

Van Duuren, B.L., Langseth, L., Goldschmidt, B.M. & Orris, L. (1967) Carcinogenicity of epoxides, lactones, and peroxy compounds. VI. Structure and carcinogenic action. *J. natl Cancer Inst.*, 39, 1217–1228

Voogd, C.E., van der Stel, J.J. & Jacobs, J.J.J.A.A. (1981) The mutagenic action of aliphatic epoxides. *Mutat. Res.*, 89, 269–282

Weast, R.C., ed. (1985) *Handbook of Chemistry and Physics*, 66th ed., Cleveland, OH, CRC Press, p. C-165

Weil, C.S., Condra, N., Haun, C. & Striegel, J.A. (1963) Experimental carcinogenicity and acute toxicity of representative epoxides. *Am. ind. Hyg. Assoc. J.*, 24, 305–325

Weinstein, D., Katz, M. & Kazmer, S. (1981) Use of a rat/hamster S–9 mixture in the Ames mutagenicity assay. *Environ. Mutagenesis*, 3, 1–9

Williams, G.M., Laspia, M.F. & Dunkel, V.C. (1982) Reliability of the hepatocyte primary culture/DNA repair test in testing of coded carcinogens and noncarcinogens. *Mutat. Res.*, 97, 359–370

RESIN MONOMERS AND RELATED COMPOUNDS

BIS(2,3–EPOXYCYCLOPENTYL)ETHER

1. Chemical and Physical Data

1.1 Synonyms

Chem. Abstr. Services Reg. No.: 2386–90–5
Chem. Abstr. Name: 2,2'–Oxybis(6–oxabicyclo[3.1.0]hexane)

1.2 Structural and molecular formulae and molecular weight

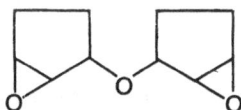

$C_{10}H_{14}O_3$

Mol. wt: 182.22

1.3 Chemical and physical properties of the pure substance

From Union Carbide Corp. (1985) unless otherwise noted
(a) *Description*: Homogeneous liquid at 43 °C and above; mixture of liquids and solids at lower temperatures
(b) *Boiling–point*: 203°C at 100 mm Hg
(c) *Freezing–point*: 29.7°C
(d) *Solubility*: Miscible with acetone (Holland *et al.*, 1979); < 0.3% by weight at 30°C in water
(e) *Volatility*: Vapour pressure, 0.01 mg Hg at 20°C; evaporation rate, 0.0017 (*n*–butyl acetate = 1.0)
(f) *Flash–point*: 138°C (open–cup)
(g) *Reactivity*: Reacts with amines, bases, acids and alcohols

(h) *Conversion factor*: mg/m³ = 7.45 × ppm[1]

1.4 Technical products and impurities

Trade names: ERL® 4205; W95

2. Production, Use, Occurrence and Analysis

2.1 Production and use

(a) *Production*

This compound was produced in the USA from the mid–1960s until 1985 but is not be-lieved to have been produced commercially elsewhere. No data on production levels were available.

(b) *Uses*

Bis(2,3–epoxycyclopentyl)ether was developed as a high–performance component and modifier of epoxy resins (Hine *et al.*, 1981).

(c) *Regulatory status and guidelines*

No data were available to the Working Group.

2.2 Occurrence

Bis(2,3–epoxycyclopentyl)ether is not known to occur as a natural product. No data on its presence in the environment or on occupational exposure to this compound were avail-able to the Working Group.

2.3 Analysis

The identity and purity of bis(2,3–epoxycyclopentyl)ether have been determined by in-frared spectroscopy and nuclear magnetic resonance analysis (Holland *et al.*, 1979).

3. Biological Data Relevant to the Evaluation of Carcinogenic Risk to Humans

3.1 Carcinogenicity studies in animals

(a) *Skin application*

Mouse: A group of 30–40 C3H mice [number and sex unspecified], 13 weeks old, re-ceived applications of a 30% solution of bis(2,3–epoxycyclopentyl)ether [purity and dose un-

---

[1]Calculated from: mg/m³ = (molecular weight/24.45) × ppm, assuming standard temperature (25 °C) and pressure (760 mm Hg)

specified] in acetone three times a week for up to 21 months. Ten mice survived to month 17. No skin tumour occurred (Weil *et al.*, 1963). [The Working Group noted the incomplete reporting of the results and that no untreated control was included.]

Groups of 40 male and 40 female C3H and 20 male and 20 female C57Bl/6 mice, ten to 12 weeks of age, received applications of 50 μl (0, 15 or 75 mg per week for C3H mice and 0, 7.5 or 37.5 mg per week for C57Bl/6 mice) commercial-grade bis(2,3-epoxycyclopentyl)ether (no impurity detected by nuclear magnetic resonance analysis) on the shaved dorsal skin three times a week for 24 months. Survival of C3H mice at 24 months was 22, 25 and 22 for males and 23, 18 and 15 for females in the control, low- and high-dose groups. In the high-dose males, one skin papilloma and one skin carcinoma and in high-dose females, one skin papilloma and two skin carcinomas, were observed. No skin tumour occurred in the low-dose group of either sex, whereas a skin papilloma was found in a single control female. The incidences of lung tumours [type unspecified] were 2/22, 4/25 and 5/22 in males and 1/23, 3/18 and 9/15 in females of the control, low-dose and high-dose groups [$p = 0.02$ for positive trend in females]. Only grossly observed lung tumours were examined microscopically. Survival of C57Bl/6 mice at 24 months was 20, 16 and 13 in males and 15, 17 and 16 in females of the control, low- and high-dose groups. One papilloma and three carcinomas of the skin were seen in high-dose males, and one carcinoma was observed in a high-dose female. No skin tumour was seen in the control or low-dose groups (Holland *et al.*, 1979).

(b) Combined exposure

Mouse: Groups of 40 male and 40 female C3H mice and 20 male and 20 female C57Bl/6 mice, ten to 12 weeks of age, received 0, 15 or 75 mg per week of a mixture of equal parts of bis(2,3-epoxycyclopentyl)ether and bisphenol A diglycidyl ether (see monograph, p. 237) in acetone on the shaved back skin for 24 months. Survival of C3H mice at 24 months was 22, 20 and 23 for males and 23, 23 and 19 for females in the control, low- and high-dose groups. Skin tumours occurred in 14 low-dose males (four papillomas and ten carcinomas) and in 32 high-dose males (13 papillomas and 19 carcinomas), in five low-dose females (three papillomas and two carcinomas) and in 19 high-dose females (12 papillomas and seven carcinomas). One skin papilloma occurred among control females and no skin tumour was observed in control males. In C57Bl/6 mice, survival at 24 months was 20, 15 and four for males and 15, 14 and four for females in the respective dose groups. The difference between control and high-dose groups was statistically significant ($p < 0.05$). Skin tumours (mostly carcinomas) were observed in one low-dose (carcinoma) and 17 high-dose (carcinomas) males and in two low-dose (one papilloma, one carcinoma) and 15 high-dose (two papillomas, 13 carcinomas) females, but not in controls of either sex. When tested alone at the same dose levels, each substance revealed a much lower tumour response (see above and the monograph on bisphenol A diglycidyl ether, p. 245), indicating a synergistic effect of the compounds when tested as a mixture (Holland *et al.*, 1979).

3.2 Other relevant data

(a) Experimental systems

(i) Absorption, distribution, excretion and metabolism

No data were available to the Working Group.

(ii) Toxic effects

The single oral LD_{50} of bis(2,3-epoxycyclopentyl)ether in rats has been reported to be 2.14 ml/kg bw and the single dermal penetration LD_{50} for rabbits, >5.0 ml/kg bw. Bis(2,3-epoxycyclopentyl)ether has been reported to be only mildly irritant to rabbit skin and to produce moderate corneal injury in rabbits; it was reported not to induce sensitization reactions three weeks after eight intracutaneous injections of 0.1 ml diluted compound in guinea-pigs (Weil et al., 1963).

Male and female C3H and C57Bl/6 mice received daily skin applications of 50 µl of a 50% solution of bis(2,3-epoxycyclopentyl)ether in acetone on shaved back skin on five days per week for two weeks. No mortality or sign of toxicity was observed in C3H mice; 5/9 C57Bl/6 mice died after four applications. Gross necropsy revealed pale, swollen liver and kidney (Holland et al., 1979).

(iii) Effects on reproduction and prenatal toxicity

No data were available to the Working Group.

(iv) Genetic and related effects

Bis(2,3-epoxycyclopentyl)ether was mutagenic to Salmonella typhimurium TA98 [details not given] and TA100 in the presence and absence of an exogenous metabolic system. It also increased the frequency of sister chromatid exchange in human lymphocytes in vitro and induced micronuclei in mice in vivo (Xie & Dong, 1984).

(b) Humans

No data were available to the Working Group.

3.3 Epidemiological studies of carcinogenicity to humans

No data were available to the Working Group.

4. Summary of Data Reported and Evaluation

4.1 Exposures

Bis(2,3-epoxycyclopentyl)ether is a synthetic organic liquid which has been used as a component and modifier of epoxy resins. Measurements of occupational exposure levels have not been reported.

4.2 Experimental carcinogenicity data

Bis(2,3–epoxycyclopentyl)ether was tested for carcinogenicity by skin application in one experiment in two strains of mice, producing a small number of skin tumours in both strains; an increased incidence of lung tumours was observed in females of one strain. Another experiment by skin application in mice was inadequate for evaluation.

4.3 Human carcinogenicity data

No data were available to the Working Group.

4.4 Other relevant data

Bis(2,3–epoxycyclopentyl)ether induced sister chromatid exchange in cultured human cells and micronuclei in mice. It was mutagenic to bacteria. (See Appendix 1.)

4.5 Evaluation[1]

There is *limited evidence* for the carcinogenicity of bis(2,3–epoxycyclopentyl)ether in experimental animals.

No data were available from studies in humans on the carcinogenicity of bis(2,3–epoxycyclopentyl)ether.

Overall evaluation

Bis(2,3–epoxycyclopentyl)ether *is not classifiable as to its carcinogenicity to humans (Group 3)*.

5. References

Hine, C., Rowe, V.K., White, E.R., Darmer, K.I., Jr & Youngblood, G.T. (1981) Epoxy compounds. In: Clayton, G.D. & Clayton, F.E., eds, *Patty's Industrial Hygiene and Toxicology*, 3rd rev. ed., Vol. 2A, New York, John Wiley & Sons, pp. 240–241

Holland, J.M., Gosslee, D.G. & Williams, N.J. (1979) Epidermal carcinogenicity of bis(2,3–epoxycyclopentyl)ether, 2,2–bis(*p*–glycidyloxyphenyl)propane and *m*–phenylenediamine in male and female C3H and C57BL/6 mice. *Cancer Res., 39*, 1718–1725

Union Carbide Corp. (1985) *Material Safety Data Sheet: Bis(2,3–epoxycyclopentyl)ether*, Hahnville, LA

Weil, C.S., Condra, N., Haun, C. & Striegel, J.A. (1963) Experimental carcinogenicity and acute toxicity of representative epoxides. *Am. ind. Hyg. Assoc. J., 24*, 305–325

---

[1]For definitions of the italicized terms, see Preamble, pp. 27–30.

Xie, D. & Dong, S. (1984) Mutagenicity of bis(2,3–epoxycyclopentyl)ether (Chin.). *Acta acad. med. sin.*,
       6, 210–212

SOME GLYCIDYL ETHERS

This monograph covers bisphenol A diglycidyl ether and phenyl glycidyl ether, for which there were carcinogenicity studies in animals. Data on toxicity, genetic and related effects, as well as basic chemical information, are also included for nine other glycidyl ethers produced in moderate to high volumes.

1. Chemical and Physical Data

1.1 Synonyms

Bisphenol A diglycidyl ether

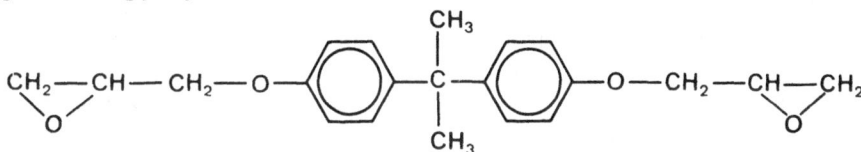

$C_{21}H_{24}O_4$ Mol. wt: 340.42

Chem. Abstr. Services Reg. No.: 1675-54-3

Chem. Abstr. Name: 2,2'-[(1-Methylethylidene)bis(4,1-phenyleneoxymethylene)]bis-(oxirane)

IUPAC Systematic Name: 2,2'-[(1-Methylethylidene)bis(4,1-phenyleneoxymethyl-ene)]bis(oxirane)

Synonyms: 4,4'-Bis(2,3-epoxypropoxy)diphenyldimethylmethane; 2,2-bis[*para*-(2,3-epoxypropoxy)phenyl]propane; 2,2-bis[4-(2,3-epoxypropoxy)phenyl]propane; bis(4-glycidyloxyphenyl)dimethylmethane; 2,2-bis(*para*-glycidyloxyphenyl)propane; 2,2-bis (4-glycidyloxyphenyl)propane; bis(4-hydroxyphenyl)dimethylmethane diglycidyl ether 2,2-bis(*para*-hydroxyphenyl)propane diglycidyl ether; 2,2-bis(4-hydroxyphenyl)pro-pane diglycidyl ether; BPDGE; dian diglycidyl ether; diglycidyl bisphenol A; diglycidyl bisphenol A ether; diglycidyl diphenylolpropane ether; diglycidyl ether of 2,2-bis-(*para*-hydroxyphenyl)propane; diglycidyl ether of 2,2-bis(4-hydroxyphenyl)propane; diglycidyl ether of bisphenol A; diglycidyl ether of 4,4'-isopropylidenediphenol; *para,para'*-dihydroxydiphenyldimethylmethane diglycidyl ether; 4,4'-dihydroxydiphenyldimethylmethane diglycidyl ether; diomethane diglycidyl ether; 4,4'-isopropylidenebis[1-(2,3-epoxypropoxy)benzene]; 4,4'-isopropylidenediphenol diglycidyl ether; oligomer 340

Phenyl glycidyl ether

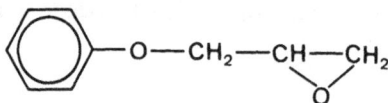

$$\text{C}_6\text{H}_5\text{—O—CH}_2\text{—CH—CH}_2\text{ (epoxide: }\text{O}\text{)}$$

$C_9H_{10}O_2$ Mol. wt: 150.18

Chem. Abstr. Services Reg. No.: 122-60-1

Chem. Abstr. Name: (Phenoxymethyl)oxirane

IUPAC Systematic Name: (Phenoxymethyl)oxirane

Synonyms: 1,2-Epoxy-3-phenoxypropane; 2,3-epoxypropoxybenzene; 2,3-epoxypropyl phenyl ether; glycidol phenyl ether; glycidyl phenyl ether; PGE; phenol glycidyl ether; 1-phenoxy-2,3-epoxypropane; 3-phenoxy-1,2-epoxypropane; phenoxypropene oxide; phenoxypropylene oxide; γ-phenoxypropylene oxide; 3-phenoxy-1,2-propylene oxide; phenyl 2,3-epoxypropyl ether; 3-phenyloxy-1,2-epoxypropane

1.2   Chemical and physical properties of the pure substances

From National Institute for Occupational Safety and Health (1978) unless otherwise noted

Bisphenol A diglycidyl ether

Bisphenol A diglycidyl ether is not produced as a pure monomer but as a mixture of monomer, dimer, trimer and tetramer; therefore, very few, if any, chemical/physical properties are reported for the pure substance.

(a)   Spectroscopy data: Electron impact mass spectral data have been reported (Brown & Creaser, 1980).

(b)   Reactivity: Glycidyl ethers, as epoxide-containing chemicals, react readily with acids, with water and with nucleophiles such as proteins and nucleic acids

(c)   Conversion factor: mg/m³ = 13.92 × ppm[1]

Phenyl glycidyl ether

(a)   Description: Colourless liquid

(b)   Boiling-point: 245°C

---

[1]Calculated from: mg/m³ = (molecular weight/24.45) × ppm, assuming standard temperature (25°C) and pressure (760 mm Hg)

(c) *Melting–point*: 3.5°C

(d) *Volatility*: Vapour pressure, 0.01 mm Hg at 25°C; relative vapour density (air = 1), 4.37 at 25°C

(e) *Specific gravity*: 1.1092 at 20°C

(f) *Refractive index*: 1.5314

(g) *Spectroscopy data*: Infrared, proton and carbon–13 nuclear magnetic resonance, ultraviolet and electron impact mass spectral data have been reported (Patterson, 1954; Shapiro, 1977; Brown & Creaser, 1980; Sadtler Research Laboratories, 1980; Pouchert, 1983, 1985).

(h) *Solubility*: Soluble in acetone and toluene; slightly soluble in octane (12.9%); nearly insoluble in water (0.24%)

(i) *Reactivity*: Glycidyl ethers, as epoxide–containing chemicals, react readily with acids, with water and with nucleophiles such as proteins and nucleic acids

(j) *Viscosity*: 6 cP at 25°C (Urquhart *et al.*, 1988)

(k) *Conversion factor*: mg/m^3 = 6.14 × ppm[1]

Other glycidyl ethers

Physical properties of selected other glycidyl ethers are given in Table 1.

1.3 Technical products and impurities

Trade names: Araldite 6005; Araldite® GY 250; Araldite® GY 6010; D.E.R.® 331; Epikote® 815; Epikote® 828; EPI–REZ® 510; EPON® 828; EPOTUF® 37–140; Epoxide A

Bisphenol A diglycidyl ether is available as a medium viscosity, unmodified liquid epoxy resin with the following typical properties: epoxy value, 0.52–0.55 equivalents/100 g; equivalent weight per epoxide, 182–192; viscosity (at 25°C), 12 000–16 000 cP (Ciba–Geigy Corp., 1984).

Trade name: Heloxy® WC–63

Phenyl glycidyl ether is available with the following properties: specific gravity (at 25°C), 1.10–1.12; viscosity (at 25°C), 4–7 cP; equivalent weight per epoxide, 155–170 (Wilmington Chemical Corp., 1987).

---

[1]Calculated from: mg/m^3 = (molecular weight/24.45) × ppm, assuming standard temperature (25˚C) and pressure (760 mm Hg)

Table 1. Physical properties of selected glycidyl ethers[a]

Compound (CAS No.)	Formula	Molecular weight	Boiling-point (°C)	Melting-point (°C)	Vapour pressure (mm Hg)
C8-C10 Alkyl glycidyl ether[b] (68609-96-1)	$CH_3(CH_2)_{7-13}-O-O-CH_2-CH-CH_2$	229	139.4 (100 mm Hg)	-12	0.08 (21°C)
C12-C14 Alkyl glycidyl ether[b] (68609-97-2)		286	215.5 (100 mm Hg)	1.7	0.06 (21°C)
Allyl glycidyl ether (106-92-3)	$H_2C-CH-CH_2-O-CH_2-CH=CH_2$	114.14	153.9	[c]	4.7 (25°C)
1,4-Butanediol diglycidyl ether (2425-79-8)	$H_2C-CH-CH_2-O-CH_2-C-CH_2-O-CH_2-CH-CH_2$	202.25	-	-	-
n-Butyl glycidyl ether (2426-08-6)	$H_2C-CH-CH_2-O-CH_2-CH_2-CH_2-CH_3$	130.21	164	-	3.2 (25°C)
tert-Butyl glycidyl ether (7665-72-7)	$H_2C-CHCH_2OC(CH_3)_3$	130.21	152	-	-
tert-Butylphenyl glycidyl ether (3101-60-8)	$H_2C-CHCH_2O$ ... $C(CH_3)_3$	206.28	294 (para-)	-	-
Cresyl glycidyl ether (26447-14-3)	$O-CH_2-CH-CH_2$	164.21	-	-	-
Neopentylglycol diglycidyl ether (17557-23-2)	$H_2C-CH-CH_2-O-(CH_2)_4-O-CH_2-CH-CH_2$	216	-	-	-

[a]From Ulbrich et al. (1964); National Institute for Occupational Safety and Health (1978); Hine et al. (1981)

[b]Two fractions of straight-chain alcohols derived from reduction of fats, namely the C8 to C10 fraction and the C12 to C14 fraction, are converted to their respective glycidyl ethers (epoxide 7 and epoxide 8).

[c]Forms a glass at about -100°C

2. Production, Use, Occurrence and Analysis

2.1 Production and use

(a) Production

The principal commercial market for glycidyl ethers is in epoxy resins, a large proportion of which are based on bisphenol A diglycidyl ether. These resins are produced by a combination of the Taffy process and the fusion process. In the Taffy process, epichlorohydrin (see IARC, 1987) and bisphenol A monomers are reacted in the presence of caustic soda to produce a mixture of bisphenol A diglycidyl ether and its low molecular-weight oligomers – primarily dimers, trimers and tetramers. As the proportion of bisphenol A in the reactant mixture is increased, the average molecular weight of the resultant epoxy resin increases. In the fusion process, low molecular-weight (liquid) bisphenol A diglycidyl ether-based epoxy resins are converted to higher molecular-weight (solid) resins by reaction with more bisphenol A (Urquhart et al., 1988).

Phenyl glycidyl ether is synthesized in a similar manner by adding phenol to epichlorohydrin in the presence of a catalyst. The intermediate chlorohydrin is not isolated and undergoes dehydrochlorination to yield a glycidyl ether (National Institute for Occupational Safety and Health, 1978).

Although specific worldwide figures are not available, relative production of the glycidyl ethers covered in this monograph was reported to be 'ultra high' for bisphenol A diglycidyl ether and 'medium' for phenyl glycidyl ether (Chemical Manufacturers' Association, 1984). Production levels of other glycidyl ethers were classified as 'medium', except for tert-butyl glycidyl ether, for which no data were available.

US production of unmodified epoxy resins increased from 81 250 tonnes in 1972 to 196 700 tonnes in 1986. The corresponding figures for modified epoxy resins (those to which reactive viscosity modifiers (usually also epoxy) have been added) were 19 600 and 125 400 tonnes. US production of these glycidyl ethers, reported separately from epoxy resins, was 1700 tonnes in 1984 and 2800 tonnes in 1985 (US International Trade Commission, 1974, 1985, 1986, 1987). Approximately 15% of US epoxy resin production in 1986 was exported (Urquhart et al., 1988).

In 1986, Japan produced approximately 92 000 tonnes, exported approximately 24 000 tonnes and imported approximately 10 000 tonnes of epoxy resins (Shikado, 1987).

(b) Use

Bisphenol A diglycidyl ether is the most common active component in epoxy resins, although other glycidyl ethers are frequently incorporated into epoxy resin systems as reactive modifiers. For most uses, liquid epoxy resins resulting from the reaction of epichlorohydrin and bisphenol A are too viscous, and modifiers, which are low viscosity liquids, are added to improve flow characteristics. Modifiers also decrease the tendency of curing agents in the resin formulation to volatilize prematurely, aid in the wetting of fillers in the resin, enhance penetration of the resin into castings, and frequently improve the mechanical properties of the product (Urquhart *et al.*, 1988).

Phenyl glycidyl ether is a monofunctional reactive modifier referred to as a 'chain stopper' or reaction inhibitor. It reduces resin system functionality (by reacting with hydroxyl groups, for example) and decreases system cross–linking density (Urquhart *et al.*, 1988).

The epoxy group of the glycidyl ethers reacts during the curing process, and glycidyl ethers are therefore generally no longer present in completely cured products. Typical curing agents include aliphatic amines (e.g., diethylenetriamine, triethylenetetramine), aromatic amines (e.g., *meta*–phenylenediamine, methylenedianiline, diaminodiphenylsulfone), catalytic curing agents (e.g., boron trifluoride–ethylamine complex) and acid anhydrides (e.g., dodecenylsuccinic anhydride, hexahydrophthalic anhydride; Harper, 1979).

Epoxy resins based on glycidyl ethers are used in a variety of applications, such as protective coatings and reinforced plastics, as well as bonding materials and adhesives, where they exhibit exceptional properties, such as toughness, chemical resistance and superior electrical properties. They are used in both decorative and protective coatings for automobiles, tins and closures, boats and ships, appliances, piping and miscellaneous metal decoration. They are also widely used in the electrical/electronic, structural/composite, adhesive and aggregate applications as adhesives, laminates, encapsulants and grouting compounds (Chemical Manufacturers' Association, 1984; Urquhart *et al.*, 1988). Table 2 presents the use pattern for epoxy resins produced in the USA (Urquhart *et al.*, 1988).

Table 2. US consumption (thousands of tonnes) of epoxy resins by use[a]

Use	1981	1982	1983	1984	1985
Bonding and adhesives	8.2	6.8	6.8	9.1	7.7
Flooring, paving and aggregates	9.1	8.2	8.6	8.1	7.7
Protective coatings	62.1	55.3	59.4	69.4	76.2
Fibre–reinforced laminates and composites	29.9	26.3	33.6	39.5	31.3
Tooling, casting and moulding resins	13.2	10.9	12.7	14.1	10.4
All other uses	9.5	11.3	13.1	13.1	14.1
Exports	18.6	17.7	15.9	18.6	19.1
Total	150.6	136.5	150.1	171.9	166.5

[a]From Urquhart *et al.* (1988)

An increase in the production of epoxy resins in Japan between 1985 and 1986 was due chiefly to the recovery in demand for electrical products such as laminated boards and sealing materials, which are major application fields in Japan and which accounted for 41% of the total epoxy resin demand (Shikado, 1987).

(c)   *Regulatory status and guidelines*

The US Food and Drug Administration (1988) permits the use of epoxy resins as components of coatings that may come into contact with food.

Occupational exposure limits for phenyl glycidyl ether in 14 countries are presented in Table 3. Exposure limits have also been set for other glycidyl ethers, including allyl (threshold limit value, 22 mg/m³; short–term exposure limit, 44 mg/m³) and n–butyl (threshold limit value, 135 mg/m³) glycidyl ethers (American Conference of Governmental Industrial Hygienists, 1988). No exposure limit has been set for bisphenol A diglycidyl ether.

Table 3. Occupational exposure limits for phenyl glycidyl ethers[a]

Substance and country	Year	Concentration (mg/m³)[b]	Interpretation[c]
Australia	1984	60	TWA
Belgium	1984	60	TWA
Denmark	1988	5	TWA
Finland	1987	S 60	STEL (15 min)
Germany, Federal Republic of	1988	S 6	TWA
Indonesia	1985	60	TWA
Mexico	1985	60	TWA
Netherlands	1986	6	TWA
Norway	1981	5	TWA
Romania	1984	75	TWA
		100	STEL
Sweden	1987	S 60	TWA
		S 90	STEL
Switzerland	1984	6	TWA
USA[d]			
OSHA	1987	60	TWA
NIOSH	1986	5	Ceiling (15 min)
ACGIH	1988	6	TWA
Yugoslavia	1984	60	TWA

[a]From Direktoratet for Arbeidstilsynet (1981); International Labour Office (1984); Arbeidsinspectie (1986); Cook (1987); National Swedish Board of Occupational Safety and Health (1987); Työsuojeluhallitus (1987); US Occupational Safety and Health Administration (1987); American Conference of Governmental Industrial Hygienists (1988); Arbejdstilsynet (1988); Deutsche Forschungsgemeinschaft (1988)

[b]S, skin notation

[c]TWA, time–weighted average; STEL, short–term exposure limit

[d]OSHA, Occupational Safety and Health Administration; NIOSH, National Institute for Occupational Safety and Health; ACGIH, American Conference of Governmental Hygienists

2.2 Occurrence

(a) Natural occurrence

Glycidyl ethers are not known to occur as natural products.

(b) Occupational exposure

On the basis of a US National Occupational Hazard Survey of 1972–74 and a US National Exposure Survey of 1981–83, the National Institute for Occupational Safety and

Health (1974, 1983) estimated that 45 700 and 13 138 workers, respectively, were potentially exposed to bisphenol A diglycidyl ether in the USA. The corresponding figures for phenyl glycidyl ether were 8554 and 1328.

During use of a powdered spray paint, levels of bisphenol A diglycidyl ether ranged from 0.005 to 0.200 mg/m³ in personal samples and from 0.002 to 0.008 mg/m³ in area samples (Hervin *et al.*, 1979). Personal time–weighted averages (TWAs) for industrial designers at a truck manufacturing plant were 0.0002–0.0004 mg/m³ bisphenol A diglycidyl ether (Boiano, 1981).

(c) Water

In the late 1970s and early 1980s, a technique was developed for lining cast–iron water pipes with epoxy resin to inhibit corrosion. In the UK, where the technique has been used for in–situ renovation of existing water distribution systems, bisphenol A diglycidyl ether and its dimer and trimer have been detected [not quantified] in drinking–water samples. The epoxy resin components were identified by high–performance liquid chromatography (HPLC) and field desorption mass spectrometry and were estimated to be present 'at the low microgram per litre concentration range or less' (Crathorne *et al.*, 1984).

2.3 Analysis

Methods have been reported for the analysis of various glycidyl ethers in air and water. More volatile glycidyl ethers (e.g., phenyl glycidyl ether) have been collected by drawing air samples through charcoal, silica gel or Tenax adsorption tubes, desorbed with an appropriate solvent and analysed by gas chromatography (Taylor, 1977, 1978; Guenier *et al.*, 1986). Bisphenol A diglycidyl ether has been determined in air by collection on a glass fibre filter (nominal pore size, 1 μm), extraction and analysis by HPLC (Taylor, 1980; Peltonen *et al.*, 1986). Bisphenol A diglycidyl ether has been quantified in water by adsorption on C_{18}–bonded silica and analysis by HPLC (Crathorne *et al.*, 1986).

3. Biological Data Relevant to the Evaluation of Carcinogenic Risk to Humans

3.1 Carcinogenicity studies in animals[1]

Bisphenol A diglycidyl ether

(a) Oral administration

Mouse: Groups of 30 male Heston A strain mice [age unspecified] were fed a diet containing 2% bisphenol A diglycidyl ether [type and quantity of impurities unspecified] or a

---

[1]The Working Group was aware of a study in progress on allyl glycidyl ether by inhalation in mice and rats (IARC, 1988).

normal control diet. The study was terminated after 11 months. The incidences of pulmonary tumours in survivors were 12/23 in the epoxy resin–treated group and 15/29 in the untreated control group. No lung tumour was observed in mice that died during the study. The other organs were not examined for tumours (Hine *et al.*, 1958). [The Working Group noted the inadequate description of the test material and that the study was not designed to investigate carcinogenicity in tissues other than the lungs.]

(b) Skin application

Mouse: Groups of 30 male C3H mice (16–18 g bw) received skin applications of 0.2 ml of a 0.3% solution weekly or a 5% solution of bisphenol A diglycidyl ether [type and quantity of impurities unspecified] in acetone once or three times weekly for 24 months. A control group of 30 male mice received 0.2 ml acetone alone, and a positive control group was treated weekly with a 0.3% solution of 20–methylcholanthrene in acetone. No skin tumour occurred in any of the treated mice. The group treated with 20–methylcholanthrene showed a high incidence of malignant skin tumours (19/20) within six months (Hine *et al.*, 1958). [The Working Group noted the inadequate description of the test material.]

A group of about 40 C3H mice [exact number and sex unspecified], aged 13 weeks, received skin applications of undiluted bisphenol A diglycidyl ether [purity and dose unspecified] on shaved back skin for life (maximum, 23 months). After 16 months of treatment, at which time 32 mice were still alive, a single skin papilloma occurred; no other skin tumour appeared during the experiment. The authors stated that in a second, similar experiment, no skin tumour was observed [details not given] (Weil *et al.*, 1963). [The Working Group noted that the amount of test substance per application was not given and that untreated controls were not included in the experiment.]

Groups of 40 male and 40 female C3H and 20 male and 20 female C57Bl/6 mice, ten to 12 weeks of age, received applications of 0, 5 or 25 mg of commercial–grade bisphenol A diglycidyl ether (containing 10% (w/w) of an epoxidized polyglycol (mol. wt, > 500) and small amounts of phenyl glycidyl ether) in acetone on shaved back skin three times a week for 24 months. At that time, 18–23 C3H mice were still alive, and no skin tumour was observed. In male C57Bl/6 mice, survival was 20, 17 and 15 control, low–dose and high–dose animals, respectively, and survival in females was 15, 15 and 13, respectively. Skin tumours occurred in 0, 1 (papilloma) and 6 (carcinomas) control, low–dose and high–dose males, and in 0, 0 and 2 (1 papilloma and 1 carcinoma) control, low–dose and high–dose females (Holland *et al.*, 1979).

Groups of 50 male and 50 female CF1 mice, six weeks old, received applications of 0, 1 or 10% [equivalent to 2 and 20 mg] Araldite GY 250 (technical grade; main component, bisphenol A diglycidyl ether; containing 4.3 mg/kg epichlorohydrin as a contaminant) in 0.2 ml acetone on shaved back skin twice a week for two years. There was no effect on survival; no skin tumour was observed at the site of application, and there was no significant difference in the occurrence of other tumours. A positive control group that received skin applications of a 2% solution of β–propiolactone showed high incidences of malignant skin tumours (Zakova *et al.*, 1985).

Groups of 50 male and 50 female CF1 mice, six weeks of age, received applications of 0.2 ml of a 1% or 10% solution of pure (analytical grade) bisphenol A diglycidyl ether or of one of two technical grades of bisphenol A diglycidyl ether (EPON 828, containing 29 mg/kg epichlorohydrin, or Epikote 828, containing 3 mg/kg epichlorohydrin) in acetone on shaved back skin twice a week for two years. Positive control groups of 50 males and 50 females received applications of 2% β-propiolactone in acetone; a control group of 100 male and 100 female mice was treated with acetone only. Survival was not affected by treatment with epoxy resins, but was considerably decreased in the β-propiolactone-treated group. In animals treated with EPON 828, one skin carcinoma occurred in high-dose males and one fibrosarcoma of the subcutis in high-dose females. In Epikote 828-treated mice, one squamous-cell papilloma of the skin was observed in low-dose males and three basal-cell carcinomas, in a low-dose and a high-dose male and a high-dose female. In addition, one sebaceous-gland adenoma was observed in a high-dose male. In the animals treated with pure epoxy resin and in the group of acetone controls, no epidermal tumour was observed. Two dermal tumours (haemangiosarcomas) were seen in the high-dose males given the pure epoxy resin. A large number of skin tumours was observed in the β-propiolactone-treated groups – 132 in 30/50 males and 63 in 13/50 females – which were generally malignant epithelial tumours and, to a lesser extent, mesenchymal tumours. The authors reported that no epidermal tumour was seen in 200 females or 100 males used as controls in that laboratory. The incidence of renal and lymphoreticular/haematopoietic tumours was increased in the treated groups. Kidney tumours, mainly carcinomas, were observed only in males: 6/99 (control), 8/50 (10% EPON 828; [$p = 0.05$]), 0/50 (1% EPON), 6/50 (1% Epikote), 2/50 (10% Epikote), 5/50 (1% epoxy resin) and 3/50 (10% epoxy resin). In female mice, an increase in the incidence of lymphoreticular/haematopoietic tumours was observed in animals treated with 1% epoxy resin (23/50) and in those treated with 10% Epikote (24/50), compared to controls (27/100). In the other groups, the incidence of these tumours was comparable to that seen in control female mice (Peristianis *et al.*, 1988).

Rabbit: Each of 16 male albino rabbits [strain and age unspecified] received skin applications [site unspecified] of 0.5 ml acetone thrice weekly, a 0.3% solution in acetone once per week, a 5% solution of bisphenol A diglycidyl ether [type and quantity of impurities unspecified] once or three times per week and a 0.3% solution of 20-methylcholanthrene in acetone. At 24 months, 13/16 rabbits were still alive. Skin tumours were seen only at 20-methylcholanthrene treated sites (Hine *et al.*, 1958). [The Working Group noted the inadequate description of the test material.]

(c) *Subcutaneous injection*

Rat: Groups of 30 male Long-Evans rats (80–100 g bw) were given three weekly subcutaneous injections of bisphenol A diglycidyl ether [type and quantity of impurities unspecified] dissolved in propylene glycol (50% solution; total dose, 2.58 g/kg bw). A negative control group was injected with propylene glycol alone, and a positive control group received three injections of 1,2,5,6-dibenzanthracene. The experiment was terminated after 24 months, at which time survival was 17, 14 and four in the negative control, epoxy resin and positive control groups, respectively. The numbers of malignant tumours at the site of injection were 0, 4 (fibrosarcomas) and 17 (mainly fibrosarcomas or sarcomas), respectively (Hine *et al.*, 1958). [The Working Gruop noted the inadequate description of the test material.]

(*d*) Combined exposure

Groups of 40 male and 40 female C3H mice and 20 male and 20 female C57Bl/6 mice, ten to 12 weeks of age, received 0, 15 or 75 mg per week of a mixture of equal parts of bisphenol A diglycidyl ether and bis(2,3–epoxycyclopentyl)ether (see monograph, p. 233) in acetone on the shaved back skin for 24 months. Survival of C3H mice at 24 months was 22, 20 and 23 for males and 23, 23 and 19 for females in the control, low and high–dose groups. Skin tumours occurred in 14 low–dose males (four papillomas and ten carcinomas) and 32 high–dose males (13 papillomas and 19 carcinomas), in five low–dose females (three papillomas and two carcinomas) and in 19 high–dose females (12 papillomas and seven carcinomas). One skin papilloma was observed in control females, and no skin tumour was seen in control males. In C57Bl/6 mice, survival at 24 months was 20, 15 and four for males and 15, 14 and four for females in the respective dose groups. The difference between control and high–dose groups was statistically significant ($p < 0.05$). Skin tumours (mostly carcinomas) were observed in one low–dose (carcinoma) and 17 high–dose (carcinomas) males and in two low–dose (one papilloma, one carcinoma) and 15 high–dose (two papillomas, 13 carcinomas) females, but not in controls of either sex. When tested alone at the same dose levels, each substance revealed a much lower tumour response (see p. 245 above and the monograph on bis(2,3–epoxycyclopentyl)ether, p. 233), indicating a synergistic effect of the compounds when tested as a mixture (Holland *et al.*, 1979).

(*e*) Carcinogenicity of metabolites

Glycidaldehyde is carcinogenic to experimental animals (IARC, 1976, 1987).

Phenyl glycidyl ether

Inhalation

Rat: Groups of 100 male and 100 female Sprague–Dawley rats, six weeks old, were exposed to 0, 1 or 12 ppm (6 or 73.5 mg/m³) phenyl glycidyl ether vapour (purity, 99.6% with trace amounts of phenol and diglycidyl ether) by inhalation for 6 h per day, on five days per week for 24 months. [No data were given on survival or body weights.] Epidermoid carcinomas occurred in the anterior parts of the nasal cavity in 1/89 male and 0/87 female controls, in 0/83 male and 0/88 female low–dose rats and in 9/85 male [$p = 0.007$] and 4/89 [$p = 0.06$] female high–dose rats. The first nasal tumour was observed in week 89. In the group receiving 12 ppm, squamous metaplasia, rhinitis, epithelial desquamation, regeneration, hyperplasia and dysplasia of the respiratory epithelium were also observed, especially in the anterior parts of the nasal cavity. No such increase in non–neoplastic changes occurred in the group receiving 1 ppm (Lee *et al.*, 1983).

3.2 Other relevant data

(a) Experimental systems

(i) Absorption, distribution, excretion and metabolism

When ^{14}C-*n-butyl glycidyl* ether was administered orally to rats and rabbits, 91 and 80% of the administered dose was recovered in urine within four days, and most of the radioactivity was excreted during day 1 (Eadsforth *et al.*, 1985a). The main urinary metabolite (23% of total) in rats receiving an oral dose of 20 mg/kg bw was 3–butoxy–2–acetylaminopropionic acid. Other metabolites were 3–butoxy–2–hydroxypropionic acid (9%) and butoxyacetic acid (10%). The same metabolites were found in rabbits, but the main metabolite (35%) was 3–butoxy–2–hydroxypropionic acid (Eadsforth *et al.*, 1985a,b).

When ^{14}C-*bisphenol A diglycidyl ether* (56 mg/kg) was applied to the shaved skin of mice, 67% and 11% of the radioactivity could be recovered from the application site after 24 h and eight days, respectively. Urinary and faecal excretion continued at low levels for at least six days following application. After oral administration of 55 mg/kg bw, 79% and 10% of the radioactivity was eliminated in the faeces and urine, respectively, over eight days; most of the excretion occurred during the first 24 h. Only 0.1% of the administered dose remained in the body after eight days (Climie *et al.*, 1981a).

Bisphenol A diglycidyl ether is rapidly metabolized in mice *via* the epoxide groups to form the corresponding bis–diol. Epoxide hydratase catalyses this reaction, but diols are also formed *via* nonenzymatic hydrolysis. Further oxidation and dealkylation reactions take place. The metabolites excreted in faeces and urine include conjugates (glucuronides and sulfates) of the bis–diol and corresponding carboxylic acids. In the dealkylation steps, glyceraldehyde and glycidaldehyde are putative intermediates (Climie *et al.*, 1981b).

Percutaneous absorption of *phenyl glycidyl ether* was high in rats and rabbits, with absorption rates of 13.5 and 4.2 mg/cm² per h, respectively (Czajkowska & Stetkiewicz, 1972).

Phenyl glycidyl ether bound to glutathione in the presence of liver microsomes from different types of birds (Wit & Snel, 1968). *Allyl glycidyl ether, butyl glycidyl ether, phenyl glycidyl ether* and *bisphenol A diglycidyl ether* (Epikote 828) bound nonenzymatically to the *N*–7 position of guanosine *in vitro* (Hemminki *et al.*, 1980).

(ii) Toxic effects

1,2-Epoxydodecane (C_{12} alkyl glycidyl ether) and epoxide 8 (C_{12}–C_{14} alkyl glycidyl ether) gave positive results in the guinea-pig skin maximization test (Thorgeirsson *et al.*, 1975; Thorgeirsson, 1978).

The oral (intragastric) LD$_{50}$ for undiluted *allyl glycidyl ether* has been reported to be 0.4 g/kg bw in mice and 1.6 g/kg bw in rats. Following intragastric administration, focal liver necrosis was observed in some animals and central nervous system depression. The percutaneous LD$_{50}$ has been reported to be 2.6 g/kg bw for rabbits. A 4–h inhalation LC$_{50}$ of 270 ppm (1270 mg/m³) was reported for mice and an 8–h inhalation LC$_{50}$ of 670 ppm (3150 mg/m³) for rats (Hine *et al.*, 1956).

In rats receiving two courses of two intramuscular injections of 400 mg/kg bw allyl glycidyl ether, separated by a four–day recovery period, necropsy and histological examination on day 12 showed thymic atrophy or loss of lymphoid tissue, focal necrosis of the pancreas and testis, and pneumonia (Kodama *et al.*, 1961).

Rats were exposed by inhalation for 7 h per day on five days per week to 260 and 400 ppm (1214 and 1868 mg/m³) allyl glycidyl ether for 50 days and to 600 and 900 ppm (2802 and 4203 mg/m³) for 25 days. The 400-ppm dose induced irritation of the eyes and respiratory tract, reduced body weight gain and increased relative kidney weight. With 600 and 900 ppm, severe irritation of the eyes and respiratory tract, decreased respiratory rate and corneal cloudiness were observed (Hine *et al.*, 1956).

When mice were exposed by inhalation (head only) for 15 min to 1.9–8.6 ppm (8.9–40.4 mg/m³) allyl glycidyl ether, a 50% decrease in respiratory rate was seen at 5.7 ppm (26.8 mg/m³). Exposure of mice by inhalation (whole body) to 7.1 ppm (33.4 mg/m³) allyl glycidyl ether for 6 h per day for four days caused necrosis of respiratory epithelium and complete erosion of olfactory epithelium without pulmonary injury, whereas exposure to 2.5 ppm (11.8 mg/m³) for 14 days caused no nasal or pulmonary injury (Gagnaire *et al.*, 1987).

The oral LD_{50} for *bisphenol A diglycidyl ether* was reported to be 19.6 ml/kg bw in rats and the dermal LD_{50} to be > 20 ml/kg bw in rabbits. Intracutaneous injection of the compound to guinea-pigs sensitized 19/20 animals (Weil *et al.*, 1963). Bisphenol A diglycidyl ether monomer sensitized all animals in a guinea-pig skin maximization test and was classified by the authors as an extreme allergen (Thorgeirsson & Fregert, 1977).

There was a dose-related increase in erythema, exfoliation/fissuring, haemorrhage and oedema of the skin at the application site following daily dermal application to rabbits of 100 or 300 mg/kg bw bisphenol A diglycidyl ether (Breslin *et al.*, 1988).

The oral and dermal LD_{50}s of *1,4-butanediol diglycidyl ether* in rats were 2.98 g/kg bw and 1.13 g/kg bw, respectively; the compound caused skin and eye irritation in rabbits (Cornish & Block, 1959). It gave positive results in the guinea-pig skin maximization test (Thorgeirsson, 1978; Clemmensen, 1984).

The LD_{50}s of n-*butyl glycidyl ether* by oral gavage were 1.5 and 2.3 g/kg bw in mice and rats, respectively (Hine *et al.*, 1956). In another study, the oral and dermal LD_{50}s in rats were 3.4 and 2.3 mg/kg bw, respectively; the compound caused skin and mild eye irritation in rabbits (Cornish & Block, 1959).

tert-*Butylglycidyl ether* gave negative results in the guinea-pig skin maximization test (Rao *et al.*, 1981).

The oral LD_{50} of *cresyl glycidyl ether* in rats was 5.1 mg/kg bw, and that in mice was 1.7 g/kg bw. The LC_{50} in rats was 282 mg/m³, and necrosis of the renal epithelium was reported (Krechkovskii *et al.*, 1985a). The subcutaneous LD_{50} in mice was 980 mg/kg bw (Söllner & Irrgang, 1965). Cresyl glycidyl ether caused sensitization in guinea-pigs previously sensitized to Epoxide 8 (C_{12}-C_{14} alkyl glycidyl ether) or butyl glycidyl ether (Thorgeirsson *et al.*, 1975).

Neopentylglycol diglycidyl ether gave positive results in the guinea-pig skin maximization test (Thorgeirsson, 1978).

The single oral LD_{50} for *phenyl glycidyl ether* has been reported to be 4.3 ml/kg bw in rats and the single skin penetration LD_{50} to be 1.5 ml/kg bw in rabbits (Weil *et al.*, 1963). The LD_{50} by gavage was 1.4 g/kg bw in mice (Hine *et al.*, 1956). Intragastric and cutaneous LD_{50} values in white rats were 2.6 and 2.1 g/kg bw, respectively (Czajkowska & Stetkiewicz, 1972). The subcutaneous LD_{50} in mice was 760 mg/kg bw (Söllner & Irrgang, 1965). Phenyl glycidyl ether caused moderate skin irritation and corneal injury in rabbits; sensitization of guinea-pigs after intracutaneous injection was low (Weil et al., 1963). Local application of phenyl glycidyl ether produced skin necrosis (Czajkowska & Stetkiewicz, 1972).

Rats and beagle dogs were exposed by inhalation to 1, 5 or 12 ppm (6, 31 or 74 mg/m³) phenyl glycidyl ether for 6 h per day on five days per week for 90 days. The only significant finding was bilateral hair loss in rats exposed to 5 or 12 ppm. Blood and urine analysis and histopathological examination of all major organs revealed no other treatment-related effect (Terrill & Lee, 1977).

(iii) *Effects on reproduction and prenatal toxicity*

Spermatogenic effects (decreased sperm motility, abnormal sperm morphology, increased numbers of tubules with desquamated epithelium) were seen in male rats, and embryolethality (especially in the pre-implantation periods) in female rats, exposed by inhalation to 0, 2.55 and 19.1 mg/m³ *cresyl glycidyl ether* (Krechkovskii et al., 1985b).

Groups of 26 pregnant New Zealand white rabbits received daily dermal applications of 0, 30, 100 or 300 mg/kg bw *bisphenol A diglycidyl ether* (purity, 99.1%) on days 6–18 of gestation. The dose was prepared as a solution in propylene glycol 400 and applied under occlusion to a shaved area on the back of the test animal at a rate of 1 ml/kg bw per day. After 6 h, the occlusive bandage was removed; the treated area was not washed. Fetuses were examined by routine teratological techniques on day 28 of gestation. No treatment-related effect was reported, except for dose-related dermatological effects in the mothers (Breslin et al., 1988).

Four groups of eight male adult Sprague-Dawley rats were exposed by inhalation to 0, 1, 5 or 12 ppm (6, 31 or 74 mg/m³) *phenyl glycidyl ether* [purity unspecified] for 6 h per day on 19 consecutive days. After the last exposure, the rats were shipped to another laboratory, where their reproductive capabilities were assessed. Male fertility was evaluated during six-weekly co-habitation periods with three untreated females, some of which were killed on day 18 of gestation and examined for corpora lutea, implantation sites and resorption sites. Other females were allowed to deliver, and the offspring were followed through production of a second generation. The percentage of females that became pregnant was significantly reduced in the high-dose group in the first breeding week. Histological analysis of the testes indicated atrophy in 1/8 males from each treatment group. [The Working Group noted that it was not stated when the examination was performed.] No other treatment related effect was reported (Terrill et al., 1982).

In a teratology study, four groups of 25 Sprague-Dawley rats were exposed by inhalation to 0, 1, 5 or 12 ppm phenyl glycidyl ether [purity unspecified] for 6 h per day between days 4 and 15 of gestation. The fetuses were examined on day 20 of gestation for external, internal and skeletal abnormalities. No clinical sign of systemic toxicity was observed in the females, and no effect on fetal viability, growth or morphological development was found (Terrill et al., 1982).

(iv) Genetic and related effects

Alkyl C_8–C_{14} glycidyl ethers

Octyl and decyl glycidyl ethers were weakly mutagenic to *Salmonella typhimurium* TA1535 and TA100 in the presence of an exogenous metabolic system but were reported not to be mutagenic to TA1537, TA1538 or TA98 [details not given]. Neither octyl nor decyl glycidyl ethers induced unscheduled DNA synthesis in cultured human WI–38 cells. Decyl glycidyl ether was reported to have a marginal effect and octyl glycidyl ether no effect in inducing mutation in mouse lymphoma L5178Y TK$^{+/-}$ cells in the absence of an exogenous metabolic system (Thompson *et al.*, 1981).

Dodecyl glycidyl ether was weakly mutagenic to *S. typhimurium* TA1535 and TA100 in the presence of an exogenous metabolic system but not to strains TA1537, TA1538 or TA98 [details not given]. Tetradecyl glycidyl ether was not mutagenic in any of the five strains tested [details not given for TA1537, TA1538 or TA98]. Neither dodecyl nor tetradecyl glycidyl ether induced unscheduled DNA synthesis in cultured human WI–38 cells or mutation in mouse lymphoma L5178Y TK$^{+/-}$ cells (Thompson *et al.*, 1981). As reported in an abstract, alkyl (C_{12}–C_{14}) glycidyl ether did not induce dominant lethal mutations in B6D2F1 mice after dermal application (Pullin, 1978).

Allyl glycidyl ether was mutagenic to *Escherichia coli* WP2 *uvrA* in the absence of an exogenous metabolic system (Hemminki *et al.*, 1980). It was mutagenic to *S. typhimurium* TA1535 and TA100 in the presence and absence of an exogenous metabolic system but not to TA1537 or TA98 (Wade *et al.*, 1979; Canter *et al.*, 1986). It induced sex–linked recessive lethal mutations in *Drosophila melanogaster* when fed at 5500 ppm (mg/kg) for three days (Yoon *et al.*, 1985).

Bisphenol A diglycidyl ether

As reported in an abstract, Epikote 828 (composed mainly of bisphenol A diglycidyl ether) induced DNA repair in *E. coli* (Nishioka & Ohtani, 1978). It was mutagenic to *E. coli* WP2 *uvrA* in the absence of an exogenous metabolic system (Hemminki *et al.*, 1980). An aqueous emulsion of Epikote 828 was mutagenic to *S. typhimurium* TA100 and TA1535 in the absence of an exogenous metabolic system; in TA1535 its mutagenicity was increased when it was tested in the presence of an exogenous metabolic system (Andersen *et al.*, 1978). Araldite 6005 and EPON 828 (composed mainly of bisphenol A diglycidyl ether) induced mutation in *S. typhimurium* in the presence and absence of an exogenous metabolic system (Ringo *et al.*, 1982) [details not given]. As reported in an abstract, Epikote 828 was mutagenic to *S. typhimurium* TA100 but not TA98 (Nishioka & Ohtani, 1978). In one study, bisphenol A diglycidyl ether was mutagenic to *S. typhimurium* TA100 and TA1535 but not TA98 or TA1537 (Canter *et al.*, 1986); however, in another study, bisphenol A diglycidyl ether was not mutagenic to *S. typhimurium* TA98 or TA100 (Wade *et al.*, 1979).

1,4–Butanediol diglycidyl ether

1,4–Butanediol diglycidyl ether was mutagenic to *S. typhimurium* TA1535, TA98 and TA100 in the presence and absence of an exogenous metabolic system. Results in TA1537 in the absence of an exogenous metabolic system and in the presence of Aroclor 1254–induced hamster liver were equivocal, but positive findings were obtained using Aroclor 1254-induced rat liver (Canter *et al.*, 1986).

Butyl glycidyl ethers

Butyl glycidyl ether [not further specified] was mutagenic to *E. coli* WP2 *uvrA* in the absence of an exogenous metabolic system (Hemminki *et al.*, 1980). *n*-Butyl glycidyl ether was mutagenic to *S. typhimurium* TA97, TA100 and TA1535 in the presence and absence of an exogenous metabolic system, but not to TA1537, TA1538 or TA98 (Wade *et al.*, 1979; Connor *et al.*, 1980a; Canter *et al.*, 1986). *tert*-Butyl glycidyl ether produced mutations in *S. typhimurium* TA1535 and TA100 in the presence and absence of an exogenous metabolic system (Canter *et al.*, 1986). As reported in an abstract, *n*-butyl and *tert*-butyl glycidyl ether induced dose-related DNA damage in cultured human lymphocytes, as determined by scintillation counting and autoradiography; *tert*-butyl glycidyl ether did not produce micronuclei or chromosomal aberrations in mice treated *in vivo* (Connor *et al.*, 1980b). *n*-Butyl glycidyl ether did not induce morphological transformation of BALBc/3T3 clone A31-1-13 cells. The urine of mice that received oral and dermal administration at various doses and for various times of *n*-butyl glycidyl ether was not mutagenic to *S. typhimurium* TA1535 or TA98. *n*-Butyl glycidyl ether produced micronuclei in bone marrow of female BDF mice when administered intraperitoneally (225–900 mg/kg bw) but not when administered orally (200 mg/kg bw; Connor *et al.*, 1980a). *n*-Butyl glycidyl ether was tested for its ability to induce dominant lethal mutations in BDF hybrid mice at doses of 0.375, 0.750 and 1.5 g/kg bw by topical application. No significant dose-related change in either pregnancy rates or in average number of implants per pregnant female was observed, but a significant increase in fetal death rates occurred at the end of the first week of administration of the highest dose (Whorton *et al.*, 1983).

tert-Butylphenyl glycidyl ether

tert-Butylphenyl glycidyl ether was mutagenic to *S. typhimurium* TA100 in the absence, but not in the presence, of an exogenous metabolic system, but was not mutagenic to TA1535, TA1537 or TA98 (Neau *et al.*, 1982; Canter *et al.*, 1986).

Cresyl glycidyl ether

ortho-Cresyl glycidyl ether was mutagenic to *S. typhimurium* TA1535 and TA100 in the absence of an exogenous metabolic system, and *para*-cresyl glycidyl ether was mutagenic to the same strains in the presence and absence of an exogenous metabolic system. These compounds were not mutagenic to TA1537 or TA98 (Canter *et al.*, 1986). As reported in an abstract, dermal application of *ortho*-cresyl glycidyl ether caused a significant reduction in the mean number of implants per pregnancy in a dominant lethal mutation assay in B6D2F1 mice (Pullin, 1978).

Neopentylglycol diglycidyl ether

Neopentylglycol diglycidyl ether was mutagenic to *S. typhimurium* TA1535, TA97 and TA100, but not to TA98, in the presence and absence of an exogenous metabolic system (Canter *et al.*, 1986). As reported in an abstract, it significantly reduced the mean number of implants per pregnancy in a dominant lethal mutation assay in B6D2F1 mice after dermal application (Pullin, 1978).

Phenyl glycidyl ether

As reported in an abstract, phenyl glycidyl ether gave positive results in assays for DNA repair in *E. coli* (Nishioka & Ohtani, 1978). It was mutagenic to *E. coli* WP2 *uvrA* in the absence of an exogenous metabolic system (Hemminki *et al.*, 1980). Phenyl glycidyl ether was mutagenic to *S. typhimurium* TA1535, TA97 and TA100 but not to TA1537, TA1538 or TA98 in the presence and absence of an exogenous metabolic system (Greene *et al.*, 1979; Ivie *et al.*, 1980; Seiler, 1984; Canter *et al.*, 1986). At an oral dose of 2500 mg/kg, it was active in the host–mediated assay using C57Bl/6 × C3H mice and *S. typhimurium* TA1535, in two out of five animals tested. It did not induce 6–thioguanine–resistant cells in Chinese hamster ovary cells with or without an exogenous metabolic system (Greene *et al.*, 1979). Phenyl glycidyl ether did not induce chromosomal aberrations in Chinese hamster ovary cells (Seiler, 1984). It induced morphological transformation in secondary Syrian hamster embryo cells and enhanced transformation by SA7 virus in primary Syrian hamster embryo cells (Greene *et al.*, 1979). It did not induce micronuclei in the bone marrow of ICR mice after oral administration of 400–1000 mg/kg bw (Seiler, 1984). No evidence of dominant lethal mutation was observed in Sprague–Dawley rats following inhalation of 2–11 ppm (12.3–67.5 mg/m³) phenyl glycidyl ether, and chromosomal aberrations were not induced in bone–marrow cells of exposed animals (Terrill *et al.*, 1982).

(b) Humans

(i) Absorption, distribution, excretion and metabolism

No data were available to the Working Group.

(ii) Toxic effects

Butyl glycidyl ether was found to be a strong contact sensitizer in the skin maximization test in humans (Kligman, 1966).

Phenyl glycidyl ether has also been recognized as a contact allergen using a patch test in a study of persons with dermatitis from occupational contact with epoxy resins (Rudzki & Krajewska, 1979; Rudzki *et al.*, 1983). It was also an allergen in a patch test on eight of 15 workers in a cable production plant who developed dermatosis; phenyl glycidyl ether was present in the plastic insulation material as a stabilizer (Zschunke & Behrbohm, 1965).

Thirty–four workers with occupational dermatitis on the hands, arms and occasionally on the face, all of whom had worked with low molecular–weight epoxy resins in different factories, were given a patch test with *bisphenol A diglycidyl ether* after symptoms had regressed; all gave positive reactions (Fregert & Thorgeirsson, 1977).

Eleven workers handling epoxy resins based mainly on Epikote 815, containing bisphenol A diglycidyl ether (89% w/w) and butyl glycidyl ether (11%), in the production of transformers for television sets developed a scleroderma–like dermatosis and other symptoms, including muscle and joint disease and central nervous system and respiratory disturbances (Tomizawa *et al.*, 1979).

Epoxy dermatitis due to exposure to bisphenol A diglycidyl ether was reported among insulation workers at an electric power station (Niinimäki & Hassi, 1983). Allergic contact dermatitis was also observed on the upper lip of a patient who had used a nasal oxygen cannula made of epoxy resin; analysis indicated the presence of a chemical reported to be related to bisphenol A (Wright & Fregert, 1983).

Out of 20 resin workers, 19 developed contact allergy to epoxy resins; three of them reacted in a patch test to allyl glycidyl ether, two to n-butyl glycidyl ether and 14 to phenyl glycidyl ether (Fregert & Rorsman, 1964).

Three workers employed in a brush factory developed a contact allergy from a two-component glue containing epoxy resin (37% w/w bisphenol A diglycidyl ether). The reactive diluents, allyl glycidyl ether, 1,4-butanediol diglycidyl ether, n-butyl glycidyl ether, ortho-cresyl glycidyl ether, neopentylglycol diglycidyl ether and phenyl glycidyl ether, were recognized as sensitizers in the patch test (Jolanki et al., 1987).

(iii) Effects on fertility and on pregnancy outcome

No data were available to the Working Group.

(iv) Genetic and related effects

Cytogenetic evaluation was performed in peripheral lymphocytes from 18 workers currently exposed to bisphenol A diglycidyl ether-type epoxy resins. Nine workers had been exposed to a low molecular-weight product (less than 900; main oligomer, MW340-bisphenol A diglycidyl ether) for five to 16 years (median, 6.5 years) and nine to high molecular-weight (about 2000) epoxy resins (for three to ten years; median, seven years). The results were compared with those for an equal number of control individuals matched for sex and age. There was no difference in the frequency of chromosomal aberrations between controls and the groups exposed to epoxy resins (Mitelman et al., 1980).

Chromosomal aberrations were not more frequent in the peripheral lymphocytes of 22 employees in an epoxy resin plant than in those of ten persons from the medical department at the manufacturing site, who were used as control subjects and matched for sex, age and smoking habits. The workers had been exposed to epoxy resins (a mixture of bisphenol A diglycidyl ether and its higher homologues), epichlorohydrin, butyl glycidyl ether and cresyl glycidyl ether for one to four years (11 men) or for than ten years (11 men). Exposure of workers to epichlorohydrin varied from 0.1 to 1.6 mg/m³; exposure to both butyl glycidyl ether and cresyl glycidyl ether was below 0.07 mg/m³, and concentrations of a urinary metabolite of epoxy resin (the bis-diol of bisphenol A diglycidyl ether) were below the analytical limit of detection of 0.1 μg/ml (de Jong et al., 1988).

3.3 Epidemiological studies of carcinogenicity to humans

No data were available to the Working Group.

4. Summary of Data Reported and Evaluation

4.1 Exposures

Glycidyl ethers are basic components of epoxy resins which have been commercially available since the late 1940s. Bisphenol A diglycidyl ether and its oligomers are major components of epoxy resins. Other glycidyl ethers, including phenyl glycidyl ether, are frequently incorporated into epoxy resin systems as reactive modifiers. Epoxy resins based on bisphenol A diglycidyl ether are widely used in protective coatings, including paints, in reinforced plastic laminates and composites, in tooling, casting and moulding resins, in bonding materials and adhesives, and in floorings and aggregates. Occupational exposure to bisphenol A diglycidyl ether and phenyl glycidyl ether may occur during their production, during the production of epoxy products and during various uses of epoxy products, but data on exposure levels are sparse.

4.2 Experimental carcinogenicity data

Bisphenol A diglycidyl ether of various technical grades was tested by skin application in mice in five studies. In one of the studies, an increased incidence of epidermal tumours was found in one of two strains tested. In another study, a small increase in the incidence of epidermal tumours and small increases in the incidences of kidney tumours in male mice and of lymphoreticular/haematopoietic tumours in female mice were observed. No increase in the incidence of skin tumours was observed in two further studies, and the other study was inadequate for evaluation. Following subcutaneous injection of technical-grade bisphenol A diglycidyl ether to rats, a small number of local fibrosarcomas was observed. Following application of technical-grade bisphenol A diglycidyl ether to the skin of rabbits, no skin tumour was observed.

Pure bisphenol A diglycidyl ether was tested in one experiment by skin application in mice; no epidermal but a few dermal tumours were observed in males, and there was a small increase in the incidence of lymphoreticular/haematopoietic tumours in females.

Pure phenyl glycidyl ether was tested for carcinogenicity by inhalation exposure in male and female rats of one strain, producing carcinomas of the nasal cavity in animals of each sex.

4.3 Human data

No data were available to the Working Group.

4.4 Other relevant data

Some glycidyl ethers have been shown to cause allergic contact dermatitis in humans. Glycidyl ethers generally cause skin sensitization in experimental animals. Necrosis of the mucous membranes of the nasal cavities was induced in mice exposed to allyl glycidyl ether.

Prenatal toxicity was not induced in rats exposed by inhalation to phenyl glycidyl ether or in rabbits exposed dermally to bisphenol A diglycidyl ether.

One study of workers exposed to bisphenol A diglycidyl ether showed no increase in the incidence of chromosomal aberrations in peripheral lymphocytes. A study of workers with mixed exposures was inconclusive with regard to the effects of specific glycidyl ethers. Phenyl glycidyl ether, but not n–butyl glycidyl ether, induced morphological transformation in mammalian cells *in vitro*. n–Butyl glycidyl ether induced micronuclei in mice *in vivo* following intraperitoneal but not oral administration. Phenyl glycidyl ether did not induce micronuclei or chromosomal aberrations *in vivo* or chromosomal aberrations in animal cells *in vitro*. Alkyl C_{12} or C_{14} glycidyl ether did not induce DNA damage in cultured human cells or mutation in cultured animal cells. Allyl glycidyl ether induced mutation in *Drosophila*. The glycidyl ethers were generally mutagenic to bacteria. (See Appendix 1.)

4.5   Evaluation[1]

There is *sufficient evidence* for the carcinogenicity of phenyl glycidyl ether in experimental animals.

There is *limited evidence* for the carcinogenicity of bisphenol A diglycidyl ether in experimental animals.

No data were available from studies in humans on the carcinogenicity of glycidyl ethers.

Overall evaluation

Phenyl glycidyl ether *is possibly carcinogenic to humans (Group 2B)*.

Bisphenol A diglycidyl ether *is not classifiable as to its carcinogenicity to humans (Group 3)*.

5. References

American Conference of Governmental Industrial Hygienists (1988) *Threshold Limit Values and Biological Exposure Indices for 1988–1989*, Cincinnati, OH, pp. 11, 13, 30

Andersen, M., Kiel, P., Larsen, H. & Maxild, J. (1978) Mutagenic action of aromatic epoxy resins. *Nature*, *276*, 391–392

Arbeidsinspectie (Labour Inspection) (1986) *De Nationale MAC–Lijst 1986* [National MAC-list 1986] (*P145*), Voorburg, Ministry of Social Affairs, p. 13

---

[1]For definitions of the italicized terms, see Preamble, pp. 27–30.

Arbjedstilsynet (Labour Inspection) (1988) *Graensevaerdier for Stoffer og Materialer* [Limit Values for Substances and Materials] (*At–anvisning No. 3.1.0.2*), Copenhagen, p. 27

Boiano, J.M. (1981) *International Harvester, Truck Engineering and Design Center, Ft Wayne, Indiana (Health Hazard Evaluation Determination Report No. 80–165–907)*, Cincinnati, OH, National Institute for Occupational Safety and Health

Breslin, W.J., Kirk, H.D. & Johnson, K.A. (1988) Teratogenic evaluation of diglycidyl ether of bisphenol A (DGEBPA) in New Zealand white rabbits following dermal exposure. *Fundam. appl. Toxicol., 10*, 736–743

Brown, R.M. & Creaser, C.S. (1980) The electron impact and chemical ionization mass spectra of some glycidyl ethers. *Org. Mass Spectrom., 15*, 578–581

Canter, D.A., Zeiger, E., Haworth, S., Lawlor, T., Mortelmans, K. & Speck, W. (1986) Comparative mutagenicity of aliphatic epoxides in *Salmonella. Mutat. Res., 172*, 105–138

Chemical Manufacturers' Association (1984) *Comments in Response to Advance Notice of Proposed Rule-making on Testing of Glycidol and its Derivatives*, submitted to the US Environmental Protection Agency by the CMA Epoxy Resins Program Panel, Washington DC

Ciba-Geigy Corp. (1984) *Product Data: Araldite® Gy 6010*, Hawthorne, NY, Resins Department

Clemmensen, S. (1984) Cross-reaction patterns in guinea pigs sensitized to acrylic monomers. *Drug chem. Toxicol., 7*, 527–540

Climie, I.J.G., Hutson, D.H. & Stoydin, G. (1981a) Metabolism of the epoxy resin component 2,2-bis[4-(2,3-epoxypropoxy)phenyl]propane, the diglycidyl ether of bisphenol A (DGEBPA) in the mouse. Part I. A comparison of the fate of a single dermal application and of a single oral dose of ^{14}C–DGEBPA. *Xenobiotica, 11*, 391–399

Climie, I.J.G., Hutson, D.H. & Stoydin, G. (1981b) Metabolism of the epoxy resin component 2,2-bis[4-(2,3-epoxypropoxy)phenyl]propane, the diglycidyl ether of bisphenol A (DGEBPA) in the mouse. Part II. Identification of metabolites in urine and faeces following a single oral dose of ^{14}C–DGEBPA. *Xenobiotica, 11*, 401–424

Connor, T.H., Ward, J.B., Jr, Meyne, J., Pullin, T.G. & Legator, M.S. (1980a) The evaluation of the epoxide diluent, *n*-butylglycidyl ether, in a series of mutagenicity assays. *Environ. Mutagenesis, 2*, 521–530

Connor, T.H., Pullin, T.G., Meyne, J., Frost, A.F. & Legator, M.S. (1980b) Evaluation of the mutagenicity of *n*-BGE and *t*-BGE in a battery of short-term assays (Abstract No. Ec–12). *Environ. Mutagenesis, 2*, 284

Cook, W.A. (1987) *Occupational Exposure Limits – Worldwide*, Washington DC, American Industrial Hygiene Association

Cornish, H.H. & Block, W.D. (1959) The toxicology of uncured epoxy resins and amine curing agents. *Arch. environ. Health, 20*, 390–398

Crathorne, B., Fielding, M., Steel, C.P. & Watts, C.D. (1984) Organic compounds in water: analysis using coupled-column high-performance liquid chromatography and soft-ionization mass spectrometry. *Environ. Sci. Technol., 18*, 797–802

Crathorne, B., Palmer, C.P. & Stanley, J.A. (1986) High-performance liquid chromatographic determination of bisphenol A diglycidyl ether and bisphenol F diglycidyl ether in water. *J. Chromatogr., 360*, 266–270

Czajkowska, T. & Stetkiewicz, J. (1972) Evaluation of acute toxicity of phenyl glycidyl ether with special regard to percutaneous absorption (Pol.). *Med. Prac., 23*, 363–371

Deutsche Forschungsgemeinschaft (German Research Society) (1988) *Maximale Arbeitsplatzkonzentrationen und Biologische Arbeitsstofftoleranzwerte 1988* [Maximal Concentrations at the Workplace and Biological Tolerance Values for Working Materials 1988] (*Report No. XXIV*), Weinheim, VCH Verlagsgesellschaft, p. 51

Direktoratet for Arbeidstilsynet (Directorate for Labour Inspection) (1981) *Administrative Normer for Forurensning i Arbeidsatmosfaere 1981* [Administrative Norms for Pollution in the Work Atmosphere] (*No. 361*), Oslo, p. 11

Eadsforth, C.V., Hutson, D.H., Logan, C.J. & Morrison, B.J. (1985a) The metabolism of *n*-butyl glycidyl ether in the rat and rabbit. *Xenobiotica, 15*, 579–589

Eadsforth, C.V., Logan, C.J., Page, J.A. & Regan, P.D. (1985b) *n*-Butylglycidyl ether: the formation of a novel metabolite of an epoxide. *Drug Metab. Dispos., 13*, 263–264

Fregert, S. & Rorsman, H. (1964) Allergens in epoxy resins. *Acta allergol., 19*, 269–299

Fregert, S. & Thorgeirsson, A. (1977) Patch testing with low molecular oligomers of epoxy resins in humans. *Contact Derm., 3*, 301–303

Gagnaire, F., Zissu, D., Bonnet, P. & De Ceaurriz, J. (1987) Nasal and pulmonary toxicity of allyl glycidyl ether in mice. *Toxicol. Lett., 39*, 139–145

Greene, E.J., Friedman, M.A., Sherrod, J.A. & Salerno, A.J. (1979) In vitro mutagenicity and cell transformation screening of phenylglycidyl ether. *Mutat. Res., 67*, 9–19

Guenier, J.P., Lhuillier, F. & Muller, J. (1986) Sampling of gaseous pollutants on silica gel with 1400 mg tubes. *Ann. occup. Hyg., 30*, 103–114

Harper, C.A. (1979) Embedding. In: Mark, H.F., Othmer, D.F., Overberger, C.G., Seaborg, G.T. & Grayson, M., eds, *Kirk–Othmer Encyclopedia of Chemical Technology*, 3rd ed., Vol. 8, New York, John Wiley & Sons, pp. 877–899

Hemminki, K., Falck, K. & Vainio, H. (1980) Comparison of alkylation rates and mutagenicity of directly acting industrial and laboratory chemicals. Epoxides, glycidyl ethers, methylating and ethylating agents, halogenated hydrocarbons, hydrazine derivatives, aldehydes, thiuram and dithiocarbamate derivatives. *Arch. Toxicol., 46*, 277–285

Hervin, R.L., Frederick, L. & McQuilkin, S. (1979) *Greenheck Fan Corporation, Schofield, Wisconsin* (*Health Hazard Evaluation Determination Report No. 79–7–639*), Cincinnati, OH, National Institute for Occupational Safety and Health

Hine, C.H,. Kodama, J.K., Wellington, J.S., Dunlap, M.K. & Anderson, H.H. (1956) The toxicology of glycidol and some glycidyl ethers. *Am. med. Assoc. Arch. ind. Health, 14*, 250–264

Hine, C.H., Guzman, R.J., Coursey, M.M., Wellington, J.S. & Anderson, H.H. (1958) An investigation of the oncogenic activity of two representative epoxy resins. *Cancer Res., 18*, 20–26

Hine, C., Rowe, V.K., White, E.R., Darmer, K.I., Jr & Youngblood, G.T. (1981) Epoxy compounds. In: Clayton, G.D. & Clayton, F.E., eds, *Patty's Industrial Hygiene and Toxicology*, 3rd rev. ed., Vol. 2A, New York, John Wiley & Sons, pp. 2141–2257

Holland, J.M., Gosslee, D.G. & Williams, N.J. (1979) Epidermal carcinogenicity of bis(2,3–epoxycyclopentyl)ether, 2,2–bis(p–glycidyloxyphenyl)propane, and m–phenylenediamine in male and female C3H and C57Bl/6 mice. *Cancer Res., 39*, 1718–1725

IARC (1976) *IARC Monographs on the Evaluation of Carcinogenic Risk of Chemicals to Man*, Vol. 11, *Cadmium, Nickel, Some Epoxides, Miscellaneous Industrial Chemicals and General Considerations on Volatile Anaesthetics*, Lyon, pp. 175–181

IARC (1987) *IARC Monographs on the Evaluation of Carcinogenic Risks to Humans*, Suppl. 7, *Overall Evaluations of Carcinogenicity: An Updating of* IARC Monographs *Volumes 1 to 42*, Lyon

IARC (1988) *Information Bulletin on the Survey of Chemicals Being Tested for Carcinogenicity*, No. 13, Lyon, p. 212

International Labour Office (1984) *Occupational Exposure Limits for Airborne Toxic Substances*, 2nd rev. ed. (*Occupational Safety and Health Series No. 37*), Geneva, pp. 172–173

Ivie, G.W., MacGregor, J.T. & Hammock, B.D. (1980) Mutagenicity of psoralen epoxides. *Mutat. Res.*, *79*, 73–77

Jolanki, R., Estlander, T. & Kanerva, L. (1987) Contact allergy to an epoxy reactive diluent: 1,4–butanediol diglycidyl ether. *Contact Derm.*, *16*, 87–92

de Jong, G., van Sittert, N.J. & Natarajan, A.T. (1988) Cytogenetic monitoring of industrial populations potentially exposed to genotoxic chemicals and of control populations. *Mutat. Res.*, *204*, 451–464

Kligman, A.M. (1966) The identification of contact allergens by human assay. III. The maximization test – a procedure for screening and rating contact sensitizers. *J. invest. Dermatol.*, *47*, 393–409

Kodama, J.K., Guzman, R.J., Dunlap, M.K., Loquvam, G.S., Lima, R. & Hine, C.H. (1961) Some effects of epoxy compounds on the blood. *Arch. environ. Health*, *2*, 50–61

Krechkovskii, E.A., Anisimova, I.G., Shevchuk, R.M. & Zaprivoda, L.P. (1985a) Toxicity of cresyl glycidyl ether (Russ.). *Gig. Tr. prof. Zabol.*, *3*, 50

Krechkovskii, E.A., Anisimova, I.G., Vislova, S.V., Zaprivoda, L.P. & Zhchkova, N.V. (1985b) Experimental study of the remote effects on reproduction of cresyl glycidyl ether (Russ.). *Gig. Sanit.*, *10*, 81–83

Lee, K.P., Schneider, P.W., Jr & Trochimowicz, H.J. (1983) Morphologic expression of glandular differentiation in the epidermoid nasal carcinomas induced by phenylglycidyl ether inhalation. *Am. J. Pathol.*, *111*, 140–148

Mitelman, F., Fregert, S., Hedner, K. & Hillbertz–Nilsson, K. (1980) Occupational exposure to epoxy resins has no cytogenetic effect. *Mutat. Res.*, *77*, 345–348

National Institute for Occupational Safety and Health (1974) *US National Exposure Survey 1972–77*, Cincinnati, OH

National Institute for Occupational Safety and Health (1978) *NIOSH Criteria for a Recommended Standard ... Occupational Exposures to Glycidyl Ethers (DHEW (NIOSH) Publ. No. 78–166)*, Cincinnati, OH

National Institute for Occupational Safety and Health (1983) *US National Exposure Survey 1981–83*, Cincinnati, OH

National Swedish Board of Occupational Safety and Health (1987) *Hygienska Gränsvärden* [Hygienic Limit Values] (*Ordinance 1987:12*), Solna, p. 22

Neau, S.H., Hooberman, B.H., Frantz, S.W. & Sinsheimer, J.E. (1982) Substituent effects on the mutagenicity of phenyl glycidyl ethers in *Salmonella typhimurium*. *Mutat. Res.*, *93*, 297–304

Niinimäki, A. & Hassi, J. (1983) An outbreak of epoxy dermatitis in insulation workers at an electric power station. *Dermatosen*, *31*, 23–25

Nishioka, H. & Ohtani, H. (1978) Mutagenicity of epoxide resins; constituents and commercial adhesives in bacterial test systems (Abstract No. 21). *Mutat. Res.*, *54*, 247–248

Patterson, W.A. (1954) Infrared absorption bands characteristic of the oxirane ring. *Anal. Chem.*, *26*, 823–835

Peltonen, K., Pfäffli, P., Itkonen, A. & Kalliokoski, P. (1986) Determination of the presence of bisphenol–A and the absence of diglycidyl ether of bisphenol–A in the thermal degradation products of epoxy powder paint. *Am. ind. Hyg. Assoc. J.*, *47*, 399–403

Peristianis, G.C., Doak, S.M.A., Cole, P.N. & Hend, R.W. (1988) Two–year carcinogenicity study on three aromatic epoxy resins applied cutaneously to CF1 mice. *Food chem. Toxicol.*, *26*, 611–624

Pouchert, C.J., ed. (1983) *The Aldrich Library of NRM Spectra*, 2nd ed., Vol. 1, Milwaukee, WI, Aldrich Chemical Co., pp. 195D, 858C, 859C

Pouchert, C.J., ed. (1985) *The Aldrich Library of FT–IR Spectra*, Vol. 1, Milwaukee, WI, Aldrich Chemical Co., p. 232C, 233D, 234C, 1064C, 1065B

Pullin, T.G. (1978) Mutagenic evaluation of several industrial glycidyl ethers (Abstract). *Diss. Abstr. int. (B)*, 4795–B – 4796–B

Rao, K.S., Betso, J.E. & Olson, K.J. (1981) A collection of guinea pig sensitization test results – grouped by chemical class. *Drug Chem. Toxicol.*, *4*, 331–351

Ringo, D.L., Brennan, E.F. & Cota-Robles, E.H. (1982) Epoxy resins are mutagenic: implications for electron microscopists. *J. ultrastruct. Res.*, *80*, 280–287

Rudzki, E. & Krajewska, D. (1979) Contact sensitivity to phenyl glycidyl ether. *Dermatosen*, *27*, 42–44

Rudzki, E., Rebandel, P., Grzywa, Z. & Jakiminska, B. (1983) Dermatitis from phenyl glycidyl ether. *Contact Derm.*, *9*, 90–91

Sadtler Research Laboratories (1980) *Standard Spectra Collection, 1980 Cumulative Index*, Philadelphia, PA

Seiler, J.P. (1984) The mutagenicity of mono– and di-functional aromatic glycidyl compounds. *Mutat. Res.*, *135*, 159–167

Shapiro, M.J. (1977) Conformational analysis of 1-substituted 2,3–epoxypropanes. A carbon–13 nuclear magnetic resonance study. *J. org. Chem.*, *42*, 1434–1436

Shikado, R. (1987) Plastics. In: *Japan Chemical Annual 1987/1988, Japan's Chemical Industry and Engineering*, Tokyo, The Chemical Daily Co., pp. 50–55

Söllner, K. & Irrgang, K. (1965) Studies on the pharmacological action of aromatic glycidyl ethers (Ger.). *Arzneimittel-Forsch.*, *15*, 1355–1357

Taylor, D.G. (1977) *NIOSH Manual of Analytical Methods*, 2nd ed., Vol. 2 *(DHEW (NIOSH) Publ. No. 77–157–B)*, Washington DC, US Government Printing Office, pp. S74–1–S74–8, S81–1–S81–8

Taylor, D.G. (1978) *NIOSH Manual of Analytical Methods*, 2nd ed., Vol. 4 *(DHEW (NIOSH) Publ. No. 78–175)*, Washington DC, US Government Printing Office, pp. S346–1–S346–9

Taylor, D.G. (1980) *NIOSH Manual of Analytical Methods*, 2nd ed., Vol. 6 *(DHEW (NIOSH) Publ. No. 80–125)*, Washington DC, US Government Printing Office, pp. 333–1–333–6

Terrill, J.B. & Lee, K.P. (1977) The inhalation toxicity of phenyl glycidyl ether. I. 90–Day inhalation study. *Toxicol. appl. Pharmacol.*, *42*, 263–269

Terrill, J.B., Lee, K.P., Culik, R. & Kennedy, G.L., Jr (1982) The inhalation toxicity of phenylglycidyl ether: reproduction, mutagenic, teratogenic, and cytogenetic studies. *Toxicol. appl. Pharmacol.*, *64*, 204–212

Thompson, E.D., Coppinger, W.J., Piper, C.E., McCarroll, N., Oberly, T.J. & Robinson, D. (1981) Mutagenicity of alkyl glycidyl ethers in three short-term assays. *Mutat. Res.*, *90*, 213–231

Thorgeirsson, A. (1978) Sensitization capacity of epoxy reactive diluents in the guinea pig. *Acta dermatol. venerol.*, *58*, 329–331

Thorgeirsson, A. & Fregert, S. (1977) Allergenicity of epoxy resins in the guinea pig. *Acta dermatovener.*, *57*, 253–256

Thorgeirsson, A., Fregert, S. & Magnusson, B. (1975) Allergenicity of epoxy-reactive diluents in the guinea-pig. *Berufsdermatosen*, *23*, 178–183

Tomizawa, T., Yamaguchi, J., Horiguchi, M. & Anzai, T. (1979) Scleroderma-like skin changes observed among workers handling epoxy resins. In: Gonzalez-Ochoa, A., Dominguez-Soto, L. & Ortiz, Y., eds, *Proceedings of the XV International Congress of Dermatology, Mexico City, 16–21 October 1977*, Amsterdam, Excerpta Medica, pp. 271–275

Työsuojeluhallitus (National Finnish Board of Occupational Safety and Health) (1987) *HTP–Arvot 1987* [TLV Values 1987] (*Safety Bulletin No. 25*), Helsinki, Valtion Painatuskeskus, p. 15

Ulbrich, V., Makeš, J. & Jureček, M. (1964) Identification of glycidyl ether bis-phenyl of glycerine and of bis-α-naphthylurethane of α-alkyl(aryl)ether of glycerine (Ger.). *Collect. Czech. chem. Commun.*, 29, 1466–1475

Urquhart, G.A., Riordan, B.J., Orrell, J.K. & Mackie, M. (1988) *Economic Impact Analysis of Proposed Test Rule for Glycidol and its Derivatives*, Washington DC, US Environmental Protection Agency, Economic and Technology Division, Office of Toxic Substances

US Food and Drug Administration (1988) Resinous and polymeric coatings. *US Code fed. Regul., Title 21*, Part 175.300, pp. 144–160

US International Trade Commission (1974) *Synthetic Organic Chemicals, US Production and Sales, 1972* (*USITC Publication 681*), Washington DC, US Government Printing Office, p. 134

US International Trade Commission (1985) *Synthetic Organic Chemicals, US Production and Sales, 1984* (*USITC Publication 1745*), Washington DC, US Government Printing Office, pp. 135, 259

US International Trade Commission (1986) *Synthetic Organic Chemicals, US Production and Sales, 1985* (*USITC Publication 1892*), Washington DC, US Government Printing Office, pp. 139, 268

US International Trade Commission (1987) *Synthetic Organic Chemicals, US Production and Sales, 1986* (*USITC Publication 2009*), Washington DC, US Government Printing Office, pp. 106, 213

US Occupational Safety and Health Administration (1987) Labor. *US Code fed. Regul., Title 29*, Part 1910.1000, pp. 676–682

Wade, M.J., Moyer, J.W. & Hine, C.H. (1979) Mutagenic action of a series of epoxides. *Mutat. Res.*, 66, 367–371

Weil, C.S., Condra, N., Haun, C. & Striegel, J.A. (1963) Experimental carcinogenicity and acute toxicity of representative epoxides. *Am. ind. Hyg. Assoc. J.*, 24, 305–325

Whorton, E.B., Jr, Pullin, T.G., Frost, A.F., Onofre, A., Legator, M.S. & Folse, D.S. (1983) Dominant lethal effects of *n*-butyl glycidyl ether in mice. *Mutat. Res.*, 124, 224–233

Wilmington Chemical Corp. (1987) *Technical Summary Chart: HELOXY® Specialty Epoxy Resins and Reactive Diluents*, Wilmington, DE

Wit, J.G. & Snel, J. (1968) Enzymatic glutathione conjugations with 2,3–epoxyphenylpropylether and diethylmaleate by wild bird liver supernatant. *Eur. J. Pharmacol.*, 3, 370–373

Wright, R.C. & Fregert, S. (1983) Allergic contact dermatitis from epoxy resin in nasal canulae. *Contact Derm.*, 9, 387–389

Yoon, J.S., Mason, J.M., Valencia, R., Woodruff, R.C. & Zimmering, S. (1985) Chemical mutagenesis testing in *Drosophila*. IV. Results of 45 coded compounds tested for the National Toxicology Program. *Environ. Mutagenesis*, 7, 349–367

Zakova, N., Zak, F., Froehlich, E. & Hess, R. (1985) Evaluation of skin carcinogenicity of technical 2,2–bis(*p*-glycidyloxyphenyl)propane in CF1 mice. *Food chem. Toxicol.*, 23, 1081–1089

Zschunke, E. & Behrbohm, P. (1965) Eczema due to phenoxypropenoxide and similar glycide ethers (Ger.). *Dermatol. Wochenschr.*, 151, 480–484

PHENOL

1. Chemical and Physical Data

1.1 Synonyms

Chem. Abstr. Services Reg. No.: 108–95–2
Chem. Abstr. Name: Phenol
IUPAC Systemic Name: Hydroxybenzene
Synonyms: Carbolic acid; monohydroxybenzene; oxybenzene; phenic acid; phenyl alcohol; phenyl hydrate; phenyl hydroxide; phenylic alcohol; phenylic acid

1.2 Structural and molecular formulae and molecular weight

C_6H_6O

Mol. wt: 94.11

1.3 Chemical and physical properties of the pure substance

(a) *Description*: White, crystalline solid that liquefies on absorption of water from air; acrid odour; sharp burning taste (Hawley, 1981)

(b) *Boiling-point*: 181.7°C at 760 mm Hg, 70.9°C at 10 mm Hg (Weast, 1985)

(c) *Melting-point*: 43°C (Weast, 1985)

(d) *Density*: 1.06 at 20°/4°C (Weast, 1985)

(e) *Spectroscopy data*: Infrared, ultraviolet and nuclear magnetic resonance spectral data have been reported (Sadtler Research Laboratories, 1980; Pouchert, 1981, 1983, 1985).

(f) *Refractive index*: 1.5408 at 41°C (Weast, 1985)

(g) *Solubility*: Soluble in water (82 g/l at 15°C; Considine, 1974), acetone, benzene, ethanol, diethyl ether, chloroform, glycerol, carbon disulfide and aqueous alkalies (Hawley, 1981; Windholz, 1983; Weast, 1985)

(h) *Volatility*: Vapour pressure: 0.357 mm Hg at 20°C (Dow Chemical Co., 1988)

(i) *Flash-point*: 80°C (closed cup); 85°C (open cup); mixture of air and 3–10% phenol vapour is explosive (Deichmann & Keplinger, 1981)

(j) *Reactivity*: Hot phenol is incompatible with aluminium, magnesium, lead, and zinc. Iron and copper catalyse discolouration. Contact with strong oxidizers and calcium hypochlorite must be avoided (Dow Chemical Co., 1988).

(k) *Octanol/water partition coefficient*: log P = 1.46 (Verschueren, 1983)

(l) *Conversion factor*: mg/m³ = 3.85 × ppm[1]

1.4 Technical products and impurities

Trade name: ENT 1814

Phenol is available in commercial grades of 82–84%, 90–92% (Considine, 1974) and 95%. Typical impurities from cumene–derived phenol include small amounts of acetol, acetone, acetophenone, *sec*-butyl alcohol, cumene, cyclohexanol, α,α–dimethylphenyl carbinol, isopropyl alcohol, mesityl oxide, 2-methylbenzofuran, α–methylstyrene and 2-phenyl–2-butene (Dow Chemical Co., 1986).

2. Production, Use, Occurrence and Analysis

2.1 Production and use

(a) Production

Phenol was first isolated from coal–tar in the 1830s. A relatively small but steady supply of phenol is recovered as a by–product of metallurgical coke manufacture. By–product coal–tar is fractionally distilled and the phenolic fraction extracted with aqueous alkali. Coal–tar was the only source of phenol until the First World War, when sulfonation of benzene and hydrolysis of the sulfonate led to the production of the first synthetic phenol (Considine, 1974; Thurman, 1982).

Other synthetic routes to phenol have involved the hydrolysis of chlorobenzene (diphenyl ether and *ortho*- and *para*-hydroxydiphenyl occur as by–products) and oxidation of toluene (see monograph, p. 79) to benzoic acid followed by oxydecarboxylation to phenol after purification. The chlorobenzene process and, to a lesser extent, the toluene–based process have been of major importance in phenol production in the past and are still used in some facilities (Thurman, 1982).

More than 98% of the phenol currently produced in the USA is derived from cumene (isopropylbenzene). This method is also the most commonly used method worldwide due to

---

[1]Calculated from: mg/m³ = (molecular weight/24.45) × ppm, assuming standard temperature (25°C) and pressure (760 mm Hg)

its high yield and economy. In this process, cumene is formed from benzene and propylene, then oxidized to the hydroperoxide, which is cleaved with sulfuric acid to yield phenol and acetone. Purification is achieved by distillation or ion–exchange resin separation. The total phenol yield from this process is about 93%, based on cumene and 84% based on benzene (Thurman, 1982).

Annual production of phenol by several countries is given in Table 1.

Table 1. Annual production of phenol (in thousand tonnes)[a]

Country	1980	1981	1982	1983	1984	1985	1986
Brazil	87.1	72.9	91.5	95.8	99.6	NA	NA
Czechoslovakia	43.9	42.1	45.9	45.3	43.6	42.6	46.3
Finland	1.0	2.9	29.9	NA	NA	NA	NA
India	14.1	10.3	15.0	18.0	20.0	NA	NA
Japan	215	214	211	271	272	262	260
Mexico	21.0	23.0	20.7	22.0	24.5	27.1	NA
Romania	65.7	66.0	71.7	88.4	99.6	87.9	NA
Spain	47.9	41.3	37.0	39.7	64.3	73.8	69.7
Sweden[b]	4.2	5.2	5.2	7.1	6.5	6.7	7.6
Turkey	0.1	0.1	NA	NA	NA	NA	NA
UK	NA	109.7	136.8	143.2	184.4	117.5	52.9
USA[c]	1164.8	1169.4	917.6	1196.6	1310.4	1288.7	1412.9
USSR[d]	496.0	497.0	459.0	484.0	511.0	502.0	515.0
Yugoslavia	NA	NA	NA	NA	NA	NA	2.3

[a]From Anon. (1984, 1987, 1988); US International Trade Commission (1981, 1982, 1983, 1984, 1985, 1986, 1987, 1988); NA, not available
[b]Phenol and phenol alcohols
[c]Does not include data from coke ovens and gas retorts
[d]Synthetic and crystallized from coal

(b) Use

Phenol is the basic feedstock from which a number of commercially important materials are made, including phenolic resins, bisphenol A and caprolactam (see IARC, 1987) as well as chlorophenols such as pentachlorophenol (see IARC, 1986). The products made in largest volume are the phenolic resins, derived by condensation of phenol and substituted phenols with aldehydes, particularly formaldehyde (see IARC, 1987). Phenolic resins are used as adhesives in plywood and particle board, as binders for fibreglass, mineral wool and other insulating products, for impregnating and laminating wood and plastic agents, and as moulding compounds and foundry resins (Greek, 1983; Mannsville Chemical Products Corp., 1985).

Bisphenol A is the second most important product of phenol. It is derived by reaction of phenol with acetone and is used mainly in the manufacture of epoxy and polycarbonate resins for plastic mouldings, protective coatings such as paints (see monograph on occupational exposures in paint manufacture and painting) and adhesive applications (epoxy resins), as well as in automotive, appliance, electronic, glazing and other types of applications.

Bisphenol A may also be used to produce phenoxy, polysulfone and polyester resins. Caprolactam, prepared from phenol *via* cyclohexanone as an intermediate, is used to make Nylon–6 fibres, moulding resins and plastic film (Mannsville Chemical Products Corp., 1985).

Phenol is also converted to alkyl phenols, which are used as surface–active agents, emulsifiers, antioxidants and lubricating oil additives (nonylphenols) and to make plasticizers, resins and synthetic lubricants (by conversion to adipic acid). In 1982, a US chemical company initiated the production of aniline from phenol. Another phenol derivative, 2,6–xylenol, is used to make polyphenylene oxide (Mannsville Chemical Products Corp., 1985).

Phenol was widely used in the 1800s as a wound treatment, antiseptic and local anaesthetic; the medical uses of phenol today include incorporation into lotions, salves and ointments. It is also used in the manufacture of disinfectants and antiseptics, paint and varnish removers, lacquers, paints, rubber, ink, illuminating gases, tanning dyes, perfumes, soaps and toys (National Institute for Occupational Safety and Health, 1976; Deichmann & Keplinger, 1981).

In 1986, an estimated 45% of the phenol produced in the USA was used to make phenolic resins, 25% to make bisphenol A, 15% to make caprolactam, 4% to make alkyl phenols, 4% for xylenols, and 7% for miscellaneous uses (Mannsville Chemical Products Corp., 1985).

(c)   Regulatory status and guidelines

Occupational exposure limits for phenol in 32 countries or regions are presented in Table 2.

Table 2. Occupational exposure limits for phenol[a]

Country or region	Year	Concentration (mg/m³)[b]	Interpretation[c]
Australia	1984	S 19	TWA
Austria	1985	S 19	TWA
Belgium	1985	S 19	TWA
Brazil	1985	15	TWA
Bulgaria	1985	S 5	TWA
Commission of the European Communities	1986	19	TWA
		95	Maximum
Chile	1985	S 15.2	TWA
China	1985	S 5	TWA
Czechoslovakia	1985	20	Average
		40	Maximum
Denmark	1988	S 19	TWA
Finland	1987	S 19	TWA
		S 38	STEL (15 min)
France	1986	S 19	TWA
Germany, Federal Republic of	1988	S 19	TWA
German Democratic Republic	1985	S 20	TWA

Table 2 (contd)

Country or region	Year	Concentration (mg/m³)[b]	Interpretation[c]
Hungary	1985	S 5	TWA
		S 10	STEL
India	1985	S 19	TWA
		S 38	STEL
Indonesia	1985	S 19	TWA
Italy	1985	8	TWA
Japan	1988	S 19	TWA
Mexico	1985	S 19	TWA
Netherlands	1986	S 19	TWA
Norway	1981	S 19	TWA
Poland	1985	S 10	TWA
Romania	1985	S 10	Average
		S 15	Maximum
Sweden	1987	S 4	TWA
		S 8	STEL
Switzerland	1985	S 19	TWA
Taiwan	1985	S 19	TWA
UK	1987	S 19	TWA
		S 38	Stel (10 min)
USA[d]			
OSHA	1985	19	TWA
NIOSH	1983	20	TWA
		60	Ceiling (15 min)
ACGIH	1988	S 19	TWA
USSR	1986	S 0.3	Ceiling
Venezuela	1985	S 19	TWA
		S 38	Ceiling
Yugoslavia	1985	S 5	TWA

[a]From Direktoratet for Arbeidstilsynet (1981); International Labour Office (1984); Arbeidsinspectie (1986); Commission of the European Communities (1986); Institut National de Recherche et de Sécurité (1986); Cook (1987); Health and Safety Executive (1987); National Swedish Board of Occupational Safety and Health (1987); Työsuojeluhallitus (1987); American Conference of Governmental Industrial Hygienists (1988); Arbejdstilsynet (1988); Deutsche Forschungsgemeinschaft (1988)

[b]S, skin notation

[c]TWA, time-weighted average; STEL, short-term exposure limit

[d]OSHA, Occupational Safety and Health Administration; NIOSH, National Institute for Occupational Safety and Health; ACGIH, American Conference of Governmental Hygienists

2.2 Occurrence

(a) Natural occurrence

Phenol is a constituent of coal–tar and is formed during the natural decomposition of organic materials (Cleland & Kingsbury, 1977).

(b) Occupational exposure

On the basis of a US National Occupational Exposure Survey, the National Institute for Occupational Safety and Health (1983) estimated that 193 000 workers were potentially exposed to phenol in the USA in 1981–83.

Airborne phenol concentrations in area samples ranged from nondetected to 12.5 mg/m^3 in a bakelite factory in Japan (Ohtsuji & Ikeda, 1972). Exposure levels of 5–88 mg/m^3 have been reported for employees in the USSR who quenched coke with waste-water containing 0.3–0.8 g/l phenol (Petrov, 1960). Occupational exposure to 5 ppm (19 mg/m^3) in a synthetic fibre plant in Japan corresponded to a urinary phenol level of 251 mg/g creatinine (Ogata et al., 1986). Air levels of phenol were correlated with urinary excretion rates in workers at five plants producing phenol, phenol resins and caprolactam in the USSR. The mean personal air and urinary phenol levels at two phenol resin plants were 0.6 mg/m^3 and 33.4 mg/l and 3.0 mg/m^3 and 34.2 mg/l, respectively; those of workers in another plant, who manufactured phenol from chlorobenzene, were 1.2 mg/m^3 and 91.3 mg/l. In a plant for the manufacture of caprolactam, the mean urinary phenol level in workers was 34.0 mg/l (air levels were not determined); and mean personal air and urinary phenol levels in workers in a plant that produced phenol from cumene were 5.8 mg/m^3 in air and 28.5 mg/l in urine (Mogilnicka & Piotrowski, 1974). Phenol levels in the air of 19 Finnish plywood plants ranged from < 0.01 to 0.5 ppm (< 0.04–1.9 mg/m^3; Kauppinen, 1986), and those at a plant in the USA that manufactured fibrous glasswool were 0.01–0.35 ppm (0.05–1.3 mg/m^3), with a mean of 0.11 ppm (Dement et al., 1973).

(c) Air

Phenol was detected in urban air (0.55–1.01 ppb; 2–4 µg/m^3), in exhaust from cars (0.233–0.320 ppm; 0.9–1.2 mg/m^3) and in tobacco smoke (312–436 µg/cigarette) collected in Osaka, Japan (Kuwata et al., 1980).

(d) Water and sediment

Phenols may occur in domestic and industrial waste waters, natural waters and potable water supplies. Chlorination of such waters may produce chlorophenols, giving the water an objectionable smell and taste. Processes for the removal of phenol include superchlorination, chlorine dioxide or chloramine treatment, ozonation, and activated carbon adsorption (American Public Health Association–American Waterworks Association–Water Pollution Control Federation, 1985). Phenol was found at a level of 1 µg/l in a domestic water supply in the USA (Ramanathan, 1984). It has been detected in US river water at 0.02–0.15 mg/l (Verschueren, 1983) and in industrial waste waters at average concentrations of up to 95 mg/l (US Environmental Protection Agency, 1983).

(e)   *Soil and plants*

In studies of environmental fate, phenol has been reported to biodegrade completely in soil within two to five days (Baker & Mayfield, 1980; Verschueren, 1983). When high soil concentrations are produced by a spill, the compound may destroy the degrading bacterial population and leach through to groundwater (Delfino & Dube, 1976; Baker & Mayfield, 1980; Ehrlich *et al.*, 1982).

(f)   *Food*

Phenol has been found to taint the taste of fish and other organisms when present at concentrations of 1.0–25 mg/l in the marine environment (Verschueren, 1983). It has been detected in smoked summer sausage (7 mg/kg) and in smoked pork belly (28.6 mg/kg; US Environmental Protection Agency, 1980).

2.3   Analysis

In the presence of other phenolic compounds, phenol is readily determined by conversion to the corresponding bromophenol by reaction with bromine. The minimal detectable amount of bromophenol by gas chromatography is about 0.01 ng (Hoshika & Muto, 1979).

Phenol present in polluted air (industrial emissions, automobile exhaust, tobacco smoke) may be collected by drawing the air through a 0.1M solution of sodium hydroxide and determined by reversed–phase high–performance liquid chromatography after derivatization with *para*–nitrobenzene diazonium tetrafluoroborate, with a detection limit of 0.05 ppb (0.2 μg/m^3) for 150 l of gas sample (Kuwata *et al.*, 1980). Phenol collected similarly can also be measured by gas chromatography with flame ionization detection. This method has been validated in the range of 10–38 mg/m^3 in 100-l samples (Eller, 1984). Air samples can be collected on a solid sorbent (e.g., resin; Cummins, 1981), and this method has been used to determine phenol in industrial waste and in natural and potable waters (American Public Health Association–American Waterworks Association–Water Pollution Control Federation, 1985).

Phenol can be determined in water samples by steam distillation followed by reaction with 4–aminoantipyrine in the presence of potassium ferricyanide to form a coloured antipyrine dye, which is determined spectrophotometrically. The sensitivity of this method is 1 μg/l (American Public Health Association–American Waterworks Association–Water Pollution Control Federation, 1985). It can also be detected at levels of approximately 0.2 μg/l by gas chromatography with flame ionization and electron capture detection following derivitization with pentafluorobenzyl bromide (US Environmental Protection Agency, 1986a).

In urine samples, phenol can be determined by acidification, diethyl ether extraction, and analysis by gas chromatography with flame–ionization detection. The limit of detection is estimated to be 0.5 μg phenol/l urine (Eller, 1985).

Environmental samples can also be analysed by gas chromatography/mass spectrometry using either packed or capillary columns. The practical quantitative limit is approximately 1 mg/kg (wet weight) for soil/sediment samples, 1–200 mg/kg for wastes and 10 μg/l for ground water samples (US Environmental Protection Agency, 1986b,c).

Colorimetric systems have been developed for detecting phenol in air (ENMET Corp., undated; Matheson Gas Products, undated; Roxan, Inc., undated; The Foxboro Co., 1983; Sensidyne, 1985; National Draeger, Inc., 1987; SKC Inc., 1988).

3. Biological Data Relevant to the Evaluation of Carcinogenic Risk to Humans

3.1 Carcinogenicity studies in animals

(a) Oral administration

Mouse: Groups of 50 male and 50 female B6C3F1 mice, five to six weeks old, were administered drinking-water containing 0, 2500 or 5000 ppm (mg/l) phenol for 103 weeks, and surviving animals were killed at weeks 104–106. Three batches of phenol were used, one with a purity of 98.47%, one with 1.36% impurities, and a homogeneous batch with one impurity. Throughout most of the study period, there was a dose–related reduction in mean body weight in both males and females and treatment–related reduction in water consumption. At weeks 104–106, 42/50 control, 45/50 low–dose and 48/50 high–dose males were still alive; at weeks 105–106, 41/50 control, 40/50 low–dose and 42/50 high–dose females survived. No treatment–related increase in the incidence of tumours was observed in mice of either sex. The incidence of uterine endometrial polyps in females was 1/50 in controls, 0/48 at the low dose and 5/48 at the high dose (National Toxicology Program, 1980).

Rat: Groups of 50 male and 50 female Fischer 344 rats, five to six weeks old, were administered drinking-water containing 0, 2500 or 5000 ppm (mg/l) phenol (purity, see above) for 103 weeks, and surviving animals were killed at weeks 104–105. After about 20 weeks of study, there was a reduction in mean body weight in both males and females which coincided with a reduction in water consumption. At weeks 104–105, 26/50 control, 22/50 low–dose and 30/50 high–dose males, and 38/50 control, 39/50 low–dose and 37/50 high–dose females were still alive. Low–dose males had increased incidences of phaeochromocytomas of the adrenal medulla (control, 13/50; low–dose, 22/50, $p = 0.046$; high–dose, 9/50), leukaemias or lymphomas (control, 18/50; low–dose 31/50, $p = 0.08$; high–dose, 22/50) and C–cell carcinomas of the thyroid (control, 0/50; low–dose, 5/49; high–dose, 1/50; National Toxicology Program, 1980).

(b) Skin application

In a wide range of studies using the two–stage mouse skin model, phenol was investigated as an initiator and a promoter with a number of polycyclic hydrocarbons. The following studies exemplify this approach.

Mouse: Three groups of 30 and one of 22 female albino mice [strain unspecified], nine weeks old, received single initiating applications of 75 μg dimethylbenz[a]anthracene (DMBA), given as 0.25 μl of a 0.3% solution in benzene, followed by applications of 25 μl benzene twice a week (group 1); 25 μl of a 5% solution of phenol (purified reagent grade) in benzene twice a week (group 2); 25 μl of a 10% solution of phenol in benzene twice a week (group 3); or 25 μl of a 0.5% solution of croton oil in benzene twice a week (group 4). A further three groups of 30 mice received no DMBA treatment but applications of 25 μl of the 5% phenol solution twice a week (group 5); 25 μl of the 10% phenol solution twice a week (group 6); or 25 μl of the 0.5% croton oil solution twice a week (Group 7). Secondary treatments continued for 51 weeks, at which time the study was terminated. At week 20, the average numbers of papillomas per mouse were 6.0, 3.3 and 0.25 in groups 4, 3 and 2, respectively. Groups without DMBA pretreatment developed papillomas more slowly, and at week 36 there was a 25% incidence in groups 6 and 7 but only one mouse with papillomas in group 5. No papilloma was seen in group 1, and no skin carcinoma was reported in mice of group 1 or in those receiving 5 or 10% phenol alone (groups 5 and 6). There was a dose-related increase in the incidence of skin carcinomas in mice receiving DMBA and phenol (groups 2 and 3), with an incidence of 47% at 40 weeks in group 3 (Boutwell & Bosch, 1959).

Groups of 30 female Swiss (Millerton) mice, six weeks old, received either 75 μg DMBA in acetone (group 1); applications of a 5% purified phenol solution in acetone three times a week (group 2); applications of a 10% phenol solution in acetone three times a week (group 3); applications of a 10% phenol solution in acetone three times a week (group 4); 75 μg DMBA in acetone followed one week later by applications of the 5% phenol solution in acetone three times a week (group 5); 75 μg DMBA in acetone followed one week later by applications of the 10% phenol solution in acetone twice a week (group 6); or 75 μg DMBA in acetone followed one week later by applications of the 10% phenol solution in acetone three times a week (group 7). Treatment of groups 2–7 continued for 51 weeks, and the study was terminated at 15 months. At this time, the percentages of mice with papillomas of the skin were: group 1, 10; group 2, 0; group 3, 7; group 4, 3; group 5, 33; group 6, 87; and group 7, 80. The percentages of those with skin carcinomas were: group 1, 7; group 2, 0; group 3, 3; group 4, 0; group 5, 10; group 6, 70; and group 7, 47 (Wynder & Hoffmann, 1961).

Other, similar studies with similar results that were reviewed by the Working Group but are not summarized here are those of Salaman and Glendenning (1957), Van Duuren *et al.* (1968) and Van Duuren and Goldschmidt (1976).

(c)   *Administration with known carcinogens*

Groups of female C57Bl mice, 12–14 g in weight, received 20 instillations of 1 mg benzo[a]pyrene in 0.1 ml triethyleneglycol twice a week by gavage simultaneously with a solution of phenol in water (1 mg in 0.1 ml). An increased incidence of forestomach tumours, including carcinomas was seen: 15/43 and 2/43 total tumours and carcinomas, respectively, in the group receiving benzo[a]pyrene alone *versus* 16/22 and 6/22, respectively, in the group receiving combined treatment. A lower dose of phenol (0.02 mg) did not influence the benzo[a]pyrene-induced carcinogenesis of the forestomach (8/23 and 1/23 total tumours and carcinomas, respectively). When administration of benzo[a]pyrene was followed by phenol, forestomach carcinogenesis was inhibited: 5/21 and 0/21 total tumours and carcinomas, respectively. An inhibitory effect was also observed when phenol treatment preceded benzo[a]pyrene: 6/24 and 0/24 total tumours and carcinomas, respectively (Yanysheva *et al.*,

1988). [The Working Group noted that the duration of the study and survival were not specified.]

Three groups of 28–40 female Swiss (Millerton) mice, six weeks old, received skin applications of approximately 5 µg of a 0.005% solution of benzo[a]pyrene (purified) in acetone
three times a week; two of the groups also received applications of either a 5 or 10% solution
of phenol in acetone alternatively with benzo[a]pyrene twice a week. Treatment was continued for 52 weeks, and animals were observed for 15 months. At 12 months, the percentages
of mice with skin papillomas were: benzo[a]pyrene alone, 58; benzo[a]pyrene plus 5% phenol, 83; benzo[a]pyrene plus 10% phenol, 80. The percentages of mice with skin carcinomas
were: benzo[a]pyrene alone, 47; benzo[a]pyrene plus 5% phenol, 77; benzo[a]pyrene plus
10% phenol, 70 (Wynder & Hoffmann, 1961).

Groups of 20 female ICR/Ha mice, six to eight weeks old, received skin applications of
0.1 ml acetone, 5 µg purified benzo[a]pyrene in 0.1 ml acetone or 5 µg benzo[a]pyrene together with 3 mg purified phenol in 0.1 ml acetone three times a week for 508, 460 or 460
days, respectively. No skin tumour was observed in the group that received acetone. In the
benzo[a]pyrene-treated group, 8/20 papillomas and 1/20 skin carcinomas developed, compared to 3/20 papillomas and 1/20 skin carcinomas in the group treated with benzo[a]pyrene
and phenol (Van Duuren et al., 1971).

Groups of 50 female ICR/Ha mice, seven weeks old, received skin applications of 0.1
ml acetone containing 5 µg purified benzo[a]pyrene, 5 µg benzo[a]pyrene together with 3 mg
purified phenol, 3 mg phenol or acetone alone three times a week for 52 weeks. At the end of
the treatment period, survival was 42/50 and 39/50 in the benzo[a]pyrene-treated and benzo[a]pyrene plus phenol-treated groups, respectively. In the benzo[a]pyrene-treated
group, 13/50 papilloma–bearing mice had a total of 14 papillomas, and 10/50 mice had squamous–cell carcinomas of the skin. In the group treated with benzo[a]pyrene and phenol,
7/50 papilloma–bearing mice had a total of nine papillomas and 3/50 mice had squamous–cell
carcinomas. In the group treated with phenol alone, 1/50 papilloma–bearing mouse had one
papilloma; no tumour was reported at 63 weeks in the group given acetone alone (Van Duur-
en et al., 1973; Van Duuren & Goldschmidt, 1976).

3.2   Other relevant data

The toxicology of phenol has been reviewed (National Institute for Occupational Safety and Health, 1976; Bruce et al., 1987).

(a)   Experimental systems

(i)   Absorption, distribution, excretion and metabolism

Phenol is absorbed through the lungs (Deichmann & Keplinger, 1981) and the alimentary tract in various species (Capel et al., 1972). In addition, phenol in solution was absorbed
through the clipped skin of rabbits (Freeman et al., 1951; Deichmann et al., 1952) and through
excised clipped skin of rats at a rate directly related to the concentration of phenol up to 3%
(Roberts et al., 1974).

The concentration of phenol in the body of rabbits 15 min after oral administration of a
5% aqueous solution was highest in the liver, followed by the kidneys, lungs, brain and spinal

cord, and blood (Deichmann, 1944). Following oral administration of 207 mg/kg bw ^{14}C-phenol to rats, the highest concentration ratios between tissues and plasma were found in the liver followed by the kidneys, spleen, adrenal glands, thyroid gland and lungs (Liao & Oehme, 1981). Radioactivity was high in the lungs, kidneys and small intestines in autoradiograms of rats killed 2 h after intravenous injection of 0.6 mg/kg bw ^{14}C-phenol (Greenlee *et al.*, 1981).

Urinary excretion is generally rapid in various species; the percentage of radioactivity excreted within 24 h after oral administration of ^{14}C-phenol was highest in rats (95%) and lowest in squirrel monkeys (31%) among 18 species of animals tested. While absorbed phenol is excreted in the urine as conjugated phenol and a small fraction as conjugated quinol, a marked species difference is observed in the ratio of sulfate to glucuronide. Among 18 species of animals given 25 mg/kg bw ^{14}C-phenol orally, cats excreted phenyl (87%) and quinol (13%) sulfates but no glucuronide, and pigs excreted only phenyl glucuronide and no sulfate, whereas other species excreted substantial amounts of both phenyl and quinol sulfates and glucuronides (Capel *et al.*, 1972). In a similar experiment in which sheep, pigs and rats were given 25 mg/kg bw ^{14}C-phenol orally, glucuronides accounted for 49, 83 and 42% of the total urinary metabolites in the three species, respectively, and sulfates accounted for 32, 1 and 55%. Less than 7% was excreted as quinol conjugates. Only in sheep, 12% of the urinary metabolites were conjugated with phosphate (Kao *et al.*, 1979). Phenyl sulfates (80%) and quinol sulfate (20%) were detected in the urine of cats given 20 mg/kg bw ^{14}C-phenol intraperitoneally (Miller *et al.*, 1976). The ratio between sulfation and glucuronidation is dose-dependent; preferential formation of sulfate occurs at lower doses (Williams, 1959; Ramli & Wheldrake, 1981).

Phenol is metabolized in the liver and other tissues. In an experiment with an isolated gut preparation, it was shown that phenol is transported from the intestinal lumen not in the free form but conjugated. Formation and urinary excretion of phenol conjugates occur in rats even after removal of the liver and gastrointestinal tract (Powell *et al.*, 1974). In rats with cannulae in the left jugular vein and left carotid artery, about 60% of ^{14}C-phenol administered *via* the venous cannula was extracted by the lungs on the first pass following administration (Cassidy & Houston, 1980). ^{14}C-Phenol was extensively metabolized to phenyl sulfate and phenyl glucuronide in a whole rat–lung preparation (Hogg *et al.*, 1981). In comparison, a study in which phenol was administered to rats *via* the duodenal lumen and jugular and hepatic portal veins showed that intestinal and hepatic conjugation are comparable at low doses (< 1 mg/kg bw), although the capacity of the hepatic enzyme is readily saturated, whereas intestinal conjugation far exceeds the contribution of the hepatic and pulmonary enzymes after high doses (> 5 mg/kg bw; Cassidy & Houston, 1984).

^{14}C-Phenol binds irreversibly to calf thymus DNA in the presence of horseradish peroxidase and hydrogen peroxide (Subrahmanyam & O'Brien, 1985a). One of the products formed by the oxidation of phenol, *ortho,ortho'*-biphenol (but not *para,para'*-biphenol), readily binds to DNA following peroxidase–catalysed oxidation (Subrahmanyam & O'Brien, 1985b).

Phenol did not bind covalently to rat haemoglobin *in vivo* (Pereira & Chang, 1981) but was associated with plasma protein in rats *in vivo* (Liao & Oehme, 1981) and with human serum *in vitro*, where the binding occurred predominantly (48.7%) in the albumin fraction (Judis, 1982).

(ii) *Toxic effects*

The majority of the LD_{50} values for phenol in several species fall within one order of magnitude (except for dermal application), cats being the most sensitive and pigs the most resistant species. This difference in sensitivity to the toxicity of phenol can be attributed to quantitative and qualitative differences in phenol metabolism (glucuronidation *versus* sulfatation) between species (Oehme & Davis, 1970 [abstract]; Miller *et al.*, 1973).

The oral LD_{50} for phenol in male mice was about 300 mg/kg bw (von Oettingen & Sharpless, 1946). The oral LD_{50} in different strains of rats ranged from 340–650 mg/kg bw (Deichmann & Oesper, 1940; Deichmann & Witherup, 1944; Flickinger, 1976). Dermal LD_{50}s of 1400 mg/kg bw in rabbits (Vernot *et al.*, 1977) and of 0.625 ml (660 mg)/kg bw in rats (Conning & Hayes, 1970) have been described.

On the basis of mortality during a 14–day post–exposure period, the single–dose LD_{50} for skin penetration in male albino rabbits was estimated to be 850 mg/kg bw. Phenol (500 mg) also produced necrosis after a maximal period of 24 h in the intact skin of exposed rabbits. Application of 100 mg phenol into the eyes of male albino rabbits resulted in inflamed conjunctiva and opaque corneas; 24 h after exposure, the eyes showed severe conjunctivitis, corneal opacities and corneal ulcerations, with no improvement during further observation (Flickinger, 1976). Following dermal application of phenol, rats developed severe skin lesions with oedema followed by necrosis (Conning & Hayes, 1970). Rats exposed to 900 mg/ m^3 phenol–water aerosol for 8 h developed ocular and nasal irritation, loss of coordination, tremors and prostration (Flickinger, 1976).

In mice exposed to phenol vapour at concentrations of 5 ppm (19 mg/m^3) for 8 h per day on five days per week for 90 days, increased stress endurance but no significant difference in any other parameter studied (haematology, urine analysis, blood chemistry, kidney function, rate of weight gain, pathological examination) was observed. In rats exposed to phenol vapour at concentrations of 5 ppm (19 mg/m^3) for 8 h per day on five days per week for 90 days and to 100–200 mg/m^3 for 7 h per day on five days per week for 53 days over 74 days, no change in the same parameters was noted, except for a slight weight gain compared to controls. Essentially the same results were obtained in monkeys exposed to phenol vapours at concentrations of 5 ppm (19 mg/m^3) for 8 h per day on five days per week for 90 days. In guinea–pigs exposed to phenol vapours of 100–200 mg/m^3 for 7 h per day on five days per week, toxicological changes observed included weight loss, respiratory difficulties and hind–quarter paralysis. Histological examination revealed myocardial necrosis, acute lobular pneumonia and liver and kidney damage. Extensive mortality (4/12) was found in guinea–pigs after 20 exposures over 28 days. In rabbits exposed similarly, damage was in general less severe than that found in guinea–pigs after 88 days (Deichmann *et al.*, 1944; National Institute for Occupational Safety and Health, 1976).

Significant effects on the central nervous system (grasping reflex and vestibular function) in rats were observed after continuous exposure to 100 mg/m^3 phenol for 15 days; activities of serum liver enzymes were also increased, indicative of liver damage (Dalin & Kristof-fersson, 1974).

(iii) *Effects on reproduction and prenatal toxicity*

As reported in an abstract, groups of 23 CD rats were exposed by oral intubation to 0, 30, 60 or 120 mg/kg bw phenol per day on days 6–15 of gestation and the fetuses examined at term for growth, viability and malformations. There was no evidence of maternal toxicity or teratogenicity, but fetal growth was retarded at the highest dose (Price *et al.*, 1986).

As reported in an abstract, groups of CD-1 mice were exposed by oral intubation to 0, 70, 140 and 280 mg/kg bw phenol per day on days 6–15 of gestation. Fetuses were examined for growth, viability and malformations. Maternal and fetal toxicity but no significant evidence of teratogenicity were observed. Greater maternal toxicity as well as cleft palates in the fetus were reported at the high dose level (Price *et al.*, 1986).

Phenol was one of a series of chemicals used in a structure–activity developmental toxicology study reported in an abstract. The chemicals were administered [route unspecified] to groups of Sprague–Dawley rats on day 11 of gestation at four dose levels between 0 and 1000 mg/kg or added to embryos of the same developmental age in whole embryo culture *in vitro*. *In vivo*, phenol induced hind–limb and tail defects. *In vitro*, phenol was the least potent of seven congeners tested; the activity, however, was increased following co–culture with primary hepatocytes (Kavlock *et al.*, 1987).

(iv) *Genetic and related effects*

Phenol was mutagenic to *Escherichia coli* B/Sd-4 at highly toxic doses only (survival level, 0.5–1.7%; Demerec *et al.*, 1951). It did not induce filamentation in the *lon⁻* mutant of *E. coli* (Nagel *et al.*, 1982). It was not mutagenic to *Salmonella typhimurium* TA1535, TA1537, TA1538, TA98 or TA100 in the presence or absence of an exogenous metabolic system from Aroclor-induced rat and hamster livers (Cotruvo *et al.*, 1977; Epler *et al.*, 1979; Florin *et al.*, 1980; Gilbert *et al.*, 1980; Kinoshita *et al.*, 1981; Thompson & Melampy, 1981; Pool & Lin, 1982; Haworth *et al.*, 1983; Kazmer *et al.*, 1983; Ludewig & Glatt, 1986 [abstract]). It was mutagenic to *S. typhimurium* TA98 only in the presence of an exogenous metabolic system when the assay was performed using a modified medium (ZLM) instead of the standard Vogel–Bonner medium (Gocke *et al.*, 1981).

Phenol weakly induced mitotic segregation in *Aspergillus nidulans* (Crebelli *et al.*, 1987). It induced C–mitosis in the root tips of *Allium cepa* but only rarely induced chromosomal fragmentation (Levan & Tjio, 1948). It induced chromosomal aberrations in maize and wheat [details not given] (Chebotar *et al.*, 1975).

Phenol did not increase the frequency of recessive lethal mutations in *Drosophila melanogaster* (Sturtevant, 1952). Feeding or injection of phenol did not induce sex–linked recessive lethal mutations in meiotic or postmeiotic germ–cell stages of adult male *Drosophila* (Gocke *et al.*, 1981; Woodruff *et al.*, 1985).

Phenol did not induce DNA single–strand breaks in mouse lymphoma L5178YS cells (Pellack-Walker & Blumer, 1986). It was reported in an abstract that phenol induced DNA strand breaks in mouse lymphoma cells, as measured by the alkaline unwinding technique followed by elution through hydroxylapatite (Garberg & Bolcsfoldi, 1985). It was reported in a further abstract that phenol did not induce strand breaks, as measured by the alkaline elution technique, in rat germ-cell DNA after either acute or subchronic treatment (Skare & Schrotel, 1984).

Phenol induced mutations at the *hprt* locus of Chinese hamster V79 cells in the absence of an exogenous metabolic system from the livers of phenobarbital-induced mice (Paschin & Bahitova, 1982).

Phenol was reported to inhibit DNA synthesis in HeLa cells (Dobashi, 1974; Painter & Howard, 1982) and to inhibit repair of radiation-induced chromosomal breaks in human leucocytes (Morimoto *et al.*, 1976). However, it only slightly inhibited DNA repair synthesis and DNA replication synthesis in WI-38 human diploid fibroblasts (Poirier *et al.*, 1975). Phenol induced sister chromatid exchange in human lymphocytes (Morimoto & Wolff, 1980a,b; Erexson *et al.*, 1985a,b); the number of sister chromatid exchanges was further increased by the presence of an exogenous metabolic system from rat livers (Morimoto *et al.*, 1983). In another study, phenol was reported to be incapable of inducing sister chromatid exchange in human lymphocytes (Jansson *et al.*, 1986).

Administration of phenol either intraperitoneally (two doses of 188 mg/kg bw; Gocke *et al.*, 1981) or orally (250 mg/kg bw; Gad-el Karim *et al.*, 1986) to female and male NMRI or male CD-1 mice did not induce micronuclei in bone marrow. However, phenol induced micronuclei in the bone marrow of pregnant CD/1 mice after a single administration of 265 mg/kg bw by gastric intubation; micronuclei were not seen in the liver of fetuses (Ciranni *et al.*, 1988). As reported in abstract, phenol induced micronuclei in male and female mice at doses of 150 and 200 mg/kg bw (Sofuni *et al.*, 1986). It was reported in another abstract that phenol induced chromosomal aberrations in bone marrow of mice *in vivo* [details not given] (Lowe *et al.*, 1987). Phenol did not inhibit intercellular communication (as measured by metabolic cooperation) in Chinese hamster V79 cells (Chen *et al.*, 1984; Malcolm *et al.*, 1985).

(b) Humans

(i) *Absorption, distribution, excretion and metabolism*

Studies in human volunteers have shown that 70–80% of inhaled phenol vapour is retained (Piotrowski, 1971) and that phenol is absorbed almost quantitatively through the alimentary tract (Capel *et al.*, 1972). Phenol in lotion, ointment (Rogers *et al.*, 1978) and vapour form (Piotrowski, 1971) can penetrate the skin. When absorbed, almost all of the dose is excreted in the urine within one day (Piotrowski, 1971; Capel *et al.*, 1972). In male volunteers given 0.01 mg/kg bw ^{14}C-phenol orally, 90% of the dose was excreted within 24 h, mainly as phenyl sulfate (77% of 24-h excretion) and phenyl glucuronide (16%), together with very small amounts of quinol sulfate and glucuronide (Capel *et al.*, 1972).

Exposure-dependent increases in the concentration of phenol in urine have been observed among factory workers occupationally exposed to phenol vapour (Ohtsuji & Ikeda, 1972; Knapik *et al.*, 1980; Gspan *et al.*, 1984). The increase was attributable entirely to conjugated phenol, and no significant change in the concentration of free phenol was observed, regardless of the intensity of exposure to phenol vapour (up to 13 mg/m^3 in workroom air; Ohtsuji & Ikeda, 1972).

(ii) *Toxic effects*

Phenol poisoning occurs by skin absorption, vapour inhalation or ingestion. Phenol has a marked corrosive effect on all tissues and, on contact with skin, causes whitening of the exposed area followed by severe chemical burns; long after cessation of contact, progressive areas of depigmentation may develop (Pardoe *et al.*, 1976).

Application of a bandage containing 2% phenol to the umbilicus of a newborn baby resulted in death after 11 h. Another newborn baby treated with 30% phenol:60% camphor for a skin ulcer experienced circulatory failure, cerebral intoxication and methaemoglobinaemia but recovered after a blood transfusion (Hinkel & Kintzel, 1968). An accidental spill in industry resulting in cutaneous absorption also caused death (Griffiths, 1973).

After an acute percutaneous intoxication of a chemical worker with phenol, local effects on the skin were seen in conjunction with several effects due to systemic intoxication, including massive intravascular haemolysis, tachycardia, respiratory depression, and renal and liver damage. The latter was concluded from the increased activities of liver enzymes in the serum (Schaper, 1981). Ingestion of 10–56 ml phenol caused severe irritation in the gastrointestinal tract, cardiovascular collapse, respiratory depression and seizures (Bennett *et al.*, 1950; Stajduhar-Caric, 1968).

As reported in a review, exposure by inhalation to low concentrations of phenol (0.004 ppm; 0.015 mg/m^3) six times for 5 min produced increased sensitivity to light in three volunteers adapted to the dark. Exposures to 0.006 ppm (0.02 mg/m^3) phenol for 15 sec resulted in the formation of conditioned electrocortical reflexes in four volunteers (Bruce *et al.*, 1987).

'Phenol marasmus', described as an occupational hazard resulting from chronic exposure to phenol, involves anorexia, weight loss, headache, vertigo, salivation and dark urine (Merliss, 1972).

Repeated oral exposure for several weeks (estimated intake, 10–240 mg/day) due to contamination of groundwater after an accidental spill of phenol resulted in mouth sores (burning of the mouth), diarrhoea and dark urine. Examination six months after the exposure revealed no residual effect (Baker *et al.*, 1978).

(iii) *Effects on fertility and on pregnancy outcome*

No data were available to the Working Group.

(iv) *Genetic and related effects*

As reported in an abstract, increased frequencies of chromosomal aberrations were found in peripheral lymphocytes of 50 workers occupationally exposed to formaldehyde, styrene and phenol, as compared to 25 controls (Mierauskienė & Lekevičius, 1985).

3.3 Epidemiological studies of carcinogenicity in humans

Wilcosky *et al.* (1984) performed a case–control study in a cohort of rubber workers (see the monograph on some petroleum solvents, p. 69). Exposure to phenol was associated with an increased risk for stomach cancer (relative risk, 1.4; six cases). [The Working Group noted that the number of cases in each category is small and multiple exposures were evaluated independently of other exposures.]

Kauppinen *et al.* (1986) conducted a case–control study of 57 male cases of 'respiratory' tumours, defined as cancers originating in organs in direct contact with chemical agents, such as the tongue, mouth, pharynx, nose, sinuses, larynx, epiglottis, trachea and lung; approximately 90% were of the lung and trachea. Three control subjects for each case (171 men) were selected from the same cohort of 3805 men who had started working in one of 19 Finnish plywood, particle–board, sawmill and formaldehyde glue plants in 1944–65, had worked for at least one year and had been followed up from 1957 to 1981. Exposure histories were assessed for each control until the month of diagnosis of his matched case. A job–exposure matrix was used to determine exposures, in which the emphasis was on wood dust exposure and chlorophenols; other exposures were determined qualitatively (yes/no) and as a function of exposure time. Smoking histories were obtained. The relative risks for exposure to phenol, adjusted for smoking, were 4.0 (12 cases; $p < 0.05$) and 2.9 with a requirement of ten years' latency (seven cases, $p > 0.05$). The relative risks for 'phenol in wood dust' were also increased but diminished after the requirement of ten years of latency time. Relative risks for exposure to phenol did not increase with duration of exposure, and the authors noted confounding by exposure to pesticides.

4. Summary of Data Reported and Evaluation

4.1 Exposures

Phenol is a basic feedstock for the production of phenolic resins, bisphenol A, caprolactam, chlorophenols and several alkylphenols and xylenols. Phenol is also used in disinfectants and antiseptics. Occupational exposure to phenol has been reported during its production and use, as well as in the use of phenolic resins in the wood products industry. It has also been detected in automotive exhaust and tobacco smoke.

4.2 Experimental carcinogenicity data

Phenol was tested for carcinogenicity by oral administration in drinking–water in one strain of mice and one strain of rats. No treatment–related increase in the incidence of tumours was observed in mice or in female rats. In male rats, an increase in the incidence of leukaemia was observed at the lower dose but not at the higher dose. Phenol was tested extensively in the two–stage mouse skin model and showed promoting activity.

4.3 Human carcinogenicity data

In one case–control study of workers in various wood industries, an increased risk was seen for tumours of the mouth and respiratory tract in association with exposure to phenol; however, the number of cases was small and confounding exposures were inadequately controlled.

4.4 Other relevant data

In humans, phenol poisoning can occur after skin absorption, inhalation of vapours or ingestion. Acute local effects are severe tissue irritation and necrosis. At high doses, the most prominent systemic effect is central nervous system depression. Phenol causes irritation, dermatitis, central nervous system effects and liver and kidney toxicity in experimental animals.

Phenol induced micronuclei in female mice and sister chromatid exchange in cultured human cells. It did not inhibit intercellular communication in cultured animal cells. It induced mutation but not DNA damage in cultured animal cells. It did not induce recessive lethal mutation in *Drosophila*. It had a weak effect in inducing mitotic segregation in *Aspergillus nidulans*. Phenol did not induce mutation in bacteria. (See Appendix 1.)

4.5 Evaluation[1]

There is *inadequate evidence* for the carcinogenicity of phenol in humans.

There is *inadequate evidence* for the carcinogenicity of phenol in experimental animals.

Overall evaluation

Phenol *is not classifiable as to its carcinogenicity to humans (Group 3)*.

5. References

American Conference of Governmental Industrial Hygienists (1988) *Threshold Limit Values and Biological Exposure Indices for 1988–1989*, Cincinnati, OH, p. 30

American Public Health Association–American Waterworks Association–Water Pollution Control Federation (1985) *Standard Methods for the Examination of Water and Wastewater – Phenol*, 16th ed., Washington DC, American Public Health Association, pp. 556–570

Anon. (1984) Facts and figures for the chemical industry. *Chem. Eng. News, 62*, 32–74

Anon. (1985) Facts and figures for the chemical industry. *Chem. Eng. News, 63*, 22–66

Anon. (1987) Facts and figures for the chemical industry. *Chem. Eng. News, 65*, 24–76

Anon. (1988) *CHEM–INTELL Database*, Chemical Intelligence Service, Dunstable, UK, Reed Telepublishing Limited

Arbeidsinspectie (Labour Inspection) (1986) *De Nationale MAC–Lijst 1986* [National AAC List 1986] (*P145*), Voorburg, Ministry of Social Affairs, p. 13

Arbejdstilsynet (Labour Inspection) (1988) *Graensevaerdir for Stoffer og Materialer* [Limit Values for Substances and Materials] (*At–anvisning no. 3.1.0.2*), Copenhagen, p. 27

Baker, E.L., Landrigan, P.J., Bertozzi, P.E., Field, P.H., Basteyns, B.J. & Skinner, H.G. (1978) Phenol poisoning due to contaminated drinking water. *Arch. environ. Health, 83*, 89–94

---

[1]For definitions of the italicized terms, see Preamble, pp. 27–30.

Baker, M.D. & Mayfield, C.I. (1980) Microbial and non-biological decomposition of chlorophenols and phenol in soil. *Water Air Soil Pollut.*, *13*, 411–424

Bennett, I.L., Jr, James, D.F. & Golden, A. (1950) Severe acidosis due to phenol poisoning: report of two cases. *Ann. intern. Med.*, *32*, 324–327

Boutwell, R.K. & Bosch, D.K. (1959) The tumor-promoting action of phenol and related compounds for mouse skin. *Cancer Res.*, *19*, 413–424

Bruce, R.M., Santodonato, J. & Neal, M.W. (1987) Summary review of the health effects associated with phenol. *Toxicol. ind. Health*, *3*, 535–568

Capel, I.D., French, M.R., Millburn, P., Smith, R.L. & Williams, R.T. (1972) The fate of [¹⁴C]phenol in various species. *Xenobiotica*, *2*, 25–34

Cassidy, M.K. & Houston, J.B. (1980) Phenol conjugation by lung *in vivo*. *Biochem. Pharmacol.*, *29*, 471–474

Cassidy, M.K. & Houston, J.B. (1984) In vivo capacity of hepatic and extrahepatic enzymes to conjugate phenol. *Drug Metab. Disposition*, *12*, 619–624

Chebotar, A.A., Kaptar, S.G., Suruzhiu, A.I. & Bukhar, B.I. (1975) Chromosomal and nucleoplasmic changes in maize and wheat induced by hexachlorocyclohexane, naphthalene and phenol. *Dokl. Akad. Nauk. SSSR, Ser. Biol.*, *223*, 320–321

Chen, T.-H., Kavanagh, T.J., Chang, C.C. & Trosko, J.E. (1984) Inhibition of metabolic cooperation in Chinese hamster V79 cells by various organic solvents and simple compounds. *Cell Biol. Toxicol.*, *1*, 155–171

Ciranni, R., Barale, R., Marrazzini, A. & Loprieno, N. (1988) Benzene and the genotoxicity of its metabolites. I. Transplacental activity in mouse fetuses and in their dams. *Mutat. Res.*, *208*, 61–67

Cleland, J.G. & Kingsbury, G.L. (1977) *Multimedia Environmental Facts for Environmental Assessment*, Vol. II, *MEG Charts and Background Information EPA–60017–77–136b*, Washington DC, US Environmental Protection Agency

Commission of the European Communities (1986) Occupational limit values. *Off. J. Eur. Commun.*, *C164*, 7

Conning, D.M. & Hayes, M.J. (1970) The dermal toxicity of phenol: an investigation of the most effective first-aid measures. *Br. J. ind. Med.*, *27*, 155–159

Considine, D.M., ed. (1974) *Chemical and Process Technology Encyclopedia*, New York, McGraw Hill, pp. 297–302, 864–866

Cook, W.A. (1987) *Occupational Exposure Limits – Worldwide*, Washington DC, American Industrial Hygiene Association, pp. 32, 124, 150, 207

Cotruvo, J.A., Simmon, V.F. & Spanggord, R.J. (1977) Investigation of mutagenic effects of products of ozonation reactions in water. *Ann. N.Y. Acad. Sci.*, *298*, 124–140

Crebelli, R., Conti, G. & Carere, A. (1987) On the mechanism of mitotic segregation induction in *Aspergillus nidulans* by benzene hydroxy metabolites. *Mutagenesis*, *2*, 235–238

Cummins, K. (1981) Phenol and cresol. In: *Analytical Methods Manual*, Salt Lake City, UT, Organic Methods Evaluation Branch, Occupational Safety and Health Administration

Dalin, N.-M. & Kristoffersson, R. (1974) Physiological effects of sublethal concentration of inhaled phenol on the rat. *Ann. Zool. fenn.*, *11*, 193–199

Deichmann, W.B. (1944) Phenol studies. V. The distribution, detoxification, and excretion of phenol in the mammalian body. *Arch. Biochem.*, *3*, 345–355

Deichmann, W.B. & Keplinger, M.L. (1981) Phenols and phenolic compounds. In: Clayton, G.D. & Clayton, F.E., eds, *Patty's Industrial Hygiene and Toxicology*, Vol. 2A, 3rd rev. ed., New York, John Wiley & Sons, pp. 2567–2584

Deichmann, W.B. & Oesper, P. (1940) Ingestion of phenol. Effects on the albino rat. *Ind. Med.*, *9*, 296–298

Deichmann, W.B. & Witherup, S. (1944) Phenol studies. VI. The acute and comparative toxicity of phenyl and *o*–, *m*–, and *p*–cresols for experimental animals. *J. Pharmacol. exp. Ther.*, *80*, 233–240

Deichmann, W.B., Kitzmiller, K.V. & Witherup, S. (1944) Phenol studies. VII. Chronic phenol poisoning, with special reference to the effects upon experimental animals of the inhalation of phenol vapor. *Am. J. clin. Pathol.*, *14*, 273–277

Deichmann, W.B., Witherup, S. & Dierker, M. (1952) Phenol studies. XII. The percutaneous and alimentary absorption of phenol by rabbits with recommendations for the removal of phenol from the alimentary tract or skin of persons suffering exposure. *J. Pharmacol. exp. Therap.*, *105*, 265–272

Delfino, J.J. & Dube, D.J. (1976) Persistent contamination of ground water by phenol. *J. environ. Sci. Health*, *11*, 345–355

Dement, J., Wallingford, K. & Zumwalde, R. (1973) *Industrial Hygiene Survey of Owens–Corning Fiberglas, Kansas City, Kansas (Report No. IW 35.16)*, Cincinnati, OH, National Institute for Occupational Safety and Health

Demerec, M., Bertani, G. & Flint, J. (1951) A survey of chemicals for mutagenic action on *E. coli. Am. Nat.*, *85*, 119–136

Deutsche Forschungsgemeinschaft (German Research Society) (1988) *Maximale Arbeitsplatzkonzentrationen und Biologische Arbeitsstofftoleranzwerte 1988* [Maximal Concentrations at the Workplace and Biological Tolerance Values for Working Materials 1988] (*Report No. XXIV*), Weinheim, VCH Verlagsgesellschaft mbH, p. 51

Direktoratet for Arbeidstilsynet (Directorate for Labour Inspection) (1981) *Administrative Normer for Forurensning i Arbeidsatmosfaere 1981* [Administrative Norms for Pollution in Work Atmosphere 1981] (*No. 361*), Oslo, p. 11

Dobashi, Y. (1974) Influence of benzene and its metabolites on mitosis of cultured human cells (Jpn). *Jpn. J. ind. Health*, *16*, 453–461

Dow Chemical Co. (1986) *Organic Impurities in Phenol Synthetic*, Midland, MI

Dow Chemical Co. (1988) *Material Data Safety Sheet: Phenol*, Midland, MI

Ehrlich, G.G., Goerlitz, D.F., Godsy, E.M. & Hult, M.F. (1982) Degradation of phenolic contaminants in ground water by anaerobic bacteria: St Louis Park, Minnesota. *Ground Water*, *20*, 703–710

Eller, P.M. (1984) *NIOSH Manual of Analytical Methods*, 3rd ed., Vol. 1 (*DHHS (NIOSH) Publ. No. 84–100*), Washington DC, US Government Printing Office, pp. 3502-1–3502-3

Eller, P.M. (1985) *NIOSH Manual of Analytical Methods*, 3rd ed., 1st Suppl. (*DHHS (NIOSH) Publ. No. 84–100*), Washington DC, US Government Printing Office, pp. 8305-1–8305-4

ENMET Corp. (undated) *ENMET–Kitagawa Toxic Gas Detector Tubes*, Ann Arbor, MI

Epler, J.L., Rao, T.K. & Guerin, M.R. (1979) Evaluation of feasibility of mutagenic testing of shale oil products and effluents. *Environ. Health Perspect.*, *30*, 179–184

Erexson, G.L., Wilmer, J.L. & Kligerman, A.D. (1985a) Sister chromatid exchanges induction in human lymphocytes exposed to benzene and its metabolites *in vitro. Cancer Res.*, *45*, 2471–2477

Erexson, G.L., Wilmer, J.L. & Kligerman, A.D. (1985b) Sister chromatid exchanges (SCE) induction in human lymphocytes exposed *in vitro* to benzene or its metabolites (Abstract). *Environ. Mutagenesis*, *7*, 66–67

Flickinger, C.E. (1976) The benzenediols: catechol, resorcinol and hydroquinone – a review of the industrial toxicology and current industrial exposure limits. *Am. ind. Hyg. Assoc. J.*, *37*, 596–606

Florin, I., Rutberg, L., Curvall, M. & Enzell, C.R. (1980) Screening of tobacco smoke constituents for mutagenicity using the Ames' test. *Mutat. Res.*, *18*, 219–232

The Foxboro Co. (1983) *Chromatographic Column Selection Guide for Century Organic Vapor Analyzer*, Foxboro, MA

Freeman, M.V., Draize, J.H. & Alvarez, E. (1951) Cutaneous absorption of phenol. *J. Lab. clin. Med.*, *38*, 262–266

Gad–el Karim, M.M., Ramanujam, V.M.S. & Legator, M.S. (1986) Correlation between the induction of micronuclei in bone marrow by benzene exposure and the excretion of metabolites in urine of CD-1 mice. *Toxicol. appl. Pharmacol.*, *85*, 464–477

Garberg, P. & Bolcsfoldi, G. (1985) Evaluation of a genotoxicity test measuring DNA strandbreaks in mouse lymphoma cells by alkaline unwinding and hydroxylapatite chromatography (Abstract). *Environ. Mutagenesis*, *7*, 73

Gilbert, P., Rondelet, J., Poncelet, F. & Mercier, M. (1980) Mutagenicity of *p*–nitrosophenol. *Food Cosmet. Toxicol.*, *18*, 523–525

Gocke, E., King, M.-T., Eckhardt, K. & Wild, D. (1981) Mutagenicity of cosmetics ingredients licensed by the European Communities. *Mutat. Res.*, *90*, 91–109

Greek, B.F. (1983) Phenol, vinyl acetate head for moderate pickup in 1984. *Chem Eng. News*, *61*, 7–10

Greenlee, W.F., Gross, E.A. & Irons, R.D. (1981) Relationship between benzene toxicity and the disposition of ^{14}C-labelled benzene metabolites in the rat. *Chem.–biol. Interactions*, *33*, 285–299

Griffiths, G.J. (1973) Fatal acute poisoning by intradermal absorption of phenol. *Med. Sci. Law*, *13*, 46–48

Gspan, P., Jeršic, A. & Čadež, E. (1984) Phenol concentration in urine of exposed workers as a function of phenol concentration in the workplace (Ger.). *Staub–Reinhalt. Luft*, *44*, 314–316

Hawley, G.H. (1981) *Condensed Chemical Dictionary*, 10th ed., New York, Van Nostrand Reinhold, pp. 796

Haworth, S., Lawlor, T., Mortelmans, K., Speck, W. & Zeiger, E. (1983) *Salmonella* mutagenicity test results for 250 chemicals. *Environ. Mutagenesis, Suppl. 1*, 3–142

Health and Safety Executive (1987) *Occupational Exposure Limits 1987* (*Guidance Note EH 40/87*), London, Her Majesty's Stationary Office, p. 19

Hinkel, G.K. & Kintzel, H.-W. (1968) Phenol poisoning in newborns by cutaneous resorption (Ger.). *Dtsch. Gesund.*, *23*, 2420–2422

Hogg, S.I., Curtis, C.G., Upshall, D.G. & Powell, G.M. (1981) Conjugation of phenol by rat lung. *Biochem. Pharmacol.*, *30*, 1551–1555

Hoshika, Y. & Muto, G. (1979) Sensitive gas chromatographic determination of phenols as bromophenols using electron capture detection. *J. Chromatogr.*, *179*, 105–111

IARC (1986) *IARC Monographs on the Evaluation of the Carcinogenic Risk of Chemicals to Humans*, Vol. 41, *Some Halogenated Hydrocarbons and Pesticide Exposures*, Lyon, pp. 319–356

IARC (1987) *IARC Monographs on the Evaluation of Carcinogenic Risks to Humans*, Suppl. 7, *Overall Evaluations of Carcinogenicity: An Updating of* IARC Monographs *Volumes 1 to 42*, Lyon, pp. 211–216

Institut National de Recherche et de Sécurité (National Institute for Research and Safety) (1986) *Valeurs Limites pour les Concentrations des Substances Dangereuses Dans l'Air des Lieux de Travail* [Limit Values for Concentrations of Dangerous Substances in the Air of Work Places] (*ND 1609-125-86*), Paris, p. 574

International Labour Office (1984) *Occupational Exposure Limits for Airborne Toxic Substances*, 2nd rev. ed. (*Occupational Safety and Health Series No. 37*), Geneva, pp. 172–173

Jansson, T., Curvall, M., Hedin, A. & Enzell, C.R. (1986) In vitro studies of biological effects of cigarette smoke condensate. II. Induction of sister chromatid exchanges in human lymphocytes by weakly acidic, semivolatile constituents. *Mutat. Res.*, 169, 129–139

Judis, J. (1982) Binding of selected phenol derivatives to human serum proteins. *J. pharm. Sci.*, 71, 1145–1147

Kao, J., Bridges, J.W. & Faulkner, J.K. (1979) Metabolism of [^{14}C]phenol by sheep, pig and rat. *Xenobiotica*, 9, 141–147

Kauppinen, T. (1986) Occupational exposure to chemical agents in the plywood industry. *Ann. occup. Hyg.*, 30, 19–29

Kauppinen, T.P., Partanen, T.J., Nurminen, M.M., Nickels, J.I., Hernberg, S.G., Hakulinen, T.R., Pukkala, E.I. & Savonen, E.T. (1986) Respiratory cancers and chemical exposures in the wood industry: a nested case–control study. *Br. J. ind. Med.*, 43, 84–90

Kavlock, R.J., Oglesby, L., Copeland, M.F. & Hall, L.L. (1987) Structure–activity relationships in the developmental toxicity of phenols (Abstract). *Teratology*, 36, 19A

Kazmer, S., Katz, M. & Weinstein, D. (1983) The effect of culture conditions and toxicity on the Ames *Salmonella*/microsome agar incorporation mutagenicity assay. *Environ. Mutagenesis*, 5, 541–551

Kinoshita, T., Santella, R., Pulkrabek, P. & Jeffrey, A.M. (1981) Benzene oxide: genetic toxicity. *Mutat. Res.*, 91, 99–102

Knapik, Z., Hańczyc, H., Lubczyńska-Kowalska, W., Menzel-Lipińska, M., Cader, J., Paradowski, L. & Borówka, Z. (1980) Assessing the subclinical forms of toxic action of phenol (Ger.). *Z. ges. Hyg.*, 26, 585–587

Kuwata, K., Uebori, M. & Yamazaki, Y. (1980) Determination of phenol in polluted air as *p*-nitrobenzeneazophenol derivative by reversed phase high performance liquid chromatography. *Anal. Chem.*, 52, 857–860

Levan, A. & Tjio, J.H. (1948) Induction of chromosome fragmentation by phenols. *Hereditas*, 34, 453–484

Liao, T.F. & Oehme, F.W. (1981) Tissue distribution and plasma protein binding of [^{14}C]phenol in rats. *Toxicol. appl. Pharmacol.*, 57, 220–225

Lowe, K.W., Holbrook, C.J., Linkous, S.L. & Roberts, M.R. (1987) Preliminary comparison of three cytogenetic assays for genotoxicity in mouse bone marrow cells (Abstract No. 160). *Environ. Mutagenesis*, 9 (*Suppl. 8*), 63

Ludewig, G. & Glatt, H.R. (1986) Mutations in bacteria and sister chromatid exchanges in cultured mammalian cells are induced by different metabolites of benzene (Abstract No. 82). *Naunyn-Schmiedeberg's Arch. Pharmacol.*, 332 (*Suppl.*), R21

Malcolm, A.R., Mills, L.J. & McKenna, E.J. (1985) Effects of phorbol myristate acetate, phorbol dibutyrate, ethanol, dimethylsulfoxide, phenol, and seven metabolites of phenol on metabolic cooperation between Chinese hamster V79 lung fibroblasts. *Cell Biol. Toxicol.*, 1, 269–283

Mannsville Chemical Products Corp. (1985) *Chemical Products Synopsis: Phenol*, Cortland, NY

Matheson Gas Products (undated) *The Matheson–Kitagawa Toxic Gas Detector System*, East Rutherford, NJ

Merliss, R.R. (1972) Phenol marasmus. *J. occup. Med.*, *14*, 55–56

Mierauskienė, J.R. & Lekevičius, R.K. (1985) Cytogenetic studies of workers occupationally exposed to phenol, styrene and formaldehyde (Abstract No. 60). *Mutat. Res.*, *147*, 308–309

Miller, J.J., Powell, G.M., Olavesen, A.H. & Curtis, C.G. (1973) The metabolism and toxicity of phenols in cats. *Biochem. Soc. Trans.*, *1*, 1163–1165

Miller, J.J., Powell, G.M., Olavesen, A.H. & Curtis, C.G. (1976) The toxicity of dimethoxyphenol and related compounds in the cat. *Toxicol. appl. Pharmacol.*, *38*, 47–57

Mogilnicka, E.M. & Piotrowski, J.K. (1974) The exposure test for phenol in the light of field study (Hung.). *Med. Prac.*, *25*, 137–141

Morimoto, K. & Wolff, S. (1980a) Benzene metabolites increase sister chromatid exchanges and disturb cell division kinetics in human lymphocytes (Abstract N. Ea–6). *Environ. Mutagenesis*, *2*, 274–275

Morimoto, K. & Wolff, S. (1980b) Increase of sister chromatid exchanges and perturbations of cell division kinetics in human lymphocytes by benzene metabolites. *Cancer Res.*, *40*, 1189–1193

Morimoto, K., Koizumi, A., Tachibana, Y. & Dobashi, Y. (1976) Inhibition of repair of radiation–induced chromosome breaks. Effect of phenol in cultured human leukocytes. *Jpn. J. ind. Health*, *18*, 478–479

Morimoto, K., Wolff, S. & Koizumi, A. (1983) Induction of sister–chromatid exchanges in human lymphocytes by microsomal activation of benzene metabolites. *Mutat. Res.*, *119*, 355–360

Nagel, R., Adler, H.I. & Rao, T.K. (1982) Induction of filamentation by mutagens and carcinogens in a *lon⁻* mutant of *Escherichia coli*. *Mutat. Res.*, *105*, 309–312

National Draeger, Inc. (1987) *Detector Tube Products for Gas and Vapor Detection*, Pittsburgh, PA

National Institute for Occupational Safety and Health (1976) *Criteria For a Recommended Standard ... Occupational Exposure to Phenol*, Washington DC, US Department of Health, Education, and Welfare

National Institute for Occupational Safety and Health (1983) *US National Occupational Exposure Survey 1981–1983*, Cincinnati, OH

National Swedish Board of Occupational Safety and Health (1987) *Hygienska Gränsvärden* [Hygienic Limit Values] (*Ordinance 1987:12*), Solna, p. 22

National Toxicology Program (1980) *Bioassay of Phenol for Possible Carcinogenicity (CAS No. 108–95–2) (NCI–CG–TR–203: NTP No. 80–15)*, Research Triangle Park, NC, US Department of Health and Human Services

Oehme, F.W. & Davis, L.E. (1970) The comparative toxicity and biotransformation of phenol (Abstract No. 28). *Toxicol. appl. Pharmacol.*, *17*, 283

von Oettingen, W.F. & Sharpless, N.E. (1946) The toxicity and toxic manifestations of 2,2-bis(p–chlorophenyl)–1,1,1-trichloroethane (DDT) as influenced by chemical changes in the molecule. *J. Pharmacol. exp. Ther.*, *88*, 400–413

Ogata, M., Yamasaki, Y. & Kawai, T. (1986) Significance of urinary phenyl sulfate and phenyl glucuronide as indices of exposure to phenol. *Int. Arch. occup. environ. Health*, *58*, 197–202

Ohtsuji, H. & Ikeda, M. (1972) Quantitative relationship between atmospheric phenol vapour and phenol in the urine of workers in bakelite factories. *Br. J. ind. Med.*, *29*, 70–73

Painter, R.B. & Howard, R. (1982) The HeLa DNA–synthesis inhibition test as a rapid screen for mutagenic carcinogens. *Mutat. Res.*, *92*, 427–437

Pardoe, R., Minami, R.T., Sato, R.M. & Schlesinger, S.L. (1976) Phenol burns. *Burns*, *3*, 29–41

Paschin, Y.V. & Bahitova, L.M. (1982) Mutagenicity of benzo[a]pyrene and the antioxidant phenol at the HGPRT locus of V79 Chinese hamster cells. *Mutat. Res.*, *104*, 389–393

Pellack-Walker, P. & Blumer, J.L. (1986) DNA damage in L5178YS cells following exposure to benzene metabolites. *Mol. Pharmacol.*, *30*, 42–47

Pereira, M.A. & Chang, L.W. (1981) Binding of chemical carcinogens and mutagens to rat hemoglobin. *Mutat. Res.*, *33*, 301–305

Petrov, V.I. (1960) Cases of phenol vapor poisoning during coke slaking with phenol water (Russ.). *Gig. Sanit.*, *25*, 60–62

Piotrowski, J.K. (1971) Evaluation of exposure to phenol: absorption of phenol vapour in the lungs and through the skin and excretion of phenol in urine. *Br. J. ind. Med.*, *28*, 172–178

Poirier, M.C., De Cicco, B.T. & Lieberman, M.W. (1975) Nonspecific inhibition of DNA repair synthesis by tumor promoters in human diploid fibroblasts damaged with N-acetoxy-2-acetylaminofluorene. *Cancer Res.*, *35*, 1392–1397

Pool, B.L. & Lin, P.Z. (1982) Mutagenicity testing in the *Salmonella typhimurium* assay of phenolic compounds and phenolic fractions obtained from smokehouse smoke condensates. *Food. chem. Toxicol.*, *20*, 383–391

Pouchert, C.J., ed. (1981) *The Aldrich Library of Infrared Spectra*, 3rd ed., Milwaukee, WI, Aldrich Chemical Co., p. 644A

Pouchert, C.J., ed. (1983) *The Aldrich Library of NMR Spectra*, 2nd ed., Vol. 1, Milwaukee, WI, Aldrich Chemical Co., p. 867A

Pouchert, C.J., ed. (1985) *The Aldrich Library of FT–IR Spectra*, Vol. 1, Milwaukee, WI, Aldrich Chemical Co., p. 1069A

Powell, G.M., Miller, J.J., Olavesen, A.H. & Curtis, C.G. (1974) Liver as major organ of phenol detoxication? *Nature*, *252*, 234–235

Price, C.J., Ledoux, T.A., Reel, J.R., Fisher, P.W., Paschke, L.L., Mann, M.C. & Kimmel, C.A. (1986) Teratologic evaluation of phenol in rats and mice. *Teratology*, *33*, 92C–93C

Ramanathan, M. (1984) Water pollution. In: Mark, H.F., Othmer, D.F., Overberger, C.G., Seaborg, G.T. & Grayson, M., eds, *Kirk–Othmer Encyclopedia of Chemical Technology*, 3rd ed., Vol. 24, New York, John Wiley & Sons, p. 299

Ramli, J.B. & Wheldrake, J.F. (1981) Phenol conjugation in the desert hopping mouse, *Notomys alexis*. *Comp. Biochem. Physiol.*, *69C*, 379–381

Roberts, M.S., Shorey, C.D., Arnold, R. & Anderson, R.A. (1974) The percutaneous absorption of phenolic compounds. I. Aqueous solutions of phenol in the rat. *Aust. J. pharm. Sci.*, *NS3*, 81–91

Rogers, S.C.F., Burrows, D. & Neill, D. (1978) Percutaneous absorption of phenol and methyl alcohol in Magenta Paint BPC. *Br. J. Dermatol.*, *98*, 559–560

Roxan, Inc. (undated) *Precision Gas Detector*, Woodland Hills, CA

Sadtler Research Laboratories (1980) *Standard Spectra Collection, 1980 Cumulative Index*, Philadelphia, PA

Salaman, M.H. & Glendenning, O.M. (1957) Tumour promotion in mouse skin by sclerosing agents. *Br. J. Cancer*, *11*, 434–444

Schaper, K.-A. (1981) Acute phenol intoxication – a report on clinical experience (Ger.). *Anaesthesiol. Reanimat.*, *6*, 73–79

Sensidyne (1985) *The First Truly Simple Precision Gas Detector System*, Largo, FL

Skare, J.A. & Schrotel, K.R. (1984) Detection of strand breaks in rat germ cell DNA by alkaline elution and criteria for the determination of a positive response (Abstract No. Gb–3). *Environ. Mutagenesis, 6,* 445

SKC Inc. (1988) *Comprehensive Catalog and Guide,* Eighty Four, PA

Sofuni, T., Hayashi, M., Shimada, H., Ebine, Y., Matsuoka, A., Sawada, S. & Ishidate, M., Jr (1986) Sex difference in the micronucleus induction of benzene in mice (Abstract No. 51). *Mutat. Res., 164,* 281

Štajduhar–Carić, Z. (1968) Acute phenol poisoning. Singular findings in a lethal case. *J. forens. Med., 15,* 41–42

Sturtevant, F.M., Jr (1952) Studies on the mutagenicity of phenol in *Drosophila melanogaster. J. Hered., 43,* 217–220

Subrahmanyam, V.V. & O'Brien, P.J. (1985a) Peroxidase–catalysed binding of [U–¹⁴C]phenol to DNA. *Xenobiotica, 15,* 859–871

Subrahmanyam, V.V. & O'Brien, P.J. (1985b) Phenol oxidation product(s), formed by a peroxidase reaction, that bind to DNA. *Xenobiotica, 15,* 873–885

Thompson, E.D. & Melampy, P.J. (1981) An examination of the quantitative suspension assay for mutagenesis with strains of *Salmonella typhimurium. Environ. Mutagenesis, 3,* 453–465

Thurman, C. (1982) Phenol. In: Mark, H.F., Othmer, D.F., Overberger, C.G., Seaborg, G.T. & Grayson, M., eds, *Kirk–Othmer Encyclopedia of Chemical Technology,* 3rd ed., Vol. 17, New York, John Wiley & Sons, pp. 373–384

Työsuojeluhallitus (National Finnish Board of Occupational Safety and Health) (1987) *HTP–Arvot 1987* [Limit Values 1987] (*Safety Bulletin 25*), Helsinki, Valtion Painatuskeskus, p. 15

US Environmental Protection Agency (1980) *Ambient Water Criteria for Phenol (PB–81–117772),* Washington DC

US Environmental Protection Agency (1983) *Treatability Manual,* Vol. 1, *Treatability of Data (ORD US EPA–600/2–82–001a),* Washington DC

US Environmental Protection Agency (1986a) Method 8010: phenols. In: *Test Methods for Evaluating Solid Waste – Physical/Chemical Methods,* 3rd ed. (*EPA No. SW–846*), Washington DC, Office of Solid Waste and Emergency Response, pp. 8040–1–8040–17

US Environmental Protection Agency (1986b) Method 8250: gas chromatography/mass spectrometry for semi–volatile organics: packed column technique. In: *Test Methods for Evaluating Solid Waste – Physical/Chemical Methods,* 3rd ed. (*EPA No. SW–846*), Washington DC, Office of Solid Waste and Emergency Response, pp. 8250–1–8250–32

US Environmental Protection Agency (1986c) Method 8270: gas chromotagraphy/mass spectrometry for semi–volatile organics: capillary column technique. In: *Test Methods for Evaluating Solid Waste – Physical/Chemical Methods,* 3rd ed. (*EPA No. SW–846*), Washington DC, Office of Solid Waste and Emergency Response, pp. 8270–1–8270–34

US International Trade Commission (1981) *Synthetic Organic Chemicals, US Production and Sales, 1980 (USITC Publ. 1183),* Washington DC, US Government Printing Office

US International Trade Commission (1982) *Synthetic Organic Chemicals, US Production and Sales, 1981 (USITC Publ. 1292),* Washington DC, US Government Printing Office

US International Trade Commission (1983) *Synthetic Organic Chemicals, US Production and Sales, 1982 (USITC Publ. 1422),* Washington DC, US Government Printing Office

US International Trade Commission (1984) *Synthetic Organic Chemicals, US Production and Sales, 1983 (USITC Publ. 1588),* Washington DC, US Government Printing Office

US International Trade Commission (1985) *Synthetic Organic Chemicals, US Production and Sales, 1984 (USITC Publ. 1745)*, Washington DC, US Government Printing Office

US International Trade Commission (1986) *Synthetic Organic Chemicals, US Production and Sales, 1985 (USITC Publ. 1892)*, Washington DC, US Government Printing Office

US International Trade Commission (1987) *Synthetic Organic Chemicals, US Production and Sales, 1986 (USITC Publ. 2009)*, Washington DC, US Government Printing Office

US International Trade Commission (1988) *Synthetic Organic Chemicals, US Production and Sales, 1987 (USITC Publ. 2118)*, Washington DC, US Government Printing Office

Van Duuren, B.L. & Goldschmidt, B.M. (1976) Cocarcinogenic and tumor-promoting agents in tobacco carcinogenesis. *J. natl Cancer Inst.*, *56*, 1237–1242

Van Duuren, B.L., Sivak, A., Langseth, L., Goldschmidt, B.M. & Segal, A. (1968) Initiators and promoters in tobacco carcinogenesis. *Natl Cancer Inst. Monogr.*, *28*, 173–180

Van Duuren, B.L., Blazej, T., Goldschmidt, B.M., Katz, C, Melchionne, S. & Sivak, A. (1971) Cocarcinogenesis studies on mouse skin and inhibition of tumor induction. *J. natl Cancer Inst.*, *46*, 1039–1044

Van Duuren, B.L., Katz, C. & Goldschmidt, B.M. (1973) Cocarcinogenic agents in tobacco carcinogenesis. *J. natl Cancer Inst.*, *51*, 703–705

Vernot, E.H., MacEwen, J.D., Haun, C.C. & Kinkead, E.R. (1977) Acute toxicity and skin corrosion data for some organic and inorganic compounds and aqueous solutions. *Toxicol. appl. Pharmacol.*, *42*, 417–423

Verschueren, K. (1983) *Handbook of Environmental Data on Organic Chemicals*, 2nd ed., New York, Van Nostrand Reinhold Co., pp. 973–982

Weast, R.C., ed. (1985) *Handbook of Chemistry and Physics*, 66th ed., Cleveland, OH, CRC Press, p. C–406

Wilcosky, T.C., Checkoway, H., Marshall, E.G. & Tyroler, H.A. (1984) Cancer mortality and solvent exposures in the rubber industry. *Am. ind. Hyg. Assoc. J.*, *45*, 809–811

Williams, R.T. (1959) *Detoxication Mechanisms*, London, Chapman & Hall, pp. 278–317

Windholz, M., ed. (1983) *The Merck Index*, 10th ed., Rahway, NJ, Merck & Co., p. 1043

Woodruff, R.C., Mason, J.M., Valencia, R. & Zimmering, S. (1985) Chemical mutagenesis testing in *Drosophila*. V. Results of 53 coded compounds tested for the National Toxicology Program. *Environ. Mutagenesis*, *7*, 677–702

Wynder, E.L. & Hoffmann, D. (1961) A study of tobacco carcinogenesis. VIII. The role of acidic fractions as promoters. *Cancer*, *14*, 1306–1315

Yanysheva, N.Y., Balenko, N.V., Chernichenko, I.A., Babiy, V.F., Bakanova, G.N. & Lemeshko, L.P. (1988) Manifestations of carcinogenesis after combined treatment with benzo[a]pyrene and phenol depending on the schedule of administration (Russ.). *Gig. Sanit.*, *41*, 29–33

SOME PIGMENTS

ANTIMONY TRIOXIDE AND ANTIMONY TRISULFIDE

1. Chemical and Physical Data

1.1 Synonyms

Antimony trioxide

Chem. Abstr. Services Reg. No.: 1309–64–4 – Antimony oxide
1317–98–2 – Valentinite
12412–52–1 – Senarmontite
Chem. Abstr. Name: Antimony oxide
IUPAC Systemic Name: Diantimony trioxide
Synonyms: Antimonious oxide; antimony (III) oxide; antimony sesquioxide; antimony white; AP 50; flowers of antimony; CI 77052; CI Pigment White 11; senarmontite; valentinite

Antimony trisulfide

Chem. Abstr. Services Reg. No.: 1345–04–6 – Antimony sulfide
1317–86–8 – Stibnite
Chem. Abstr. Name: Antimony sulfide
IUPAC Systematic Name: Diantimony trisulfide
Synonyms: Antimonous sulfide; antimony glance; antimony needles; antimony orange; antimony sesquisulfide; antimony trisulfide colloid; antimony vermilion; black antimony; CI 77060; CI Pigment Red 107; crimson antimony sulfide; needle antimony; stibnite

1.2 Molecular formulae and molecular weights

Sb_2O_3 antimony trioxide – Mol. wt: 291.50

Sb_2S_3 antimony trisulfide – Mol. wt: 339.68

Antimony trioxide is a dimorphic crystalline solid existing in an orthorhombic configuration as the mineral valentinite and in cubic form as senarmontite (Weast, 1985). Antimony trisulfide in its mineral form, stibnite, is an orthorhombic–bipyramidal crystalline structure (Roberts *et al.*, 1974).

1.3   Chemical and physical properties of the pure substance

Antimony trioxide

(a) *Description*: White, odourless, crystalline powder (Asarco, Inc., 1988a)

(b) *Melting–point*: 656°C (Weast, 1985)

(c) *Boiling–point*: Sublimes (Weast, 1985)

(d) *Density*: Valentinite, 5.7 g/cm³; senarmontite, 5.2 g/cm³ (Weast, 1985)

(e) *Reactivity*: Reacts with strong alkalis to form antimonates (Mannsville Chemical Products Corp., 1981)

(f) *Solubility*: Very slightly soluble in water; soluble in potassium hydroxide, hydrochloric acid and acetic acid (Weast, 1985); insoluble in organic solvents (Freedman *et al.*, 1978)

(g) *Spectroscopy data*: X–ray diffraction patterns for valentinite and senarmontite have been reported (Roberts *et al.*, 1974).

(h) *Refractive index*: Valentinite: 2.18, 2.35, 2.35; senarmontite: 2.087 (Weast, 1985)

Antimony trisulfide

(a) *Description*: Purified antimony trisulfide is usually a yellow–red amorphous powder (Weast, 1985). In its natural form (stibnite), antimony trisulfide is commonly found as well–formed crystals, sometimes very large and solid and at other times slender and fragile. The crystals are often vertically striated, bent or twisted. They are pale to dark lead–grey, but may appear tarnished, iridescent, bluish or blackish (Roberts *et al.*, 1974).

(b) *Density*: amorphous, 4.12 g/cm³; stibnite, 4.64 g/cm³ (Weast, 1985)

(c) *Solubility*: Very slightly soluble in water; soluble in hydrochloric acid and ethanol (Weast, 1985)

(d) *Spectroscopy*: The X–ray diffraction pattern for stibnite has been reported (Roberts *et al.*, 1974)

(e) *Refractive index*: Stibnite: 3.194, 4.064, 4.303 (Weast, 1985)

1.4   Technical products and impurities

Antimony trioxide

Trade names: Amspec-KR; Anzon-TMS; Asarco antimony oxide (LT, HT, VHT); A 1582; A 1588 LP; Blue Star; Dechlorane A-O; Exitelite; Extrema; Laurel (formerly Chemtron) Fire Shield; Thermoguard B; Thermoguard S; Twinkling Star; White Star

Antimony trioxide is available in several product grades of varying particle and tint. All are of at least 99.0% purity. Lead, arsenic and iron are common contaminants of the product in quantities of ≤ 2, ≤ 0.5 and < 0.01 wt%, respectively (Mansville Chemical Products Corp., 1985; Anzon, Inc., 1988; Asarco, Inc., 1988a,b).

Antimony trisulfide

Trade name: Lymphoscan

No data on technical-grade antimony trisulfide or its impurities were available to the Working Group.

2. Production, Use, Occurrence and Analysis

2.1 Production and use

(a) Production

Antimony trioxide is typically produced by roasting stibnite ores, which are reported to contain 55% antimony. The production of antimony trioxide occurs as a vapour–phase reaction at temperatures in excess of 1550°C. The stoichiometric addition of oxygen to the feed ore produces the desired product. Tetra- and pentoxides produce precipitate from the vapour as white and yellow powders, respectively. Antimony trioxide has also been isolated for many years as a by-product of lead smelting and production. Approximately 10–15% of US production occurs *via* this route. Antimony trioxide can be purified through serial vapour phase recrystallization (Freedman *et al.*, 1978; Mannsville Chemical Products Corp., 1981, 1985; Asarco, Inc., 1988a).

The following countries produce antimony trioxide: Belgium, Bolivia, China, France, Guatemala, Mexico, South Africa, the UK, the USA and Yugoslavia. Total US production has tripled since 1960 (Mannsville Chemical Products Corp., 1981, 1985; Palencia & Mishra, 1986) and was approximately 19 000 tonnes in 1987 (Llewellyn & Isaac, 1988).

Antimony trisulfide pigment is prepared by the addition of sodium thiosulfate solution to a solution of antimony potassium tartrate (tartar emetic) and tartaric acid or to a solution of another suitable antimony salt (LeSota, 1978).

(b) Use

The largest end use for antimony trioxide is as a fire retardant in plastics, rubbers, textiles, paper and paints (Mannsville Chemical Products Corp., 1981). This application represents approximately 60–75% of total US consumption (Mannsville Chemical Products Corp., 1981, 1985; Llewellyn & Isaac, 1988).

Antimony trioxide by itself is not a fire retardant. It is used as a synergist, typically at 2–10% by weight, with organochlorine and brominated compounds to diminish the inflammability of a wide range of plastics and textiles. Antimony trioxide also filters ultraviolet radiation which cause textile fibres to deteriorate (Drake, 1980; Lyons, 1980; Mannsville Chemical Products Corp., 1981, 1985).

When added to ceramic products, antimony trioxide imparts opacity, hardness and acid resistance, for instance, to sanitary ware and enamels. In the preparation of optical and ruby glass, antimony trioxide is used as a bubble remover. In other glasses, it is incorporated as a colour stabilizer to protect against the weathering effects of the sun. Its use as a stabilizer and as a catalyst accounted for an estimated 15% of total US consumption in 1980 (Mannsville Chemical Products Corp., 1981; Windholz, 1983).

Antimony trioxide is used as a catalyst in the production of polyester resins and in the decomposition of hydrogen bromide. When used with tin dioxide at 480°C, antimony trioxide catalyses the partial oxidation of propylene (Samsonov, 1982).

Antimony trisulfide is used as a primer in ammunition and smoke markers, in the production of vermilion or yellow pigment and antimony salts such as antimony oxide and chloride, and in the manufacture of ruby glass (Mannsville Chemical Products Corp., 1985; Palencia & Mishra, 1986; Hawley, 1981).

(c)   *Regulatory status and guidelines*

An occupational exposure limit of 0.5 mg/m³ as an 8–h time–weighted average (TWA) has been set for antimony and its compounds (measured as antimony) in many countries (Direktoratet for Arbeidstilsynet, 1981; Arbeidsinspectie, 1986; Institut National de Recherche et de Sécurité, 1986; Health and Safety Directorate, 1987; National Swedish Board of Occupational Safety and Health, 1987; Työsuojeluhallitus, 1987; US Occupational Safety and Health Administration, 1987; American Conference of Governmental Industrial Hygienists, 1988; Arbejdstilsynet, 1988; Deutsche Forschungsgemeinschaft, 1988). An exposure limit of 0.5 mg/m³ (8–h TWA) has also been set specifically for the handling and use of antimony trioxide (measured as antimony), in some cases with a carcinogen or skin sensitivity notation (International Labour Office, 1984; American Conference of Governmental Industrial Hygienists, 1988). Lower exposure limits (0.05 mg/m³, or no permissible exposure) have been set for the production of antimony trioxide because these compounds are classified as carcinogenic in several countries, e.g., Belgium, Finland, Italy, Sweden and the USA (International Labour Office, 1984; National Swedish Board of Occupational Safety and Health, 1987; Työsuojeluhallitus, 1987; American Conference of Governmental Industrial Hygienists, 1988).

2.2 Occurrence

(a)   *Natural occurrence*

Antimony trioxide occurs in nature as the minerals valentinite and senarmontite. The orthorhombic valentinite and cubic senarmontite are secondary minerals formed by the geologic alteration of stibnite (Sb_2S_3) and other antimony minerals (Roberts et al., 1974).

Antimony trisulfide occurs in nature as the mineral stibnite. It is formed as a low temperature deposit from hot solutions often associated with arsenic minerals and cinnabar. Significant deposits occur in Algeria, Borneo, Canada, China, Czechoslovakia, the Federal Republic of Germany, France, Italy, Japan, Mexico and Peru (Pough, 1960; Roberts et al., 1974; Palencia & Mishra, 1986).

(b)   *Occupational exposure*

On the basis of US National Occupational Exposure Surveys, the National Institute for Occupational Safety and Health (1974, 1983) estimated that 28 957 workers were potentially exposed to antimony trioxide in the USA in 1972–74 and 85 650 in 1981–83.

In a UK plant where antimony ore was processed, levels of antimony oxide in the air of work areas were reported to be highest during the short periods when tapping operations (pouring molten metal) at the furnace were under way; the mean value at these times was 37 mg/m³. Air levels in other areas of the plant were 0.53–5.3 mg/m³ (McCallum, 1963).

Personal TWA exposure to antimony trioxide at a glass–producing factory in the Federal Republic of Germany ranged from less than 50 μg/m³ to 840 μg/m³, with corresponding blood levels of 0.4–3.1 μg/l and a median of 1.0 μg/l (Lüdersdorf *et al.*, 1987).

A study of two major US antimony producing companies in which imported antimony sulfide ore was roasted to produce antimony trioxide showed personal TWA exposures to range from 0.21 to 3.2 mg/m³ (mean, 1.32 mg/m³) and 2.7 to 8.7 mg/m³ (mean, 5.2 mg/m³; Donaldson & Cassady, 1979).

At a smelting plant in Yugoslavia, concentrations of dust consisting of 36–90% antimony trioxide ranged from 16 to 248 mg/m³ (Karajovic *et al.*, 1960). At a plant in the USA where antimony sulfide ore was smelted, personal TWA airborne concentrations of antimony were 0.92–70.7 mg/m³; the author postulated a predominance of antimony trioxide (Renes, 1953). At another US antimony smelting plant, workers were exposed to antimony ore dust containing primarily antimony trioxide at concentrations (area samples) ranging from 0.08 to 138 mg/m³ (Cooper *et al.*, 1968). In a plant in the USA manufacturing resinoid grinding wheels, occupational exposures to antimony trisulfide were reported to range from 0.6 to 5.5 mg/m³ (Brieger *et al.*, 1954).

2.3   Analysis

No information was available to the Working Group on standard methods for the quantitative determination of antimony trioxide or antimony trisulfide in environmental samples. Antimony can be quantified in environmental matrices by a variety of methods, including atomic absorption spectrophotometry, inductively coupled plasma emission and X-ray fluorescence spectrometry, neutron activation analysis, anodic stripping voltammetry and various titrimetric and colorimetric methods (Freedman *et al.*, 1978; US Environmental Protection Agency, 1983; Eller, 1985; US Environmental Protection Agency, 1986; Lodge, 1989).

3. Biological Data Relevant to the Evaluation of Carcinogenic Risk to Humans

3.1 Carcinogenicity studies in animals[1]

Inhalation exposure

Rat: Groups of 49–51 female Fischer rats (CDF from Charles River), 19 weeks old, were exposed by inhalation to 0, 1.6 ± 1.5, or 4.2 ± 3.2 mg/m³ commercial grade *antimony trioxide* (measured as antimony; purity, 99.4%; arsenic, 0.02%; particle size, 0.4 μm ± 2.13 (for the high concentration) and 0.44 μm ± 2.23 (for the lower concentration)) for 6 h per day on five days per week for 13 months. Mean body weights were increased in both treated groups during the exposure period but did not differ significantly from that of controls at the end of the study. Groups of rats were sacrificed and examined histologically after three, six, nine and 12 months of exposure and two months after the end of treatment [numbers of rats sacrificed and numbers of early deaths for each period unspecified]. At 12 months after the end of treatment, 13 control, 17 low–dose and 18 high–dose rats were sacrificed and selected tissues were examined. Lung tumours localized in the bronchioloalveolar region occurred in 14/18 high–dose rats (three adenomas, nine scirrhous carcinomas ($p < 0.01$ [test unspecified]) and two squamous–cell carcinomas). One bronchioloalveolar adenoma occurred in a low–dose rat, and no lung tumour was observed in the control group at terminal sacrifice. Scirrhous carcinomas were also observed in 5/7 and 1/9 high–dose rats that died or were sacrificed between two months after the end of treatment and terminal sacrifice or between the end of treatment and two months after the end of treatment. One bronchioloalveolar adenoma occurred among six control rats that died or were sacrificed between two months after the end of treatment and terminal sacrifice. There was no significant difference in the number of other tumours occurring in treated and control groups (Watt, 1983).

Groups of 90 male and 90 female Wistar rats, eight months old, were exposed by inhalation to 0 or 45 mg/m³ (TWA) *antimony trioxide* (purity, ≥95%; arsenic, 0.004%; titanium, < 3%) for 7 h per day on five days per week for 52 weeks. Five males and females were killed at six, nine and 12 months after exposure was initiated; the remainder of the animals were killed 18–20 weeks after the end of the exposure period. There was no significant difference in survival between treated and control groups of either sex; 39 control and 31 treated females survived until terminal sacrifice [estimated survival in males was 21 and 21]. Non–neoplastic lesions (interstitial fibrosis, alveolar–cell hyperplasia and metaplasia) of the lung occurred with similar frequency in male and female rats but were slightly less severe in males. The first lung tumours were seen in two (one adenoma and one squamous–cell carcinoma) of

---

[1]The Working Group was aware of a study in progress of intracheal administration of antimony trioxide to hamsters (IARC, 1988).

five treated female rats sacrificed at 53 weeks; 19/70 (27%) treated females surviving at the time the first tumour was observed developed lung tumours. No lung tumour was seen in treated males or in male or female controls. The lung tumours that were found in treated females were nine squamous–cell carcinomas, five scirrhous carcinomas and 11 bronchioloalveolar adenomas or carcinomas [numbers of benign and malignant bronchioloalveolar tumours not specified]. The incidence of other tumours was not different between treated and control rats (Groth *et al.*, 1986).

Groups of 90 male and 90 female Wistar rats, eight months old, were exposed by inhalation to 0 or 36–40 mg/m³ (TWA) antimony ore concentrate (containing 46% antimony, principally as *antimony trisulfide*; titanium, < 4%; aluminium, 0.5%; tin, 0.2%; lead, 0.3%; iron, 0.3%; arsenic, 0.08%) for 7 h per day on five days per week for up to 52 weeks. Five males and five females were killed six, nine and 12 months after exposure had been initiated; the remainder of the animals were killed 18–20 weeks after the end of the exposure period. There was no significant difference in survival between treated and control groups of either sex; 39 control and 33 treated females survived until terminal sacrifice [estimated survival in males was 23 and 21]. Non-neoplastic lesions (interstitial fibrosis, alveolar–cell hyperplasia and metaplasia) of the lung occurred at similar frequency in male and female rats but were slightly less severe in males. The first lung tumour was seen in a treated female that died 41 weeks after the beginning of treatment; 17/68 (25%) treated females surviving at the time the first tumour was observed developed lung tumours. No lung tumour was seen in treated males or in male or female control rats, and there was no difference in the incidences of other tumours between treated and control groups of either sex. The lung tumours that occurred in treated females were nine squamous–cell carcinomas, four scirrhous carcinomas and six bronchioloalveolar adenomas or carcinomas [numbers of benign and malignant bronchioloalveolar tumours not specified] (Groth *et al.*, 1986).

3.2 Other relevant data

The toxicology of antimony compounds has been reviewed (National Institute for Occupational Safety and Health, 1978).

(a) Experimental systems

(i) Absorption, distribution, excretion and metabolism

After administration of 2% *antimony trioxide* to rats in the diet for eight months, very high levels were found in the thyroid, while retention was much lower (in decreasing order) in the liver, spleen, kidney, heart and lungs (Gross *et al.*, 1955a). After administration of 1% antimony trioxide to rats in the diet for 12 weeks, the highest antimony concentrations were found (in decreasing order) in the blood, spleen, lungs, kidneys, hair, liver and heart; 12 weeks after the end of treatment, levels in the blood, lungs and kidneys had decreased to about 50%, but the spleen still contained about 75% of the concentration observed at termination of exposure (Hiraoka, 1986).

In rats exposed for two to 14 months by inhalation to 100–125 mg/m³ antimony trioxide, pulmonary retention increased with increasing length of exposure; following cessation of exposure, pulmonary levels declined slowly (Gross *et al.* 1955b).

Exposure of rats by inhalation to 119 mg/m³ antimony trioxide dust (geometric mean particle size, 1.3 μm) for 80 h resulted in total urinary excretion of less than 40 μg antimony trioxide within four days (Gross et al., 1955a).

A single administration to rats by stomach tube of 0.2 g antimony trioxide suspended in water resulted in total urinary excretion of 3.2% of the dose during the subsequent eight days. Faecal excretion was detected in rats for several weeks following cessation of administration of 2% antimony trioxide in the diet; after three weeks, faecal excretion was lower than urinary excretion levels (Gross et al., 1955a).

Exposure of female dogs by inhalation to about 5.5 mg/m³ *antimony trisulfide* dust from a smelter (particle size, <2 μm) for ten weeks resulted in urinary excretion of up to 16–18 mg/l antimony (Brieger et al., 1954).

(ii) *Toxic effects*

The oral LD$_{50}$ for *antimony trioxide* in rats is above 20 g/kg bw; however, reduced growth and other nonspecific effects were seen with 1 g/kg bw (Smyth & Carpenter, 1948). In contrast, administration of 16 g/kg bw antimony trioxide to rats by stomach tube resulted in no apparent ill effects within a 30–day observation period (Gross et al., 1955a). Reduced weight gain, a slight reduction in absolute weight of the spleen and heart and a slight increase in absolute and relative weight of the lungs were observed in rats given 1% antimony trioxide in the diet for 12 weeks (Hiraoka, 1986). Vomiting and gastrointestinal disturbances occurred in dogs after daily ingestion of approximately 0.15 g/kg bw antimony trioxide or more; vomiting was induced by similar doses in a cat, and continued daily doses caused significant weight loss (Flury, 1927).

Intratracheal instillation of 50 mg antimony trioxide–containing smelter dust to rats did not result in lung fibrosis, although thin argyrophilic fibres were seen (Potkonjak & Vishnjich, 1983). No fibrosis was described after long–term, repeated exposure of rats and rabbits by inhalation to 100–125 mg/m³ and 89 mg/m³ antimony trioxide (average particle size, 0.6 μm), respectively. However, at these dose levels, rapid mortality occurred, particularly in rabbits, due primarily to pneumonia (Gross et al., 1955b).

Increased lung weight, focal fibrosis, adenomatous hyperplasia, multinucleated giant cells and pigmented macrophages were observed in female rats exposed to 1.6 and 4.2 mg/m³ antimony trioxide (commercial grade; average particle size, 0.4 μm) for one year by inhalation. Similar doses caused no exposure–related change in miniature pigs (Watt, 1983). Increased death rates due to pneumonia were observed in guinea–pigs following inhalation exposure to antimony trioxide; cloudy swelling of the liver cells occurred in almost half of the animals exposed, but there was no other indication of systemic toxicity (Dernehl et al., 1945).

Rats exposed to 3.1 mg/m³ *antimony trisulfide* by inhalation for six weeks developed electrocardiographic changes, notably with flattened T–waves; on autopsy, the heart was found to be dilated, with signs of degenerative changes; focal haemorrhage and congestion in the lungs were considered to be secondary to heart failure. Similar pathological effects were seen in rabbits exposed to 5.6 mg/m³ for six weeks. Cardiotoxic changes were observed in two dogs exposed to 5.6 mg/m³ for ten weeks, but not in two dogs exposed to 5.3 mg/m³ for seven weeks (Brieger et al., 1954).

Degenerative changes in the liver and in the tubular epithelium of the kidney were observed in rabbits exposed to 27.8 mg/m³ antimony trisulfide for five days (Brieger *et al.*, 1954).

(iii) *Effects on reproduction and prenatal toxicity*

Female rats were exposed by inhalation for 4 h per day for 1.5–2 months to 0 or 250 mg/m³ *antimony trioxide*. They were then mated, and exposures continued until days 3–5 before expected delivery. Pregnancy was obtained in 16/24 treated females and in 10/10 controls. Litter size and weight of offspring at birth and weaning were not altered by exposure to antimony trioxide (Belyaeva, 1967).

Pregnant female rats (six to seven per group) were exposed by inhalation to 0, 0.027, 0.082 or 0.27 mg/m³ antimony trioxide for 24 h per day for 21 days. Fetal growth and viability were assessed at the end of gestation. Maternal body weight gain was not affected by exposure, but, at the high–dose level, increased pre- and postimplantation death of embryos was observed. At the mid–dose level, preimplantation loss and fetal growth retardation were evident (Grin *et al.*, 1987).

No data on *antimony trisulfide* were available to the Working Group.

(iv) *Genetic and related effects*

Antimony trioxide produced differential killing in DNA repair–proficient compared to repair–deficient strains of *Bacillus subtilis*. In a spot test, it was not mutagenic to *Escherichia coli* B/r WP2 or to *Salmonella typhimurium* TA1535, TA1537, TA1538, TA98 or TA100 [details not given] (Kanematsu *et al.*, 1980).

No data on *antimony trisulfide* were available to the Working Group.

(b) *Humans*

(i) *Absorption, distribution, excretion and metabolism*

Three workers with pulmonary changes related to exposure to antimony trioxide excreted 425, 480 and 680 μg/l antimony in urine, while another patient with antimony pneumoconiosis had urinary levels of 55 and 28 μg/l seven months and four years after retirement, respectively (McCallum, 1963). High levels [not given separately] of antimony were detected in the urine of workers exposed for several years to antimony trioxide; high excretion levels were also found after one month's cessation of exposure (Klučík & Kemka, 1960). High excretion levels were also seen in antimony production workers examined by Cooper *et al.* (1968). X–Ray spectrometry has demonstrated that inhaled antimony dust may be retained in the lung for long periods (McCallum, 1967; McCallum *et al.*, 1971). The amount of antimony retained in the lungs tended to rise with duration of employment at an antimony smelter, suggesting that accumulation may take place (McCallum *et al.*, 1971).

Among antimony trisulfide workers exposed to antimony levels generally higher than 3 mg/m³, urinary excretion of antimony was 0.8–9.6 mg/l (Brieger *et al.*, 1954).

(ii) *Toxic effects*

Accidental oral intake of antimony trioxide leached from enamel or ceramic glaze into acid beverages was reported to result in a burning sensation in the stomach, colic, nausea, vomiting and, occasionally, collapse (Monier–Williams, 1934). Complete recovery occurs after several days (Dunn, 1928).

Smelter workers exposed to antimony trioxide frequently complained of symptoms related to mucous membrane irritation, such as rhinitis (with cases of septal perforation and loss of smell), pharyngitis, laryngitis (sometimes with aphonia), gastroenteritis, bronchitis and pneumonitis. Other symptoms, less often encountered, included weight loss, nausea, vomiting, abdominal cramps and diarrhoea (Renes, 1953).

In other studies, only skin irritation (Oliver, 1933) or no indication of systemic toxicity (Potkonjak & Pavlovich, 1983) was found in smelter workers. Skin lesions ('antimony spots') in workers exposed to antimony trioxide develop mainly in areas exposed to heat and where sweating occurs (Stevenson, 1965). Occasionally, positive patch tests with antimony trioxide have been recorded (Paschoud, 1964). Antimony trisulfide has not been reported to cause dermatitis (National Institute for Occupational Safety and Health, 1978).

Radiographic changes (rounded opacities) in smelter workers exposed to antimony trioxide were first described in 1960 in Yugoslavia (Karajović et al., 1960). The smelter dust contained antimony trioxide and some pentoxide, with low concentrations of silica and arsenic oxide. In a follow-up study of these smelter workers, the earliest changes were seen only after at least nine years of exposure, and no evidence was found of progression after cessation of exposure. Some evidence was found of mixed restrictive as well as obstructive changes in bronchi and small airways (Potkonjak & Pavlovich, 1983).

McCallum (1963) also noted in the UK that smelter workers with pneumoconiosis, a condition he termed 'antimony pneumoconiosis', were generally symptomless. The degree of radiographic abnormalities was correlated with the amount of antimony retained in the lungs and with duration of exposure; early changes were recorded in these workers after only a few years of employment (McCallum et al., 1971). In a cross-sectional study of 274 smelter workers in 1965–66, 26 new cases of antimony pneumoconiosis were found and 18 were already under observation (McCallum, 1967). A subsequent study included 113 men, 46 of whom had radiographic abnormalities; six had severe abnormalities (McCallum et al., 1971). Additional cases of pneumoconiosis with rounded opacities were seen in antimony smelter workers in the USA; the radiographic abnormalities were not associated with changes in pulmonary function (Cooper et al., 1968). Possible antimony pneumoconiosis has also been recorded in two chemical workers exposed to antimony trioxide dust (Guzman et al., 1986).

Following several deaths, possibly related to heart disease, among workers exposed to 0.6–5.5 mg/m³ antimony trisulfide, a study revealed electrocardiographic changes, mainly in T-waves, in 37/75 workers; after cessation of exposure, the changes persisted in 12/56 workers who were re-examined. Gastrointestinal disturbances were also reported (Brieger et al., 1954). Electrocardiographic changes were also observed in smelter workers by Klučík and Ulrich (1960).

(iii) *Effects on fertility and on pregnancy outcome*

Belyaeva (1967) described the reproductive outcomes of 318 women in the USSR working with dusts containing metallic antimony, antimony trioxide and antimony pentasulfide and of 115 control women. The women exposed to antimony dusts more frequently had premature births (3.4% *versus* 1.2%, respectively) and spontaneous abortions (12.5% *versus* 4.1%, respectively). The average birth weights of the 70 children of the exposed women were similar to those of the 20 children of the control women (3360 g and 3350 g, respectively); however, at one year of age, the children of the exposed women were significantly lighter than the control children (8960 g and 10 050 g, respectively). [The Working Group noted that the numbers of premature births and spontaneous abortions were not stated.]

(iv) *Genetic and related effects*

No data were available to the Working Group.

3.3 Case reports and epidemiological studies of carcinogenicity to humans

In a report which quoted an unpublished statement issued in 1973 that ten cases of lung cancer had occurred among antimony process workers in the UK in 1969–71, with 8.0 expected, it was stated that no data were given on smoking nor on the methods used to calculate expected rates (National Institute for Occupational Safety and Health, 1978).

4. Summary of Data Reported and Evaluation

4.1 Exposure data

Antimony trioxide is produced from stibnite ores (antimony trisulfide) or as a by-product of lead smelting and production. It is used mainly in fire-retardant formulations for plastics, rubbers, textiles, paper and paints. It is also used as an additive in glass and ceramic products and as a catalyst in the chemical industry. Occupational exposure may occur during mining, processing and smelting of antimony ores, in glass and ceramics production, and during the manufacture and use of products containing antimony trioxide.

Antimony trisulfide is used in the production of explosives, pigments, antimony salts and ruby glass. Occupational exposure may occur during these processes and also during the mining, processing and smelting of ores containing antimony trisulfide.

4.2 Experimental carcinogenicity data

Antimony trioxide was tested for carcinogenicity by inhalation exposure in male and female rats of one strain and in female rats of another strain, producing a significant increase in the incidence of lung tumours (scirrhous and squamous–cell carcinomas and bronchioloalveolar tumours) in females in both studies. No lung tumour was seen in male rats.

Antimony ore concentrate (mainly antimony trisulfide) was tested for carcinogenicity by inhalation exposure in male and female rats of one strain, producing a significant increase in the incidence of lung tumours (scirrhous and squamous–cell carcinomas and bronchioloalveolar tumours) in females. No lung tumour was seen in males.

4.3 Human carcinogenicity data

The available data were inconclusive.

4.4   Other relevant data

Antimony trioxide causes pneumoconiosis in humans. One study of women exposed to dusts containing metallic antimony, antimony trioxide and antimony pentasulfide suggested that they may have had an excess incidence of premature births and spontaneous abortions and that their children's growth may have been retarded.

Antimony trioxide induced DNA damage in bacteria. (See Appendix 1.)

4.5   Evaluation[1]

There is *inadequate evidence* for the carcinogenicity of antimony trioxide and antimony trisulfide in humans.

There is *sufficient evidence* for the carcinogenicity of antimony trioxide in experimental animals.

There is *limited evidence* for the carcinogenicity of antimony trisulfide in experimental animals.

Overall evaluations

Antimony trioxide *is possibly carcinogenic to humans (Group 2B)*.

Antimony trisulfide *is not classifiable as to its carcinogenicity to humans (Group 3)*.

5. References

American Conference of Governmental Industrial Hygienists (ACGIG) (1988) *Threshold Limit Values and Biological Exposure Indices for 1988–1989*, Cincinnati, OH, p. 11

Anzon, Inc. (1988) *Product Bulletin: Antimony Trioxide, TMS*, Philadelphia, PA

Arbeidsinspectie (Labour Inspection) (1986) *De Nationale MAC–Lijst 1986* [National MAC–list 1986] (*P145*), Voorburg, Ministry of Social Affairs, p. 7

Arbejdstilsynet (Labour Inspection) (1988) *Graensevaerdier for Stoffer og Materialer* [Limit Values for Materials] (*At–anvisning No. 3.1.0.2*), Copenhagen, p. 11

Asarco, Inc. (1988a) *Product Bulletin: Antimony Oxide*, New York

Asarco, Inc. (1988b) *Material Data Safety Sheet: Antimony Oxide*, New York

Belyaeva, A.P. (1967) The effect produced by antimony on the generative function (Russ.). *Gig. Tr. prof. Zabol.*, *11*, 32–37

Brieger, H., Semisch, C.W., III, Stasney, J. & Piatnek, D.A. (1954) Industrial antimony poisoning. *Ind. Med. Surg.*, *23*, 521–523

---

[1]For definitions of the italicized terms, see Preamble, pp. 27–30.

Cooper, D.A., Pendergrass, E.P., Vorwald, A.J., Mayock, R.L. & Brieger, H. (1968) Pneumoconiosis among workers in an antimony industry. *Am. J. Roentgenol. Radium Ther. nucl. Med.*, *103*, 495–508

Dernehl, C.U., Nau, C.A. & Sweets, H.H. (1945) Animal studies on the toxicity of inhaled antimony trioxide. *J. ind. Hyg. Toxicol.*, *27*, 256–262

Deutsche Forschungsgemeinschaft (German Research Society) (1988) *Maximale Arbeitsplatzkonzentrationen und Biologische Arbeitsstofftoleranzwerte 1988* [Maximal Concentrations in the Workplace and Biological Tolerance Values for Working Materials 1988] *(Report No. XXIV)*, Weinheim, VCH Verlagsgesellschaft mbH, p. 19

Direktoratet for Arbeidstilsynet (Directorate for Labour Inspection) (1981) *Administrative Normer for Forurensning i Arbeidsatmosfaere 1971* [Administrative Norms for Pollution in Work Atmosphere 1981], *(No. 361)*, Oslo, p. 7

Donaldson, H. & Cassady, M. (1979) *Environmental Exposure to Airborne Contaminants in the Antimony Industry 1975–1976 (DHEW (NIOSH) Publication No. 79–140)*, Cincinnati, OH, National Institute for Occupational Safety and Health

Drake, G.J., Jr (1980) Flame retardants for textiles. In: Mark, H.F., Othmer, D.F., Overberger, C.G., Seaborg, G.T. & Grayson, N., eds, *Kirk–Othmer Encyclopedia of Chemical Technology*, 3rd ed., Vol. 10, New York, John Wiley & Sons, pp. 420–444

Dunn, J.T. (1928) A curious case of antimony poisoning. *Analyst*, *53*, 532–533

Eller, P.M., ed. (1985) *NIOSH Manual of Analytical Methods*, 3rd ed., 1st Suppl. *(DHHS (NIOSH) Publ. No. 84–100)*, Washington DC, US Government Printing Office, pp. 8005-1–8005-5

Flury, F. (1927) Toxicology of antimony (Ger.). *Nauyn–Schmiedeberg's Arch. exp. Pathol. Pharmakol.*, *126*, 87–103

Freedman, L.D., Doak, G.O. & Long, G.G. (1978) Antimony compounds. In: Mark, H.F., Othmer, D.F., Overberger, C.G., Seaborg, G.T. & Grayson, N., eds, *Kirk–Othmer Encyclopedia of Chemical Technology*, 3rd ed., Vol. 3, New York, John Wiley & Sons, pp. 105–128

Grin, N.V., Govorunova, N.N., Bessmertny, A.N. & Pavlovich, L.V. (1987) Experimental study of embryotoxic effect of antimony oxide (Russ.). *Gig. Sanit.*, *10*, 85–86

Gross, P., Brown, J.H.U., Westrick, M.L., Srsic, R.P., Butler, N.L. & Hatch, T.F. (1955a) Toxicologic study of calcium halophosphate phosphors and antimony trioxide. I. Acute and chronic toxicity and some pharmacologic aspects. *Arch. ind. Health*, *11*, 473–478

Gross, P., Westrick, M.L., Brown, J.H.U., Srsic, R.P., Schrenk, H.H. & Hatch, T.F. (1955b) Toxicologic study of calcium halophosphate phosphors and antimony trioxide. II. Pulmonary studies. *Arch. ind. Health*, *11*, 479–486

Groth, D.H., Stettler, L.E., Burg, J.R., Busey, W.M., Grant, G.C. & Wong, L. (1986) Carcinogenic effects of antimony trioxide and antimony ore concentrate in rats. *J. Toxicol. environ. Health*, *18*, 607–626

Guzman, J., Costabel, U., Orlowska, M., Schmitz-Schumann, M. & Freudenberg, N. (1986) Lung fibrosis after exposure to antimony trioxide dust (Abstract; Ger.). *Zbl. allg. Pathol. Anat.*, *131*, 278

Hawley, G.G. (1981) *Condensed Chemical Dictionary*, 10th ed., New York, Van Nostrand Reinhold Co., pp. 81–82

Health and Safety Executive (1987) *Occupational Exposure Limits 1987 (Guidance Note EH 40/87)*, London, Her Majesty's Stationery Office, p. 9

Hiraoka, N. (1986) The toxicity and organ–distribution of antimony after chronic administration to rats. *J. Kyoto pref. med. Coll.*, *95*, 997–1017

IARC (1988) *Information Bulletin on the Survey of Chemicals Being Tested for Carcinogenicity*, No. 13, Lyon, p. 144

Institut National de Recherche et de Sécurité (National Institute for Research and Safety) (1986) *Valeurs Limites pour les Concentrations des Substances Dangereuses Dans l'Air des Lieux de Travail* [Limit Values for Concentrations of Dangerous Substances in the Air of Work Places] *(ND 1609-125-86)*, Paris, p. 557

International Labour Office (1984) *Occupational Exposure Limits for Airborne Toxic Substances*, 2nd rev. ed. *(Occupational Safety and Health Series No. 37)*, Geneva, pp. 44-45

Kanematsu, N., Hara, M. & Kada, T. (1980) Rec assay and mutagenicity studies on metal compounds. *Mutat. Res.*, 77, 109-116

Karajović, D., Potkonjak, V. & Gospavić, J. (1960) Silicoantimonosis (Ger.). *Arch. Gewerbepathol. Gewerbehyg.*, 17, 651-665

Klučík, I. & Kemka, R. (1960) The excretion of antimony in workers in antimony metallurgical works (Czech.). *Prac. Lek.*, 12, 133-138

Klučík, I. & Ulrich, L. (1960) Electrocardiographic examinations of workers in an antimony metallurgical plant (Czech.). *Prac. Lek.*, 12, 236-243

LeSota, S. (1978) *Paint/Coatings Dictionary*, Philadelphia, PA, Federation of Societies for Coatings Technology, p. 33

Llewellyn, T.O. & Isaac, E. (1988) *Antimony Quarterly, Mineral Industry Surveys*, Washington DC, US Department of the Interior, Bureau of Mines, pp. 1-5

Lodge, J.P., Jr, ed. (1989) *Methods of Air Sampling and Analysis*, 3rd ed., Chelsea, MI, Lewis Publishers, pp. 83-92, 143-150, 226-230, 357-360, 623-638

Lüdersdorf, R., Fuchs, A., Mayers, P., Skulsuksai, G. & Schäcke, G. (1987) Biological assessment of exposure to antimony and lead in the glass-producing industry. *Int. Arch. occup. environ. Health*, 59, 469-474

Lyons, J.W. (1980) Flame retardants. In: Mark, H.F., Othmer, D.F., Overberger, C.G., Seaborg, G.T. & Grayson, N., eds, *Kirk-Othmer Encyclopedia of Chemical Technology*, 3rd ed., Vol. 10, New York, John Wiley & Sons, pp. 348-354

Mannsville Chemical Products Corp. (1981) *Chemical Products Synopsis: Antimony Oxide*, Cortland, NY

Mannsville Chemical Products Corp. (1985) *Chemical Products Synopsis: Antimony Oxide*, Cortland, NY

McCallum, R.I. (1963) The work of an occupational hygiene service in environmental control. *Ann. occup. Hyg.*, 6, 55-64

McCallum, R.I. (1967) Detection of antimony in process workers' lungs by X-radiation. *Trans. Soc. occup. Med.*, 17, 134-138

McCallum, R.I., Day, M.J., Underhill, J. & Aird, E.G.A. (1971) Measurement of antimony oxide dust in human lungs *in vivo* by X-ray spectrophotometry. In: Walton, W.H., ed., *Inhaled Particles III*, Old Woking, UK, Unwin Bros, pp. 611-619

Monier-Williams, G.W. (1934) *Antimony in Enamelled Hollow-ware*, London, His Majesty's Stationery Office

National Institute for Occupational Safety and Health (1974) *US National Occupational Exposure Survey 1972-74*, Cincinnati, OH

National Institute for Occupational Safety and Health (1978) *Criteria for a Recommended Standard ... Occupational Exposure to Antimony*, Cincinnati, OH

National Institute for Occupational Safety and Health (1983) *US National Occupational Exposure Survey 1981-1983*, Cincinnati, OH

National Swedish Board of Occupational Safety and Health (1987) *Hygienska Gränsvärden* [Hygienic Limit Values] *(Ordinance 1987:12)*, Solna, p. 12

Oliver, T. (1933) The health of antimony oxide workers. *Br. med. J., i*, 1094–1095

Palencia, C.M. & Mishra, C.P. (1986) *Antimony Availability – Market Economy Countries, A Minerals Availability Appraisal (Bureau of Mines Information Circular 9098)*, Washington DC, US Department of the Interior

Paschoud, J.-M. (1964) Clinical notes on occupational contact eczemas from arsenic and antimony (Fr.). *Dermatologica, 129*, 410–415

Potkonjak, V. & Pavlovich, M. (1983) Antimonosis: a particular form of pneumoconiosis. I. Etiology, clinical and X-ray findings. *Int. Arch. occup. environ. Health, 51*, 199–207

Potkonjak, V. & Vishnjich, V. (1983) Antimoniosis: a particular form of pneumoconiosis. II. Experimental investigation. *Int. Arch. occup. environ. Health, 51*, 299–303

Pough, F.H. (1960) *A Field Guide to Rocks and Minerals*, 3rd ed., Boston, MA, Houghton Mifflin Co., p. 102

Renes, L.E. (1953) Antimony poisoning in industry. *Arch. ind. Hyg. occup. Med., 7*, 99–108

Roberts, W.L., Rapp, G.R., Jr & Weber, J. (1974) *Encyclopedia of Minerals*, New York, Van Nostrand Reinhold Co., p. 582

Samsonov, G.W., ed. (1982) *The Oxide Handbook* (Engl. Transl.), New York, IFI/Plenum, p. 375

Smyth, H.F., Jr & Carpenter, C.P. (1948) Further experience with the range finding test in the industrial toxicology laboratory. *J. ind. Hyg. Toxicol., 30*, 63–68

Stevenson, C.J. (1965) Antimony spots. *Trans. St Johns Hosp. dermatol. Soc., 51*, 40–45

Työsuojeluhallitus (National Finnish Board of Occupational Safety and Health) (1987) *HTP–Arvot 1987* [TLV Values 1987] (*Safety Bulletin 25*), Helsinki, Valtion Painatuskeskus, p. 10

US Environmental Protection Agency (1983) *Methods for the Chemical Analysis of Water and Wastes (US EPA–600/4–79–020)*, Cincinnati, OH, Environmental Monitoring and Support Laboratory

US Environmental Protection Agency (1986) *Test Methods for Evaluating Solid Waste. Volume 1A: Laboratory Manual Physical/Chemical Methods (SW–846)*, 3rd ed., Washington, DC, Office of Solid Waste and Emergency Response

US Occupational Safety and Health Administration (1987) Toxic and hazardous substances. *US Code Fed. Regul. Title 29*, Part 1910.1000, p. 677

Watt, W.D. (1983) *Chronic Inhalation Toxicity of Antimony Trioxide: Validation of the Threshold Limit Value*, Detroit, MI, Wayne State University, PhD Thesis

Weast, R.C., ed. (1985) *CRC Handbook of Chemistry and Physics*, 66th ed., Cleveland, OH, CRC Press, p. B-74

Windholz, M., ed. (1983) *The Merck Index*, 10th ed., Rahway, NJ, Merck & Co., p. 104

TITANIUM DIOXIDE

1. Chemical and Physical Data

1.1 Synonyms

Chem. Abstr. Services Reg. No.: 13463-67-7 – Titanium dioxide
1317-70-0 – Anatase titanium dioxide
1317-80-2 – Rutile titanium dioxide
Chem. Abstr. Name: Titanium dioxide
IUPAC Systematic Name: Titanium dioxide
Synonyms: CI 77891; E 171; NCI-CO4240; Pigment White 6; titania; titanium (IV) oxide

1.2 Molecular formula and molecular weight

TiO_2 Mol. wt: 79.90

1.3 Chemical and physical properties of the pure substance

(a) *Description*: Fine white powder (Windholz, 1983); crystal structure is tetragonal with the titanium ion octahedrally bonded to six oxygen ions; position of octahedra in the lattice and number of molecules per unit cell differ for anatase and rutile forms: anatase titanium dioxide contains four molecules per unit cell, rutile contains only two (Schiek, 1982).

(b) *Density*: Anatase, 3.84 g/cm³; rutile, 4.26 g/cm³ (Weast, 1985)

(c) *Spectroscopy*: X-ray diffraction patterns for anatase and rutile titanium dioxide have been reported (Roberts *et al.*, 1974).

(d) *Refractive index*: Anatase, 2.554, 2.493; rutile, 2.616, 2.903 (Weast, 1985)

(e) *Solubility*: Soluble in sulfuric acid and alkalis; insoluble in water (Weast, 1985)

1.4 Technical products and impurities

Trade names: A-Fil Cream; Atlas white titanium dioxide; Austiox; Bayertitan; Calcotone White T; Cosmetic White C47-5175; Cosmetic White C47-9623; C-Weiss 7; Flamenco;

Hombitan; Horse Head A-410; Horse Head A-420; Horse Head R-710; KH 360; Kronos titanium dioxide; Levnox White RKB; Rayox; Runa RH20; Rutile; Tichlor; Tiofine; Tiona T.D.; Tioxide; Tipaque; Ti-Pure; Titafrance; Titandioxid; Titanox; Titanox 2010; Trioxide(s); Tronox; Unitane products (various); 1700 White; Zopaque

The technical products that incorporate titanium dioxide require a component that is essentially free from coloured impurities in order to produce the desired whitening and opacifying effect. International standards have been established for four types of titanium dioxide pigments. Type I (minimum, 94% titanium dioxide) is an anatase, freely chalking pigment used in white interior and exterior house paints, chalking being the formation of a layer of loose pigment powder on the surface of weathered paint film (Schurr, 1981). Type II (minimum, 92% titanium dioxide) is a rutile pigment with medium chalking resistance used in varying amounts in all types of interior paints, enamels and lacquers. Type III (minimum, 80% titanium dioxide) is also a rutile pigment with medium chalking resistance, used principally in alkyd and emulsion flat wall paints. Type IV (minimum, 80% titanium dioxide) is another rutile pigment, but with high chalking resistance; it is used in exterior paints and has excellent durability and gloss retention (American Society for Testing and Materials, 1988). Typical median particle sizes for anatase and rutile pigments range from 0.2–0.3 μm (LeSota, 1978; Schurr, 1981).

Aluminium, silicon and zinc oxides may be added to either form of the pigment to enhance specific properties such as increased dispersibility and ultraviolet light resistance. Titanium dioxide pigments must remain free of extenders such as barium sulfate, clay, magnesium silicate and calcium carbonate; however, the pigment may be extended by blending with anhydrous calcium sulfate (Lowenheim & Moran, 1975; American Society for Testing and Materials, 1988).

2. Production, Use, Occurrence and Analysis

2.1 Production and use

(a) Production

Titanium dioxide is the principal white pigment used commercially, due to its high refractive index, its ease of dispersion into a variety of matrices, and its inertness towards those matrices during processing and throughout product life (Considine, 1974).

Two main processes exist for making titanium dioxide pigments: the sulfate process and the chloride process. The sulfate process, the older of the two, was first used in Europe and the USA around 1930 and was the primary process until the early 1950s, when the chloride process was developed. By 1981, the chloride process accounted for approximately 78% of US production and by 1982 for 68% of world titanium dioxide pigment production (Considine, 1974; Lynd & Lefond, 1983).

In the sulfate process, rutile or anatase titanium dioxide is produced by digesting ilmenite (iron titanate) or titanium slag with sulfuric acid. The major concern in selecting the starting material for use in the process is that it contain as little as possible of impurities such as chromium, vanadium, manganese, niobium and phosphorus, which impair pigment properties. Ilmenite containing as little as 40% titanium dioxide can be used to produce pigment-grade titanium dioxide. Titanium slag may also be used as the starting material. Slag is produced by smelting ilmenite in an electric furnace and typically contains about 70% titanium dioxide, although concentrations may reach 85% (Considine, 1974; Lowenheim & Moran, 1975; Lynd & Lefond, 1983). Some typical materials used for pigment production throughout the world are shown in Table 1.

The sulfate process is a batch process in which concentrated sulfuric acid is added to ground ilmenite or titanium slag in proportions of 1.5:1 (acid to ore). An organic flocculent or antimony oxide may be added to induce aggregation of suspended titanyl and iron sulfates into a solid porous cake. The cake is dissolved in a dilute acid solution to release the sulfate agglomeration into solution. If necessary, scrap iron is added to reduce iron [III] to iron [II]. Also during this step, small amounts of titanium [IV] are reduced to titanium [III] to prevent later oxidation of iron [II]. The solution is clarified by settling and filtration. The resulting mother liquor is concentrated and subjected to steam for 6 h. Seed crystals may be added to aid nucleation. About 95% of the titanium in the mother liquor is hydrolysed to titanium hydrate or metatitanic acid (H_2TiO_3), which is collected on a filter and washed. The final filter cake is calcined at 900–1000°C to form titanium dioxide. The product is ground, quenched and dispersed in water; the coarse particles are separated in the thickener, reground and filtered; and the cake is dried in a rotary steam dryer and pulverized. The resulting product is anatase titanium dioxide. The rutile form is made by seeding the mother liquor with rutile seed crystals and conditioning the precipitated pigment with phosphates, potassium, antimony, aluminium or zinc compounds prior to calcination. The recovery of pigment-grade titanium dioxide in sulfate process plants is approximately 80% (Lowenheim & Moran, 1975; Lynd & Lefond, 1983; Lynd, 1985).

The chloride process is a continuous process that requires ores with a high content of titanium dioxide or concentrates such as natural or synthetic rutile. Natural rutile contains approximately 95% titanium dioxide; synthetic rutile or ilmenite (iron titanate) concentrates must have a minimum titanium dioxide content of 60% to produce economical yields of pigment in this process. The titanium dioxide content in ilmenite may be increased by reducing iron to its elemental form, followed by chemical or physical separation. Another method of enrichment is reducing iron [III] to iron [II] and chemically leaching it out of the mineral. A third method involves prior selective chlorination to remove iron and other impurities (Considine, 1974; Lowenheim & Moran, 1975; Lynd & Lefond, 1983).

In the chloride process, ore is ground and mixed with coke in a fluidized or static bed reactor and chlorinated at temperatures of 850–1000°C. Titanium tetrachloride is produced, along with chlorides of impurities present in the starting material, which include chlorides of iron, vanadium and silicon; these are removed chemically and through fractional distillation. Hydrogen chloride and carbon dioxide are present after chlorination and are vented prior to fractional distillation. Conversion to titanium dioxide is accomplished by burning titanium tetrachloride with air or oxygen at temperatures of 1200–1370°C. The resulting fine-grained oxide is sometimes calcined at about 500–600°C to remove any residual

Table 1. Composition of typical commercial ilmenite concentrates and titaniferous slag[a] (weight percent)

Material	USA		Australia		Norway	India		Malaysia (Amang[c])	Canada (Québec)	South Africa (Richards Bay slag)
	New York	Florida	Company A	Company B		Quilon deposit[b]	MK deposit			
TiO_2 (total)	46.1	64.00	54.4	55.4	45.0	60.6	54.2	53.1	70–74	85.0 (min)
Ti_2O_3									10–15	25.0 (max)
Fe_2O_3	6.7	28.48	19.0	11.1	12.5	24.2	14.2	8.7	12–15	
FeO	39.3	1.33	19.8	22.5	34.0	9.3	26.6	33.6	4–6	
Al_2O_3	1.4	1.23	1.5		0.6	1.0	1.3		3.5–5	
SiO_2	1.5	0.28	0.7	1.4	2.8	0.7				
CaO	0.5	0.007	0.04		0.25				1.2 (max)	0.15 (max)
MgO	1.9	0.20	0.45		5.0	0.9	1.0		4.5–5.5	1.3 (max)
Cr_2O_3	0.009		0.2	0.03	<0.076	0.12	0.07	0.005	0.25	0.3 (max)
V_2O_5	0.05		0.12	0.13	0.16	0.15	0.16	0.02	0.5–0.6	0.6 (max)
ZrO_2	0.01					0.9	0.8			
S	0.6		<0.01		<0.05	0.21	0.12			
P_2O_5	0.008	0.12	0.02		<0.04			0.085	0.03–0.10	
MnO	0.5		1.4		0.25	0.4	0.4	4.0	0.025 (max)	
H_2O (loss on ignition)	1.3		0.4			2.0	0.3		0.2–0.3	2.5 (max)
Rare earths						trace	0.12			
C	0.22				<0.055				0.03–0.10	

[a]From Lynd & Lefond (1983)

[b]MK, Manavalakwuchi area

[c]Amang, a crude mixture of heavy minerals that must be treated further to recover ilmenite

chlorine or hydrogen chloride. These gases are separated, and the chlorine is collected and recirculated to the chlorinator. Approximately 90% of the chlorine may be recycled. Aluminium chloride is added to titanium tetrachloride to assure near-total conversion to rutile titanium dioxide. A typical yield from this process is 90% (Lowenheim & Moran, 1975; Lynd & Lefond, 1983; Lynd, 1985).

Current worldwide demand for titanium dioxide is about 2.8 million tonnes (Anon., 1988a,b,c). Production volumes in 1978–86 in several countries are given in Table 2.

Table 2. Titanium dioxide pigment production by country in 1978–86 (thousand tonnes)[a]

Country	1978	1979	1980	1981	1982	1983	1984	1985	1986
Brazil	NA	NA	29.9	32.8	30.9	45.3	45.0	NA	NA
Czechoslavakia	19.1	18.8	16.6	16.7	20.0	21.0	20.4	NA	NA
Finland	NA	270.7	355.2	402.1	366.3	420.7	502.2	NA	NA
India	9.9	NA	NA	NA	NA	NA	NA	NA	NA
Italy[b]	60.0	59.0	–	–	–	–	–	–	–
Japan	171.4	185.4	172.8	176.2	184.0	195.9	204.7	217.7	222.9
Korea, Republic of	9.2	5.1	8.0	9.8	11.1	14.4	NA	NA	NA
Mexico	28.5	35.0	39.1	40.0	37.5	40.5	44.7	NA	NA
Spain	53.9	66.1	51.4	66.6	67.5	64.9	70.0	74.7	NA
UK	205.3	192.9	186.7	169.6	172.3	193.9	206.0	219.1	230.0
USA	635.8	673.1	659.4	690.2	598.6	689.3	757.3	780	844.7
USSR	7.1	7.1	6.0	4.7	5.0	5.0	6.4	NA	NA
Yugoslavia	19.2	19.6	19.6	21.9	21.4	25.6	21.7	NA	NA

[a]From Anon. (1988d,e); NA, not available
[b]Not produced after 1979

(b)   Use

The principal use of titanium dioxide is as a whitening and opacifying agent in paints, varnishes, lacquers, paper, plastics, ceramics, rubber and printing ink (Table 3). These industries accounted for 93–95% of the titanium dioxide pigment used in the USA during 1982–86 (Mannsville Chemical Products Corp., 1983; Lynd & Hough, 1986).

Titanium dioxide is the most common white synthetic pigment used in the paint industry. Modern paint plants often handle titanium dioxide as a slurry for convenience and to avoid airborne particulates (Schurr, 1981). In the paper industry, both anatase and rutile titanium dioxides are used to improve opacity. The total amount of pigment varies with the grade of paper but may comprise 2–40% of the final sheet, of which as much as 25% may be titanium dioxide. The pigment is dispersed during the pulping process and retained in the sheet during formation with organic polymers. The optical efficiency of titanium pigment is improved by synthetic silicas and silicates (Baum *et al.*, 1981). Paper is waterproofed by coating the sheets with polyethylene resins whitened with titanium dioxide (Locker, 1982).

Table 3. Percent distribution of titanium dioxide pigment used within US industries, 1982–86[a]

Industry	1982	1983	1984	1985	1986
Paint, varnish, lacquer	48.1	48.9	54.8	54.3	52.6
Paper	27.4	27.3	19.9	20.5	20.7
Plastics	12.7	13.2	15.4	16.2	15.8
Ceramics	1.2	1.0	1.0	0.7	2.2
Rubber	2.6	1.8	2.0	1.7	2.0
Printing inks	1.0	1.1	1.2	1.0	1.4
Other	7.0	6.7	5.7	5.6	5.3

[a]From Lynd & Hough (1986)

Titanium oxide is incorporated into various plastic products to confer opacity and whiteness, and also because it resists degradation by ultraviolet light and is chemically inert (Lynd, 1985).

Commercial use of anatase titanium dioxide as an opacifier for ceramic and enamel products began in 1946. Rutile pigments may also be used for this purpose (Friedberg, 1980; Lynd, 1985).

Titanium dioxide is also incorporated into rubber tyres to make whitewall tyres (Lynd, 1985). In printing inks, titanium dioxide is mixed with coloured pigments to add opacity or lighten the hue. In a typical nitrocellulose ink formulation, titanium pigment is present at 35 wt%. An acrylic–based ink formulation contained 30 wt% titanium dioxide (Burachinsky *et al.*, 1981).

Other uses for titanium dioxide are as a catalyst in the production of alcohol fuels (Klass, 1984), as a delusterant in a variety of synthetic fibres at up to 2 wt% (Davis & Hill, 1982), as a component of flame–retardant formulations for wood (Wegner *et al.*, 1984), in cosmetics as a physical sunscreening agent (Isacoff, 1979), as a component of gums, resins and waxes used for making dental impressions (Paffenbarger & Rupp, 1979), and as a pigment in floor coverings, leather products and soaps (Mannsville Chemical Products Corp., 1983).

(c)   Regulatory status and guidelines

Occupational exposure limits for titanium dioxide in 13 countries or regions are presented in Table 4.

Titanium dioxide may be used as a food colour additive with the following specifications limiting impurities: antimony compounds, < 100 mg/kg; zinc compounds, < 50 mg/kg; soluble barium compounds, < 5 mg/kg; and hydrochloric acid–soluble compounds, < 3.4 g/kg (Commission of the European Communities, 1962).

Table 4. Occupational exposure limits for titanium dioxide[a]

Country or region	Year	Concentration[b] (mg/m^3)	Interpretation[c]
Austria	1985	8	TWA
Denmark	1988	6	TWA
Finland	1987	10	TWA
German Democratic Republic	1985	5	TWA
		10	STEL
Germany, Federal Republic of	1988	6[d]	TWA
Netherlands	1986	10	TWA
Norway	1981	10	TWA
Switzerland	1985	6	TWA
Taiwan	1985	S 10	TWA
UK	1987		
Total inhalable dust		10	TWA
Respirable dust		5	TWA
USA[e]			
OSHA	1985	15	TWA
ACGIH	1988	10[f]	TWA
USSR	1986	10	TWA
Venezuela	1985	20	Ceiling

[a]From Direktoratet for Arbeidstilsynet (1981); Arbeidsinspectie (1986); Institut National de Recherche et de Sécurité (1986); Cook (1987); Health and Safety Executive (1987); Työsuojeluhallitus (1987); American Conference of Governmental Industrial Hygienists (1988); Arbejdstilsynet (1988); Deutsche Forschungsgemeinschaft (1988)

[b]S, skin notation

[c]TWA, 8-h time–weighted average; STEL, short–term exposure limit

[d]Total dust containing no asbestos and < 1% free silica

[e]OSHA, Occupational Safety and Health Administration; ACGIH, American Conference of Governmental Industrial Hygienists

[f]Measured as fine dust

2.2 Occurrence

(a) Natural occurrence

Titanium dioxide occurs naturally in three crystalline forms: anatase, rutile and brookite. Only anatase and rutile are of commercial importance. Rutile can be mined directly, but anatase is obtained through the processing of ilmenite, a natural iron titanate (Lynd, 1985).

Rutile is a widespread accessory mineral found in high–grade metamorphic gneisses and schists and in igneous rocks. It also occurs in black sand deposits in many parts of the world. Typically, the composition of rutile is 95% titanium dioxide, the remaining 5% being silicon, chromium, vanadium, aluminium and iron oxides (Lynd & Lefond, 1983).

Ilmenite is a common accessory grain in igneous rocks such as anorthosites, gabbros and basic lavas; it may also occur as an intergrowth in magnetite and haematite. Weathering and alterations along grain boundaries result in the removal of iron from the ilmenite lattice, producing leucoxene or pseudorutile. Ilmenite is also found in sand deposits (Lynd & Lefond, 1983).

The principal producing countries for rutile and ilmenite are Australia, Brazil, Canada, China, Finland, India, Malaysia, Mexico, Norway, Sierra Leone, South Africa, Sri Lanka and the USA (Lynd & Lefond, 1983).

(b) Occupational exposure

On the basis of a US National Occupational Exposure Survey, the National Institute for Occupational Safety and Health (1983) estimated that 1 270 000 workers were potentially exposed to titanium dioxide in the USA in 1981-83.

Although it is apparent that occupational exposure to titanium dioxide is extensive, there are few data on levels and sources of exposure (Santodonato et al., 1985), and the data available in the literature are reported as total dust or nuisance dust and not as titanium dioxide. Concentrations ranged from 10 to 400 mg/m^3 during the grinding of titanium dioxide pigment, but documentation of these levels was not provided (Elo et al., 1972). Long-term exposures to titanium dioxide dust in a titanium pigment production factory occasionally exceeded 10 mg/m^3, and exposures greater than 10 mg/m^3 were common during the repair of production machinery (Rode et al., 1981).

2.3 Analysis

No information was available to the Working Group on methods for the quantitative determination of titanium dioxide in environmental samples. Occupational exposures to titanium dioxide have been estimated gravimetrically as total or respirable dust (Lee et al., 1986).

3. Biological Data Relevant to the Evaluation of Carcinogenic Risk to Humans

3.1 Carcinogenicity studies in animals

(a) Oral administration

Mouse: Groups of 50 male and 50 female B6C3F1 mice, five weeks old, were fed diets containing 0, 2.5% or 5.0% titanium dioxide (anatase; purity, $\geq$98%) daily for 103 weeks. Mice were killed at 109 weeks of age, one week after exposure was stopped. At terminal sacrifice, there was no significant difference in survival between treated and control males: 32, 40 and 40 controls, low-dose and high-dose males were still alive at that time. In female mice, a dose-related trend for decreased survival was significant in treated groups ($p = 0.001$, Tarone test); survival at terminal sacrifice was 45, 39 and 33 among control, low-dose and high-dose females, respectively. No difference in body weights between treated and control groups and no significant increase in the incidence of tumours was observed in treated mice of either sex (National Cancer Institute, 1979).

Rat: Groups of 50 male and 50 female Fischer 344 rats, nine weeks old, were fed diets containing 0, 2.5% or 5.0% titanium dioxide (anatase; purity, $\geq 98\%$) daily for 103 weeks. Rats were killed at 113 weeks of age, one week after exposure was stopped. There was no significant difference in survival between treated and control groups of either sex; survival of males at terminal sacrifice was 31, 37 and 36 among control, low–dose and high–dose groups, respectively, and that of females was 36, 36 and 34 in control, low–dose and high–dose groups, respectively. No difference in body weights between treated and control groups and no significant increase in the incidence of tumours was observed in treated mice of either sex (National Cancer Institute, 1979).

(b) Inhalation/intratracheal administration

Rat: Groups of 50 male and 50 female Sprague–Dawley rats, eight weeks of age, were exposed by inhalation to 0 or 15.95 mg/m^3 titanium dioxide [purity unspecified] for 6 h per day on five days per week for 12 weeks. After 140 weeks (128 weeks after the end of exposure), all surviving rats were sacrificed. Average survival was 116 and 113 weeks for control and treated males and 114 and 120 weeks for control and treated females, respectively. At terminal sacrifice, 39 control males, 44 treated males, 45 control females and 45 treated females were still alive. No difference in body weights between treated and control groups and no significant increase in the incidence of tumours was observed in treated mice of either sex (Thyssen *et al.*, 1978). [The Working Group noted the short duration of exposure and the relatively low exposure level.]

Groups of 100 male and 100 female CD rats, five weeks of age, were exposed by inhalation to 0, 10, 50 or 250 mg/m^3 titanium dioxide (rutile; 99% pure; 84% of dust particles of respirable size) for 6 h per day on five days per week for two years. At three, six and 12 months, five, five and ten rats of each sex at each dose, respectively, were removed for interim kills. No difference in mortality, body weight or clinical signs was observed. Nasal cavities were examined histologically, and no tumour was observed. Lung tumours were observed primarily in high–dose rats of each sex. The incidences of lung adenomas were: males – control, 2/79; low–dose, 1/71; mid–dose, 1/75; and high–dose, 12/77; females – control, 0/77; low–dose, 0/75; mid–dose, 0/74; and high–dose, 13/74. The incidences of cystic keratinizing squamous–cell carcinomas were: males – 0/79, 0/71, 0/75 and 1/77; females – 0/77, 1/75, 0/74 and 13/74. One anaplastic carcinoma occurred in a low–dose male. The lung tumours occurred in the bronchioloalveolar region, and no evidence of metastasis was observed. The authors noted difficulty in distinguishing between the squamous–cell carcinomas and keratinizing squamous metaplasia (Lee *et al.*, 1985a,b, 1986).

Hamster: Groups of 24 male and 24 female Syrian golden hamsters, six to seven weeks old, received intratracheal administrations of 0 or 3 mg titanium dioxide ([purity unspecified] particle size, 97% < 5 μm) in 0.2 ml saline once a week for 15 weeks. Animals were observed until spontaneous death and all control and treated hamsters died by 120 and 80 weeks, respectively, after the beginning of treatment. The respiratory tract and other organs with gross lesions were examined microscopically; no respiratory tract tumour occurred in treated hamsters, but two tracheal papillomas were found in untreated controls (Stenbäck *et al.*, 1976).

(c)   *Subcutaneous injection*

Rat: Groups of 20 male and 20 female Sprague-Dawley rats, 13 weeks old, received a single subcutaneous injection of 1 ml saline or 30 mg of one of three preparations of titanium dioxide ($\geq$99%, $\geq$95% or $\geq$85% pure) in 1 ml saline into the flank. All rats were observed until spontaneous death, which occurred as late as 136, 126, 146 and 133 weeks in the control and titanium dioxide-treated groups, respectively. No tumour was observed at the site of the injection in any group (Maltoni *et al.*, 1982). [The Working Group noted the inadequate reporting of the study.]

(d)   *Intraperitoneal injection*

Mouse: Groups of 30 or 32 male Marsh-Buffalo mice, five to six months old, received a single intraperitoneal injection of 0 or 25 mg titanium dioxide (purity, $\geq$98%; manually ground) in 0.25 ml saline. All survivors (ten control and 13 treated mice) were killed 18 months after treatment. No difference in the incidence of local or distant tumours was observed between treated and control animals (Bischoff & Bryson, 1982). [The Working Group noted the small number of animals used.]

Rat: As part of a large study on various dusts, three groups of female Wistar rats, nine, four and five weeks of age, received intraperitoneal injections of granular titanium dioxide [purity unspecified] in 2 ml 0.9% sodium chloride solution. The first group received a total dose of 90 mg per animal by five weekly injections; the second group received a single injection of 5 mg per animal; and the third group received three weekly injections of 2, 4 and 4 mg per animal. One concurrent group of 32 five-week-old Wistar rats received a single injection of saline alone. Average lifespans were 120, 102, 130 and 120 weeks. No intra-abdominal tumour was reported in 47 and 32 rats from the second and third groups that were examined; six of 113 rats examined from the first group had sarcomas, mesotheliomas or carcinomas of the abdominal cavity [numbers unspecified]. Two controls had abdominal tumours. In a similar experiment with female Sprague-Dawley rats given single intraperitoneal injections of 5 mg per animal titanium dioxide, 2/52 rats developed abdominal tumours (average lifespan, 99 weeks). Controls were not available for comparison (Pott *et al.*, 1987).

(e)   *Administration with known carcinogens*

Hamster: Groups of 24 male and 24 female Syrian golden hamsters, six to seven weeks old, received intratracheal administrations of 3 mg titanium dioxide ([purity unspecified] particle size, 97% < 5 μm) plus benzo[*a*]pyrene in 0.2 ml saline, or benzo[*a*]pyrene in saline (controls) once a week for 15 weeks. Animals were observed until spontaneous death; all control and treated hamsters had died by 100 and 70 weeks, respectively. In the 48 hamsters [number of tumours per sex unspecified] treated with titanium dioxide plus benzo[*a*]pyrene, tumours occurred in the larynx (11 papillomas, five squamous-cell carcinomas), trachea (three papillomas, 14 squamous-cell carcinomas, one adenocarcinoma) and lung (one adenoma, one adenocarcinoma, 15 squamous-cell carcinomas, one anaplastic carcinoma). Two papillomas occurred in the larynx of benzo[*a*]pyrene-treated controls. In the same study, ferric oxide and benzo[a]pyrene induced a similar spectrum of tumours as that induced by the combination with titanium dioxide; aluminium oxide and benzo[a]pyrene produced no increase in tumour incidence compared with benzo[a]pyrene-treated controls (Stenbäck et al., 1976).

3.2 Other relevant data

(a) Experimental systems

The toxicology of titanium dioxide has been reviewed (World Health Organization, 1982).

(i) Absorption, distribution, excretion and metabolism

The pattern of deposition of titanium dioxide (anatase) dust in the lungs of rats was similar to that seen with other particles (Ferin et al., 1983). After rats were exposed to 16.5 mg/m³ anatase and 19.3 mg/m³ rutile aerosols (mass median aerodynamic diameters, about 1 μm) for 7 h, the half-lives of pulmonary clearance were 51 and 53 days, respectively (Ferin & Oberdörster, 1985). Pulmonary clearance of titanium dioxide was significantly decreased after long-term exposure to 1 ppm (2.6 mg/m³) sulfur dioxide (Ferin & Leach, 1973).

Following intratracheal instillation of titanium dioxide to mice and rapid partial removal of particles by ciliary clearance within the first 15 min, subsequent pulmonary clearance had a half-life of about 20 days; the slow phase was assumed to be due to uptake of the particles and removal by macrophages (Finch et al., 1987).

After exposure of rats by inhalation to 10, 50 and 250 mg/m³ titanium dioxide (rutile; mass median aerodynamic diameter, 1.5–1.7 μm) for 6 h per day on five days per week for two years, inhaled particles were found to be engulfed by macrophages, and dense accumulation of dust-laden macrophages was seen in perivascular and peribronchial lymphoid tissue. Massive accumulation of these dust-containing cells was also seen in tracheobronchial lymph nodes and, to a lesser extent, in cervical lymph nodes. The occurrence of dust particles in mesenteric lymph nodes, in the liver and in the spleen suggests that small amounts could have entered the general circulation from the lungs (Lee et al., 1985a). Also after intravenous injection to rats, titanium dioxide particles were found to accumulate in abdominal lymph nodes (coeliac nodes) which drain the liver (Huggins & Froehlich, 1966). Pulmonary accumulation increased considerably in rats exposed by inhalation to 250 mg/m³ as compared to 10 and 50 mg/m³, indicating that pulmonary clearance mechanisms can be overloaded at the highest dose (Lee et al., 1986).

(ii) Toxic effects

Titanium dioxide pigments (anatase and others) caused significant mobilization of peritoneal macrophages when injected into the peritoneal cavity of mice (Nuuja et al., 1982). They decreased the level of acid phosphatase in mouse peritoneal macrophages in vitro; silica-coated particles were the most effective (Nuuja & Arstila, 1982). In both studies, titanium dioxide had minimal cytotoxicity in comparison to quartz and asbestos. In vivo, titanium dioxide (anatase and rutile) failed to induce proline hydroxylase in the lungs of rats four weeks after they had been exposed by inhalation (Zitting & Skyttä, 1979).

Anatase, rutile and rutile containing trace amounts of nickel or chromium had no fibrotic potential in either an in-vitro test for cell viability using bovine alveolar macrophages
or following instillation into rat lungs *in vivo*; α–quartz gave positive results in both assays
(Richards *et al.*, 1985). Similarly, injection of an untreated anatase form of titanium dioxide
(particle size, 0.8–16 μm) into the pleura of rats caused no pleural effusion and only a few
strands of connective tissue surrounding the collections of macrophages (Grasso *et al.*, 1983).
Additional studies *in vivo* have all failed to demonstrate any fibrotic potential in rats or rabbits (Grandjean *et al.*, 1956; Christie *et al.*, 1963; Dale, 1973; Sethi *et al.*, 1973; Ferin &
Oberdörster, 1985); however, one study showed that intratracheal instillations of 3 mg titanium dioxide to hamsters once a week for 15 weeks resulted in slight pulmonary inflammation and, subsequently, interstitial fibrosis (Stenbäck *et al.*, 1976). Inflammatory changes and
formation of collagen fibres were also seen after intratracheal instillation into rats of ilmenite (iron titanate) or titanium dioxide dusts (Shevtsova, 1968).

Following overloading of lung clearance in rats exposed to 250 mg/m^3 rutile for 6 h per
day on five days per week for two years, there was massive accumulation of dust-laden macrophages; free particles and cellular debris were found in the alveoli, and alveolar proteino-
sis and cholesterol granulomas developed. Lung weights increased, and white patches of
accumulated dusts were seen in the lungs at autopsy (Lee *et al.*, 1985b, 1986).

(iii) *Effects on reproduction and prenatal toxicity*

No data were available to the Working Group.

(iv) *Genetic and related effects*

Titanium dioxide did not induce differential killing in DNA repair-proficient compared to repair–deficient strains of *Bacillus subtilis* rec$^{+/-}$ [details not given] (Kada *et al.*,
1980).

Titanium dioxide was not mutagenic to *Salmonella typhimurium* TA1535, TA1537,
TA1538, TA97, TA98 or TA100 (Dunkel *et al.*, 1985; Zeiger *et al.*, 1988) or to *Escherichia coli*
WP2, either in the presence or absence of an exogenous metabolic system from the livers of
uninduced and Aroclor–induced rats, mice and Syrian hamsters (Dunkel *et al.*, 1985).

It did not induce morphological transformation in Syrian hamster embryo cells (Di Paolo & Casto, 1979) and did not enhance transformation of Syrian hamster embryo cells by the
SA7 adenovirus (Casto *et al.*, 1979).

(b) *Humans*

(i) *Absorption, distribution, excretion and metabolism*

Pulmonary retention of titanium dioxide dusts has been documented in several studies.
The lungs usually contain higher levels of titanium than any other organ analysed (Schroeder
et al., 1963). In lung tissue obtained from one autopsy and two thoracotomies performed on
workmen employed in a factory that processed titanium dioxide, titanium dioxide was found
in lysosomes of alveolar macrophages and in lymphatic macrophages (Elo *et al.*, 1972). Dust-
laden macrophages were seen in sputum samples from current titanium dioxide workers and
from retired workers who had left the factory three years earlier; in lung biopsies from retired workers, macrophages were loaded with titanium dioxide and with silicon, which possi-

bly originated from the titanium dioxide coating (Määttä & Arstila, 1975). Massive deposition of rutile was also reported at autopsy in the lungs of a man who had been employed at a titanium dioxide production plant until four years before his death (Rode *et al.*, 1981). Titanium dioxide deposition has also been reported in regional lymph nodes (Schmitz–Moormann *et al.*, 1964; Yamadori *et al.*, 1986).

The pattern of regional deposition of titanium dioxide in the lungs and the relative contents in lymph nodes suggest that titanium dioxide is removed relatively slowly *via* the lymphatic system. The lungs of 35 miners contained an average of 71 mg titanium dioxide, but a total of only 4 mg was found in the regional lymph nodes (Einbrodt & Liffers, 1968).

(ii) *Toxic effects*

At an ilmenite (iron titanate) extraction plant, where workers were also exposed to rutile, chest X-rays (70 mm) were abnormal in 3/136 workers, of whom 24 had had more than ten years' exposure. The prevalence was similar to that (4/170) seen in a control group (Uragoda & Pinto, 1972). [The Working Group noted that the X-ray technique used would have detected only cases with severe pneumoconiosis.]

A man who was employed to pack titanium dioxide into cans developed pneumoconiosis after nine years of exposure; lung tissue examined five years later revealed slight fibrosis of interstitial tissue surrounding bronchioles and alveolar spaces (Yamadori *et al.*, 1986). In a study of lung tissue obtained from an autopsy and from two thoracotomies performed on titanium dioxide workers, deposits of titanium dioxide in interstitial lung tissue were also associated with cell destruction and slight fibrosis (Elo *et al.*, 1972). Määttä and Arstila (1975) suggested that the fibrogenic effect might be due to the presence of silicon compounds associated with the titanium dioxide dust. Another worker in titanium pigment production, however, had massive deposition of rutile in the lungs but no inflammation or fibrosis (Rode *et al.*, 1981). The absence of fibrotic changes has also been reported in a case after 15 years of exposure at a titanium dioxide mill (Schmitz–Moormann *et al.*, 1964).

As reported in an abstract, in a cross-sectional study of 207 workers producing titanium dioxide from ilmenite ore, the predominant signs were obstructive airway changes. In 26 workers, chest X-rays showed irregular or nodular opacities of limited extent; eight of these workers were known also to have been exposed to silica or asbestos (Daum *et al.*, 1977). [The Working Group noted that when titanium dioxide is produced from ilmenite ore the dried ore is digested by sulfuric acid, causing exposure to both sulfuric acid mist and titanium dioxide dust. The possible role of titanium dioxide could not be assessed.]

In a cross-sectional study of 209 titanium metal production workers, including 78 workers involved in the reduction process who were exposed to titanium tetrachloride vapour, titanium oxychloride and titanium dioxide particles, reductions in lung function (forced expiratory volume in one minute) were found. The authors noted that this finding could be due to exposure to titanium tetrachloride, which reacts violently with water to liberate heat and produce hydrochloric acid, titanium oxychloride and titanium dioxide. Pleural disease with plaques and pleural thickening was observed in 36 of the 209 workers, including eight of the 78 reduction process workers. Some cases were probably caused by prior exposure to asbestos; however, among workers not known to have been exposed to asbestos, the risk for pleural disease after more than ten years of employment was 3.8 times the risk in those who had been employed for fewer than five years. The authors noted that past exposure to asbestos at the titanium production facility may have contributed to the risk (Garabrant *et al.*, 1987).

A chest X-ray study of 336 workers at two titanium dioxide–producing plants showed 19 cases of pleural abnormalities (thickening/plaques), as compared to 3/62 among unexposed workers at the same plants. The odds ratio for chest X-ray abnormality associated with titanium dioxide exposure was 1.4. No lung fibrosis was observed. Exposures at the plants included titanium tetrachloride, potassium titanate and asbestos (Chen & Fayerweather, 1988).

A case of granulomatous lung disease was reported in a worker with possible exposure to titanium dioxide at an aluminium smelting plant where he worked near a firebrick furnace. A lymphocyte transformation test showed proliferative response to titanium chloride but not to any other metal tested, suggesting a possible link with titanium hypersensitivity (Redline *et al.*, 1986).

(iii) *Effects on fertility and on pregnancy outcome*

No data were available to the Working Group.

(iv) *Genetic and related effects*

No data were available to the Working Group.

3.3 Case reports and epidemiological studies of carcinogenicity to humans

Titanium dioxide was found in the lungs of three patients who died of cancers at various sites (Schmitz-Moormann *et al.*, 1964; Ophus *et al.*, 1979; Rode *et al.*, 1981; Yamadori *et al.*, 1986), one of which was a papillary adenocarcinoma in the right lung. The latter had been engaged in packing titanium dioxide for about 13 years prior to his death at the age of 53 and had smoked about 17 cigarettes per day for 40 years; he had been diagnosed at the age of 48 as having pneumoconiosis.

Chen and Fayerweather (1988) studied mortality and cancer incidence among 1576 male employees who had been exposed to titanium dioxide for more than one year in two plants in the USA. Information on cancer incidence was obtained from the company cancer registry, which was started in 1956. Information on deaths among active and pensioned employees was obtained from the company mortality registry, which was started in 1957: vital status was determined for about 94% of the cohort, and death certificates were available for about 94% of those known to be deceased. Observed numbers of incident cancer cases were compared with expected numbers based on company rates, and the observed numbers of deaths were compared with both company and US rates. Mortality from all cancers was lower than expected. For lung cancer, nine deaths were observed, with 17.3 expected on the basis of national rates (standardized mortality ratio (SMR), 52; [95% confidence interval, 24–99]) and 15.3 expected on the basis of company rates (59; [27–112]). There was a slight excess of incident cancer cases, 38 *versus* 32.6 (117; [83–160]), due mainly to ten cases of tumours of the genitourinary system *versus* 6.3 expected (159; [76–292]); there were eight cases of lung cancer with 7.7 expected (104; [45–205]). [The Working Group noted that details of exposure to titanium dioxide and other factors were not described, that cancer mortality and specific cancer sites were not reported in detail, that incident cancer cases only in actively employed persons were used for both observed and company reference rates, and that the numbers of incident cases were compared only with company rates.]

4. Summary of Data Reported and Evaluation

4.1 Exposures

Titanium dioxide is a white pigment produced mainly from ilmenite (iron titanate) and natural rutile (titanium dioxide). It is widely used in paints, paper, plastics, ceramics, rubber, inks and a variety of other products. Occupational exposure to titanium dioxide during its production, the production of paints, in painting trades and during other industrial use is likely to be extensive, but there is a paucity of data on levels, both occupational and environmental.

4.2 Experimental carcinogenicity data

Titanium dioxide was tested for carcinogenicity by oral administration in one strain of mice and in one strain of rats, by inhalation in two strains of rats, by intratracheal administration in one strain of hamsters, by subcutaneous injection in one strain of rats and by intraperitoneal administration in one strain of male mice and two strains of female rats. Increased incidences of lung adenomas in rats of both sexes and of cystic keratinizing lesions diagnosed as squamous–cell carcinomas in female rats were observed in animals that had inhaled the high but not the low doses of titanium dioxide. Oral, subcutaneous, intratracheal and intraperitoneal administration did not produce a significant increase in the frequency of any type of tumour in any species. Intratracheal administration of titanium dioxide in combination with benzo[a]pyrene to hamsters resulted in an increase in the incidence of benign and malignant tumours of the larynx, trachea and lungs over that in benzo[a]pyrene–treated controls.

4.3 Human carcinogenicity data

The only available epidemiological study provided inconclusive results.

4.4 Other relevant data

Titanium dioxide did not induce morphological transformation in mammalian cells or mutation in bacteria. (See Appendix 1.)

4.5   Evaluation[1]

There is *inadequate evidence* for the carcinogenicity of titanium dioxide in humans.

There is *limited evidence* for the carcinogenicity of titanium dioxide in experimental animals.

Overall evaluation

Titanium dioxide *is not classifiable as to its carcinogenicity to humans (Group 3).*

5. References

American Conference of Governmental Industrial Hygienists (1988) *Threshold Limit Values and Biological Exposure Indices for 1988–1989*, Cincinnati, OH, p. 36

American Society for Testing and Materials (1988) Standard specification for titanium dioxide pigments. In: Storer, R.A., Cornillit, J.L., Savini, D.F., Hart, S.H., Richardson, D., O'Brien, K.W., Stanton, A.L., Loux, L.A., Craig, R.E., Fanelle, C.S., Fazio, P.C., Winfree, M.S., Gutman, E.L., McGee, P.A., Wise, C.J. & Shupak, H.J., eds, *1988 Annual Book of ASTM Standards: Paint–Pigments, Resins, and Polymers*, Philadelphia, PA, pp. 100–101

Anon. (1988a) TiO$_2$ makers move to balance market. *Chem. Mark. Rep.*, *234*, 3, 29–30

Anon. (1988b) Prime pigments – titanium dioxide. *Chem. Mark. Rep.*, *234*, 30

Anon. (1988c) NL boosting TiO$_2$ stake with grassroots plant in US. *Chem. Mark. Rep.*, *234*, 3

Anon. (1988d) Facts & figures for the chemical industry. *Chem. Eng. News*, *66*, 34–82

Anon. (1988e) *CHEM–INTELL Database*, Chemical Intelligence Service, Dunstable, UK, Reed Telepublishing Limited

Arbeidsinspectie (Labour Inspection) (1986) *De Nationale MAC–Lijst 1986* [National MAC-list 1986] (*P145*), Voorburg, Ministry of Social Affairs and Work Inspection, p. 21

Arbejdstilsynet (Labour Inspection) (1988) *Graensevaerdier for Stoffer og Materialer* [Limit Values for Materials] (*At–anvisning No. 2.1.0.2*), Copenhagen, p. 31

Baum, G.A., Malcolm, E.W., Wahren, D., Swanson, J.W., Easty, D.B., Litvay, J.D. & Dugal, H.S. (1981) Paper. In: Mark, H.F., Othmer, D.F., Overberger, C.G., Seaborg, G.T. & Grayson, N., eds, *Kirk–Othmer Encyclopedia of Chemical Technology*, 3rd ed., Vol. 16, New York, John Wiley & Sons, pp. 777–778

Bischoff, F. & Bryson, G. (1982) Tissue reaction to and fate of parenterally administered titanium dioxide. I. The intraperitoneal site in male Marsh–Buffalo mice. *Res. Commun. chem. Pathol. Pharmacol.*, *38*, 279–290

Burachinsky, B.V., Dunn, H. & Ely, J.K. (1981) Inks. In: Mark, H.F., Othmer, D.F., Overberger, C.G., Seaborg, G.T. & Grayson, N., eds, *Kirk–Othmer Encyclopedia of Chemical Technology*, 3rd ed., Vol. 13, New York, NY, John Wiley & Sons, pp. 377, 391

---

[1]For definitions of the italicized terms, see Preamble, pp. 27–30.

Casto, B.C., Meyers, J. & DiPaolo, J.A. (1979) Enhancement of viral transformation for evaluation of the carcinogenic or mutagenic potential of inorganic metal salts. *Cancer Res.*, *39*, 193–198

Chen, J.L. & Fayerweather, W.E. (1988) Epidemiologic study of workers exposed to titanium dioxide. *J. occup. Med.*, *30*, 937–942

Christie, H., MacKay, R.J. & Fisher, A.M. (1963) Pulmonary effects of inhalation of titanium dioxide by rats. *Am. ind. Hyg. Assoc. J.*, *24*, 42–46

Commission of the European Communities (1962) Council Directive concerning the uniformity of regulations of member states concerning colouring agents that could be used in food products for human consumption. *Off. J. Eur. Commun.*, *115*, 2645–2664

Considine, D.M. (1974) *Chemical and Process Technology Encyclopedia*, New York, McGraw-Hill, pp. 1102–1104

Cook, W.A. (1987) *Occupational Exposure Limits – Worldwide*, Washington DC, American Industrial Hygiene Association

Dale, K. (1973) Early effects of quartz and titanium dioxide dust on pulmonary function and tissue. An experimental study on rabbits. *Scand. J. respir. Dis.*, *54*, 168–184

Daum, S., Anderson, H.A., Lilis, R., Lorimer, W.V., Fischbein, S.A., Miller, A. & Selikoff, I.J. (1977) Pulmonary changes among titanium workers (Abstract). *Proc. R. Soc. Med.*, *70*, 31–32

Davis, G.W. & Hill, E.S. (1982) Polyester fibers. In: Mark, H.F., Othmer, D.F., Overberger, C.G., Seaborg, G.T. & Grayson, N., eds, *Kirk–Othmer Encyclopedia of Chemical Technology*, 3rd ed., Vol. 18, New York, John Wiley & Sons, p. 531

Deutsches Forschungsgemeinschaft (German Research Society) (1988) *Maximale Arbeitsplatzkonzentrationen und Biologische Arbeitsstofftoleranzwerte 1988* [Maximal Concentrations in the Workplace and Biological Tolerance Values for Working Materials 1986] (*Report No. XXIV*), Weinheim, VCH Verlagsgesellschaft, p. 58

Di Paolo, J.A. & Casto, B.C. (1979) Quantitative studies of in vitro morphological transformation of Syrian hamster cells by inorganic metal salts. *Cancer Res.*, *39*, 1008–1013

Direktoratet for Arbeidstilsynet (Directorate for Labour Inspection) (1981) *Administrative Normer for Forurensning i Arbeidsatmosfaere 1981* [Administrative Norms for Pollution in Work Atmosphere] (*No. 361*), Oslo, p. 21

Dunkel, V.C., Zeiger, E., Brusick, D., McCoy, E., McGregor, D., Mortelmans, K., Rosenkranz, H.S. & Simmon, V.F. (1985) Reproducibility of microbial mutagenicity assays. II. Testing of carcinogens and noncarcinogens in *Salmonella typhimurium* and *Escherichia coli*. *Environ. Mutagenesis*, *7* (Suppl. 5), 1–248

Einbrodt, H.J. & Liffers, R. (1968) Studies of lymphatic transport of titanium dioxide from human lung (Ger.). *Arch. Hyg.*, *152*, 2–6

Elo, R., Määttä, K., Uksila, E. & Arstila, A.U. (1972) Pulmonary deposits of titanium dioxide in man. *Arch. Pathol.*, *94*, 417–424

Ferin, J. & Leach, L.J. (1973) The effect of SO_2 on lung clearance of TiO_2 particles in rats. *Am. ind. Hyg. Assoc. J.*, *34*, 260–263

Ferin, J. & Oberdörster, G. (1985) Biological effects and toxicity assessment of titanium dioxides: anatase and rutile. *Am. ind. Hyg. Assoc. J.*, *46*, 69–72

Ferin, J., Mercer, T.T. & Leach, L.J. (1983) The effect of aerosol charge on the deposition and clearance of TiO_2 particles in rats. *Environ. Res.*, *31*, 148–151

Finch, G.L., Fisher, G.L. & Hayes, T.L. (1987) The pulmonary effects and clearance of intratracheally instilled Ni_3S_2 and TiO_2 in mice. *Environ. Res.*, *42*, 83–93

Friedberg, A.L. (1980) Enamels, porcelain or vitreous. In: Mark, H.F., Othmer, D.F., Overberger, C.G., Seaborg, G.T. & Grayson, N., eds, *Kirk–Othmer Encyclopedia of Chemical Technology*, 3rd ed., Vol. 9, New York, John Wiley & Sons, pp. 14–15

Garabrant, D.H., Fine, L.J., Oliver, C., Bernstein, L. & Peters, J.M. (1987) Abnormalities of pulmonary function and pleural disease among titanium metal production workers. *Scand. J. Work Environ. Health, 13*, 47–51

Grandjean, E., Turrian, H. & Nicod, J.L. (1956) The fibrogenic action of quartz dusts. Biological methods for their quantitative evaluation in rats. *Arch. ind. Health, 14*, 426–441

Grasso, P., Mason, P.L., Cameron, W.M. & Sharratt, M. (1983) Tissue reaction to the intrapleural injection of polyvinyl chloride powder, α–quartz and titanium dioxide. *Ann. occup. Hyg., 27*, 415–425

Health and Safety Executive (1987) *Occupational Exposure Limits, 1987 (Guidance Note EH 40–87)*, London, Her Majesty's Stationery Office, p. 21

Huggins, C.B. & Froehlich, J.P. (1966) High concentration of injected titanium dioxide in abdominal lymph nodes. *J. exp. Med., 124*, 1099–1105

Institut National de Recherche et de Sécurité (National Institute for Research and Safety) (1986) *Valeurs Limites pour les Concentrations des Substances Dangereuses Dans l'Air des Lieux de Travail* [Limit Values for the Concentrations of Toxic Substances in the Air of Work Places] *(ND 1609–125–86)*, Paris, p. 579

Isacoff, H. (1979) Cosmetics. In: Mark, H.F., Othmer, D.F., Overberger, C.G., Seaborg, G.T. & Grayson, N., eds, *Kirk–Othmer Encyclopedia of Chemical Technology*, 3rd ed., Vol. 7, New York, John Wiley & Sons, p. 152

Kada, T., Hirano, K. & Shirasu, Y. (1980) Screening of environmental chemical mutagens by the rec-assay system with *Bacillus subtilis*. *Chem. Mutagens, 6*, 149–173

Klass, D.L. (1984) Alcohol fuels. In: Mark, H.F., Othmer, D.F., Overberger, C.G., Seaborg, G.T. & Grayson, N., eds, *Kirk–Othmer Encyclopedia of Chemical Technoplogy*, 3rd ed., Suppl., New York, John Wiley & Sons, p. 19

Lee, K.P., Trochimowicz, H.J. & Reinhardt, C.F. (1985a) Transmigration of titanium dioxide (TiO_2) particles in rats after inhalation exposure. *Exp. mol. Pathol., 42*, 331–343

Lee, K.P., Trochimowicz, H.J. & Reinhardt, C.F. (1985b) Pulmonary response of rats exposed to titanium dioxide (TiO_2) by inhalation for two years. *Toxicol. appl. Pharmacol., 79*, 179–192

Lee, K.P., Henry, N.W., III, Trochimowicz, H.J. & Reinhardt, C.F. (1986) Pulmonary response to impaired lung clearance in rats following excessive TiO_2 dust deposition. *Environ. Res., 41*, 144–167

LeSota, S. (1978) *Paint/Coatings Dictionary*, Philadelphia, PA, Federation of Societies for Coatings Technology, p. 424

Locker, D.J. (1982) Photography. In: Mark, H.F., Othmer, D.F., Overberger, C.G., Seaborg, G.T. & Grayson, N., eds, *Kirk–Othmer Encyclopedia of Chemical Technology*, 3rd ed., Vol. 17, New York, John Wiley & Sons, pp. 629–630

Lowenheim, F.A. & Moran, M.K. (1975) *Faith, Keyes, and Clark's Industrial Chemicals*, 4th ed., New York, John Wiley & Sons, pp. 814–821

Lynd, L.E. (1985) Titanium. In: *Mineral Facts and Problems (Bureau of Mines Bulletin 675)*, Washington DC, US Government Printing Office, pp. 1–21

Lynd, L.E. & Hough, R.A. (1986) Titanium. In: *Bureau of Mines Mineral Yearbook*, Vol. 1, Washington DC, Bureau of Mines, pp. 1–17

Lynd, L.E. & Lefond, S.J. (1983) Titanium minerals. In: Lefond, S.J., Bates, R., Bradbury, J.C., Buie, B.F., Foose, R.M., Hoy, R.B., Husted, J.E., McCarl, H.N., Roe, L., Rooney, L.F. & Stokowski, S., eds, *Industrial Minerals and Rocks (Nonmetallics other than Fuels)*, 5th ed., Vol. 2, New York, Society of Mining Engineers of the American Institute of Mining, Metallurgical and Petroleum Engineers, Inc., pp. 1303–1362

Määttä, K. & Arstila, A.U. (1975) Pulmonary deposits of titanium dioxide in cytologic and lung biopsy specimens. Light and electron microscopic X-ray analysis. *Lab. Invest., 33*, 342–346

Maltoni, C., Morisi, L. & Chieco, P. (1982) Experimental approach to the assessment of the carcinogenic risk of industrial inorganic pigments. In: Englund, A., Ringen, K. & Mehlman, M.A., eds, *Advances in Modern Environmental Toxicology, Vol. 2, Occupational Health Hazards of Solvents*, Princeton, NJ, Princeton Scientific Publishers, pp. 77–92

Mannsville Chemical Products Corp. (1983) *Chemical Products Synopsis: Titanium Dioxide*, Cortland, NY

National Cancer Institute (1979) *Bioassay of Titanium Dioxide for Possible Carcinogenicity (Tech. Rep. Ser. No. 97)*, Bethesda, MD

National Institute for Occupational Safety and Health (1983) *US National Occupational Exposure Survey, 1981–83*, Cincinnati, OH

Nuuja, I.J.M. & Arstila, A.U. (1982) On the response of mouse peritoneal macrophages to titanium dioxide pigments *in vitro. Environ. Res., 29*, 174–184

Nuuja, I.J.M., Ikkala, J., Määttä, K. & Arstila, A.U. (1982) Effects of titanium dioxide pigments on mouse peritoneal macrophages *in vivo. Bull. environ. Contam. Toxicol., 28*, 208–215

Ophus, E.M., Rode, L.E., Gylseth, B., Nicholson, D.G. & Saeed, K. (1979) Analysis of titanium pigments in human lung tissue. *Scand. J. Work Environ. Health, 5*, 290–296

Paffenbarger, G.C. & Rupp, N.W. (1979) Dental materials. In: Mark, H.F., Othmer, D.F., Overberger, C.G., Seaborg, G.T. & Grayson, N., eds, *Kirk–Othmer Encyclopedia of Chemical Technology*, 3rd ed., Vol. 7, New York, John Wiley & Sons, pp. 507, 512

Pott, F., Ziem, U., Reiffer, F.-J., Huth, F., Ernst, H. & Mohr, U. (1987) Carcinogenicity studies on fibres, metal compounds, and some other dusts in rats. *Exp. Pathol., 32*, 129–152

Redline, S., Barna, B., Tomashefski, J.F., Jr & Abraham, J.L. (1986) Granulomatous disease associated with pulmonary deposition of titanium. *Br. J. ind. Med., 43*, 652–656

Richards, R.J., White, L.R. & Eik-Nes, K.B. (1985) Biological reactivity of different crystalline forms of titanium dioxide *in vitro* and *in vivo. Scand. J. Work Environ. Health, 11*, 317–320

Roberts, W.L., Rapp, G.R., Jr & Weber, J. (1974) *Encyclopedia of Minerals*, New York, Van Nostrand Reinhold, pp. 21–22, 530

Rode, L.E., Ophus, E.M. & Gylseth, B. (1981) Massive pulmonary deposition of rutile after titanium dioxide exposure. Light-microscopical and physico-analytical methods in pigment identification. *Acta pathol. microbiol. scand., A89*, 455–461

Santodonato, J., Bosch, S., Meylan, W., Becker, J. & Neal, M. (1985) *Monograph on Human Exposure to Chemicals in the Workplace: Titanium Dioxide (Report No. SRC–TR–84–804)*, Syracuse, NY, Center for Hazard Assessment, Syracuse Research Corporation

Schiek, R.C. (1982) Pigments (inorganic). In: Mark, H.F., Othmer, D.F., Overberger, C.G., Seaborg, G.T. & Grayson, N., eds, *Kirk–Othmer Encyclopedia of Chemical Technology*, 3rd ed., Vol. 17, New York, John Wiley & Sons, pp. 793–801

Schmitz-Moormann, P., Hörlein, H. & Hanefeld, F. (1964) Lung alterations after titanium dioxide dust exposure (Ger.). *Beitr. Silkoseforsch., 80*, 1–17

Schroeder, H.A., Balassa, J.J. & Tipton, I.H. (1963) Abnormal trace metals in man: titanium. *J. chron. Dis.*, *16*, 55–69

Schurr, G.G. (1981) Paint. In: Mark, H.F., Othmer, D.F., Overberger, C.G., Seaborg, G.T. & Grayson, N., eds, *Kirk–Othmer Encyclopedia of Chemical Technology*, 3rd ed., Vol. 16, New York, John Wiley & Sons, p. 742

Sethi, S., Hilscher, W. & Flasbeck, R. (1973) Tissue response to a single intraperitoneal injection of various substances in rats. *Zbl. Bakt. Hyg., I. Abt. Orig. B*, *157*, 131–144

Shevtsova, V.M. (1968) Data on the effect of titanium and zirconium ore dust on workers and on animals in chronic experiments (Russ.). *Gig. Tr. prof. Zabol.*, *12*, 24–29

Stenbäck, F., Rowland, J. & Sellakumar, A. (1976) Carcinogenicity of benzo[a]pyrene and dusts in the hamster lung (instilled intratracheally with titanium oxide, aluminium oxide, carbon and ferric oxide). *Oncology*, *33*, 29–34

Thyssen, J., Kimmerle, G., Dickhaus, S., Emminger, E. & Mohr, U. (1978) Inhalation studies with polyurethane foam dust in relation to respiratory tract carcinogenesis. *J. environ. Pathol. Toxicol.*, *1*, 501–508

Työsuojeluhallitus (National Finnish Board of Occupational Safety and Health) (1987) *HTP–Arvot 1987* [TLV Values 1987] (*Safety Bulletin No. 25*), Helsinki, Valtion Painatuskeskus, p. 14

Uragoda, C.G. & Pinto, M.R.M. (1972) An investigation into the health of workers in an ilmenite extracting plant. *Med. J. Aust.*, *1*, 167–169

Weast, R.C., ed. (1985) *Handbook of Chemistry and Physics*, 66th ed., Cleveland, OH, CRC Press, pp. B-154–B-155

Wegner, T.H., Baker, A.J., Bendtsen, B.A., Brenden, J.J., Eslyn, W.E., Harris, J.F., Howard, J.L., Miller, R.B., Pettersen, R.C., Rowe, J.W., Rowell, R.M., Simpson, W.T. & Zinkel, D.F. (1984) Wood. In: Mark, H.F., Othmer, D.F., Overberger, C.G., Seaborg, G.T. & Grayson, N., eds, *Kirk–Othmer Encyclopedia of Chemical Technology*, 3rd ed., Vol. 24, New York, John Wiley & Sons, pp. 594–595

Windholz, M., ed. (1983) *The Merck Index*, 10th ed., Rahway, NJ, Merck 26, p. 1356

World Health Organization (1982) *Titanium* (*Environmental Health Criteria 24*), Geneva

Yamadori, I., Ohsumi, S. & Taguchi, K. (1986) Titanium dioxide deposition and adenocarcinoma of the lung. *Acta pathol. jpn.*, *36*, 783–790

Zeiger, E., Anderson, S., Haworth, S., Lawlor, T. & Mortelmans, K. (1988) *Salmonella* mutagenicity tests: IV. Results from the testing of 300 chemicals. *Environ. mol. Mutagenesis*, *11* (*Suppl. 12*), 1–158

Zitting, A. & Skyttä, E. (1979) Biological activity of titanium dioxides. *Int. Arch. occup. environ. Health*, *43*, 93–97

OCCUPATIONAL EXPOSURES IN
PAINT MANUFACTURE AND PAINTING

OCCUPATIONAL EXPOSURES IN PAINT MANUFACTURE AND PAINTING

1. Historical Perspectives and Description of Painting Trades

1.1 Description of paint products

The term *organic coating* encompasses conventional paints, varnishes, enamels, lacquers, water–emulsion and solution finishes, nonaqueous dispersions (organosols), plastisols and powder coatings. The following definitions have been used commonly, although they have not always been strictly applied.

Paint is a suspension of finely divided pigment particles in a liquid composed of a binder (resin) and a volatile solvent, sometimes with additives to impart special characteristics. The volatile solvent evaporates from the drying film after application, while the binder holds the pigment in the dry film, causing it to adhere to the substrate. Some high quality, hard gloss paints are referred to as *enamels* (Piper, 1965; Schurr, 1981).

Lacquer is defined as a coating that dries primarily by evaporation rather than by oxidation or polymerization. Because the solvents used in lacquers are relatively volatile and no chemical change is required for formation of the film, lacquers dry very rapidly (Piper, 1965; Hamilton & Early, 1972).

Varnish is defined as a homogeneous, transparent or translucent liquid that is converted to a solid, transparent film after being applied as a thin layer (Schurr, 1981).

The basic components of paints may have a widely varying chemical composition, depending on the colour, durability and other properties required from the paint.

(a) Pigments

Pigments can be classified as (i) inorganic, (ii) organic and (iii) earth pigments, such as ochre. They can also be classified into whites, colours, metallic flake pigments and powders. They are generally added in considerable proportion by weight (20–60%) and are used in paints to provide colour, opacity and sheen and also affect the viscosity, flow, toughness, durability and other physical properties of the coating. The physical properties of pigments, such as particle shape and size, vary; the diameter of the particles is generally <3 μm. The particles in dry pigment powders (0.5–10 μm) are partially in the range of respirable dust (Krivanek, 1982).

(i) Inorganic pigments

Inorganic pigments are an integral part of numerous decorative, protective and functional coating systems, such as automobile finishes, marine paints, industrial coatings, traffic paints, maintenance paints, and exterior and interior oil, alkyd and latex house paints. Inorganic pigments belong to numerous chemical classes, primarily including elements, oxides, carbonates, chromates, phosphates, sulfides and silicates (Schiek, 1982).

Many forms of lead have been used for more than 200 years in pigments; these include carbonate (white lead), oxides (litharge, red lead), sulfate, oxychloride (Turner's yellow), acetate, borate and chromates (IARC, 1980a,b; Schiek, 1982).

Zinc chromate, little known before 1914, was widely employed during the Second World War to inhibit rust on all sorts of equipment (Brunner, 1978). Other chromium pigments that have been used in paint for many years include lead chromates, barium chromate and chromium oxide (IARC, 1980b). Other inorganic pigments include cadmium sulfide, cadmium sulfoselenide and antimony trioxide (see monograph, p. 291). Various grades of naturally occurring ferric oxide provide yellow, red and brown pigments (Schiek, 1982).

White pigments constitute over 90% of all pigments used (Krivanek, 1982). Until the nineteenth century, white lead in linseed oil was used primarily, and prior to 1920 the available white pigments were basic carbonate white lead, basic sulfate white lead, zinc oxide, leaded zinc oxide and lithopone (Martens, 1964; Federation of Societies for Paint Technology, 1973). Increasing awareness of the toxic hazards of white lead stimulated the development of other pigments, which became available to paint manufacturers in the early part of the twentieth century (Brunner, 1978).

The most common pigment employed in paint is the white pigment titanium dioxide (see monograph, p. 307), produced in two different crystal forms – rutile and anatase. Although it was introduced shortly after 1918, it was not used widely because of its high cost. The first titanium dioxide pigment was a composite of 30% titanium dioxide (anatase crystal structure) and 70% barium sulfate. A major gain was made by the production of titanium dioxide with the rutile crystal structure, which has almost 25% greater opacity than the anatase form. Because of the chemical inertness of titanium dioxide, its extreme whiteness, excellent covering power and lack of toxicity, compared to white lead, it soon dominated in the manufacture of white paint and, by 1945, represented 80% of white pigment on the market. Concomitantly, the use of white lead in paints fell during 1900–45 from nearly 100% to less than 10%. The share of lithopone, a coprecipitate of 28–30% zinc sulfide and 70–72% barium sulfate (Schiek, 1982), introduced before the First World War, rose to 60% by about 1928 but fell to 15% by 1945 (Brunner, 1978). Calcium carbonate and aluminium silicate have also been used as white pigments.

The most common metallic dusts and powders used in paint are aluminium powder, zinc dust (Schiek, 1982) and bronze powders, which consist of metals in a finely divided state; e.g., gold bronzes are alloys of copper with varying proportions of zinc or aluminium.

Materials used as extender pigments include barium sulfate (barytes), calcium carbonate (ground limestone and chalk), silica (diatomaceous and amorphous; see IARC, 1987a,b), clays (hydrated aluminium silicate), talc (hydrated magnesium silicate; see IARC, 1978c,d) and mica (hydrated potassium aluminium silicate). These minerals are often added to paint to reduce cost, improve physical characteristics and increase resistance to wear; their effects are largely governed by the average particle size (Martens, 1964).

(ii) *Organic pigments*

Hundreds of organic pigments, comprising a broad spectrum of structural classes, are used in the paint industry. Organic pigments may be classified as azo or nonazo pigments.

Azo pigments are formed by successive diazotization of a primary amine and coupling. Monoazo and diazo pigments contain, respectively, one and more than one chromophore ($-N = N-$) group and are subdivided into two types, the pigment dyes and the precipitated azo dyes. The most important and established uses for pigment products include the coloration of surface–coating compositions for interior, exterior, trade and automotive applications, including oil and water emulsion paints and lacquers (Fytelson, 1982).

Prior to the discovery of Perkins' mauve in 1856, colour was obtained from natural sources, i.e., woad, madder, indigo, cochineal and log wood. The development of synthetic colouring materials continued with the discovery of fuchsin in 1858 and of other triphenyl-methane dyes, such as alkali blue, methyl violet and malachite green. Lakes of these dyes were used as the first synthetic organic pigments. The largest single advance in pigment technology after the First World War was the discovery in the 1930s of phthalocyanine blue (Monastral blue) and, later, its halogenated derivatives (Monastral greens) which were widely used in automotive finishes (Brunner, 1978; Fytelson, 1982). Other main categories of nonazo organic dyes and pigments used in paints and related products include quinacridones, thioindigos, perinones, perylenes and anthraquinone (Fytelson, 1982).

(iii) Earth pigments

Iron oxides are the most widely used of the coloured pigments derived from natural sources. Natural iron oxides are processed from several different ores, including haematite (see IARC, 1972, 1987e), limonite, siderite and magnetite, and provide a range of reds, yellows, purples, browns and blacks (Schiek, 1982).

(b) Binders (resins)

The vehicle portion of paints contains components collectively termed 'binders', which hold the pigment in the dry film and cause it to adhere to the surface to be painted. Almost all binders in modern paint films are composed of polymer materials, such as resins, and drying oils, whose main functions are to provide film hardness, gloss, surface adhesion and resistance of the film to acids, alkalis and other agents (Krivanek, 1982). A large variety of both natural and synthetic resins has been employed in paints. Natural resins have been used in paints for centuries, while synthetic resins have been commercially available since the early 1900s.

(i) Natural resins and oils

From early times, various natural resins have been used to reinforce linseed oil and other drying oils, since paints based only on pigment and oil yield only very soft films. Shellac (Brunner, 1978) and insect exudations are natural oleoresins that have been used in paints for centuries. Oleoresins from tree saps (shellac) are a mixture of single- and fused–ring compounds with various oxygenated groups possessing a wide range of molecular weights, solubilities and chemical and physical properties.

Kauri, a fossil resin, was used widely but had been replaced by the beginning of the twentieth century by Congo copal, which is a much harder resin but which requires prolonged heat treatment, known as 'gum running'. The term 'copal' is a generic name covering a number of fossilized and recent resins found in many tropical and subtropical parts of the world, which include all the harder resins used for oil varnishes (Brunner, 1978).

Another useful natural resin is rosin (colophony), which is obtained as a residue after distilling pine oleoresin for the production of turpentine. Rosin consists of about 85% rosin acids and 15% neutral substances and can be classified into two main types – gum rosin and wood rosin. Rosin has been used in paints (principally alkyd resins) for many years. They are often upgraded to yield higher quality resins by chemical reactions, including liming (calcium rosinate), esterification of rosin with glycerol, and reactions with trimethylolpropane, phthalic anhydride, maleic anhydride, adipic acid and sebacic acid (Krivanek, 1982).

Vegetable and fish oils have long been used as binders in traditional paints and varnishes. White linseed oil has been the most important oil in standard exterior paints, despite its moderately slow drying rate. It is infrequently employed in interior paints because of yellowing. Other important oils include castor oil, tall oil, soya bean oil, coconut oil, cottonseed oil, tung oil and various fish oils (Brunner, 1978; Lowell, 1984).

Although raw oils are useful as paint binders, it has been advantageous to use them in conjunction with refined oils and oils treated with heat to increase viscosity (heat-bodied oils) which isomerize the oil and improve the drying rate of the films. Oleoresinous varnishes are made by cooking oils with natural or synthetic resin, resulting in more rapid drying and a harder film (Lanson, 1978).

(ii) *Synthetic resins*

A wide variety of synthetic resins has been commercially available since the early 1900s. Those that have been most frequently employed in paints, varnishes and lacquers include cellulosic, phenolic, alkyd, vinyl, acrylic and methacrylic, and polyurethane resins, chlorinated rubber derivatives, styrene–butadiene and silicone oils (Martens, 1964; Krivanek, 1982).

Mixtures of different synthetic resins are often incorporated into a paint to furnish certain properties not provided by a single resin. While the amount of resin in paint varies, values of 20–35 wt% are common (Krivanek, 1982). The choice of a resin(s) for a particular application depends on factors such as appearance, ease of application, cost and resistance to chemicals, heat and wear.

Some resins (polyurethanes, epoxys; see IARC, 1976a) are blended immediately before use with cross–linking agents between the individual polymer chains, resulting in a hard, serviceable film.

Alkyd, acrylic, polyurethane and polyester resins have broad areas of use in paints, including paints for houses, automobiles, furniture and appliances, as well as in the protection of metal surfaces, e.g., in chemical plants and oil refineries.

Phenolic resins: The first synthetic resins used in paints were phenolic resins, which were introduced in the 1920s and are made from formaldehyde (see IARC, 1982a, 1987f) and phenol (see monograph, p. 263) or substituted phenols in the presence of alkaline or acid catalysts. Depending on the type and proportion of reactants, and on the reaction conditions, the resins may be heat–reactive or not. The first product of the reaction is methylol phenol, which reacts further. With an excess of formaldehyde under alkaline conditions, methylol groups react slowly with phenol, are retained in the reaction product and can act as reactive sites in varnish preparations or for cross–linking in finished products (Lowell, 1984).

The early phenolic resins developed between 1905 and 1910 were based on unsubstituted phenols, e.g., cresols, *para*–phenylphenol and *para*–*tert*–butylphenol, which are oil–soluble, and constituted a new type of varnish with superior hardness and resistance to water, solvents, chemicals and heat. Heat–sensitive phenolic resins that are insoluble in oil may be dissolved in solvents and employed as the sole vehicle for metal coatings (Lowell, 1984).

Alkyd resins: The advent of alkyd resins is considered to be a major breakthrough in modern paint technology. Alkyds are oil–modified polyester resins produced by the condensation reaction of polyhydric alcohols, polybasic acids and monobasic fatty acids, e.g., linseed or soya fatty acids (Lowell, 1984). The specific definition, which has gained wide acceptance, is that alkyds are polyesters modified with monobasic fatty acids (Lanson, 1978). The alkyds used initially were principally products of the chemical reaction of phthalic anhydride and glycerol with certain vegetable oils or their corresponding fatty acids (Brunner, 1978).

In recent years, the terms 'non–oil' and 'oil–free' alkyd have been used to describe polyesters formed by the reaction of polybasic acids with polyhydric alcohols in excess of stoichiometric amounts. These products are best described as saturated polyesters containing unreacted –OH or –COOH groups (Lanson, 1978).

Monobasic acids modify the properties of alkyd resins by controlling functionality, and thus polymer growth, as well as by the nature of their inherent physical and chemical properties. The majority of monobasic acids used in alkyd resins are derived from natural glyceride oils and are in varying degrees of unsaturation. The most common fatty acids present in these oils include lauric, palmitic, stearic, oleic, linoleic, linolenic, eleostearic, ricinoleic and licanic acids. Alkyd resins with relatively high fatty acid contents are called 'long–oil' alkyds; when the oil percentage is relatively low, they are known as 'short–oil' alkyds (Lanson, 1978).

Although glycerol and pentaerythritol are the major polyhydric alcohol components of alkyd resins, a number of other polyols are employed to a lesser degree, including sorbitol, trimethylolethanol, trimethylol propane, dipentaerythritol, tripentaerythritol, neopentylglycol and diethylene glycol. The principal polybasic acid for alkyd resins is phthalic acid, which is prepared and used as the anhydride. Isophthalic anhydride is also employed to yield somewhat faster drying and tougher, more flexible films than the analogous *ortho*–phthalic resins.

Long–oil alkyds are soluble in mineral spirits and are widely employed in architectural brushing enamels, exterior trim paints and wall paints, and their flexibility and durability have made them useful for top–side marine paints, metal maintenance paints and as a clear varnish. Medium–oil alkyds are the most versatile of the alkyd class, and their superior air drying, flexibility and durability allow for their use in maintenance paints, metal primers and a variety of general–purpose enamels. Short–oil alkyds are either drying or nondrying, require a strong aromatic solvent, such as toluene or xylene (see monographs, p. 79 and p. 125), and have been employed principally as industrial baking finishes.

Many polymeric materials and reactive functional materials can be used to produce suitably designed alkyds and to impart improved and/or special film–forming properties, in-

cluding nitrocellulose, polyisocyanates, urea–formaldehyde resins, silicones, melamine-
formaldehyde resins, reactive monomers, phenolic resins, cellulose acetobutyrate, chlorinated rubber, phenolic varnishes, vinyl resins, polyamides, chlorinated paraffin, natural
resins, epoxy resins and monobasic aromatic acids (Lanson, 1978).

Short–oil alkyd resins with a phthalic anhydride content of 38–45% contain a higher
proportion of hydroxyl groups, which provide reactive sites for alkylated urea–formaldehyde
and melamine–formaldehyde resins. These alkyds are generally based on tall–oil fatty acids,
soya bean oil or fatty acids. Amino–alkyd resins are widely used in industrial baking enamels
(Lanson, 1978).

The compatibility of alkyd resins with nitrocellulose extends up to 55% nitrocellulose
content, and nitrocellulose lacquers are produced in large quantities. Alkyds modified with
short–chain acids, such as those from coconut oil and castor oil, are widely used in high-
grade furniture lacquers (Lanson, 1978).

The general effect of the alkyd modified resins has been to upgrade the gloss, adhesion
and durability of nitrocellulose lacquers. Alkyd resins have been used in protective coatings
for over 40 years, and they still rank as the most important synthetic coating resin, constituting about 35% of all resins used in organic coatings (Lanson, 1978). The largest market for
alkyds in product finishes includes machines and equipment and wood and metal furniture
and fixtures (Connolly et al., 1986).

Vinyl resins: Vinyl polymers and copolymers were among the first synthetic polymers
and are widely employed in trade paints. Although synthesis of polyvinyl chloride (see
IARC, 1979a) was first reported in 1872 and that of polyvinyl acetate (see IARC, 1979b) in
1913, neither was developed commercially until the mid–1920s (Powell, 1972).

Vinyl monomers can be induced to polymerize readily by the addition of initiators, such
as peroxides and azo compounds, which decompose at reactor temperature to generate free
radicals. Polymerization processes involve radical formation, initiation, propagation, including chain transfer, and termination (Powell, 1972). The principal vinyl resins of importance
in the paint industry are polyvinyl chloride, polyvinyl acetate and polyvinyl butyrate, which
are available in a range of different compositions for specific uses and in grades that can be
handled as true solutions in organic solvents, as high–solid dispersions ('organosols' or 'plastisols'), as dry powders or as water–borne latexes. Polyvinyl acetate is extensively used in
emulsion paints, providing exceptional flexibility, toughness and water and chemical resistance. Vinyl chloride copolymer coatings are used in coil coatings and in industrial and marine coatings (Lowell, 1984; Connolly et al., 1986).

Water emulsions of high molecular–weight polyvinyl acetate have been widely used in
interior house paints. Copolymers of vinyl acetate with acrylic monomers are also employed
in exterior emulsion house paints. Latexes of vinyl chloride polymers and copolymers have
been commercially important for a number of years, e.g., as copolymers in exterior house
paints, which often include a vinyl chloride–acrylic ester copolymer modified with a specially
designed alkyd resin. Polyvinyl acetate and vinyl acetate copolymers are used in latex–based
interior and exterior paints (Powell, 1972).

The principal modifying monomers that have been used with vinyl acetate include dibutyl maleate and fumarate, butyl-, 2-ethylhexyl- and isodecyl acrylates and higher vinyl esters (Powell, 1972). Copolymers of the acrylates and vinyl acetate are commonly called vinyl acrylics and generally contain 15% acrylic monomer by weight (Connolly *et al.*, 1986).

Acrylic and methacrylic ester resins: Acrylic resins have been divided into four specific types: water-based, solvent-based thermoplastic (lacquer types), solvent-based thermosetting and powder coating resins (Connolly *et al.*, 1986).

Commercial acrylic and methacrylic polymers are made from a variety of acrylic and methacrylic monomers (see IARC, 1979c). The major monomers used are the methyl, ethyl, butyl and 2-ethylhexyl esters of acrylic and methacrylic acids, which readily undergo polymerization in the presence of free-radical initiators, such as peroxides, to yield high molecular-weight polymers (Allyn, 1971; Lowell, 1984; Connolly *et al.*, 1986).

Monomeric acrylic esters are produced commercially by several processes based on ethylene cyanohydrin, acetylene, β-propiolactone (see IARC, 1974a) and ethylene oxide (see IARC, 1985a, 1987g). The acetone-cyanohydrin process is the major method for the production of monomeric methacrylate esters (Allyn, 1971).

Several types of functionality can be incorporated into acrylic and methacrylic monomers. These are principally the amide, carboxyl, hydroxyl and epoxy types and are used to confer cross-linking capabilities and thermosetting properties on the resulting polymers (Allyn, 1971). Other monomers are used in conjunction with the acrylic monomers to achieve different properties, including vinyl acetate (see IARC, 1986a), styrene (see IARC, 1979d), vinyl toluene, acrylonitrile (see IARC, 1979a, 1987h) and methylacrylamide (Allyn, 1971).

Acrylic and methacrylic polymers are used in the formulation of clear and pigmented lacquers. Dispersions in water and in organic solvents provide latex and organosol coatings, respectively. Polymethacrylates are harder and less flexible than the corresponding acrylates (Lowell, 1984).

Although thermoplastic acrylic emulsions have been commercially available since 1925, they were not widely used in coatings until 1953 when new grades specifically designed for paints were introduced (Allyn, 1971). In the late 1950s, lacquers of greatly improved durability, based on polymethylmethacrylate or thermosetting acrylic enamels were adopted by the automobile industry (Lowell, 1984). By the 1960s, the use of acrylic emulsion polymers had been firmly established in exterior coatings for wood surfaces, a field long dominated by oil paints (Allyn, 1971).

Epoxy resins: Epoxy resins were first derived from bisphenol A and epichlorohydrin (see IARC, 1976b, 1987i) and introduced into the paint industry in the late 1940s. Two major types of epoxy resin exist – glycidyl ether epoxy resins (see monograph, p. 237) and epoxidized olefins – the former of which is the most important. Epoxy resins based on bisphenol A and epichlorohydrin are the most prominent of the glycidyl ether category and are produced by a condensation reaction in which bisphenol A and epichlorohydrin are reacted in the presence of alkali (Allen, 1972). The resultant diglycidyl ether resin has a functionality of two reactive epoxy groups per molecule. Epoxy resins can be polymerized through their reactive epoxy group using amines or polyamides (Allen, 1972; Lowell, 1984).

Epoxy resins of a second major type, epoxidized olefins, are based on epoxidation of the carbon–carbon double bond. Coating compositions derived from epoxidized olefin have better weathering characteristics than analogous systems based on bisphenol A diglycidyl ether resins (Allen, 1972).

In order to proceed from the relatively low molecular weight of the coating composition, as applied, to the high molecular–weight polymer necessary for optimal film properties, a 'curing' or polymerization must take place, which can involve either the epoxide or free hydroxy groups in the resin, or a combination of the two. Some of the principal reactions that have been used include chemical cross–linking *via* the amine–epoxide reaction, an anhydride–epoxide reaction, reaction with methylol groups, e.g., between the secondary hydroxyl groups of the higher molecular–weight resins and the methylol groups of phenol–formaldehyde and urea–formaldehyde resins, cross–linking *via* the isocyanate–hydroxyl reaction and esterification reactions between solid–grade epoxy resins and carboxyl–containing compounds, particularly drying–oil fatty acids (Allen, 1972).

Solid–grade glycidyl ether resins are readily soluble in polar solvents, such as ketones, esters and ether–alcohols, as well as in chlorinated hydrocarbon solvents (Allen, 1972).

Glycidyl ether resins of high molecular weight (number average[1], about 7000; weight average[2], about 200 000) are unique among epoxy coatings in that they form coatings by solvent evaporation alone (Allen, 1972). Because of their toughness, adhesion and corrosion resistance, epoxy resins are used in many applications, including industrial maintenance, automobile primers and coatings for appliances and steel pipes. Epoxys combined with phenolic resins and thermosetting acrylic resins yield high bake finishes with hardness, flexibility and resistance to chemicals and solvents (Lowell, 1984).

Polyurethane resins: Although polyurethanes were synthesized in 1937, the utility of weather–resistant polyurethane coatings became manifest only in the 1960s. Polyurethanes are obtained from the reaction of polyhydric alcohols and isocyanates. Nonreactive polymers can be prepared by terminating the polymer chains with monofunctional isocyanates or alcohols. Cross–linked polymers are formed from polyfunctional isocyanates or alcohols (Lowell, 1984). Isocyanates that have been employed include toluene diisocyanate (see IARC, 1986b) and hexamethylenediisocyanate.

Because of the wide range of physical properties obtained through variations in formulating polyurethane coatings, they are used in industrial and maintenance coatings, as well as in coatings for wood, concrete and flexible structures (Lowell, 1984). Polyurethane coatings are being used increasingly for automobiles and aircraft. Urethane ester–type resins (also

---

[1]Molecular weight value from number of molecules each multiplied by molecular weight and total divided by number of molecules

[2]Molecular weight value from sum of number of grams of material with a particular molecular weight each multiplied by its molecular weight and total divided by total number of grams

called urethane alkyds or uralkyds) are used primarily in architectural coatings. Two–component systems are used as high–performance coatings for maintenance and product finishes (Connolly et al., 1986).

Silicone resins: Silicones are characterized by a siloxane backbone, e.g., –Si–O–Si–O–, with organic groups which determine the properties of the final polymer attached to the silicon atoms. The monomeric precursors of silicone polymers are mono-, di- and trisubstituted halosilanes (usualy chlorosilanes). Monosubstituted silanols condense to highly cross-linked polymers, which are chiefly used in coatings. The degree of cross-linking and consequent physical properties are controlled by adjusting the ratio of mono- and disubstituted chlorosilanes. Alkyd resins with terminal hydroxyl groups can be condensed with silicones to produce hybrid polymers (Lowell, 1984).

Silicone resins are used to waterproof masonry and are blended with alkyds to formulate industrial maintenance coatings for storage tanks and other metal structures (Lowell, 1984).

Cellulose derivatives: Cellulose nitrate, commonly misnamed nitrocellulose, is the oldest cellulose derivative, first prepared in 1838 from cotton linters using a nitrating mixture of nitric and sulfuric acids (Jones, 1938; Sears, 1974). Before the early 1920s, only very high molecular-weight cellulose nitrate was available, which had limited utility in lacquers. The development of stable cellulose nitrate with lower viscosity after the First World War resulted in fast–drying lacquer coatings which were used extensively in automobile and furniture production. In the USA, three types of commercially available cellulose nitrates are distinguished by their nitrogen content and solubility. Each of these types is available in a variety of viscosity grades, which are a measure of the polymer chain length (Hamilton & Early, 1972; Brewer & Bogan, 1984).

Cellulose nitrate lacquers have also been formulated to contain resins, plasticizers, solvents and thinners. The resins that were employed initially with cellulose nitrate lacquers include shellac, sandarc, mastic and ester gums, which were added in amounts of about two–thirds of the weight of cellulose nitrate. Plasticizers are usually added at about 10% of the weight of cellulose nitrate. Camphor, which was first used in the USA in the mid–1800s, was replaced by castor oil; by the late 1920s, plasticizers such as triphenyl phosphate, tricresyl phosphate, dibutyl phthalate and butyl tartrate were being used increasingly. The principal solvents used initially with cellulose nitrate included ethyl, butyl and amyl acetates, acetone, 'diacetone alcohol', industrial spirit, ethanol and mixtures of alcohol with benzene or toluene and of alcohol with esters (Heaton, 1928; Hamilton & Early, 1972).

Another cellulose derivative, ethyl cellulose, is made by treating cellulose from wood pulp or cotton with a solution of sodium hydroxide to obtain primarily what is commonly referred to as 'alkali' or 'soda' cellulose. Further treatment with ethyl chloride under heat and pressure yields ethyl cellulose, which can be made in different viscosities. It is widely used in clear, dyed or pigmented lacquers for flexible substrates. Although less often used in paints, it has been formulated with silicone coatings to prevent pigment settling and sagging (Singer, 1957; Hamilton & Early, 1972).

Cellulose acetate is a linear high polymer that is obtained by first pretreating cellulose with a reduced amount of acetic acid to cause a certain amount of swelling, and then reacting it with acetic anhydride in the presence of sulfuric acid. Cellulose acetate lacquers are stable to light and heat and have good resistance to oils, greases and weak acids (Singer, 1957).

Methylcellulose, carboxymethyl cellulose and hydroxyethyl cellulose are water–soluble polymers that are used as thickeners in latex–based coatings. Cellulose acetate–butyrate is used as a resin modifier in automobile lacquers based on polymethylmethacrylate.

(c)   Solvents (see also the monographs on solvents, pp. 43 et seq.)

Solvents are widely used to keep paints in liquid form so that they can be applied easily. Until the late nineteenth century, turpentine and alcohol were the only solvents of any importance. Since the early 1900s, the number of solvents has increased considerably to encompass initially a broad range of petroleum and coal–tar distillates, alcohols, esters, ketones, glycols and halogenated hydrocarbons and, more recently, synthesized glycol ethers and esters. A large variety of mixtures of these classes is also employed.

Solvents in the turpentine category are derived mainly from the resinous exudations of various species of pine and other conifers and consist essentially of mixtures of various terpenes such as α– and β–pinene (Heaton, 1928).

A petroleum distillate, known as 'white spirits' (see the monograph on some petroleum solvents, p. 43), which consists mainly of aliphatic, alicyclic and aromatic hydrocarbons, was introduced as a solvent in the paint industry in 1885. For many years it was regarded as a cheap adulterant for turpentine but, as its use developed, it attained recognition as a different solvent. In addition to white spirits, several other paint solvents are prepared from petroleum and coal–tars (see IARC, 1985b, 1987j). Coal–tar distillates were the original source of commercial quantities of solvents such as benzol (a mixture containing mainly benzene, with smaller amounts of toluene, light hydrocarbons and carbon disulfide), benzene (see IARC, 1982b, 1987k), toluene, xylene and solvent naphtha (Heaton, 1928).

(d)   Additives

Additives are defined as those chemicals that perform a special function or impart a special property to paint. They are present at low concentrations, generally 0.2–10%, and include driers, thickeners, anti–skinning agents, plasticizers, biocides, surfactants and dispersing agents, antifoam agents and catalysts (Krivanek, 1982).

(i)   Surfactants

Surfactants, which are classified into anionic, cationic, amphoteric and nonionic categories, are used as pigment dispersants in both nonaqueous and aqueous systems. Dispersants employed in nonaqueous systems include lecithin, zinc naphthenate, calcium naphthenate, copper oleate and oleic acid. Ionizable dispersants that are usually employed in aqueous coatings include tripotassium polyphosphate, tetrapotassium pyrophosphate, sodium salts of arylalkyl–sulfonic acids and sodium salts of carboxylic acids (Lowell, 1984).

In addition to pigment dispersion, surfactants are used in paints as emulsifying agents, protective colloids, wetting agents, thickeners and antifoaming agents. A number of water-soluble resins and gums have been used as protective colloids or thickeners in emulsion paints. Water–soluble hydrophilic colloids include agents such as gum arabic, gum traga-canth, starch, sodium alginate, methyl cellulose, hydroxyethyl cellulose, polyvinyl alcohol, ammonium caseinate and sodium polyacrylate. The acrylate salts, casein and cellulosics, have been widely used in acrylic paints, while the major thickeners for styrene–butadiene paints have been alkali–soluble proteins (soya bean proteins). Methyl cellulose and hydroxy-ethyl cellulose are common thickeners for polyvinyl acetate paints (Martens, 1964).

Noncellulosic thickeners used in latex paints include maleic anhydride copolymers, mineral fillers, such as colloidal attapulgite (see IARC, 1987l,m) and treated magnesium montmorillonite clays, natural products (e.g., alginic acid, casein and soya bean protein), polyacrylamides, polyacrylic acid salts and acid–containing cross–linked acrylic emulsion co-polymers (Connolly et al., 1986).

A recent partial list of surfactants employed in water–borne paints includes aluminium stearate, cellulose ethers, polydimethyl siloxanes, polyethylene, alkali metal phosphates and sodium dioctyl sulfosuccinate (Hansen et al, 1987).

A variety of other surface–active agents are added to paints to control flow, levelling, sagging, settling and viscosity, including hydrogenated castor oils, lecithin, metallic soaps (e.g., linoleates, palmitates and stearates), treated montmorillonite clays, peptized oil gels, polyolesters, silicas and soap solutions (Connolly et al., 1986).

(ii) *Driers*

Driers (siccatives) that have been used in water–borne paints containing unsaturated polymers (e.g., alkyds) to accelerate curing are principally metal salts (lead, calcium, cobalt, manganese, zirconium, barium, zinc and cerium–lanthanum) of naphthenic acid, tall oil acid, 2-ethylhexanolic acid and neodecanoic acid, generally at levels ranging from 0.3 to 0.8% (w/w; Hansen et al., 1987). Cobalt–based driers are the most commonly used commercially and are active catalysts in both air drying and heat cure systems. Manganese is another major active drier. Other metal driers serve as auxiliary driers and are usually used in combination with cobalt and manganese. Lead (see IARC, 1980a, 1987n) driers were at one time the ma-jor auxiliary driers; however, legislation limiting the amount of lead that can be used in sur-face coatings has resulted in a sharply reduced use. The most suitable replacements for lead appear to be zirconium, calcium and cobalt–zirconium compounds (Connolly et al., 1986). In addition, 1,10–phenanthroline has been employed at levels of 0.02% (w/w; Hansen et al., 1987).

(iii) *Plasticizers*

The early use of plasticizers is illustrated by the incorporation of castor oil and glycerine into alcoholic spirit varnishes and of camphor into spirit varnishes and lacquers, as well as into cellulose ester enamels and lacquers in the late nineteenth and early twentieth centu-ries (Heaton, 1928). In 1912, triphenyl phosphate began to replace camphor for the plastici-zation of cellulose nitrate; later, tricresyl phosphate was used. The use of plasticizers was generally expanded by the mid 1920s with the introduction of di(2–ethylhexyl)phthalate (see IARC, 1982c) and dibutylphthalate in the mid–1930s (Sears, 1974).

Plasticizers that are generally added in quantities of up to about 2% include dibutyl-, diethyl-, diethylhexyl- and dioctylphthalates and, to a lesser extent, the low molecular-weight esters of adipic and sebacic acid, tributyl phosphate and castor oil. Polyester resins, including maleic residues, sulfonamides, triorthocresyl phosphate and chlorinated diphenyls, have been used occasionally (Piper, 1965; Krivanek, 1982).

(iv) Biocides (fungicides, preservatives and 'mildewcides')

Biocides are generally added to paint at low concentrations – less than 1% – for preservation in the tin. Each biocide formulation can contain several agents.

The function of a preservative is to retard the enzymatic degradation of cellulosic and other thickeners in latex paints in the tin. The function of a mildewcide is to retard the growth of fungi on applied exterior latex and solvent-based paints. These compounds are often the same, but are used in different quantities. Much less preservative is needed to preserve a latex paint than the amount of mildewcide required to retard mildew growth on an exterior paint (Connolly et al., 1986).

Phenylmercury compounds (e.g., acetate, propionate, benzoate, dodecyl succinate and oleate) were previously used extensively as mildewcides. Although mercuric compounds are no longer employed in solvent-based paints in the USA because of legislative restrictions, they are still permitted in water-based paints. Biocides that have been employed include tributyltin oxide, chlorothalonil (see IARC, 1983a), 1,2-benzisothiazolin-3-one, carbendazim, benzyl alcohol mono(poly) hemiformal, 1-(3-chloroallyl)-tetra-aza-adamantane hydrochloride, 5-chloro-2-methyl-4-isothiazolin-3-one, dodecyl dimethylammonium chloride, 5,8,11,13,16,19-hexaoxatricosane, 3-iodopropynyl butyl carbamate, 2-methyl-4-isothiazoline-3-one, formaldehyde, sodium nitrite and sodium benzoate (Connolly et al., 1986; Hansen et al., 1987).

(v) Antiskinning agents

Antiskinning agents are added to paints to retard the formation of skin on the surface of the liquid coating, in either closed or open tins, without retarding the drying of the product. The principal antiskinning agents are oximes or phenol derivatives; the major oxime used is methyl ethyl ketoxime. Smaller quantities of butyraldoxime and cyclohexanone oxime are used. The phenol derivatives used are mainly methoxyphenol, ortho-aminophenol and polyhydroxyphenol. Minor quantities of cresols, guaiacol, hydroquinone (see IARC, 1977a), isobutoxysafrol and lignocol have also been used as antiskinning agents (Connolly et al., 1986).

(vi) Miscellaneous additives

Other additives are employed in paints, including polymerization initiators such as benzoyl peroxide (see IARC, 1985c) and azobisbutyronitrile; antioxidants such as hydroquinones, phenols and oximes; ultraviolet light absorbers, luminescent and fluorescent materials and heat stabilizers (Connolly et al., 1986).

1.2 History of the manufacture of paints and related products

(a) *History of paint manufacture*

The history of paints and other related products has been reviewed (Heaton, 1928; Jones, 1938; Singer, 1957; Martens, 1964).

Paint has existed from the earliest times – literally from the beginning of history. Water-based paint was used for pictorial and decorative purposes in caves in France and Spain as early as 30 000–15 000 BC. The earliest efforts of cavemen were expressed by daubing coloured mud on the walls of their caves. The addition of crushed berries, blood, milk, eggs, dandelion or milkweed sap and other natural materials, such as chalk, earth colours, charcoal and ashes, to early paints improved both adhesion and utility.

The earliest pigments were natural ores. By 6000 BC, calcined (fired) mixtures of inorganic components and organic pigments were employed in China. By 1500 BC, the Egyptians were using dyes such as indigo and madder to prepare blue and red pigments, and by 1000 BC they had developed a varnish from gum arabic that contributed to the permanence of their art. Although paint and varnish have existed since earliest times, it was not until the 1700s that commercial manufacture of paints began in Europe and the USA. The early manufacturers of paint ground their pigments on a stone table with a round stone.

Whitewash, which is essentially water–slaked lime, was used during the early history of many countries. Various materials, such as Portland cement, were added to improve the material and it gradually evolved into the present–day cement paints. Skimmed milk and, later, casein were added to improve adhesion and durability. The addition of pigments, whiting and clay and, finally, the replacement of lime traced the evolution from whitewash to casein paint. Drying oils were later added to casein paints to improve water resistance. Casein paints were continually improved and, by the 1930s, they contained highly refractive index pigments similar to those used in oil paints.

In the late 1800s, grinding and mixing machines were developed to enable manufacturers to produce large volumes of paint. However, it was not until the twentieth century, during the period from the mid–1920s to prior to the Second World War, when improvements in paint technology began to parallel more closely advances in chemistry, that paint became sophisticated with regard to the process of manufacture, its use and the methods of its application. The development of highly efficient paint–making equipment and an expanding scope of areas of utility, e.g., automobile, marine and architectural areas, contributed to these advances. Paint technology advanced most rapidly in countries where industrialization developed fastest, i.e., in western Europe and the USA, the greatest advances occurring after the Second World War (Brunner, 1978).

For centuries, it was common practice for painters to prepare their own products from pigments and oils (principally linseed oil). Little 'ready–mixed' paint was available until just before the beginning of the twentieth century, when a large number of factories were established, principally in Europe and the USA (Brunner, 1978).

Mixing was originally conducted in an apparatus known as a 'pug mill', the general construction of which was a cylinder fitted with a vertical shaft carrying arms. Until the introduction of machinery, grinding was generally accomplished by rubbing down the mixed paint with a large 'muller' of glass or granite on a slab of similar material. Cone mills were later employed which were made in various sizes – small ones for hand use and larger sizes for factory use. Grinding machinery was introduced into paint manufacture in Europe around the late 1870s and developed enormously in design and efficiency during the twentieth century. Two– and, later, three–roller grinding mills were first introduced which greatly enhanced the speed and efficiency of grinding. Multiple–roll mills, e.g., four and five as well as combinations of edge–runner, horizontal mixers and roller mills, were all used in the 1920s for paint production and were employed until the advent of more modern plant methods. The conversion of paste paints to the liquid or 'ready–mixed' condition required a step called 'thinning' or the addition of a thinning vehicle. Vertical mixers were used for this purpose (Heaton, 1928).

(b)   History of lacquer manufacture

The use of lacquer dates from about 500 BC when the Chinese and Japanese used the sap of the tree, *Rhus vernicefera*, to prepare lacquer for the ornamentation of woods and metals (Singer, 1957).

In the early manufacturing processes of the late eighteenth and early nineteenth centuries, the most frequent method of mixing the components of lacquers was by use of a slowly revolving churn that contained baffles. In older works, these were merely wooden barrels, but later they were specially designed and made of aluminium, since other metals are likely to affect the colour of clear lacquers due to the acidity of the solvents or resins used. At larger installations, a method that involved lifting up a container by means of a special lifting carriage to a battery of paddle stirrers was also used. After completion of dissolution, the container was covered and lifted by crane to its storage bench as standardized lacquers or stock solutions for further mixing (Heaton, 1928; Jones, 1938).

(c)   History of varnish manufacture

The use of oil varnishes and resins was already well understood at the time of Theophilus the monk, who wrote about them in the eleventh century. The preparation of a drying oil by treating linseed oil with lime and litharge was described at an even earlier date by Eraclius in the ninth century (Barry, 1939). The travels of Marco Polo and the discovery of the New World brought a great variety of exotic gums and resins to Europe.

Towards the end of the eighteenth century, paints and varnishes were based on natural resins, e.g., copal and amber, and vegetable oils, e.g., linseed, walnut, hempseed and poppy seed oils. By the nineteenth century, fossil and semi–fossil gums replaced amber, which was by then scarce and expensive. These principal substitutes included gum arabic, gum elastic, copal, mastic and shellac (Singer, 1957). Early in the twentieth century, the first practical phenol–formaldehyde resins were developed. Research over the following two decades provided the basis for greatly improved varnishes.

Varnishes can be divided into three main types – oil, spirit or alcohol and water – of which the oil varnishes are the most important, such that the term 'varnish' generally implies oil varnish. The manufacture of oil varnishes involves the following operations: (i) 'gum running' or melting the resin; (ii) boiling the oil and mixing with the melted 'gum'; (iii) boiling the varnish; (iv) introducting driers; (v) thinning the varnish; and (vi) maturing the varnish (Heaton, 1928).

1.3 Construction painting and paint products used

Paints that are used on architectural structures (indoor and outdoor surfaces) are comprised of primers or undercoats for walls and woodwork and mat, semigloss or gloss finishing coats. The primers and finishing coats differ primarily in pigment/vehicle balance and in additives and vehicle type. Primers (usually called 'primers/sealers') are used to seal the variable porosity of the substrate (e.g., wood) and to adhere to the substrate and to subsequent coats of paints. For most architectural uses, an alkyd–based primer/sealer is used (Schurr, 1981).

(a) *Exterior house paints*

Casein paints, which have been used since the mid–1800s, were continually improved so that, by the 1930s, they contained high–refractive index pigments. Later, a drying oil was added to casein to produce an emulsion paint (Martens, 1964). Traditionally, linseed oil and oleoresinous vehicles have accounted for the bulk of architectural (house) paints. Several other oils have been used, but to a much lesser degree than linseed oil, and often in conjunction with linseed oil. The more important have been tung oil, perilla oil, soya bean oil, fish oils, safflower oil and dehydrated castor oil. Modern oil–based house paints generally contain combinations of untreated drying oil (unbodied oil) and drying oil treated (polymerized) so that its viscosity is increased (bodied oil) (Schurr, 1974).

A wide variety of thinners and solvents has been employed traditionally – principally turpentine, white spirits, benzene and solvent naphtha. Turpentine was the accepted thinner until the 1930s, at which time white spirits were introduced for reasons of cost and odour (Schurr, 1974).

The classic house paint consisted of ~80% basic lead carbonate – white lead – and 11% raw or boiled linseed oil which contained a small amount of drier in the form of metallic soaps. The paint was thinned with additional oil for application. Gradually, other primer pigments and inert fillers were introduced into the basic lead–in–oil formula. These included zinc oxide, leaded zinc oxide, lithopone and, finally, titanium dioxide (see monograph, p. 307). The use of lead pigments has been increasingly curtailed because of legislation (Schurr, 1974, 1981).

In the late 1930s and 1940s, alkyd paints were gradually introduced, particularly in the dark colours, with a marked improvement in properties such as colour and gloss retention. In the 1960s, white alkyd house paints without lead pigments were marketed by a few paint companies; they usually contained phenylmercury oleate and other arylmercury derivatives as fungicides. By the 1970s, alkyd house paints were replacing oil house paints, the faster drying time of the alkyds being obtained with cobalt and calcium soaps.

Around 1957, the first exterior water–based house paints were introduced. Most of these were based on acrylic-type latexes, and the paint had excellent colour retention on exterior exposure. Since that time, water–soluble and emulsified linseed oil house paints have been marketed which combine the advantage of an oil paint and a water–type paint in one product (Martens, 1964). Because of ease of application, cleaning ability with soap and water and good service, latex paints comprise most of the exterior paint market. Among the more common latexes are the acrylics, polyvinylacetate–dibutylmaleate copolymers, ethylene copolymers and acrylate copolymers (Schurr, 1974). Table 1 lists the ingredients of typical white house paints with oil and latex binders (Fisher, 1987).

Table 1. Examples of formulations of white house paints from the 1980s[a]

Type of paint and ingredients	Weight (%)
Oil–based	
Titanium dioxide (anatase) } Titanium dioxide (rutile)	12.7
Zinc oxide (acicular)	19.8
Water–ground mica	6.2
Magnesium silicate	23.2
Refined linseed oil	23.4
Bodied linseed oil	7.8
Lead and manganese soap solution	1.3
Mineral spirits	5.7
Acrylic latex	
Titanium dioxide (rutile)	20.2
Titanium dioxide (anatase)	0.8
Water–ground mica	2.5
Magnesium silicate	8.4
Calcium carbonate	6.9
Acrylic latex	41.9
Water	7.0
Cellulosic thickener	8.0
Nonionic emulsifier	0.9
Alkyl–aryl surfactant	0.2
Commercial defoamer	0.2
Ethylene glycol	2.1
Ammonium hydroxide	0.2
Organic mercurial fungicide	0.2
Pine oil	0.6

[a]From Fisher (1987)

The main categories of organic pigments used in interior and exterior construction or architectural paints include phthalocyanine and monoazo dyes (Volk & Abriss, 1976).

(b)   *Interior paints*

Until the early 1930s, the vehicles employed in interior paints were based on oils treated to increase viscosity (bodied oils) and heat-treated oils, usually combined with rosin, ester gum or other natural gums. Most solvent–type paints for interior use contain some oil and dry by oxidation. Oils that have been used in interior compositions include bleached linseed oil, dehydrated castor oil, soya bean oil, tung oil and oiticica oil. The principal pigments used for interior white paints are titanium dioxide, zinc oxide and various carbonate and siliceous extenders which are used to control pigment volume and gloss. From about 1927, with the development of alkyd resins, a variety of architectural enamels for interior and exterior use was based on these resins. However, the bulk of enamels produced for interior use contained oil treated to increase viscosity (bodied oil) and/or varnish as the binder until after 1945 (Volk & Abriss, 1976).

Early water–based interior paints were alkyd–resin emulsions stabilized with large amounts of casein and other stabilizers. The alkyds used were generally long–oil vehicles and the paints generally had poor emulsion stability. Three types of latex polymers are used most commonly in the manufacture of latex paints: styrene–butadiene types, polyvinyl acetate types and acrylics. Copolymer blends of styrene and acrylate have also been employed, combining the most durable features of each monomer into a single polymer (Volk & Abriss, 1976).

After the Second World War, the excess capacity for manufacturing styrene–butadiene rubber (see IARC, 1979d) was adapted to make styrene–butadiene latexes that could be used in paint. These water–based latexes appeared in the USA around 1948 in interior wall finishes (Martens, 1964; Schurr, 1981). Although polyvinyl acetate latexes have been in existence since the late 1930s, they were used as adhesives rather than in paints until after the Second World War (Volk & Abriss, 1976).

The general categories of extender pigments that are used in latex paints include clays, calcium carbonates, silicates, diatomaceous earths, silicas, barytes and talcs (see IARC, 1987a,b,c,d). Along with latex, surfactants, pigments and several other additives are usually incorporated into the formulation to obtain a stable and satisfactory product, including thickeners, defoaming agents, freeze–thaw stabilizers, coalescents and pH adjusters. Although natural thickeners, such as casein, were formerly used, their use has decreased appreciably in recent years. The thickeners employed most commonly are cellulosics – principally hydroxyethyl cellulose and methyl cellulose – polyacrylates, polyacrylamide, polyvinyl alcohols and many others. Ethylene and propylene glycols serve as freeze–thaw stabilizers. Coalescents are additives designed to optimize the coalescence of latex particles (Volk & Abriss, 1976) and include hexylene glycol, butyl cellosolve and butyl carbitol.

(c) Masonry paints

Casein–based paints applied to masonry and plaster surfaces were used in early construction work. These paints usually contained about 10% casein with some lime to insolubilize the casein after application (Martens, 1964). Latex–based primers/sealers are now often used for masonry surfaces. The latex vehicle is generally more resistant to alkali and permits evaporation of water from masonry surfaces without disruption of the film. Both alkyd and latex vehicles adequately seal porous surfaces (Schurr, 1981). Also, oil paints and styrene–butadiene copolymer, polyvinyl acetate emulsion, resin–emulsion and chlorinated rubber paints have all been used extensively on masonry surfaces.

Concrete floor coatings must possess good water resistance and adhesion over damp surfaces. Powdery concrete is first covered with a solvent primer. A satisfactory floor paint can be formulated using a styrene–butadiene latex fortified with an epoxy ester (Martens, 1964). Two examples of concrete floor enamels are presented in Table 2. Acrylic emulsion paints are widely used outdoors on concrete, stucco and cinder block because of their durability, adhesion and flexibility (Allyn, 1971).

Table 2. Examples of formulations of grey concrete floor enamels from the 1950s[a]

Type of paint and ingredients	Weight (%)[b]
Polystyrene–butadiene–based	
Titanium dioxide (rutile)	17.6
Lampblack	0.5
Organic ester	0.2
Polystyrene–butadiene copolymer resin	27.0
Raw linseed oil	1.9
40% Chlorinated paraffins	4.7
High–flash naphtha	25.0
White spirits	24.4
Chlorinated rubber–based	
Titanium dioxide (rutile)	17.2
Lampblack	0.5
Organic ester	0.3
Chlorinated rubber	13.8
40% Chlorinated paraffins	3.2
Thermolysed tung oil	3.5
Alkyd resin	7.8
Soya lecithin	0.4
Dipentene	7.5
High–flash naphtha	14.1
Aromatic high–solvency petroleum solvent	24.6
White spirits	6.4
Antiskinning agent	0.1

[a]From Singer (1957)
[b]Calculated by the Working Group

Although cement paints are used on all types of masonry, they tend to be brittle and to 'powder off'. Cement paint typically contains white Portland cement, gypsum, calcium chloride and hydrated lime added to water. Fortified cements have been prepared using latexes such as styrene–butadiene, polyvinyl acetate and acrylic esters added to Portland cement in amounts ranging from 10 to 40% latex on a solid basis (Martens, 1964).

(d) Waterproofing paints

These paints are applied on the outside of unpainted concrete, brick, stucco, etc., and have been formulated in a variety of ways to include components such as wax, aluminium stearate and silicone resins. The earliest, simplest waterproofing formulations were of the wax type and consisted of paraffin wax, raw chinawood oil and white spirits. Stearate formulations consisted of aluminium stearate in white spirits, with occasional addition of paraffin wax. A significant advance in the manufacture of waterproofing paints in the mid–1950s involved the use of silicone resins. A typical formulation of silicone waterproofing contains silicone resin and xylene. Another commonly used product contains special silicones, such as sodium methyl siliconate, in aqueous solution (Singer, 1957).

1.4 Surface coating in the wood industry and products used

Shellac and other gums or resins, such as elemi, sandarac, manila and benzoin, dissolved in alcohol or spirits of wine were introduced by the French in the seventeenth century and have been used for a long time in Europe for finishing wood grain and for their quick-drying properties. By the beginning of the eighteenth century, use of varnishes on furniture increased rapidly, particularly in France and England, although the manufacture of varnishes was still incidental to the work of painters, decorators and gilders. By the mid- and late eighteenth century, varnishing of furniture was well established in Europe (Jones, 1938; Barry, 1939).

Four properties are considered to be essential in furniture varnish: quick, hard, tough drying (3–4 h); good sanding and polishing properties; good resistance to water, acids, alkalis, etc.; and good heat resistance (Singer, 1957). Table 3 gives the formulation of a best–grade furniture varnish used in the 1950s.

Table 3. Formulation of a furniture coating from the 1950s[a]

Type of product and ingredients	Weight (%)[b]
Phenolic resin–based furniture varnish	
para-Phenyl phenol, pure phenolic resin	16.5
Modified phenolic resin, hard–oil type	11.0
China wood oil	17.3
Dehydrated castor oil	4.4
Xylene	16.5
White spirits	35.8
Cobalt naphthenate	0.3
Lead naphthenate	0.7
Antiskinning agent	0.3

[a]From Singer (1957)
[b]Calculated by the Working Group

Four categories of wood stains were employed in the late nineteenth and early twentieth centuries: (i) water stains which consisted of water-soluble dyes or colouring agents (e.g., potassium permanganate, potassium bichromate, pigments such as Venetian Brown, Bismark Brown); (ii) oil stains (e.g., brownish resins, bitumens, asphalts or pitches in white spirits, or petroleum tinted with oil–soluble dyes); (iii) naphtha stains, which consisted of naphtha–soluble dyes dissolved in coal–tar naphtha and containing a little resin such as coumarone or ester gum as a binder; (iv) and spirit varnish stains which consisted of methylated spirit–soluble dyes with a little alcohol–soluble gum and usually contained manila as a binder (Jones, 1938).

The range of organic dyes that are found in wood stains (see IARC, 1981) include rosaniline [magenta; see IARC, 1974b], nitrosine, indigo, amaranths, carmoisine (see IARC, 1975), croceine, rhodamine (see IARC, 1978a) and several CI solvent and acid dyes (Krivanek, 1982).

Finishing operations for wood include staining, wash coating, filling (if necessary), sealing, sanding, application of one or two lustre coats and polishing. Two types of oil stains – soluble and suspended pigment type – impart the desired colour to wood. Wood stains are dissolved in a vehicle that enables the stain to soak into the wood rather than stick to its surface as a film. After the staining operation, a clear, thin coat of lacquer is often applied before application of a filler, which is called a 'wash coat'; it stiffens the protruding fine wood fibres and can be removed by light sanding. In some procedures, filler is used to fill the depressions before the sealer and finish coats are applied. These finishes are cured by solvent evaporation; finish coats usually contain cellulose nitrate (Lowell, 1984). Some typical sanding sealer formulations that have been used contain cellulose nitrate in ethanol, zinc stearate paste, maleic resin solution, castor oil, butyl and ethyl acetates, toluene and petroleum lacquer diluent (Singer, 1957).

Oil stains based on linseed oil and tung oil have been used in the finishing of furniture, since they protect against staining without leaving an apparent film on the surface. Varnishes based on urethane oils rather than oil resins are being used increasingly (Wicks, 1984).

Formulations of paint utilized for furniture are principally dependent on the end–use. Nursery furniture, for example, requires extremely hard, tough coatings containing non–toxic pigments. A wide variety of coatings has been used on furniture, including low–bake finishes based on urea–formaldehyde resins, polyurethane paints based on diphenylmethane diisocyanate and hexamethylene diisocyanate, and lacquers composed of ethyl cellulose or cellulose acetate butyrate combined with acrylic resins (Singer, 1957; Lowell, 1984).

1.5 Painting in the metal industry and paint products used

(a) Metal primers, finish coats and corrosion inhibition paints

Since iron and steel rust in time when in contact with moisture and oxygen, many products made with these metals are coated with rustproof primers and finish coats (Schurr, 1974).

Primers are vehicle–rich coatings intended for application as foundation and adhesion–promoting coats. Metal primers are used to form a firm adhesive bond with the surface and also serve as an impermeable barrier between the environment and metal surface. When active rust prevention is essential, rust–inhibitive pigments that retard oxidation chemically are used.

Although there are many formulae for structural steel primers, red lead (Pb_3O_4; see IARC, 1980a) in a linseed oil vehicle has been used for a long time. Other formulations of red lead include combinations with alkyd resins and with red iron pigment. A typical red lead–iron oxide primer formulation is shown in Table 4. Zinc chromate (zinc yellow; a double salt of zinc and potassium and chromic acid) was introduced during the Second World War and is still used extensively. It is usually formulated as the basic pigment with an alkyd resin or linseed oil. Less zinc chromate is required to give the same protection as red lead, and zinc chromate is often combined with red iron oxide (Singer, 1957).

Because of restrictions on the use of lead and chromates, the pigments favoured in industrial maintenance coatings are now mainly zinc metal, zinc oxide, molybdates and phosphates (Schurr, 1981).

Finish coats cover the metal primer and seal it. Some metal products are covered by enamels which contain alkyd resins and dry by oxidation. The most durable coatings available are generally used on machinery and other industrial equipment and are based on epoxy or polyurethane resins which are cured by chemical reaction. Typical formulations are shown in Table 4.

(b) Marine paints

Paints for surfaces that are continuously immersed in seawater must be formulated with antifouling properties to resist the growth of marine fauna. Diverse species of hard and soft fauna which require a permanent anchorage in order to mature and reproduce form colonies on hulls.

The Phoenicians used copper on the hulls of their ships more than 3000 years ago. During the early nineteenth century, compounds (generally oxides) of copper, tin, lead, mercury and arsenic (see IARC, 1980c, 1987o) were the biocides used in antifouling paints, since these agents are effective against the broad range of organisms encountered in the marine environment. Biocides based on lead, mercury and arsenic are now prohibited from use in many countries (Brunner, 1978; Brady et al., 1987).

Antifouling coatings based on derivatives of triphenyl or tributyl tin have been introduced during the past 15–20 years. In some coatings, an organotin compound, such as the acetate, chloride, fluoride or oxide, is simply mixed into the formulation. These coatings are known as 'free–association' coatings and are characterized by a leach rate of organotin which is quite high when the coating is new but rapidly diminishes until it is insufficient to prevent fouling. A more useful formulation is obtained when the organotin in 'copolymer' coatings is covalently bound to the resin of the coating and is released when the bond hydrolyses in seawater. Since organotin compounds do not prevent accumulation of algae on hulls, some commercial organotin coatings contain a small amount of cuprous oxide to control algae and grasses (Brady et al., 1987).

Table 4. Examples of formulations of metal paints

Type of paint and ingredients	Weight (%)
Red lead–iron oxide primer (from 1950s)[a]	
Stearated processed clay antisetting agent	0.3
Red lead (97%)	28.2
Red iron oxide (85% Fe_2O_3 min)	7.0
Magnesium silicate	14.1
Mica, white, water–ground	3.5
Alkyd	27.9
White spirits	18.3
Cobalt naphthenate (6%)	0.1
Lead naphthenate (24%)	0.4
Antiskinning agent	0.1
White epoxy powder paint (from 1980s)[b]	
Epoxy resin (1400 D)	60
Dicyanamide curing agent	4
2–Methylimidazole (accelerator)	1
Calcium carbonate (extender)	15
Titanium dioxide (pigment)	15
Acrylic polymer flow additive	1
Water–based white epoxy enamel (from 1970s)[c]	
Epoxy resin emulsion (50% solids)	28.3
Polyamide resin curing agent (65% nonvolatiles)	20.0
Titanium dioxide	22.2
Hydroxyethyl cellulose	0.1
Water	29.4
White polyurethane enamel (from 1980s)[d]	
Hydroxyl–functional resin (solids)	21.0
Dibutyl tin dilaurate (catalyst)	<0.1
Titanium dioxide (pigment)	19.0
Aromatic hydrocarbons	6.0
Propylene glycol monomethyl ether acetate	29.0
Polyisocyanate resin (solids)	8.0
Butyl acetate	1.0
Ethyl acetate	17.0

[a]From Singer (1957); weight calculated by the Working Group
[b]From Peltonen (1986); two–component product
[c]From Allen (1972); two–component product
[d]From Dupont (1988); two–component product

Most commercial antifouling paints contain a vinyl binder, although products with other binders are also available. Rosin or some other leaching agent is generally added to cuprous oxide formulations to permit its controlled release into seawater, where it is lethal to fouling larva forms. More recent developments in antifouling methods have involved use of a sheet material of black neoprene rubber impregnated with tributyl tin (Drisko, 1985). An antifouling coating used extensively by the the US Navy consists of cuprous oxide dispersed in a mixture of natural rosin and a vinyl chloride–vinyl acetate copolymer (Brady et al., 1987).

More recent strategies have focused on nontoxic alternatives to antifouling paints. These include the use of fluoropolyurethane foulant-release coatings. One such formulation consisted of Desmodur–N–75 (an aliphatic polyisocyanate), polytetrafluoroethylene (38% by volume; see IARC, 1979c), titanium dioxide (see monograph, p. 307) and solvent (Brady et al., 1987).

(c) *Automobile coatings*

The development of low–viscosity cellulose nitrate lacquers in the early 1920s revolutionized the painting of automobiles. Although these lacquers did not flow well and required an expensive buffing operation to obtain an acceptable gloss, their fast–drying characteristics permitted production line assembly and painting of automobiles for the first time (Lowell, 1984).

Cellulose nitrate lacquers were followed by the introduction of alkyd enamels to the automobile industry in the early 1930s. These compositions were usually modified with small amounts of amino resins to provide harder, more thoroughly cross–linked films. These were followed by the adoption of thermosetting acrylic enamels in which alkyds were replaced by acrylic copolymers containing hydroxyl groups which could still react with melamine modifiers (Lowell, 1984).

In the late 1950s, lacquers of greatly improved durability and gloss, based on polymethylmethylacrylate or thermosetting acrylic enamels, were adopted by the automobile industry (Lowell, 1984).

Today, many new polymers, including maleic resins, amino resins (urea–formaldehyde and melamine–formaldehyde polymers), silicones, epoxides, polyesters and polyurethanes form the basis of highly diverse coating systems. In addition, nonaqueous dispersion lacquers and acrylic enamels have been developed. Steel used in automobiles is pretreated with a conversion coating (phosphating or bonderizing) to improve corrosion resistance and adhesion. In the traditional procedure, which is still employed to some extent, the solvent–borne primer was sprayed onto the automobile body shell followed by a surfacer which could be sanded. The primer and surfacer were often combined into a single adhesive formulation which could be sanded. The vehicles of primer-surfacers were combinations of oxidizing alkyd, epoxy and amino–formaldehyde resins, alkyd– and rosin–modified phenolic resins and others. Applications of the colour coats followed applications of the primer–surfacer combinations (Lowell, 1984).

Solvent-borne primers have been almost completely replaced by water-borne electrodeposited primers. The original anodic type has been largely replaced by the cathodic type which is superior in corrosion protection. The binders for cathodic deposition are typically acid salts of amino-treated epoxy. The formulations contain polyepoxides or blocked poly-isocyanates which cross-link the coating when it is baked. Prior to application of the top coat, a coat of solvent or water-borne epoxyester primer-surfacer is applied (Lowell, 1984).

Very solid top coats are being used increasingly, thus eliminating lacquer-type formulations. Versions of conventional thermosetting acrylic enamels that can be applied in about 40–50% volume solids are now available (Lowell, 1984).

A broad range of organic pigments is employed in automotive finishes. These include Hansa yellows (prepared from chloro- and nitroanilines and acetoanilides), diarylide yellow, nickel azo yellow (nickel (see IARC, 1976c, 1987p) chelate of diazotized 4-chloroaniline and 2,4-dihydroxy quinoline), lithol reds (precipitated azo pigments comprised of a family of the sodium, barium, calcium and strontium salts of the coupling product from diazotized 2-naphthylamine-1-sulfonic acid and 2-naphthol), yellow BON-maroon (manganese salt of the coupling product of diazotized 4-chloroanthranilic acid with 3-hydroxy-2-naphthoic acid), and naphthol reds and maroons (monoazo pigments such as the copper precipitation product from the coupling of diazotized 4-nitroanthranilic acid with Naphthanil RC). Other classes of nonazo organic pigments that have been employed in automotive finishes include quinacridones, thioindigos, perinones (diimides of naphthalene-1,4,5,8-tetracarboxylic acid), perylenes (diimides of perylene-3,4,9,10-tetracarboxylic acid) and anthraquinones (Fytelson, 1982).

Table 5 gives a typical formulation of a lacquer for automobile finishing.

Table 5. Formulation of an automobile paint from the 1970s[a]

Type of paint and ingredients	Weight (%)
Blue metallic lacquer	
Acrylic resin	41.8
Methyl methacrylate (93%)	
Butyl acrylate (7%), comprising 40% nonvolatiles in toluene	
Cellulose acetate butyrate	4.4
Plasticizer	8.2
Aluminium pigment (65%)	1.8
Phthalocyanine blue pigment dispersion	1.4
Acetone	10.5
Methyl ethyl ketone	10.0
Xylene	4.0
Ethylene glycol monoethyl ether acetate	13.0
Mixed methyl esters of adipic and glutaric acid	4.9

[a]From Williams (1977)

(*d*)   *Aluminium paints*

Aluminium paints sold today in ready-mixed form are used as protective coatings on iron and steel, aluminium, magnesium and other metals, providing high resistance to moisture penetration and heat and a high reflectance for ultraviolet radiation. These paints contain aluminium pigment in finely divided form and are formulated in two types – non–leafing and leafing. The former consists of aluminium ground to a powder that is dispersed in vehicles to give a metallic–grey finish. 'Leafing' is a phenomenon in which finely divided aluminium flakes rise to the surface and form a continuous metallic finish consisting of intermeshing aluminium flakes (Singer, 1957).

(e) Coil coatings

One of the most rapidly growing areas of industrial coating is coil coating. The coil stock consists of enormous rolls of thin–gauge steel or aluminium, which are coated at steel mills, aluminium mills or by specially equipped contractors. The coils are unwound, coated on high–speed roller coaters, heat cured and rewound. Binder compositions include alkyd–amino–formaldehyde combinations, vinyl chloride–vinyl acetate copolymers (see IARC, 1979a,b) and thermosetting acrylics, often modified with small amounts of epoxy, which produce coatings that are flexible, durable and adhesive. In the coil coating industry, solvent vapours are often collected and disposed of by incineration (Lowell, 1984).

1.6 Other painting trades and paint products

(a) Traffic paints

The major requirements for traffic (road) paints are fast and hard drying. The paints that are generally used contain a high pigment volume, fast–drying vehicles, such as resin combinations with low oil content or oil–free synthetic resins, and low–boiling solvents (e.g., petroleum fractions with distillation ranges of 100–150°C). Solutions of plasticized chlorinated rubber made of styrene–butadiene copolymers (containing an aromatic hydrocarbon solvent to maintain solubility) have been used. Glass spheres are added to formulations of reflecting paints (Lowell, 1984).

Conventional alkyd formulations (including both straight alkyd and alkyd/chlorinated rubber types) still account for most of the traffic paints used in the USA. However, there has been a significant increase in the use of more durable pavement marking materials, such as two–component polyester and epoxy systems and one–component hot extruded thermoplastic types (Connolly et al., 1986).

Although traffic paints can be made in any colour, the most widely used are white (titanium oxide) and yellow (chrome yellow; Connolly et al., 1986).

(b) Fire–retardant paints

Fire–retardant or intumescent paints, when applied to wood or other combustable surfaces, retard the spread of fire by foaming at elevated (but less than charring) temperatures. A number of intumescent formulae employed in the mid 1950s contained a chemical combination of polyol (e.g., pentaerythritol), a mono– or diammonium phosphate and an amide (e.g., dicyandiamide; Martens, 1964; Lowell, 1984). Certain pigments such as antimony oxide and borates are also added to enhance the fire–resistant properties of such paints. Other intumescent paint formulae have included polyvinyl acetate and acrylic latexes (Martens, 1964).

(c)   Aerosol colours

A large variety of paints have been packaged in aerosol tins for touching up and painting small areas, hobby aircrafts and other such objects. The principal types of paint used are of alkyd composition, are thinned out to a low viscosity (generally with ketones and aromatic hydrocarbons) to allow atomization, and contain a gaseous propellant which is liquid under pressure (e.g., propane, butane and isobutane or dichloromethane, which has replaced dichlorofluoromethane in many countries). Other aerosol paint compositions include acrylic and cellulosic lacquers and epoxyester systems (Cannell, 1967; Lowell, 1984; Connolly *et al.*, 1986).

(d)   Paint and varnish strippers

Before the advent of chemically–resistant synthetic paint vehicles, simple organic solvents, mixtures of solvents or solutions of caustic alkalis could be used to soften and strip most paints and various films (Downing, 1967).

Dichloromethane (see IARC, 1986c, 1987q) is the most widely used paint stripper base of the organic group. Other chlorinated hydrocarbons that have been used with dichloromethane are, in order of decreasing effectiveness, 1,2–dichloroethane (see IARC, 1979f), propylene dichloride, dichloroethyl ether and *ortho*–dichlorobenzene (see IARC, 1982d, 1987r). Other solvents that can soften paint films are, in (approximate) decreasing order of effectiveness, ketones (e.g., methyl ethyl ketone), esters, aromatic hydrocarbons (toluene (see monograph, p. 79) and naphthas), alcohols and aliphatic hydrocarbons (Downing, 1967).

Phenols and chloroacetic acids have been used in certain, specific situations, such as for stripping epoxy coatings. In addition, many less common solvents have been used, some primarily for specific applications, including 2–nitropropane (see IARC, 1982e), dimethylformamide (see monograph, p. 171), dimethyl sulfoxide, tetrahydrofuran and 1,1,2–trimethoxyethane (Downing, 1967).

The main inorganic compounds that have been used as paint strippers are alkalis, principally in the form of a boiling solution of sodium hydroxide and, to a lesser degree, potassium hydroxide and lime or soda ash (anhydrous sodium carbonate). Additives such as sequestering agents (e.g., gluconic acid and alkali metal gluconates), surfactants (e.g., sodium resinate, fatty acid soaps, sodium lignin sulfonate, alkylarenesulfonates and petroleum sulfonates), water–soluble activators (e.g., phenolic compounds and their sodium salts – cresol, chlorocresol, sodium pentachlorophenate) and solvents (e.g., monoethers of ethylene glycol and diethylene glycol) are often used to increase the stripping rates of inorganic paint removers. Paint removers that are used on steel, aluminium and other nonferrous alloys often contain corrosion inhibitors such as phosphates and chromates (Downing, 1967).

Molten and fused alkali baths are also employed to salvage ferrous metal parts with defective finishes. At temperatures up to 500°C, even heavy films of epoxy and silicone coatings can be removed rapidly (Downing, 1967).

2. Production and Use of Paint Products

2.1 Production

(a) Production processes

The modern manufacture of paints, which are generally made in batches, involves three major steps: (i) mixing and grinding of raw materials; (ii) tinting (shading) and thinning; and (iii) filling operations (US Environmental Protection Agency, 1979), as illustrated in Figure 1.

To produce a batch of paints, manufacturers first load an appropriate amount of pigment, resin and various liquid chemicals into a roller mill, which is a large, hollow, rotating steel cylinder. Mills for grinding primer or dark pigments are partly filled with steel balls that measure about 1–2 cm in diameter. Mills for grinding light colours usually contain flattened ceramic spheres (pebbles) that measure about 3–4 cm in diameter. Depending on the type of mill used, the grinding process lasts about 24 h or until the pigment has been ground to a sufficiently fine paste. After the pigment has been ground, more resin and solvent are added to the paste in the mill and the paste is 'let down', that is, pumped out of the mill through a strainer which removes the grinding media to a holding tank.

Until the 1930s, drying vegetable oils, primarily linseed oil, were used as binders in paints and as liquids for grinding. Since these oils were relatively poor pigment wetters, considerable energy was required for the grinding (dispersion) steps. Earlier dispersion techniques which involved pebble, steel ball or roller mills were replaced during the 1960s by high–speed equipment which was first used to dissolve large chips of pigment dispersed in solid binders. As pigment production and wetting characteristics improved, pigments were dispersed satisfactorily in high–speed dissolvers (Schurr, 1981).

The 'tinting' step involves comparing samples in the holding tank with colour standards. Small amounts of shading pastes, which are highly concentrated blends of ground pigments, and a vehicle are added as required to match the standard. After the batch has been shaded to specifications, it is thinned to the desired viscosity by the addition of solvent, filtered and poured into containers for shipment (Schurr, 1981).

The complexity of paint technology is indicated by the numerous types and number of raw materials required. A plant that produces a broad line of trade, maintenance and industrial paints requires over 500 different raw materials and purchased intermediates, including oils, pigments, extenders, resins, solvents, plasticizers, surfactants, metallic driers and other materials (Federation of Societies for Paint Technology, 1973).

The modern manufacture of unpigmented lacquers is generally a cold–cutting or simple mixing operation. For example, cellulose nitrate solutions are made by adding the nitrated cellulose from alcohol-wet cotton to the solvent mixture and agitating for 1–2 h in a paddle or turbine blade mixer. Alkyd resins, which are supplied in solution, can be added directly to the cotton–based solution. Hard resins may be dissolved separately, usually in toluene, and added as solutions, or the lumps may be dissolved directly in the cotton–based

Fig. 1. Process for manufacturing solvent–based paints[a]

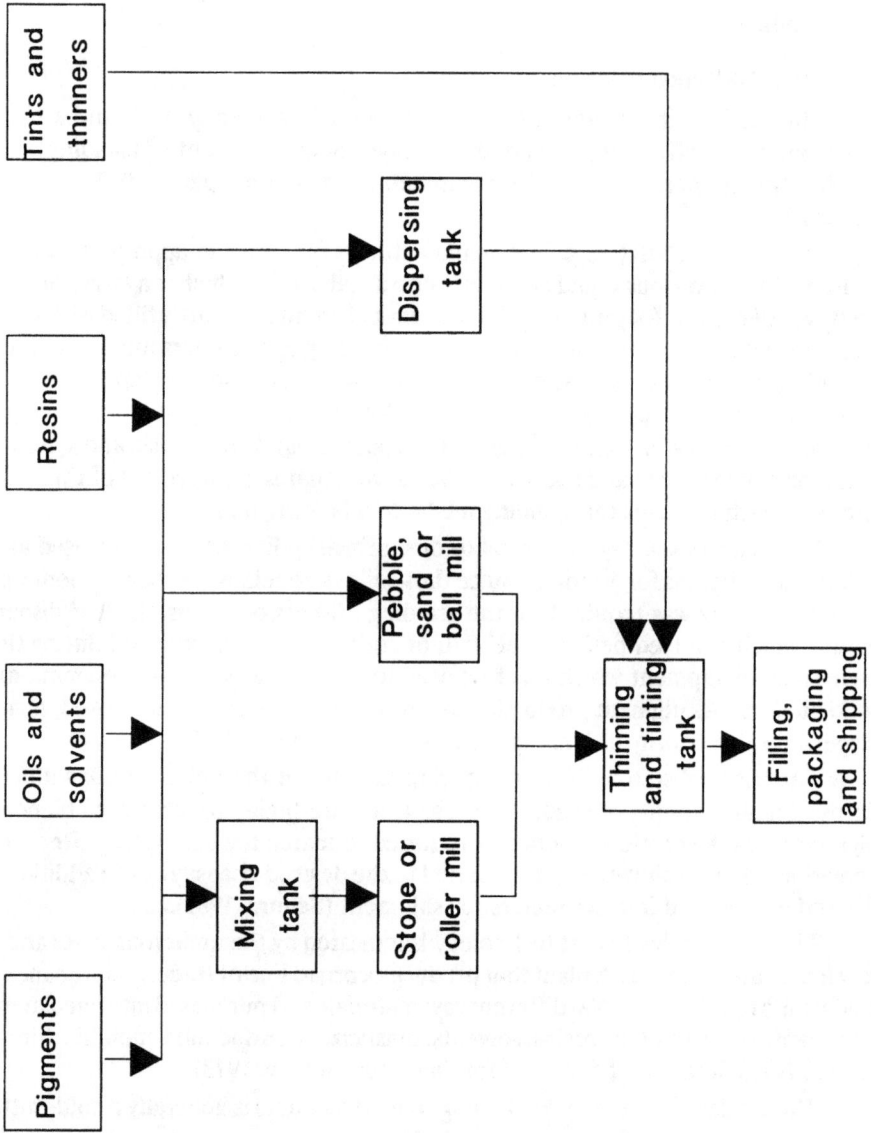

[a]From US Environmental Protection Agency (1979)

solution by stirring. Liquid plasticizers are then simply poured or pumped into the mixing tank (Hamilton & Early, 1972).

In pigmented lacquer manufacture, the pigments are first dispersed in ball mills with plasticizers, such as dibutyl phthalate, after which natural or synthetic resins are added. Cellulose nitrate (or cellulose acetate) is then added and all the components are mixed in a vertical mixer or churn before the finished product is run off into containers (Browne, 1983).

Modern manufacture of varnish is carried out in jacketed and enclosed kettles or set pots, and the required high temperature is achieved by different methods, including the use of heat–transfer media (Browne, 1983).

(b)   Production figures

Traditionally, two distinct types of coatings are produced – trade sale paints and industrial product finishes (Kline & Co., 1975).

Trade sale paints are sold through a variety of distribution channels to builders, contractors, industrial and commercial users and government units, as well as to the general public. These products are primarily for exterior and interior coatings for houses and buildings, although sizeable amounts of automobile and machinery refinishes, traffic paints and marine shelf-goods are also dispensed through trade sales outlets (Kline & Co., 1975).

Industrial product finishes or chemical coatings are produced to user specification and sold to other manufacturers for factory applications on such items as automobiles, aircraft, appliances, furniture and metal containers. They also include the category of industrial maintenance coatings, which are specially formulated and are used to maintain industrial plants and equipment (e.g,. as resistance to corrosion). Within these major product lines, the paint industry produces thousands of different products for a broad spectrum of substrates, applications and customers (Kline & Co., 1975).

World production in 1971 of surface coatings by selected countries or regions is given in Table 6. North America was the largest producer and manufactured 4.5 million tonnes (31.6%), while western Europe produced 4.2 million tonnes (29.3%) and eastern Europe produced slightly over 3 million tonnes (21.6%; Kline & Co., 1975).

A more recent listing of paint production in the USA, Japan and western Europe is presented in Table 7. Japan is now the largest paint producer outside of the USA, followed by the Federal Republic of Germany, France, the UK and Italy. In 1986, US paint production was 967 million gallons [approximately 4340 thousand tonnes] (Reisch, 1987).

Estimated consumption of various resins, pigments and solvents in the USA in 1975, 1980 and 1985 is shown in Tables 8, 9 and 10. The major resins used in the production of paints are alkyd, acrylic and vinyl resins, which account for over 60% of total resin use in the USA. The main pigment was titanium dioxide and the major solvents aliphatic hydrocarbons, toluene and xylenes (see monographs, p. 125).

The number of paint manufacturers in the USA has declined steadily from about 1500 in 1963, to about 1300 in 1972 and 1000 in 1987 (Kline & Co., 1975; Layman, 1985; Reisch, 1987). In the UK, there have been similar reductions from about 500 paint manufacturers in the 1950s to only about 250–300 in 1985 (Layman, 1985).

Table 6. World production (in thousands of tonnes) of surface coatings by selected country or region in 1971[a]

Region	Production	% Distribution
North America		
USA	4155	29.0
Canada	379	2.6
Total	4534	31.6
Europe		
Germany, Federal Republic of	1192	8.3
France	744	5.2
UK	614	4.3
Italy	420	2.9
Spain	207	1.4
Netherlands	192	1.3
Sweden	160	1.1
Belgium/Luxembourg	130	0.9
Austria	101	0.7
Denmark	84	0.6
Yugoslavia	80	0.5
Switzerland	68	0.5
Norway	66	0.5
Finland	46	0.3
Portugal	32	0.2
Turkey	27	0.2
Greece	20	0.1
Ireland	17	0.1
Total	4200	29.3
Eastern Europe (total)	3094	21.6
Asia		
Japan	1140	7.9
India	67	0.5
Israel	49	0.3
Hong Kong	26	0.2
Other[b]	304	2.1
Total	1586	11.1
Latin America[c]	484	3.4
Oceania		
Australia	179	1.2
New Zealand	45	0.3
Other	4	<0.1
Total	228	1.6

Table 6 (contd)

Region	Production	% Distribution
Africa		
South Africa	81	0.6
Other	138	1.0
Total	219	1.5
TOTAL	14 345	100.0

[a]From Kline & Co. (1975)

[b]Includes the Philippines and the Republic of Korea

[c]Includes Mexico

Table 7. Paint production (in thousands of tonnes) in selected countries in 1984[a]

Country	Prodution
USA[b]	4432
Japan[c]	1803
Germany, Federal Republic of	1321
France	809
UK[d]	670
Italy	625
Spain	278
Netherlands	253
Sweden	189
Denmark	133
Belgium	131
Austria	126
Portugal	92
Finland	82
Switzerland	82
Norway	75

[a]From Layman (1985), unless otherwise specified

[b]From Connolly et al. (1986)

[c]From Kikukawa (1986)

[d]Production in millions of litres

Table 8. Estimated consumption (in thousands of tonnes) of resins in paints and coatings in the USA[a]

Resin	1975	1980	1985
Alkyd[b]	315	295	300
Acrylic	158	203	223
Vinyl	166	183	191
Epoxy[c]	38	70	87
Urethane	33	54	58
Amino	29	33	41
Cellulosic	24	27	24
Polyester[b]	11	33	62
Phenolic	11	11	12
Chlorinated rubber	6	8	8
Styrene–butadiene	11	7	6
Natural	9	8	7
Linseed oil	23	35	36
Other resins	77	61	66
Plasticizers	20	19	17
Total	931	1047	1138

[a]From Connolly et al. (1986)

[b]Data for 1985 are not comparable to those for previous years, since it is probable that some of the resins reported as alkyds in previous years were actually polyesters.

[c]Includes unmodified epoxy resins plus hybrids (e.g., acrylate enhancement)

Table 9. Estimated consumption (in thousands of tonnes) of pigments in paints and coatings in the USA[a]

Use and pigment	1975	1980	1985
Colours			
Titanium dioxide	323	354	393
Chrome	29	33	33
Iron oxide	43	53	57
Carbon black	8	8	9
Other coloured inorganic	7	5	5
Phthalocyanine	1	2	2
Other organic	7	7	8
Aluminium	10	11	11
Fillers			
Calcium carbonate	175	175	213
Talc	136	150	136
Clay	154	159	199

Table 9 (contd)

Use and pigment	1975	1980	1985
Fillers (contd)			
Silica	59	71	83
Barytes	34	33	34
Nepheline syenite and feldspar	19	34	35
Other extenders and fillers	17	20	26
Other			
Zinc oxide	10	12	12
Zinc dust	24	32	24
Lead (corrosion inhibiting)	11	8	5
Cuprous oxide	2	4	3
Other	3	6	6
Total	1072	1177	1294

[a]From Connolly et al. (1986)

Table 10. Estimated consumption (in thousands of tonnes) of solvents in paints and coatings in the USA[a]

Solvent	1975	1980	1985
Aliphatic hydrocarbons	533	456	433
Toluene	281	265	277
Xylenes	240	213	211
Other aromatic compounds	91	79	84
Butyl alcohols	50	59	68
Ethyl alcohol	82	84	95
Isopropyl alcohol	48	53	41
Other alcohols	25	26	29
Acetone	79	79	89
Methyl ethyl ketone	145	154	152
Methyl isobutyl ketone	47	48	50
Ethyl acetate	31	35	36
Butyl acetates	55	62	71
Propyl acetates	8	8	9
Other ketones and esters	61	68	75
Ethylene glycol	23	29	34
Propylene glycol	14	12	14
Glycol ethers and ether esters	109	120	136
Chlorinated solvents	6	10	21
Miscellaneous	16	16	15
Total	1944	1876	1940

[a]From Connolly et al. (1986)

2.2 Use

(a) Uses and application methods

The various uses of paint products are described by type of resin in Table 11.

Paints are applied by direct contact or by deposition by atomization processes. The direct–contact category includes the familiar brushing and roller techniques, dipping, flow coating and electrodeposition. Deposition by atomization processes includes conventional spray, hot spray and electrostatic spray. Machine roller coating is used in the industrial application of paint to paper, plywood and metal sheets, as well as continuous coating of metal coils. Dip coating is used in large industrial operations (Browne, 1983; Lowell, 1984)

Probably the greatest advance made during the early 1900s in the field of paint technology was the introduction of the spray gun. Its advent helped in the introduction of cellulose nitrate lacquers and their application to automobile assembly line production. Electrostatic spraying was first introduced in the USA in the 1940s and later in the UK. Electrodeposition of paint, introduced during the 1960s, is an important milestone in industrial painting and has proven especially advantageous for painting automobile bodies and other parts because of its superior corrosion resistance. In this technique, the coating is an aqueous dispersion of low solid content. The binder particles carry ionized functional groups which may be positive or negative, thus having either anodic or cathodic deposition. The anodic type typically uses amino– or alkali-solubilized polycarboxylic resins and the cathodic type, salts of amine-treated resins, such as epoxy resins (Brewer, 1984; Lowell, 1984).

(b) Use patterns

Use of paints in the major markets in the USA in 1985 is presented in Table 12. Distribution of use of resins and the other main components of paints in the USA in 1985 is shown in Table 13.

2.3 Exposures in the workplace

(a) Introduction

No data on the numbers of paint production workers or painters worldwide were available to the Working Group. According to a 5% census sample of the US population in 1970 (National Institute for Occupational Safety and Health, 1970), paint manufacturers employed approximately 62 000 workers. Extrapolating from the proportion of world production of surface coatings contributed by the USA (see Table 6), it can be estimated that the total number of paint production workers in the world is approximately 200 000. According to the same US census survey, there were 362 000 construction/maintenance painters and 106 000 painters/paperhangers/decorators in the USA. On the basis of these figures, it can be similarly estimated that the number of painters worldwide must be at least several million.

Table 11. Uses of polymer systems in industrial coatings[a]

Polymer systems	Coil	Metal	Appliance	Furniture	Hardboard	Lumber and plywood	Marine	Maintenance	Automobile manufacture	Automobile refinish	Tins	Paperboard
Natural and modified polymers												
Drying oils	+			+	+	+	+	+			+	
Cellulose esters		+		+	+	+			+	+	+	+
Cellulose ethers				+						+		
Condensation systems												
Alkyd resins	+	+	+	+	+	+	+	+	+	+	+	+
Polyesters, high molecular weight	+	+	+	+	+	+	+		+	+		
Amino resins	+	+	+		+	+	+	+	+		+	+
Phenolic resins	+	+		+	+		+	+	+		+	+
Polyamides							+	+			+	
Polyurethanes	+			+	+		+	+	+		+	+
Epoxy resins	+	+	+		+		+	+	+		+	+
Silicones		+	+		+		+	+	+		+	+
Vinyl polymers and copolymers based on:												
Butadiene	+	+										
Acrylic or methacrylic ester	+	+	+	+	+		+	+	+	+	+	+
Vinyl acetate		+	+	+	+	+	+	+	+		+	
Vinyl chloride	+	+	+	+	+	+	+	+	+		+	+
Vinylidene chloride							+	+				+
Styrene		+	+	+	+			+	+	+	+	+
Vinyl acetal or butyral	+	+		+			+	+			+	
Fluorocarbons	+	+										
Resin combinations												
Acrylic and amino	+	+	+	+	+		+		+	+	+	+
Acrylic and epoxy		+	+				+		+	+	+	
Acrylic and silicone	+	+					+	+			+	
Alkyd and amino	+	+	+	+			+	+	+	+	+	+
Alkyd and acrylic		+	+	+				+	+	+	+	
Alkyd and epoxy		+	+					+	+			
Alkyd and silicone	+	+	+					+	+		+	+
Polyester and epoxy	+	+					+	+			+	+
Polyester and silicone	+			+	+							
Cellulose ester and urethane				+	+							
Alkyd, acrylic and amino					+						+	
Polyester and amino							+	+			+	+
Phenolic and epoxy											+	
Epoxy and amino							+	+			+	
Phenolic and amino											+	
Alkyd and vinyl chloride polymers							+	+				

[a]From Lowell (1984)

Table 12. Consumption of paints and coatings by major market in the USA, 1985[a]

Paints and coatings	%	% of total
Architectural coatings		46
Water–based	73	
Solvent–based	27	
Product finishes		38
Miscellaneous[b]	32	
Containers	15	
Wooden furniture	14	
Automotive	12	
Machinery	10	
Metal furniture	7	
Coils	5	
Wood flat stock	4	
Special purpose coatings		16
Traffic	28	
Auto refinishes	25	
Special maintenance	19	
Aerosols	10	
Other[c]	18	

[a]From Connolly et al. (1986)
[b]Includes appliances, other transportation, marine, paper and foil, and other
[c]Includes paints for roofs, bridges, marine shelf goods, metals and others

A wide range of potential occupational health hazards is present in relation to the manufacture and use of paints, varnishes and lacquers. Coatings are complex mixtures containing a variety of groups of substances, such as organic solvents, organic and inorganic pigments, extenders, resins and additives such as catalysts, surfactants, driers, plasticizers and biocides. Each of these categories covers a range of tens or hundreds of individual chemical compounds (Connolly et al., 1986). It has been reported that over 3000 individual paint components are used worldwide.

Occupational exposure results predominantly from the inhalation of gases and vapours, mainly organic, from solvents, binders and additives, of mainly inorganic pigment dusts and of complex inorganic and organic mixtures such as dusts from dried coatings and mists generated during the spraying of paint. The other major route of occupational exposure is through cutaneous contact with the various paint compounds, many of which can be absorbed through the skin. Ingestion related to personal work habits constitutes another potential route of entry.

Table 13. Distribution of use of components of paints in the USA in 1985[a]

Use	Proportion by use (wt% of total)					Total
	Resins	Pigments	Additives	Solvents	Water	(wt% of grand total)
Product finishes						
Wood furniture and fixtures	25.7	2.9	0.1	66	5.3	4.4
Wood mat stock	38	20	0.6	24	17	1.7
Metal furniture and fixtures	32	25	0.1	36	6.7	2.8
Containers and closures	38	6.2	0.3	43	13	4.8
Sheet, strip and coil	32	26	0.1	33	8.1	2.3
Major appliances	37	24	0.1	33	5.4	1.4
Other applicances	39	22	0.1	35	3.5	0.8
Automobile						
Topcoat	32	16	0.05	46	5.7	1.7
Primer	23	24	0.3	29	23	1.7
Underbody components and parts	26	18	0.2	35	21	1.1
Trucks and buses	31	16	0.1	27	25	1.2
Railroad	25	25	0.1	33	17	0.4
Other transportation	38	25	neg	38	neg	0.4
Machinery and equipment	30	27	0.2	38	5	4.2
Electrical insulation	51	0.0	0.0	49	neg	1.1
Paper, film and foil	39	2	0.1	43	17	1.0
Other product finishes	27	17	0.2	40	16	5.2
Total	31	17	0.2	41	11	36.8
Architectural coatings						
Interior water–based						
Mat	14	40	2	1.6	43	16.5
Semigloss and gloss	18	23	2	9.5	48	4.5
Other	18	21	1.5	0.2	57	1.3
Interior solvent–based						
Mat	21	58	0.2	30	0.0	1.1
Semigloss and gloss	27	41	0.1	32	0.0	2.6
Varnish	33	1.4	0.2	27	0.0	0.7
Other	26	21	0.2	43	0.0	0.7
Exterior water–based						
Mat (house paints)	19	35	25	4.2	39	11.2
Trim	20	23	2.6	3.3	51	0.9
Stains	17	19	1.5	3.5	59	1.4
Other	17	24	1.7	5	52	0.8
Exterior solvent–based						
Mat (house paints)	28	38	0.5	33	0.0	2.1
Enamel	42	26	0.3	32	0.0	2.4
Primer	25	37	0.7	37	0.0	0.7
Varnish	58	0.0	0.4	42	0.0	0.3
Stains	41	0.0	0.4	59	0.0	1.2
Other	24	40	0.5	36	0.0	0.3
Total	20	33	1.7	12	33	48.5

Table 13 (contd)

Use	Proportion by use (wt% of total)					Total
	Resins	Pigments	Additives	Solvents	Water	(wt% of grand total)
Special–purpose coatings						
Maintenance	36	23	0.2	33	8.2	2.8
Marine						
Pleasure	50	0.0	0.5	50	0.0	0.04
Commercial and maintenance	35	29	0.1	35	0.5	1.1
Automobile refinishing	23	22	0.2	55	neg	3.2
Traffic paints	15	59	0.1	26	0.9	5.6
Aerosol	15	5.5	0.1	80	neg	1.1
Other	21	46	0.2	32	0.1	3.2
Total	23	38	0.1	37	1.8	16.4
Grand total (thousands of tonnes)	1138	1294	42	1217	884	4536

*From Connolly *et al.* (1986); neg, negligible

Workers in the painting trades may also be exposed to a number of chemical agents originating from other operations that they or fellow workers are involved in, such as cleaning and preparing by chemical or mechanical means the object to be painted or cleaning themselves and the painting equipment.

The main substances to which workers may be exposed are listed in Table 14. The main occupational agents for which quantitative exposure data are available are presented in the following sections, covering the major paint trades.

Exposure to solvent mixtures is often described in the following sections using a summary measure, the cumulative exposure index (CEI), i.e., the sum of ratios of various measured levels to the respective occupational exposure limits. If this index exceeds unity, the combined exposure to different components of a solvent mixture is considered to exceed the recommended exposure limit. The values of the CEI are not always comparable because the exposure limits may vary with country and time.

In some painting operations, personal protective equipment is worn. However, it is common industrial hygiene practice to determine potential exposure by monitoring the breathing zone outside such protective gear. The results reported are thus not necessarily actual personal exposures.

Table 14. Main substances (and classes of substances) to which workers may be exposed in the painting trades[a]

Material	Principal uses or sources of emissions	IARC Monographs[b]
Acrylates (e.g., ethyl acrylate, methyl methacry-late)	Acrylic resins, ultraviolet curing paints	IARC (1979c)
Acrylic resins	Binders	IARC (1979c)
Alcohols, aliphatic (e.g., methanol, isopropanol, n-butanol)	Solvents (lacquers), paint removers	
Alkalis (e..g. sodium hydroxide, potassium hydroxide	Paint removers	
Alkyd resins	Binders	
Aluminium, powder	Pigment	
Amides, aliphatic (e.g., dimethylformamide)	Solvents	This volume
Amines (mono), aliphatic (e.g., diethylamine) and alkanolamines (e.g., 2-amino-2-methyl-1-propanol)	Water-based paints	
Amines (poly), aliphatic (e.g., diethylenetriamine)	Curing agents (epoxy resins)	
Amines, aromatic (e.g., meta-phenylenediamine, 4,4-methylenedianiline)	Curing agents (epoxy resins)	IARC (1978b) IARC (1986d)
Amino resins (e.g., urea-formaldehyde resins, melamine-formaldehyde resins)	Binders	IARC (1982a)
Ammonia	Water-based paints	
Anhydrides, organic (e.g., maleic anhydride, phthalic anhydride, trimellitic anhydride)	Alkyd resin synthesis, curing agents (epoxy resins)	
Antimony compounds (e.g., antimony trioxide)	Pigments, fire retardant pigments	This volume
Arsenic compounds (e.g., copper aceto-arsenate)	Antifouling agents	IARC (1980c)
Asbestos	Filler, spackling and taping compounds, talc	IARC (1977b)
Barium compounds (e.g., barium sulfate, barium carbonate)	Pigments	
Benzoyl peroxide	Catalyst	IARC (1985c)
Bisphenol A	Epoxy resins	
Cadmium compounds (e.g., cadmium sulfide, cadmium sulfoselenide)	Pigments	IARC (1976c)
Calcium compounds (e.g., calcium sulfate, calcium carbonate)	Fillers	
Camphor	Plasticizer	
Carbon black	Pigment	IARC (1984)
Cellulose ester resins (e.g., cellulose nitrate, cellulose acetate)	Binders	
Chloracetamide	Fungicide (water-based paints)	
Chromium compounds (e.g., chromic oxide, chromates)	Pigments	IARC (1980a)
Chlorofluorocarbons	Spray-can paint propellants	IARC (1986e)
Clays (e.g., bentonite)	Fillers	

Table 14 (contd)

Material	Principal uses or sources of emissions	*IARC Monographs*[b]
Coal–tar and asphalt	Special waterproof coatings (ships, tanks, pipes)	IARC (1985b)
Cobalt compounds	Pigments, driers	
Copper and copper compounds (e.g., bronze powder, cuprous oxide)	Pigments, antifouling agents	
Dyes and pigments, organic (e.g., aromatic azo dyes, phthalocyanines, rhodamine)	Pigments	IARC (1974b, 1978a)
Epichlorohydrin	Epoxy resins	IARC (1976b)
Epoxy resins	Binders	IARC (1976a)
Esters, aliphatic (e.g., ethyl acetate, isopropyl acetate)	Solvents	
Ethers, aliphatic (e.g., isopropyl ether, tetrahydrofuran) and glycol ethers (e.g., methylcellosolve)	Solvents	
Formaldehyde	Amino resin varnishes, biocide (water-based paints)	IARC (1982a)
Gasoline	Solvent	IARC (1989a)
Glycidyl ethers (e.g., *n*–butyl glycidyl ether and bisphenol A diglycidyl ether)	Epoxy resin diluents and constituents	This volume
Glycols (e.g., ethylene glycol)	Polyester resins, water–based paints	
Hydrocarbons, aliphatic (e.g., hexanes, heptanes)	Solvents (naphthas, white spirits)	
Hydrocarbons, aromatic (e.g., benzene, toluene, xylenes, trimethylbenzene)	Solvents (naphthas, white spirits), paint removers	IARC (1982b); this volume
Hydrocarbons, chlorinated (e.g., dichloromethane, 1,1,1–trichloroethane, carbon tetrachloride, trichloroethylene)	Solvents, paint removers, metal de-greasers	IARC (1979g,h,i; 1986c)
Hydrochloric acid (hydrogen chloride)	Catalyst (amino resins)	
Iron compounds (e.g., iron oxides, ferric ferrocyanide)	Pigments	IARC (1972)
Isocyanates (e.g., 1,6–hexamethylene diisocyanate, toluene diisocyanate)	Two–component polyurethane resins	IARC (1986b)
Isothiazolones (e.g., 1,2–benzisothiazolin–3–one)	Biocides in tinned foods	
Kerosene	Solvent	IARC (1989b)
Ketones, aliphatic (e.g., acetone, methyl ethyl ketone, cyclohexanone, isophorone, diacetone alcohol)	Solvents, lacquers, paint removers	
Lead compounds (e.g., lead chromate, lead oxides, basic lead carbonate, lead naphthenate)	Primers, pigments, driers	IARC (1980a)
Magnesium compounds (e.g., magnesium carbonate)	Fillers	
Manganese naphthenate	Drier	
Mercury compounds (e.g., mercuric oxide, phenyl mercuric acetate)	Fungicides (water–based paints)	
Methyl cellulose	Thickener (water–based paints)	

Table 14 (contd)

Material	Principal uses or sources of emissions	*IARC Monographs*[b]
Mica	Filler	
Molybdenum compounds (e.g., lead molybdate)	Pigments	
Nickel, metal powder	Pigment	IARC (1976c)
Nitroparaffins (e.g., nitroethane, 2-nitropropane)	Solvents	IARC (1982e)
Oils, vegetable (e.g., linseed oil, tung oil)	Binders	
Oximes (e.g., methyl ethyl ketoxime)	Anti-oxidants, anti-skinning agents	
Petroleum solvents (e.g., Stoddard solvent, VM & P naphtha)	Solvents, paint removers	This volume
Phenol	Phenol-formaldehyde resins, paint remover (formerly)	This volume
Phenol-formaldehyde resins	Binders	
Phenols, chlorinated (e.g., pentachlorophenol)	Fungicides (water-based paints)	IARC (1979j)
Phosphates, organic (e.g., tricresyl-*ortho*-phosphate, tributyl phosphate)	Plasticizers	
Phthalate esters (e.g., dibutyl phthalate, dioctyl phthalate)	Plasticizers	IARC (1982c)
Polychlorinated biphenyls	Plasticizers	IARC (1978c)
Polycyclic aromatic hydrocarbons	Special waterproof coatings (ships, tanks, pipes)	IARC (1983b)
Polyester resins	Binders	
Polyurethane resins	Binders	IARC (1979k)
Polyvinylacetate resins	Binders	IARC (1979b)
Pyrolysis fumes	Removal of paint by burning; heat-curing operations	
Rosin	Binder	
Rubber, synthetic (e.g., butyl rubber, styrene-butadiene rubber)	Binders (special paints, water-based paints)	IARC (1982f)
Shellac resin	Binder	
Silica, amorphous (e.g., diatomaceous earth)	Filler	IARC (1987b)
Silica, crystalline (e.g., quartz)	Filler, sand-blasting operation	IARC (1987b)
Silicates (e.g., sodium silicate, aluminium silicate)	Fillers	
Stearates (e.g., aluminium, zinc stearates)	Soaps, flattening agents	
Strontium compounds (e.g., strontium chromate, strontium sulfide)	Pigments	IARC (1980b)
Styrene	Polyester resins	IARC (1979d)
Styrene oxide	Diluent (epoxy resins)	IARC (1985d)
Sulfuric acid	Metal cleaner	
Talc	Filler	IARC (1987c)
Tin, metal powder	Lacquers (tinplate containers)	
Tin, organic compounds (e.g., tri-*n*-butyltin oxide, dibutyltin laurate)	Antifouling agents, catalysts	
Titanium dioxide	Pigment	This volume
para-Toluenesulfonic acid	Catalyst (amino resins)	
Turpentine	Solvent	

Table 14 (contd)

Material	Principal uses or sources of emissions	*IARC Monographs*[b]
Vinyl acetate	Polyvinylacetate resins	IARC (1986a)
Zinc and compounds (e.g., zinc metal powder, zinc oxide, zinc chromate)	Pigments, catalysts, bodying agents	IARC (1980b)

[a]From Sterner (1941); Piper (1965); Phillips (1976); O'Brien & Hurley (1981); O'Neill (1981); Dufva (1982); Krivanek (1982); Ringen (1982); Adams (1983); Selikoff (1983); National Institute for Occupational Safety and Health (1984); Swedish Work Environment Fund (1987) and previous sections
[b]See also *IARC Monographs* Supplement 7

(b) *Manufacture of paints and related products*

The manufacture of paints and related products such as varnishes, lacquers, enamels and paint removers involves the handling and processing of a complex array of raw materials, e.g., pigments, extenders, solvents, binders and additives, described in section 1, implying overall potential worker exposure to hundreds of chemicals (National Institute for Occupational Safety and Health, 1984). Furthermore, raw materials are often subjected to chemical changes such as during polymerization and cooking, thus creating a variety of new hazards.

The potential for occupational exposure depends largely on the basic types of products being manufactured, the degree of automation of the manufacturing process, the availability of exposure control measures and the nature of the specific job held. Various job classification systems have been developed for the paint manufacturing industry. Workers have thus been regrouped according to the basic product made – water–based paints, solvent–based paints, lacquer and vehicle – and to functions – pre–batch assembler, mixer, tinter, filler, tank and tub cleaner, reactor operator, varnish cooker, filter press operator (Morgan *et al.*, 1981). Additional functions are raw materials handler, laboratory personnel and others such as packagers, maintenance personnel, shippers and warehouse workers (National Institute for Occupational Safety and Health, 1984).

Heavy exposures, both by inhalation and skin contact, occur specifically in operations that can involve manual handling procedures such as weighing dry ingredients (pigments, extenders, resins, additives), loading them into mixing equipment, adding solvents to mills, and cleaning equipment (mixers, mills, reactors, kettles, tanks, filters). Additional exposure to solvents occurs in thinning, tinting and shading procedures, filling operations and filtering of varnishes. The cooking of varnishes may produce emissions of various aldehydes such as acrolein, of phenol, ketones, glycerine and fatty acids as well as dusts or vapours of maleic, phthalic and fumaric anhydrides during the loading of kettles. The production of powder coatings can be associated with significant exposure to dust from resin powders, pigments, curing agents and other additives. In the manufacture of radiation–curable coatings, exposures may occur to monomers such as ethyl acrylate, other acrylates and photoinitiators. Caustic solutions may be used in the cleaning of dispersion equipment (National Institute for Occupational Safety and Health, 1984). In general, important opportunities for exposure result from the presence of spills and the continuous spattering from machines (Adams, 1983).

(i) *Exposure to solvents*

Because of their volatility, solvents are ubiquitous air contaminants in paint manufacturing industries.

Exposure levels measured for various categories of workers in nine Swedish companies and reported as the sums of standardized concentrations are summarized in Table 15. High concentrations of solvents were found in all operations, the worst situation being manual cleaning of equipment with solvents. Local exhaust ventilation was common, and respirators were not often used. Of the 14 types of solvents monitored, the most common were xylene, toluene, butanol and esters (Ulfvarson, 1977).

Exposure to organic solvents was measured in the breathing zone of 17 Swedish male paint industry workers presumed to have the highest exposure of 47 workers employed in seven factories in Sweden, by collecting air with battery–driven syringes and analysing with two portable gas chromatographs. The median exposure values were (mg/m^3): xylene, 111 (16 persons); toluene, 11 (16 persons); isobutanol, 5 (15 persons); ethylacetate, 20 (14 persons); n–butylacetate, 14 (13 persons); ethanol, 13 (13 persons); n–butanol, 7 (13 persons); methylacetate, 12 (eight persons); dichloromethane, 719 (three persons); white spirits, 45 (three persons); and isopropanol, 129 (one person) (Haglund *et al.*, 1980).

Overall solvent exposure of workers known to be exposed to toluene was measured in seven paint manufacturing companies in New Zealand. Mean total levels of solvents ranged from 19 ppm in one company (five worker sampled) to 130 ppm in another one (three workers sampled), with individual values ranging from 7 to 297 ppm. Toluene, xylene and ethylbenzene were found in the atmosphere of all seven plants; the frequencies of other solvents were as follows: heptane, 6/7; n–hexane and methylethylketone, 5/7; acetone, 4/7; pentane, methylisobutylketone, ethanol and n–butylacetate, 3/7; and isopropanol, 1/7 (Winchester & Madjar, 1986).

In another study in Sweden, 47 employees of seven paint manufacturing industries, known to be exposed to solvents and including nine manual cleaners of paint mixing equipment, were surveyed for exposure to 12 solvents. The results are summarized in Table 16. The main exposures with regard to both frequency and weight were to xylene and toluene. Outstandingly high exposures occurred during the manual cleaning of equipment (Lundberg & Håkansson, 1985).

In a study on the effects of long–term exposure to solvents in the paint industry in Sweden, overall solvent exposure in a large paint manufacturing company was estimated for various work tasks over three historical periods. The results are presented in Table 17 in terms of the CEI, i.e., the sum of the ratios of the various exposure levels to the respective Swedish occupational standards in 1982 for the various solvents (Ørbaek *et al.*, 1985).

Table 15. Exposure levels (personal breathing–zone samples) to combined organic solvents during various paint manufacturing operations[a]

Operation	Sampling condition		Combined exposure[b]		Main solvents[c]	No. of samples in which solvent found
	No. of samples	Sampling time (min)	Mean	Range		
Charging solvents	33	4–43	2.0	0.2–16	Xylene	16
					Mesitylene	4
					Toluene	4
					Styrene	2
					Butanol	9
Pigment dispersion	18	9–66	1.5	0.2–4.4	Xylene	13
					Butanol	4
Tinting, thinning	14	15–32	0.9	0.1–2.0	Xylene	11
					Butanol	3
Can filling, paints	39	11–32	1.3	0.02–6.6	Xylene	23
					Alkanes	4
					Butanol	7
					Benzene	4
					Toluene	6
Can filling, thinners	14	9–20	1.8	0.1–7.4	Toluene	3
					Xylene	5
					Trichloroethylene	3
					Esters	2
					Acetone	1
Manual cleaning of equipment with solvents	51	3–28	5.7	0.5–30	Xylene	33
					Butanol	8
					Toluene	13
					Dichloromethane	9
					Esters	7
					Ketones	4

[a]From Ulfvarson (1977)
[b]Sum of ratios of individual solvent levels to their occupational exposure limits
[c]Solvents constituting at least one–fifth of individual combined exposure levels

The overall improvement in exposure levels over time has been attributed in large part to better control measures and to the increasing production of water–based paints. These results are corroborated by other estimates of the evolution of average solvent exposures in the Swedish paint manufacturing industry with the following values for the overall CEI: 2 in 1950–69, 1.5 in 1970–74, 0.7 in 1975–79 and 0.3 since 1980 (Lundberg, 1986). Heavy naphthas, toluene and benzene are reported to have been the most commonly used solvents during the 1930s, presumably with high exposure levels. Substitutes for aromatic hydrocarbons,

Table 16. Exposure levels (8–h time–weighted average) to organic solvents of 47 paint manufacturing workers[a]

Solvent	No. exposed	Exposure (mg/m³)	
		Median	Range
Xylene	44	82	1–6070
Toluene	43	10	1–1260
Isobutanol	36	4	1–1040
n–Butanol	35	6	1–1540
Ethanol	33	12	1–1090
Ethyl acetate	32	26	1–767
n–Butyl acetate	31	9	1–1680
White spirits	18	44	5–74
Methyl acetate	11	13	3–169
Dichloromethane	5	719	10–2420
Methyl ethyl ketone	5	39	8–124
Isopropanol	3	129	6–258

[a]From Lundberg & Håkansson (1985)

Table 17. Average combined organic solvent exposure[a] of paint industry workers in various work areas over three periods[b]

Work area	Period		
	1969 and earlier	1970–75	1976 and later
Industrial paint section			
Mixing	2	1.3	0.7
Grinding	3	1.8	0.9
Tinting–finishing	2.2	1.5	0.6
Tapping	2.2	1.2	0.6
Cleaning of vessels	4.5	3	1.5
Alkyd paint section (mixing, tinting, tapping)			0.1–0.2
Small batch manufacturing		1.4	0.7
Filler manufacturing	0.2	0.15	0.1
Storage	0.2	0.15	0.1
Cellulose paint section	2.5		
Laboratory			
Product development	0.7	0.4	0.15
Control laboratory	1	0.75	0.4
Process engineering	2	1	0.4

[a]Sum of ratios of individual solvent levels to their occupational exposure limits; solvents considered: acetone, butanol, butylacetate, ethanol, ethyl acetate, white spirits, methyl isobutyl ketone, toluene and xylene
[b]From Ørbaek et al. (1985)

including turpentine, decaline and tetraline, would have been used during the Second World War and immediately after. From 1950 until today, the most commonly used solvents would have been xylene, toluene, white spirits, ethanol, butanol, ethyl acetate and butyl acetate. While operations were largely manual before the late 1960s, improvements such as local exhaust ventilation were gradually introduced in the mid-1960s.

(ii) *Exposure to dusts*

In a Swedish investigation covering ten factories manufacturing paint and industrial coatings, dust was found during tinting, handling of bags, compressing empty bags, floor cleaning and emptying air-cleaner filters. The principal exposure to dust, however, was found during charging of raw materials. Sixty-one breathing-zone samples taken over durations of 5 min to 8 h indicated total dust exposure levels of 1.7-70 mg/m³. Raw materials charged included inorganic and organic pigments and fillers, chromium and lead compounds, talc and silica. The highest total dust levels (range, 7.7-70 mg/m³; four samples) were found in a powder coatings factory. Local exhaust ventilation was widely used, but fewer than half of the workers wore respirators. A few air samples were obtained to evaluate exposure to specific dusts during charging operation in some of the companies. Quartz was measured in five factories, with levels ranging from 0.01 to 0.9 mg/m³. Asbestos levels ranged from 0.3 to 5 fibres/cm³ (four factories). Chromium levels (as CrO_3) ranged from 0.003 to 1.6 mg/m³ (seven factories), while lead levels ranged from 0.006 to 4 mg/m³ (three factories; Ulfvarson, 1977). Blood lead concentrations monitored in 80 workers in 12 paint manufacturing companies in Finland were 5-72 μg/100 ml. The highest value was found in a spray painter (Tola *et al.*, 1976).

(iii) *Other exposures*

Exposure to ammonia was reported while charging it for use in water-based paints in the Swedish paint industry, at average levels of 50-80 ppm (35-56 mg/m³). In one case, more than 700 ppm (490 mg/m³) was measured. The levels of pentachlorophenol and phthalic anhydride were below the standards of 0.5 mg/m³ and 2 ppm (12 mg/m³), respectively (Ulfvarson, 1977). The concentration of diethylene triamine was below the detection limit (0.01 mg/m³) in the breathing zone of two workers canning epoxy paint curing agents in a Finnish paint factory (Bäck & Saarinen, 1986).

In a US paint manufacturing company, the 8-h time-weighted average (TWA) concentration of vinyl acetate ranged from 1.0 to 8.4 ppm (3.6-30.6 mg/m³; four samples). Personal and area air samples indicated concentrations of ethyl acrylate ranging from below the limit of detection to 5.8 ppm (23.8 mg/m³); concentrations of butyl acetate were all below the limit of detection (16 samples), except one sample at 0.9 ppm (4.7 mg/m³; Belanger & Coye, 1980).

(c) *Construction painting and lacquering*

Usual painters' work in the construction industry involves the use of a rather limited number of types of coatings - mainly decorative water- or solvent-based paints and wood lacquers and varnishes. The potential for exposure to a variety of substances (mainly solvents and pigments) is high, however: painting performed inside buildings, where poor ventilation opportunities, especially in confined spaces such as small rooms, cupboards, bathrooms, can lead to very high levels of contaminants; whereas when painting the outside of buildings (facades, windows, roofs), natural ventilation is usually effective. Painting of new buildings usually involves mainly water-based paints and spraying equipment; however, during renovation or maintenance, solvent-based paints are still widely used and work is usually performed by hand with a brush or roller.

Surfaces to be coated can be made of plaster- or gypsum-based wall-board composite materials, concrete, wood such as for windows, doors and flooring, and more rarely metal. Construction painters may spend a good proportion of their time in preparatory or accessory work. In a Finnish study on construction painters, 92 of 231 (40%) painters estimated that they spent more time on such work than actually painting (Riala *et al.*, 1984). Removing old paint and preparing surfaces in general may involve the use of paint strippers containing solvents such as dichloromethane, of gas-operated blow torch units or hot air guns which may generate organic pyrolysis fumes, metallic fumes and dusts from pigments containing *inter alia* chromium, lead and arsenic compounds. Other accessory tasks may be polishing, sanding or sandblasting operations, which generate old paint, quartz, concrete, plaster, wood and metal dusts. Acid or alkali washing solutions may be used, as well as steam generators for removing wallpaper, which release carbon monoxide-containing exhaust gases. Preparing surfaces also often involves filling cracks and holes using plaster, cement, sealers, spackling, taping and dry wall materials, putties and wood fillers, implying possible additional exposure to inorganic dusts and fibres (including asbestos) and solvents. Further exposure stems from the use of solvents during the cleaning of equipment as well as for personal cleaning (Ringen, 1982; Huré, 1986; Swedish Work Environment Fund, 1987).

The use of solvents in construction paints, and thus painters' exposures, has evolved radically with time. Early whitewashes and distempers contained no organic solvent, and oil paints contained only about 10% turpentine or, later, white spirits. Alkyd paints introduced in the 1960s required approximately 50% of a solvent such as white spirits. With the introduction of epoxy paints for special surfaces such as floors, other solvents such as alcohols, esters and aromatic hydrocarbons became more widely used. Water-based latex paints were introduced in the 1950s but were more widely accepted in the 1960s and 1970s, to become predominant in the 1980s; now, an estimated 60–80% of building trade coatings are water-based (Dufva, 1982; Hansen, 1982; Riala *et al.*, 1984). Vinylic and acrylic water-based paints are the most common, and these contain only a small percentage of organic solvents, mainly alcohols or glycol ethers.

The 8-h TWA exposure to solvents of 45 Dutch maintenance painters working on 12 different projects has been measured. Summed air concentrations averaged 101 mg/m³ (geometric mean) for the whole group and 59 mg/m³ for a subgroup of 20 house painters who applied only alkyd resins by brush and roller. Benzene was detected at only two of the sites and at low concentrations (up to 0.2 mg/m³). Toluene concentrations were below 4 mg/m³, except at one site where it reached 43 mg/m³. C_2- and C_3-substituted benzenes and C_8–C_{11} alkanes were found at most sites, originating mainly from the use of white spirits. Workers using chlororubber paint in a pumping station were exposed to carbon tetrachloride at levels ranging from 10 to 17 mg/m³; the highest level of toluene was also found at this site (Scheffers et al., 1985).

The exposure of Danish house painters to 13 solvents was investigated in 1974. Overall exposure, standardized to relevant occupational exposure limits, was above the permissible limit for five of 11 maximal values, reaching up to 34 times the permissible limit. Individual solvent average exposure levels were especially elevated for benzene (55 ppm [175 mg/m³]; 41 samples), believed to originate from thinners, and for trichloroethylene (91 ppm [490 mg/m³]; 33 samples). The origin of the trichloroethylene was not specified (Mølhave & Lajer, 1976).

In Finland, concentrations of Stoddard solvent during application of solvent-containing alkyd paints were 22–65 ppm (seven samples) and those during application of wood preservatives or alkyd varnishes, 68–280 (four samples). The overall solvent CEI during parquet floor varnishing using cellulose nitrate lacquers and urea–formaldehyde varnish ranged from 0.6 to 2.3, according to Finnish occupational exposure limits. Acetone, ethanol, isobutanol and butyl acetate were the main solvents used. Exposure to formaldehyde during varnishing averaged 2.8–4.5 ppm (3.4–5.5 mg/m³; Riala, 1982). The risk of formation of bis(chloromethyl)ether (see IARC, 1987s) from the reaction between formaldehyde and hydrochloric acid (used as a hardener) in urea–formaldehyde varnishes has been evoked (Dufva, 1982), but levels higher than 0.2 ppb (>0.9 μg/m³) have not been found (O'Neill, 1981).

In a study in Finland mainly of maintenance construction workers, the overall average airborne concentration of solvents during alkyd and urethane painting and varnishing, expressed as solvent naphtha exposure, was 132 ppm (77 samples); this was much higher when there was no ventilation, either natural or artificial (197 ppm; 46 samples), than with ventilation (38 ppm; 31 samples). Highest concentrations were observed during painting in small, unventilated rooms (303 ppm) and on large surfaces such as walls and ceilings with no ventilation (206 ppm with roller and brush painting and 243 ppm with spray painting). Taking into account other activities, e.g., use of water-based paints, the overall average 8-h TWA exposure level was 40 ppm (Riala et al., 1984).

Air concentrations (mg/m³) of organic vapours generated during the application of water-based paints were measured by personal sampling in Denmark, as follows: butyl acrylate, 0–2; diethylene glycol butyl ether, 4–5; diethylene glycol methyl ether, 8–32; dipropylene glycol methyl ether, 30–40; ethylene glycol butyl ether, 2–60; ethylene glycol phenyl ether, 0–0.7; propylene glycol, 2–70; 2,2,4-trimethylpentane-1,3 diol monoisobutyrate, 0.5–12; triethylamine, 4–6; and white spirits, 40–75. Concentrations of two gases, formaldehyde (at 0–0.4 mg/m³) and ammonia (at 2–12 mg/m³) were also reported (Hansen et al., 1987).

In a Swedish study of renovation spray painters, very high concentrations of white spirits (1200–1500 ppm) were measured during use of alkyd-type paints and 100–1000-times lower concentrations of solvents during use of acrylate-polyvinyl acetate-based water-borne paints. Dust concentrations, originating from paint mist, were higher during use of water-based paints (77–110 mg/m³) than solvent-based paints (17–27 mg/m³). Inorganic substances were found to represent 80% and 70–85% of the dust content, respectively. Exposures to substances such as lead (in solvent–based paints) and zinc (in both types of paints) were 10–23% and 1–2% of their respective exposure limits (Bobjer & Knave, 1977).

The mean blood lead level measured in 1962 for a group of 107 decorative and house painters in the USA was 23 μg/100 g blood, similar to that in control groups (Siegel, 1963).

(d)   Painting, varnishing and lacquering in the wood industry

Application of clear varnish or lacquer finishes on furniture represents the main use of coatings in the wood industry. Paints, varnishes and lacquers are also used in the production of various wooden raw materials (e.g., composite wood boards) and miscellaneous wooden articles (e.g., toys, tableware). Until the mid–1950s, cellulose ester–type lacquers were almost the only ones used in the furniture industry; however, amino resin–based, polyurethane and polyester coatings now constitute the main coatings in the industry (Swedish Work Environment Fund, 1987).

Workers are exposed mainly through inhalation or cutaneous absorption of solvents either from paint mist or from vapours generated by spraying operations, from vapours evolved from finished products or from auxiliary work such as mixing the coatings, cleaning equipment or applying other products such as wood fillers and sealants. The amount of exposure is influenced by the method of applying coatings; the most common are spraying, usually at low pressure, curtain and roller coating and dipping. The main categories of solvents used are aliphatic esters, ketones, alcohols and hydrocarbons as well as aromatic hydrocarbons (O'Brien & Hurley, 1981; Swedish Work Environment Fund, 1987).

Low molecular–weight resin constituents such as formaldehyde and isocyanates may be evolved during application or curing of coatings. Another possible exposure is to wood dust from the general factory environment and from preparatory work such as sanding.

Air monitoring was carried out over a ten–year period (1975–84) in 50 Finnish furniture factories, where the main coatings used were acid–cured amino resin–based paints and varnishes. The most commonly used solvents were xylene, *n*–butanol, toluene, ethanol, butyl-acetate and ethylacetate, which were present in more than 50% of 394 measurements. Mean concentrations of the solvents present were below 20 ppm, except for white spirits, which occurred at 66 ppm. Arithmetic mean solvent vapour concentrations measured during different work tasks ranged from 0.4 ppm in spray painting to 2.1 ppm during cleaning of a painting machine, with individual values varying from 0.1 to 7.4 ppm. Formaldehyde, derived from the amino resin binder, was the object of 161 short–term measurements (15–30 min) covering different work tasks. The arithmetic mean of the concentrations varied from 0.9 to 1.5 ppm (1.1–1.8 mg/m^3), with individual values ranging from 0.1 to 6.1 ppm (1.2–7.5 mg/m^3; Priha *et al.*, 1986).

The 8–h TWA exposure to formaldehyde of 38 employees in a Swedish light furniture factory applying acid–hardening clear varnishes and paints was found to average 0.4 mg/m^3 (range, 0.1–1.3) with a mean exposure to peak values (15 min) of 0.7 mg/m^3. Mean exposure to solvents was low. The dust concentration was low – usually less than one-tenth of the Swedish threshold limit value of 5 mg/m^3 (Alexandersson & Hedenstierna, 1988).

In a study of a US wood furniture company producing stereo equipment cabinets, the solvent exposure of 27 employees in spray painting and finish wiping operations was measured. In spraying jobs that involved the use of an acrylic base coat, an oil-based glaze or stain and cellulose nitrate lacquers, total exposure to paint mist (8-h TWA) varied from 0.1 to 2.5 mg/m³ (geometric means). Combined exposure to solvents (CEI) varied from 0.05 to 0.11 in base coat operations (solvents measured: methyl ethyl ketone, isopropyl acetate, xylene, isopropanol, methyl isobutyl ketone, toluene and isobutyl isobutyrate), from 0.06 to 0.10 in glaze operations (toluene, xylene, ethylene glycol monobutyl ether and petroleum distillates) and from 0.08 to 0.24 in lacquer operations (isopropanol, ethanol, isophorone, isobutyl acetate, n-butanol, toluene, xylene, ethylene glycol monobutyl ether, methyl ethyl ketone, isobutyl isobutyrate, isopropyl acetate and petroleum distillates). The overall low air concentrations of paint mist and organic solvents were attributed to adequate ventilation in paint booths and good working practices (O'Brien & Hurley, 1981).

Exposure to organic solvent vapours was also measured in 16 small-scale industries in Japan, where synthetic *urushi* lacquer was applied to wooden tableware (bowls), vases and altars, and in two furniture factories. Work involved mainly brush painting, screen painting and hand-spraying operations. Toluene, xylene, ethylbenzene and n-hexane were the recorded solvents. Average mixed solvent personal exposure (CEI) was always low (below 0.44), except in the case of an automated spray operation (1.4; Ikeda *et al.*, 1985).

The average 4-h exposure to toluene of 20 workers employed in painting and hand-finishing in an Italian art furniture factory was 27–182 mg/m³. Toluene was the principal solvent found in the work environment; other major solvents found were acetone, isobutanol, ethanol and ethyl acetate (Apostoli *et al.*, 1982).

In the Finnish plywood industry, solvent concentrations in workroom air were recorded during coating operations involving polyurethane and alkyd paints. The following ranges in ppm (mg/m³) were obtained from eight to 12 measurements: (i) polyurethane paint: methyl isobutyl ketone, 2–28 (8.2–115); butylacetate, 8–50 (38–238); xylene, 10–25 (43–108); and cyclohexane, 1–28 (3.4–95); (ii) alkyd paint: toluene, 2–3 (7.5–11.3); xylene, 7–12 (30.4–52); isobutanol, 7–11 (21–33); and trimethylbenzene, 1–9 (5–44; Kauppinen, 1986).

In a US plant where paint was stripped from wood and metal, breathing zone TWA concentrations of dichloromethane for three operators ranged from 633 to 1017 mg/m³ in seven samples (Chrostek, 1980).

(e)   Painting in the metal industry

Protection from corrosion is the primary aim of metal painting. Mild steel is thus almost always subjected to the application of a primer coat containing corrosion inhibitors such as iron and lead oxides or of zinc powder, further covered with a decorative paint. Aluminium may be covered with a zinc chromate-based primer before a decorative coat is applied.

During the preparation of metal parts, painters may be exposed to cleaning and degreasing agents, such as solvents, alkalis and acids, and to abrasive dusts, such as crystalline silica generated during blast cleaning. Depending on the industry, metal painters may be exposed to a variety of dusts, solvents, fumes and gases resulting from operations such as mixing paints, maintaining equipment, applying fillers, sealers or putty, or background metal welding or assembling operations. Most coatings used in the metal industry are solvent-based, and spray painting is the main method of application, leading to potential exposures to paint mist and solvents. Two–component paints, such as those based on epoxy and polyure-thane resins, play a major role, implying potential exposure to reactive substances such as isocyanates and epoxides. Air–drying or baking after application results in the evolution of solvents and, possibly, thermal degradation products of resins (Peterson, 1984).

(i) *Exposure to organic solvents*

Exposure of metal spray painters to a variety of solvents has been measured by the US National Institute for Occupational Safety and Health in a number of industries. The results are summarized in Table 18. Except in railroad car painting, overall exposure levels were found to be low. Toluene, xylene and petroleum distillates were among the most common solvents. Analyses of bulk air samples indicated no detectable benzene (O'Brien & Hurley, 1981).

In Finland, solvent concentrations were measured in the breathing zone of 40 car paint-ers at six garages (54 1-h samples). Mean concentrations and the upper limits of various solvents were as follows (ppm) [mg/mg^3]: toluene, 30.6 (249) [115 (940)]; xylene, 5.8 (36) [25 (156)]; butylacetate, 6.8 (128) [32 (608)]; white spirits, 4.9 (150); methyl isobutyl ketone, 1.7 (39) [7 (160)]; isopropanol, 2.9 (85) [7 (209)]; ethyl acetate, 2.6 (14) [9 (50)]; acetone, 3.1 (25) [7 (60)]; and ethanol, 2.9 (27) [6 (51)] (Husman, 1980).

A large study of Swedish car refinishing workshops showed that painters spent only 15% of their time actually spray painting, the rest being occupied with grinding, filling, mask-ing and assembling activities (60%) and colour mixing, degreasing and cleaning activities (25%). The highest overall solvent exposure was observed during spray painting, with a com-bined exposure of 0.3 (CEI; 106 samples). Toluene, xylene and ethyl acetate were present in all samples, at average levels of 39, 14 and 11 mg/m^3, respectively. Ethanol, butanol and butyl acetate were observed at very low levels in nearly half the samples. Other solvents encountered frequently in other activities, although at low levels, included styrene and white spirits. A reconstitution of working conditions in 1955 indicated that exposure levels to sol-vents were higher than in 1975, which was considered to be representative of the 1960s and 1970s. In particular, when benzene was used as a solvent in 1975–77, the combined exposure (CEI) reached 0.8 (Elofsson *et al.*, 1980).

Breathing zone samples were taken during short–term spray painting operations in a small autobody repair shop in the USA. Elevated levels of total hydrocarbons (up to 1400 ppm) were measured in winter when the spraybooth fan was turned off to conserve heat. Under these conditions, high concentrations of toluene (590 ppm; 2224 mg/m^3) were seen during lacquer spray painting and of xylene (230 ppm; 1000 mg/m^3) and benzene (11 ppm; 35 mg/m^3) during enamel spray painting. Summer conditions, when the fan was on, resulted in maximal concentrations of 330 ppm total hydrocarbons, 56 ppm (211 mg/m^3) toluene, 44 ppm (191 mg/m^3) xylene and 3.7 ppm (12 mg/m^3) benzene. Other major solvents measured were acetone, cellosolve acetate, methyl isobutyl ketone, *n*–hexane, methyl cellosolve acetate, tri-methylbenzene, ethylbenzene and *n*–butyl acetate (Jayjock & Levin, 1984).

Table 18. Painters' time–weighted average exposure levels (personal breathing–zone samples) in various metal spray–painting operations[a]

Operation	Sampling time	No. of samples	Combined exposure[b,c]	Main solvents measured	Concentration (mg/m³)[c]
Light aircraft finishing, primer spraying	25–41 min	3	0.9 ± 1.5	2–Butanone	42 ± 2.1
				Toluene	60 ± 1.2
				Ethanol	26 ± 1.6
				Isopropanol	19 ± 1.6
Light aircraft finishing, topcoat spraying	27–62 min	7	0.15 ± 1.3	Ethylacetate	77 ± 1.3
				Ethoxyethylacetate	44 ± 1.4
				Aliphatic hydrocarbons	34 ± 1.2
Light aircraft finishing, stripping operations	19–35 min	6	0.13 ± 2.5	Ethylacetate	52 ± 2.5
				Ethoxyethylacetate	30 ± 2.7
				Aliphatic hydrocarbons	73 ± 1.5
Car refinishing	15–45 min	7	0.09 ± 1.5	Toluene	39 ± 1.6
				Xylene	10 ± 1.0
				Petroleum distillates	21–63
				Other solvents	< 10
Railroad car	15–60 min	14	1.3 ± 1.4	Toluene	188 ± 1.5
				Xylene	14 ± 2.6
				Other aromatic compounds	217 ± 1.4
				Aliphatic hydrocarbons	840 ± 1.4
Heavy equipment	60 min	12	0.01–0.05	Refined solvents	21–96
				Other solvents	≤5
Metal furniture, solvent and water–borne paints	8 h	5 painters	0.10–0.46	Toluene	12–61
				Xylene	7–48
				n–Butyl acetate	22–109
				Diisobutyl ketone	< 1–23
				2–Ethoxyethyl acetate	1–14
				Aliphatic hydrocarbons	33–180
Metal furniture, high–solids paints	8 h	6 painters	0.07–0.31	Xylene	6–55
				Aromatic distillates	5–60
				Other solvents	< 10
Appliance finishing	8 h	4 painters	0.38–0.79	Toluene	88–204
				Xylene	112–225

[a]From O'Brien & Hurley (1981)

[b]Cumulative exposure index (see p. 366), based on US Occupational Safety and Health Administration permissible exposure levels

[c]Geometric means ± geometric standard deviation, unless otherwise stated

In Japan, the full–shift TWA concentrations of the following solvents measured for 13 car repair painters (ppm [mg/m^3]; mean $\pm$ standard deviation) were: xylene, 8 ± 8 [35 ± 35]; toluene, 19 ± 13 [72 ± 49]; isobutanol, 5 ± 5 [15 ± 15]; and ethyl acetate, 6 ± 4 [22 ± 14]. The overall combined exposure (CEI) was 0.38 ± 0.25. Short–term samples taken during painting showed a higher combined exposure for ten of 14 workers, toluene being the major solvent encountered (Takeuchi et al., 1982). In another Japanese study of car refinishing painters, high toluene concentrations were observed during painting in side–wall ventilated booths (410–660 ppm; 1546–2488 mg/m^3), compared with those in downdraft ventilated booths (28–87 ppm; 106–328 mg/m^3). Short actual painting periods resulted in full–shift TWA concentrations of organic solvents (toluene, xylene, methyl acetate, ethyl acetate and butyl acetate) below the exposure limits. The average hippuric acid concentration in the urine of painters (0.33 mg/ml) was slightly higher than that in controls (0.19 mg/ml; Matsunaga et al., 1983).

Exposure to toluene was investigated in 1940–41 in 106 painters in a large US airplane factory. Eight–hour TWA levels of toluene ranged from 100 to 1100 ppm (377–4147 mg/m^3); approximately 60% of workers were exposed to 200 ppm (754 mg/m^3) or more. Toluene was a major constituent of zinc chromate primers, lacquers, cellulose nitrate dope (lacquer) and brush wash (Greenburg et al., 1942).

An industrial hygiene evaluation was conducted at a commercial airline maintenance facility in the USA. Employees working in and around jet aircraft during the paint stripping process were exposed to levels of dichloromethane in the breathing zone that ranged from 79 to 950 mg/m^3 with a mean of 393 mg/m^3. During application of the prime coat, exposure to solvents was as follows (mg/m^3; mean and range): toluene, 112 (51–179); methyl ethyl ketone, 39 (8–77); butyl acetate, 72 (29–130); n–butanol, 25 (9–47); isopropanol, 51 (undetectable to 132); and cyclohexanone 10 (undetectable to 23). During application of the top coat, exposure to the solvents was: ethyl acetate, 333 (undetectable to 857); methyl ethyl ketone, 69 (undetectable to 219); methyl isobutyl ketone, 44 (nondetectable to 117); butyl acetate, 80 (undetectable to 210); xylene, 21 (undetectable to 49) and cellosolve acetate, 18 (undetectable to 46; Okawa & Keith, 1977).

Another study involved workers spray painting large commercial aircraft. Industrial hygiene measurements indicated short–term personal exposures as follows (mg/m^3; mean and range): toluene, 583 (140–1230); methyl ethyl ketone, 1436 (240–3250); ethyl acetate, 1231 (160–3520); naphtha, 44 (20–120); butyl acetate, 64 (20–150); xylene, 318 (60–1330); cellosolve acetate, 4843 (670–25 170); and dichloromethane, 654 (undetectable to 2840). Long–term exposures to the solvents were: ethyl acetate, 264 (10–1100); methyl ethyl ketone, 197 (20–440); toluene, 162 (30–450); butyl acetate, 11 (undetectable to 50); naphtha, 10 (undetectable to 160); xylene, 69 (10–270); cellosolve acetate, 640 (70–2490); and dichloromethane, 100 (undetectable to 760; Hervin & Thoburn, 1975).

Airborne concentrations (CEI) of solvent mixtures for jobs as paint mixer and spray painter ranged from 0.03 to 0.32 at a US plant manufacturing school and general purpose buses. The solvents found at the plant were petroleum naphtha, toluene, xylene, benzene, methyl ethyl ketone and n–hexane (Zey & Aw, 1984).

UK shipyard painters working in ships' accommodation and bilges were exposed to various mean TWA levels of organic solvents, depending on their job: 125 mg/m³ for three painters using a chlorinated rubber paint with white spirits as solvent, 215 mg/m³ for a worker using paint stripper with dichloromethane as the main solvent and 577 mg/m³ for four men using white interior paint with white spirits as the main solvent. Other paint solvents used frequently in dockyards are methyl-*n*-butyl ketone, *n*-butanol, trichloroethylene, xylene and cellosolve (Cherry *et al.*, 1985).

In Poland, phenol and hippuric acids were measured in 51 urine samples from shipyard painters working in small spaces of superstructures and in large holds. The average values of phenol in urine were 12.4–66.4 mg/l compared to 7.9 mg/l on average for a control group. Urinary phenol was attributed to benzene: the benzene concentration in air ranged from undetectable to 11 ppm (35 mg/m³). The average concentrations of hippuric acids in urine (sum of hippuric and methylhippuric acids) were 1812–5500 mg/l compared to 790 mg/l in a control group. Concentrations of toluene and xylene in air were 7–88 ppm (26–332 mg/m³) and 23–538 ppm (100–2335 mg/m³), respectively (Mikulski *et al.*, 1972). Elevated values of hippuric (up to 6700 mg/l) and methylhippuric acids (up to 7100 mg/l) were also measured in the urine of shipyard workers in Japan (Ogata *et al.*, 1971).

In a factory producing dump–truck bodies and earth–moving machinery in the UK, full–shift personal exposure levels to xylene and white spirits (two samples) were measured as 52 and 65 ppm (226 and 282 mg/m³) xylene and 7 and 12 ppm white spirits. After ventilation was properly adjusted, these levels dropped to 9 and 7 ppm (39 and 30 mg/m³) xylene and < 5 ppm white spirits (Bradley & Bodsworth, 1983). At a US plant where truck bodies and refuse handling equipment were manufactured, breathing zone concentrations of xylene during spray painting operations (eight samples varying from 1 to 3 h) ranged from 5 to 140 ppm (22–608 mg/m³; Vandervort & Cromer, 1975). Low exposure levels of toluene (3–18 mg/m³) and isobutyl acetate (2–44 mg/m³) were observed for Swedish spray painters in a plant manufacturing fireplaces (Hellquist *et al.*, 1983).

(ii) *Exposure to paint mists, dusts and specific metals*

Exposures of metal spray painters to paint mists, lead and chromium have been measured by the US National Institute for Occupational Safety and Health in a variety of industries. The results are summarized in Table 19. High concentrations of paint mist have been recorded in several operations, often linked with the painting of enclosed spaces and internal cavities, faulty ventilation and work practices. Substantial but short–term lead exposure was encountered in situations where lead–based pigments were used, such as for painting transportation and heavy equipment. Elevated but brief exposures to chromium were noted during the spraying of aircraft with primer. No antimony, arsenic, cadmium or mercury was encountered in these studies. Very low levels of tin (2–7 μg/m³) were recorded during the spray painting of dibutyltin dilaurate containing enamel on light aircraft (O'Brien & Hurley, 1981).

In a large study of Swedish car refinishing workshops, averages of 7 mg/m³ mist, 100 μg/m³ lead and 26 μg/m³ chromium were measured during spraying activities; during grinding activities, the corresponding values were 3 mg/m³, 20 μg/m³ and 6 μg/m³. The conditions

Table 19. Painters' exposure (personal breathing–zone samples) to paint mists, lead and chromium in various metal spray–painting operations[a]

Operation	Sampling conditions	No. of samples	Exposure level[b]		
			Paint mist (mg/m³)	Lead (μg/m³)	Total (μg/m³)
Light aircraft finishing, primer spraying	25–41 min	3	23.3 ± 1.6	ND	1600 ± 1.6
Light aircraft finishing, topcoat spraying	27–62 min	6	23.3 ± 1.7	ND	–
Light aircraft finishing, stripping operations	19–35 min	6	14.1 ± 2.0	ND–5000	–
Car refinishing	15–45 min	7	8.7 ± 1.6	52 ± 1.5	–
Car refinishing	8 h	7	5.0	30	–
Railroad car	15–60 min	13	43.3 ± 1.4	211 ± 1.7	220 ± 2.2
Heavy equipment	60 min	3	2.0–36.5	230–1300	31–230
Metal furniture	8 h	6 painters	3.7–27.6	ND–1050[c]	–
Metal furniture, high–solids paints	8 h	6 painters	0.5 –6.2	5–26	5–9
Small appliance parts, powder coating, electrostatic spraying	8 h	3	1.3 ± 1.1	–	–
Appliance finishing	8 h	4 painters	21.7–54.5	< 6–20	–

[a]From O'Brien & Hurley (1981)

[b]Geometric means ± geometric standard deviation, unless otherwise indicated

[c]Only 2.5–h samples taken one afternoon showed detectable levels (30–1050)

were thought to be representative of those in the 1960s and 1970s. Simulation of work conditions in 1955 showed low concentrations of lead during the use of all colours except red, when the Swedish exposure limit was exceeded by 70–fold. The actual exposure of painters was believed to be reduced by the use of individual protective equipment (Elofsson et al., 1980). Breathing–zone samples were taken during short–term spray painting operations in a small auto–body repair workshop in the USA. Only one of eight samples, corresponding to a red paint formula, contained significant levels of chromium (490 μg/m³) and lead (210 μg/m³); in all other measurements, the levels of chromium, lead and cadmium were below the detection limit. The concentration of total dust collected during the sanding or grinding of plastic body filler was 5–40 mg/m³ and that of respirable dust, 0.3–1.2 mg/m³ (Jayjock & Levin, 1984). In a factory producing dump–truck bodies and earth–moving machinery in the UK, full–shift personal exposure levels to total paint solids (two samples) were measured as 11.6 and 15.9 mg/m³. After ventilation was properly adjusted, these levels dropped to 1.4 and 5.2 mg/m³. The major pigments used were titanium dioxide and iron oxide (Bradley & Bodsworth, 1983). At a US plant where truck bodies and refuse handling equipment were made, breathing zone concentrations of solid contaminants measured during various spray painting operations (seven samples varying from 1 to 3 h) were 4.8–47 mg/m³ for total particulates, 20–3000 µg/m³ lead and 10–400 µg/m³ chromium (Vandervort & Cromer, 1975). Low overall exposure levels were found for Swedish spray painters working in a plant manufacturing fire-places, with a total dust level of 1.7 mg/m³, chromium oxide, 5–8 µg/m³ and zinc oxide, 20–30 µg/m³ (Hellquist et al., 1983).

At a US plant where school and general purpose buses were manufactured, employees working in and around the paint booth were reported to be exposed to hexavalent chromium and lead. The concentrations of hexavalent chromium in five personal air samples were 0.03–0.45 mg/m³, with a mean of 0.23 mg/m³; airborne lead concentrations (eight personal samples) ranged from below the laboratory limit of detection (3 µg/filter) to 2.01 mg/m³, with a mean of 0.78 mg/m³ (Zey & Aw, 1984).

A US manufacturer of large-scale weapon, electronic and aero–mechanical systems reported exposure of workers to hexavalent chromium while spraying aircraft wheels with yellow lacquer primers containing zinc chromate. In 12 personal breathing zone samples, the level of chromium[VI] ranged from 13.3 to 2900 µg/m³ with a mean of 606.7 µg/m³ (Kominsky et al., 1978).

In a US plant in which bridge girders were sprayed with lead silico–chromate paint, personal air levels of lead and chromium (as Cr) in five samples were 0.01–0.25 mg/m³ (mean, 0.08) and 0.01–0.04 mg/m³ (mean, 0.02), respectively (Rosensteel, 1974). Substantial exposure to airborne lead was demonstrated for US workers involved in scraping old lead–based paint from the metallic structure of a bridge and priming it (24–1017 µg/m³); recoating with lead–based paint implied lower exposure levels (6–30 µg/m³). Blood lead levels in these workers were 30–96 µg/100 ml, with 58% above 60 µg/100 ml (Landrigan et al., 1982). In the Netherlands, workers involved in flame–torch cutting of a steel structure coated with lead–based paints were shown to be exposed to 2–38 mg/m³ airborne lead (Spee & Zwennis, 1987).

Blood lead levels have also been measured in workers in various occupations in three Finnish shipyards. Painters were among the most heavily exposed: mean blood levels in a total of 77 painters in the three shipyards were 20–28 µg/100 ml (Tola & Karskela, 1976).

(iii) Other exposures

Use of polyurethane type paints can result in exposure to diisocyanate monomers and their oligomers. In Sweden, 43 car repair painters were exposed to a TWA of 115 µg/m³ hexamethylene diisocyanate (HDI)–biuret oligomer, with a range of 10–385 µg/m³. Very high exposure peaks (up to 13 500 µg/m³) were measured. The concentration of HDI was 1.0 µg/m³ (Alexandersson et al., 1987). In Finland, average HDI and HDI–biuret oligomer levels in four car paint shops during spray painting (ten 5–10-min personal samples, outside respirator) were 49 (± 22 SD) and 1440 (± 1130) µg/m³, respectively. The proper use of a respirator with combined charcoal and particle filters was shown to reduce exposure levels to below detection limits (Rosenberg & Tuomi, 1984). In a US car repainting shop, three short-term air samples (5–13 min) taken in the breathing zone during spray painting operations showed HDI levels of < 130 µg/m³. Similar measurements taken during various light aircraft finishing operations (7–21 min, eight samples) indicated HDI levels below approximately 70 $\mu g/m^3$, except for one operation with a level of 250 $\mu g/m^3$ (O'Brien & Hurley, 1981).

Ambient levels of HDI during the spray application of an enamel top coat at a US airline maintenance facility were < 0.04–3.20 mg/m³, with a mean of 1.1 mg/m³ (Okawa & Keith, 1977).

Epoxy paints are usually applied as reactive mixtures of epoxy resins and curing agents, leading to potential exposure to compounds containing the epoxide group. Total epoxide concentrations have been measured in area samples of aerosols collected during three painting operations involving the use of a bisphenol-A diglycidyl ether type of epoxy resin. In a facility producing military aircraft, the use of an epoxy primer did not result in detectable epoxide levels, and the authors surmised that the epoxy–amine curing reaction had probably consumed most of the epoxide group. Epoxide levels of 2–12 $\mu Eq/m^3$ epoxide functional group were recorded during the painting of a tank with coal-tar epoxy coatings and the painting of a metal ceiling using an epoxy architectural coating (Herrick *et al.*, 1988). In a US company that finished structural steel members and other fabricated steel products, the products are blasted with steel shot or sand and spray–painted with two–component epoxy paints or oil–based paints. Personal air levels of epichlorohydrin were reported to range from 2.4 to 138.9 mg/m³, with a mean TWA of 64.9 mg/m³. Bisphenol A glycidyl ethers were also detected in the workers' breathing zone at levels which ranged from below the limit of detection (0.6 μg) to 28.6 $\mu g/m^3$, with a mean of 9.8 $\mu g/m^3$ (Chrostek & Levine, 1981).

The major thermal degradation components of epoxy powder paints were identified as phenol, cresols, bisphenol-A, pyridine, 2,3–dimethylpyrazine and formaldehyde; bisphenol–A glycidyl ether was not observed. Levels in the work environment of painters were not measured (Peltonen, 1986; Peltonen *et al.*, 1986). Diethylene triamine, which is a component of curing agents of epoxy paints, was measured in three samples collected from the breathing zone of a painter during spray painting of paper machine cylinders and pulp tanks at a concentration of 0.02–0.07 mg/m³ (Bäck & Saarinen, 1986).

Operators working in eight plants where coal–tar enamel protective coating was applied to pipelines with heat were exposed to high levels of coal–tar pitch volatiles (see IARC, 1985b) at up to 24 mg/m³ of benzene–soluble matter (full–shift samples). The overall respirable concentration of benzo[a]pyrene in the plants averaged 133 $\mu g/m^3$ (Larson, 1978).

3. Biological Data Relevant to the Evaluation of Carcinogenic Risk to Humans

3.1 Carcinogenicity studies in animals

No data were available to the Working Group.

3.2 Other relevant data in humans

(a) Toxic effects

(i) Skin and eyes

Workers in the paint manufacturing industry (Pirilä, 1947; Ulfvarson, 1977) and painters (Pirilä, 1947; Högberg & Wahlberg, 1980; Winchester & Madjar, 1986) are at a considerable risk of developing an occupational dermatosis. In one study of Swedish paint industry workers, the prevalence of occupational dermatoses was about 40%, 26% of which were on the hands and arms (Ulfvarson, 1977). Among Swedish house painters, the prevalence of occupational skin disease was 4–6%, mainly affecting the hands. More than half of the dermatoses were nonallergic contact eczemas, probably mostly induced by organic solvents, mainly in atopic subjects. Allergic contact eczemas involved hypersensitivity towards chromium, nickel, epoxy resin components and formaldehyde. Several cases were seen of allergy to chloracetamide, which was widely used as a biocide in water–based paints and glues. Sensitivity to turpentine, which was formerly prevalent among painters, is now rare (Högberg & Wahlberg, 1980).

The wide variety of skin sensitizing agents in paints include some of the monomer residues from resins (e.g., phenol/formaldehyde resins, carbamide resin, melamine resin, epoxy compounds, acrylates). In addition, natural resins, such as colophony, may contain sensitizing agents. Some hardeners, such as acid anhydrides and para–toluenesulfonic acid, may cause sensitization, as may some metals used as pigments and driers in paints, e.g., cobalt and zirconium. Chromate sensitivity is rare in the painting trades, due to the low solubility of the salts used. Of other additives, several biocides (e.g., formaldehyde, chlorophenols and isothiazolinones) may have this effect. Of the solvents, only turpentine and dipentene (limonene) are known to be sensitizers (Fregert, 1981; Hansen et al., 1987).

Some organic solvents (e.g. some ketones and esters) are irritants, as are some resin monomers (e.g., butyl acrylate) and additives (e.g., amines, ammonia and organic peroxides; Hansen et al., 1987).

Corneal changes have been described in workers exposed to spray paints containing xylene (Matthäus, 1964). Changes in the lens of the eye have been recorded in car painters exposed to a mixture of solvents (Raitta et al., 1976; Elofsson et al., 1980). However, no ocular effect was noted in industrial spray painters occasionally exposed to toluene at up to 4125 mg/m^3 (Greenburg et al., 1942). Water–based paints may contain triethylamine (Hansen et al., 1987), which can cause corneal oedema (Åkesson et al., 1985, 1986).

(ii) Respiratory tract

Complaints of irritation in the upper airways were reported among paint factory workers (Winchester & Madjar, 1986) and among painters occupationally exposed to white spirits and other solvents (Cohen, 1974; Seppäläinen & Lindström, 1982; Lindström & Wickström, 1983; Pham et al., 1985; White & Baker, 1988). Hyposmia has sometimes been associated with exposure of painters to solvents (Lindström & Wickström, 1983). Histological changes of the nasal mucosa were reported among industrial spray painters (Hellquist et al., 1983).

Some painters suffer from lower airway symptoms (Schwartz & Baker, 1988; White & Baker, 1988), and there is a high prevalence of chronic phlegm bronchitis among spray painters (White & Baker, 1988) and lacquerers (Sabroe & Olsen, 1979). An obstructive ventilatory pattern was recorded after testing lung function in people who abused spray paint by inhalation (Reyes de la Rocha *et al.*, 1987). A decrease in expiratory flow rates was noted in a few workers in a printing paint factory, probably due to irritant effects, but not among car painters (Beving *et al.*, 1984a). Other studies of painters have also indicated bronchial obstruction (Pham *et al.*, 1985; Schwartz & Baker, 1988; White & Baker, 1988), and small airways disease has been noted in car painters exposed to isocyanates (Alexandersson *et al.*, 1987). In contrast, no disturbance of lung function was reported among house painters using solvent-based (Hane *et al.*, 1977; Askergren *et al.*, 1988) and water-based (Askergren *et al.*, 1988) paints. Danish painters were reported to have a high rate of disability pensions due to respiratory disease (Mikkelsen, 1980).

Painting may also entail exposure to compounds that cause allergic reactions in the airways. Isocyanates can cause both asthma and pneumonitis in painters (Nielsen *et al.*, 1985; Hagmar *et al.*, 1987). Exposure in the painting trade to isocyanates and polyisocyanates may induce antibody formation (Welinder *et al.*, 1988). Acid anhydrides (e.g., trimellitic anhydride, phthalic anhydride and its derivatives, and maleic anhydride) caused sensitization in workers producing alkyd binders (Wernfors *et al.*, 1986; Hagmar *et al.*, 1987; Nielsen *et al.*, 1988). Moreover, paints sometimes contain asthma-inducing amines (Hagmar *et al.*, 1987).

Exposure to aluminium dust and iron oxide during paint production may cause fibrosis, and exposure to iron oxide can cause pneumoconiosis (Maintz & Werner, 1988).

(iii) *Nervous system*

The neurotoxic effects of exposures to solvents have been reviewed (World Health Organization, 1985; Cranmer & Golberg, 1986; National Institute for Occupational Safety and Health, 1987). Such effects have been determined by means of questionnaires about subjective symptoms, neuropsychological testing and neurophysiological examination of central and peripheral nervous system function (Table 20), as well as in epidemiological studies of neuropsychiatric diseases.

Subjective symptoms, e.g., a feeling of intoxication, fatigue, poor concentration, emotional instability, short-term memory problems and headache, have been recorded in a series of cross-sectional studies of workers in the paint manufacturing industry, of house painters, of car and industry painters and of shipyard painters. Some of the symptoms are short or mid-term, others are persistent. However, no such symptom was recorded in house painters using mainly water-based paints (Askergren *et al.*, 1988). Neuropsychological tests have documented impairment of psychomotor performance, memory and other intellectual functions, as well as changes of mood (Table 20).

Electroencephalographic changes and a slight decrease in cerebral blood flow were recorded in paint industry workers (Ørbaek *et al.*, 1985). Electroencephalographic abnormalities have also been seen in car and industry painters (Seppäläinen *et al.*, 1978; Elofsson *et al.*, 1980). Other studies of solvent-exposed painters have failed to identify such effects (Seppäläinen & Lindström, 1982; Triebig *et al.*, 1988), and no effect on auditory-provoked

Table 20. Symptoms and neurobehavioural effects in studies of workers in the painting trade[a]

Population	Symptoms	Psychomotor performance	Short–term memory	Other intellectual functions	Mood	Reference
Paint industry workers	+	+	+			Anshelm Olson (1982)
	+ +	–	–	[+]	+ +	Ørbaek et al. (1985)
	[+]	[+]				Winchester & Madjar (1986)
House painters	[+]	+	+	+		Hane et al. (1977)
	+	+ +	+	–	–	Lindström & Wickström (1983)
	[+ +]	[-]	[-]	[-]	[-]	Fidler et al. (1987)
	–	–	–	–	+	Triebig et al. (1988)
	+ +	[+ +]	[+ +]	[+ +]	+ +	Baker et al. (1988)
		+ +	[+]	[+ +]		Mikkelsen et al. (1988)
Car/industry painters		+	+	+	+	Hänninen et al. (1976)
	+					Husman (1980)
	+	+	+	+	+	Elofsson et al. (1980)
	+				+	Struwe et al. (1980)
		[-]	[+]	[-]		Maizlish et al. (1985)
Shipyard painters	+	[+]	–	–		Cherry et al. (1985)
	[+]			+		Valciukas et al. (1985)

[a]+, exposed group differed statistically significantly from a control group; + +, there was a dose–response relationship; –, there was no statistically significant difference; [], the Working Group considered that the evidence was limited because the effect was weak or inconsistent and/or the duration and/or intensity of the exposure was low.

potential was seen in painters exposed to water–based paints (Askergren et al., 1988). In one group of house painters (Mikkelsen et al., 1988) and in a study of car and industrial painters (Elofsson et al., 1980), signs of slight atrophy were found by computed brain tomography, but another study showed no such effect (Triebig et al., 1988).

Occasional cases of clinical polyneuropathy have been described in spray painters exposed to methyl–n–butyl ketone (Mallov, 1976). In a few cross–sectional studies of car and industry painters (Elofsson et al., 1980; Husman, 1980; Maizlish et al., 1985), signs of slight neurological impairment were observed during physical examinations. Neurophysiological studies of house painters (Askergren et al., 1988) and of car and industrial painters (Seppäläinen et al., 1978) have indicated slight toxic effects on the peripheral nervous system, but other studies have not (Seppäläinen & Lindström, 1982; Cherry et al., 1985; Ørbaek et al., 1985;

Triebig *et al.*, 1988). Formerly, painters exposed to lead sometimes showed clinical effects on the peripheral nervous system, including palsy (mainly affecting the extensor muscles of the forearm) and drop (affecting the wrist; Rosen, 1953). No effect on the peripheral nervous system was observed in painters who used mainly water-based paints (Askergren *et al.*, 1988).

In a cohort of Danish painters, statistically significant two- to three-fold increases in the relative risk of being granted a disability pension due to neuropsychiatric disease was found (Mikkelsen, 1980). Similarly increased risks were observed in case–control studies of applicants for disability pensions due to neuropyschiatric disease and for nursing home accommodation due to encephalopathy, in which the occupation of 'painter or other solvent-exposed trade' was used as an indicator of exposure (Axelson *et al.*, 1976; Olsen & Sabroe, 1980; Lindström *et al.*, 1984; Rasmussen *et al.*, 1985). Significant increases in risk were not, however, seen in other case–control studies, using subjects granted a disability pension (van Vliet *et al.*, 1987), subjects who had consulted general practitioners because of minor psychiatric illness (Cherry & Waldron, 1984) and deaths from presenile dementia (O'Flynn *et al.*, 1987). [The Working Group noted that the confidence intervals were wide and that the results of the latter studies could thus be considered non–positive rather than negative.] Painters were overrepresented among cases of psychomotor epilepsy (Littorin *et al.*, 1988).

[The Working Group noted that reasons for the variable outcome include differences in exposures, i.e., identity of chemicals, intensity and duration. Also, selection bias may have occurred; and the examination methods varied, some possibly being influenced by recent rather than chronic exposures. Finally, the control groups used may have not been appropriate, so that the effects of confounders cannot be ruled out.]

(iv) *Kidneys*

At the beginning of the century, it was claimed that exposure of painters to turpentine caused glomerulonephritis; this association was not firmly established (Chapman, 1941), although the suspicion that a toxic effect of solvents caused clinical disease of the glomeruli remained. Goodpasture's syndrome has been associated with exposure to paint solvents (Klavis & Drommer, 1970; Beirne & Brennan, 1972).

Case studies of glomerulonephritis indicated a possible association with exposure to various solvents, including those in paints (Zimmerman *et al.*, 1975; Ehrenreich *et al.*, 1977; Lagrue *et al.*, 1977; Ravnskov *et al.*, 1979; Finn *et al.*, 1980), although one study that showed a relative risk (RR) of 1.1 (95% confidence interval (CI), 0.4–3.1) did not (van der Laan, 1980).

Most studies on kidney disease in the painting trade have concentrated on solvents. Several solvents are nephrotoxic (Lauwerys *et al.*, 1985). In a study of industrial spray painters exposed to toluene-containing paints, no indication of kidney disease was observed (Greenburg *et al.*, 1942). Later cross-sectional studies using more sophisticated methods revealed only minor effects. Among paint industry workers who were exposed to toluene and xylene, slight haematuria and albuminuria were observed but no effect on concentrating ability or glomerular filtration rate (Askergren, 1981; Askergren *et al.*, 1981a,b,c). These results were interpreted as being a minor effect on the glomeruli. In another study of painters exposed to toluene and xylene, indications of very slight tubular effects were reported (Franchini *et al.*, 1983). In a third study of car painters exposed to low levels of white spirits and toluene, no such effect was observed (Lauwerys *et al.*, 1985); however, a minor increase in urinary albumin excretion was reported among house painters using mainly water-based paints (Askergren *et al.*, 1988).

Kidney disease may be caused by exposure to lead in paints (Skerfving, 1987; see also IARC, 1980a).

(v) *Liver and gastrointestinal tract*

Slight effects on serum liver enzymes were noted in early studies of groups of industrial spray painters (Greenburg *et al.*, 1942). In other studies of paint industry workers (Lundberg & Håkansson, 1985), of car painters (Kurppa & Husman, 1982), of house painters (Hane *et al.*, 1977) and of subjects with suspected organic solvent poisoning (e.g., car painters; Milling Pedersen & Melchior Rasmussen, 1982), no consistent change in levels of serum liver enzymes was observed.

Lead may cause colic ('painter's colic'), and solvents and arsenic have also been claimed to cause gastrointestinal symptoms among painters and varnishers (Rosen, 1953).

(vi) *Blood and haematopoietic system*

Results obtained from haematological studies of workers in the painting trade vary. In painters who used gasoline as a solvent, a reduction in blood haemoglobin level was observed. Typical levels of aromatic hydrocarbons (one-fifth to one-tenth of the total hydrocarbon content) were 300–800 ppm (Sterner, 1941). Similarly, in later studies of car spray painters exposed mainly to xylene at rather low levels (Angerer & Wulf, 1985) and of house painters who had been exposed to various solvents (Hane *et al.*, 1977), slight decreases in haemoglobin levels were reported. In contrast, in one study of car and industrial spray painters, increased levels of haemoglobin were reported (Elofsson *et al.*, 1980).

In early studies of car spray painters, a slight decrease in white cell counts was observed, with relative lymphocytosis (Lind, 1939). This was probably due to a myelotoxic effect of benzene which was a contaminant of toluene and xylene before 1950–60. However, in one later study of patients with suspected solvent poisoning (mostly house, industrial and car painters), a slight decrease in white cell counts was reported (Milling Pedersen & Melchior Rasmussen, 1982), and in a study of car spray painters, lymphocytosis was observed (Angerer & Wulf, 1985). In contrast, another study of house painters showed no change in white cell counts (Elofsson *et al.*, 1980).

In more recent studies of paint industry workers (Beving *et al.*, 1984b; Lam *et al.*, 1985; Ørbaek *et al.*, 1985) and car painters (Beving *et al.*, 1983), slight decreases in thrombocyte counts were observed; in paint industry workers, the fatty acid composition of platelet membrane was altered (Beving *et al.*, 1988). In a further study of patients with chronic poisoning suspected to be induced by solvents (e.g., car painting), no change in thrombocyte counts was reported (Milling Pedersen & Melchior Rasmussen, 1982).

In painters, lead affects the formation of haemoglobin and red cells in bone marrow and causes haemolysis in peripheral blood (Skerfving, 1987).

(vii) *Other organs*

Some indication has been found that solvents affect muscles (raised serum creatine kinase levels), as seen during short–term exposure of volunteers to white spirits (Milling Pedersen & Cohr, 1984), in workers (e.g., house painters; Milling Pedersen & Melchior Rasmussen, 1982) and in patients with poisoning suspected to be due to solvents (mostly house, industrial and car painters; Milling Pedersen *et al.*, 1980). In the latter study, an increase in the activity of lactic dehydrogenase was observed in muscle biopsy specimens.

Case histories have been reported of subjects who suffered myocardial infarction after exposure to dichloromethane in paint removers (Stewart & Hake, 1976). However, cohort studies of paint industry workers have not indicated an increased risk for cardiovascular disease (Morgan *et al.*, 1981, 1985; Lundberg, 1986).

(viii) *Mortality from conditions other than cancer*

Many of the papers mentioned below are discussed in greater detail in section 3.3. Only statistically significant results are given here.

In one study of US paint industry workers, no increase in the total deaths from diseases of the nervous system was observed over that expected (Morgan *et al.*, 1981), and in a further study a significant decrease was observed (Matanoski *et al.*, 1986). In a study of Swedish painters, there was increased mortality from suicide (Engholm & Englund, 1982; Engholm *et al.*, 1987).

A cohort of Swedish painters showed an increase in mortality from chronic obstructive respiratory disease (Engholm & Englund, 1982), but no such increase was seen in studies of US painters (Matanoski *et al.*, 1986), of US automobile painters (Chiazze *et al.*, 1980) or of US aeroplane spray painters (Dalager *et al.*, 1980) or in two studies of workers in the paint industry (Morgan *et al.*, 1981; Lundberg, 1986).

In two further studies of paint industry workers, no increase in the total number of deaths from diseases of the genitourinary system was observed (Morgan *et al.*, 1981; Lundberg, 1986), although in one of the studies three deaths from infectious urinary tract disease were observed among cleaners in paint manufacture who had been heavily exposed to solvents, while only 0.2 were expected (Lundberg, 1986).

In two studies of paint industry workers, no increase in the total number of deaths from diseases of the gastrointestinal tract was observed (Morgan *et al.*, 1981; Lundberg, 1986). In another study of US painters, a significant decrease in the number of deaths from gastrointestinal disease was observed (Matanoski *et al.*, 1986). Increased mortality from diseases of the oesophagus and stomach has been reported in painters (Engholm & Englund, 1982). There was an indication of an increased rate of liver cirrhosis in one study (Lundberg, 1986), and, in automobile (Chiazze *et al.*, 1980) and aeroplane (Dalager *et al.*, 1980) spray painters, proportionate mortality from liver cirrhosis also appeared to be increased. Similar findings were reported in Swedish house painters (Engholm *et al.*, 1987). Danish house painters did not display an increase in the incidence of cirrhosis (Mikkelsen, 1980). [The Working Group noted that, in interpreting effects on the liver and gastrointestinal tract, the possibility that workers in the painting trade have a higher alcohol consumption than the general population must be considered.]

In one study of paint industry workers, no increase in the total number of deaths from diseases of the blood or blood-forming organs was found (Morgan *et al.*, 1981).

Studies of paint industry workers have not indicated an increased risk for cardiovascular disease (Chiazze *et al.*, 1980; Morgan *et al.*, 1981; Engholm & Englund, 1982; Morgan *et al.*, 1985; Lundberg, 1986; Matanoski *et al.*, 1986). In one study of Danish painters (Mikkelsen, 1980), deaths from diseases of the circulatory system were increased 30% as compared to the general population, but not as compared to a control group of bricklayers. Spray painters in automobile factories showed increased proportionate mortality from hypertensive heart disease (Chiazze *et al.*, 1980).

An increased number of deaths from cerebrovascular disease was observed in paint factory workers (Morgan *et al.*, 1981, 1985). Data on cerebrovascular mortality among aeroplane painters are in accordance with these results but are not significant (Dalager *et al.*, 1980). In a study of US painters, a significant decrease in the number of deaths from cerebrovascular disease was observed (Matanoski *et al.*, 1986).

[The Working Group noted that cohorts of workers in the painting trades may be subject to selection, which may bias the results of mortality studies. Also, in mortality studies, the occupational and disease categories used are broad, decreasing the specificity of the observations.]

(b) Effects on fertility and on pregnancy outcome

(i) Fertility

McDowall (1985) analysed a 10% sample of 601 526 births within marriage registered in England and Wales in 1980–82 for which the occupation of the father was recorded on the birth certificate. The standardized fertility ratio for men in each of 350 occupational units, defined by the Office of Population Censuses and Surveys, was calculated, taking the value for all occupational groups combined to be 100. Table 21 summarizes the findings in the five occupational groups in which paternal exposure to paint is likely: artists and commercial artists; coach painters; other spray painters; painters and decorators not elsewhere classified, and french polishers; and painters, assemblers and related occupations. Men with occupations classified as 'other spray painters' and 'painters and decorators, and french polishers' had significantly more children than expected on the basis of national rates (standardized fertility ratios, 129 and 141, respectively, based on 694 and 2871 births).

Rachootin and Olsen (1983) carried out a case–control study of 1069 infertile couples and 4305 fertile control couples attending Odense University Hospital, Denmark, in 1977–80. The RRs associated with occupational exposure to 'lacquer, paint or glue' were 1.2 (95% CI, 0.9–1.7) for men with sperm abnormalities, 1.1 (0.7–1.7) for women with hormonal disturbances, 1.4 (0.8–2.6) for women with idiopathic infertility and 1.1 (0.7–1.8) for men with idiopathic infertility.

Bjerrehuus and Detlefsen (1986) reported on a postal survey of 3251 male painters in Copenhagen, Denmark, and 1397 construction labourers. Approximately half responded, and 18% of the painters reported failure to conceive after two years of trials, compared with 10% of the construction workers. Telephone interview with a sample of the painters who had not responded to the postal questionnaire yielded a similar infertility rate.

Table 21. Standardized fertility ratios, sex ratio, percentage of births with birthweights of less than 2500 g, stillbirths, perinatal mortality and infant mortality, according to father's occupation, in occupational units in which exposure to paint is likely; England and Wales, 1981–82[a]

Occupational title (Office of Population Census and Surveys, 1970)	No. of births[b]	Standardized fertility ratio	Sex ratio (M:F births)	Births with birthweight <2500 g (%)	Stillbirths		Perinatal mortality		Infant mortality	
					SMR	No.	SMR	No.	SMR	No.
Artists; commercial artists	373	105	0.884	4.8	75	19	65	20	72	19
Coach painters	20	89	1.000	10.0	73	1	56	1	119	2
Other spray painters	694	129*	1.224*	8.1	112	52	120	71	111	62
Painters and decorators not elsewhere specified; french polishers	2871	141*	1.049	6.6	99	191	99	242	98	224
Painters, assemblers and related occupations	341	100	1.018	7.9	131	30	131	38	144*	38
All occupations	601 526	100	1.061	6.6	100		100		100	

aFrom McDowall (1985); SMR, standardized mortality ratio

b10% sample, except for 'all occupations'

*Differs significantly from all occupations ($p < 0.05$)

(ii) *Perinatal toxicity*

Olsen and Rachootin (1983) reported in a letter to the Editor data on 2259 couples who had had a healthy child in 1978–79 at the Odense University Hospital, Denmark. Occupational exposure to various substances was assessed prior to delivery. Exposure to 'lacquer, paint or glue' was reported by 217 mothers and 1512 of their spouses. For maternal exposures, mean birth weights were 64 g less than the average; the authors reported that, after adjustment for maternal age, smoking and drinking habits and time to conception, birth weights were 51 g less than the average ($p = 0.12$). For paternal exposures, the adjusted birth weight of the babies was 14 g above average ($p = 0.56$).

Heidam (1984a,b) carried out a postal survey of the reproductive history of women living in Funen county, Denmark. Female painters were recruited from the local divisions of the trade union, and 76 of 81 (94%) to whom questionnaires were sent replied; among these, 38 pregnancies were reported (0.5 per woman), of which five (13%) were reported to be spontaneous abortions. A 91% response rate was obtained from a reference group of 1571 employed women; among these, 843 pregnancies were reported (0.5 per woman), of which 84 (10%) were reported to be spontaneous abortions. The corresponding RR was 2.9 (95% CI, 1.0–8.8) for painters after controlling for gravidity, pregnancy order and age. When, however, the analysis was done using separate data on births and spontaneous abortions registered in hospital, the corresponding RR was 1.1 (0.4–2.9). The authors suggested that there may have been some reporting bias among the painters.

McDowall (1985) also presented data relevant to perinatal toxicity, including the sex ratio of offspring, percentage of low birth weights and standardized mortality ratios (SMRs) for stillbirths and perinatal and infant deaths, for births in England and Wales in 1980–82 (Table 21). 'Other spray painters' had an elevated ratio of male:female births; but in each of the four other occupational units with exposure to paint, the sex ratio was below average. The infant mortality rate for the offspring of 'painters, assemblers and related occupations' was higher than that of all occupations. Workers in this group would generally be classified in social class IV, in which the infant mortality rate is 115.

Daniell and Vaughan (1988) used records of live births in Washington State, USA, from 1980–83 to compare the outcome of pregnancy in various occupational groups. Among the 1299 live births for which the occupation of father of the child was described as 'painter', the sex distribution and Apgar score at 1 min and 5 min were similar to that found in the 2529 live births for whom the occupation of the father was described as 'electrician' and in 1469 'general controls'. The RR for low birth weight (< 2500 g) among the offspring of painters was 1.1 (95% CI, 0.7–1.5) compared to 'electricians' and 1.4 (0.9–2.1) when compared to 'general controls'.

(iii) *Malformations*

McDowall (1985) also reported on malformations in England and Wales in 1980–82, according to maternal and paternal occupation (Table 22). Overall, there was no excess of malformations, except in the offspring of men in occupations classified as 'painters, assemblers and related occupations'. When specific malformations were considered, there was an excess of polydactyly in the children of men and women with occupations classified as 'paint-

Table 22. Standardized malformation ratios for specified malformations according to occupation, England and Wales, 1980–82[b]

Malformation	Artists; commercial artists				Coach painters				Other spray painters				Painters and decorators not elsewhere specified; french polishers				Painters, assemblers and related occupations			
	Father		Mother		Father		Mother		Father		Mother		Father		Mother		Father		Mother	
	Ratio	No.	Ratio	No.	Ratio	No.	Ratio	No.	Ratio	No.	Ratio	No.	Ratio	No.	Ratio	No.	Ratio	No.	Ratio	No.
All malformations	91	46	100	14	102	3	–	0	97	97	100	4	89*	369	100	4	241*	117	100	140
Anencephalus	381	3	–	0	–	0	–	0	120	2	–	0	58	4	–	0	126	1	187	5
Spina bifida	–	0	–	0	–	0	–	0	64	3	–	0	99	19	495	1	314*	7	177	12
Spina bifida and/or anencephalus	102	3	–	0	–	0	–	0	82	5	–	0	84	2	372	1	275*	8	187*	17
Cleft palate and/or cleft lip	93	3	–	0	–	0	–	0	111	7	–	0	77	20	–	0	163	5	111	10
Hiatus hernia and/or diaphragmatic hernia	–	0	–	0	–	0	–	0	–	10	–	0	101	3	–	0	–	0	94	1
Tracheo–oesophageal fistula, oesophageal atresia and stenosis	–	0	–	0	–	0	–	0	138	1	–	0	99	3	–	0	–	0	92	1
Rectal and anal atresia and stenosis	–	0	559	1	–	0	–	0	80	1	–	0	97	5	–	0	166	1	55	1
Malformations of the heart and circulatory system	156	5	–	0	550	1	–	0	80	5	–	0	79	20	–	0	166	5	58	5
Hypospadias, epispadias	54	2	288	3	–	0	–	0	119	9	–	0	78	24	–	0	194	7	83	9
Polydactyly	49	1	–	0	–	0	–	0	164	7	–	0	176*	31	1081*	2	392	8	144	9
Syndactyly	105	2	–	0	–	0	–	0	57	2	–	0	150	22	763	1	459*	8	213*	10
Reduction deformities	–	0	–	0	–	0	–	0	47	1	–	0	82	7	–	0	399*	4	138	4
Exomphalos, omphalocele	–	0	692	1	–	0	–	0	–	0	–	0	39	2	–	0	357	2	223	4
Down's syndrome	90	2	170	1	–	0	–	0	35	1	505	1	60	8	–	0	123	2	72	3

[a]For fathers, standardized malformation ratios are calculated, taking all occupations as 100. For mothers, standardized proportionate mortality ratios are calculated, taking all malformations in each occupational group as 100.

[b]From McDowall (1985)

*Differs significantly from all occupations (p < 0.05)

ers and decorators not elsewhere specified, and french polishers' and in children of men whose occupations were described as 'painters, assemblers and related occupations'. Syndactyly was in excess in the offspring of men and women whose occupation was 'painters, assemblers and related occupations'. Reduction deformities were also in excess for paternal exposure but not for maternal exposure. Spina bifida and/or anencephalus were in excess in the offspring of men and women described as 'painters, assemblers, and related occupations'.

Olsen (1983) reported data from the Register for Congenital Malformations in the county of Funen, Denmark, and took details of parental occupation from birth certificates. The authors reported a relative prevalence ratio of 4.9 (95% CI, 1.4–17.1) for congenital malformations of the central nervous system in the group in which the children's fathers were entered as painters in comparison with all other occupations; the ratio for mothers in this category was 0.

(c)   Genetic and related effects

Haglund et al. (1980) studied chromosomal aberrations and sister chromatid exchanges in the lymphocytes of 17 male paint industry workers (exposed to organic solvents) who were presumed to have the highest exposure among a group of 47 paint industry workers employed in seven different factories in southern Sweden. For each exposed person, a control was chosen, matched by sex, age, place of residence (rural/urban) and smoking habits. Most of the controls were also factory workers (storeroom personnel, paint grinders, electricians, drivers, carpenters), but presumably unexposed. For analysis of both chromosomal aberrations and sister chromatid exchange, lymphocytes were cultured for 72 h; 20–25 metaphases were studied for sister chromatid exchange (17 subjects) and 100 for chromosomal aberrations (five subjects with the highest combined exposure). No difference was seen in either parameter; a significant difference in the frequency of sister chromatid exchange was observed between smokers and nonsmokers (0.202 and 0.175, respectively; $p = 0.02$). [The Working Group noted the small number of workers studied for chromosomal aberrations.]

Sister chromatid exchange was studied in the peripheral lymphocytes of 106 members of the International Brotherhood of Painters and Allied Tradesmen in two major US cities (Kelsey et al., 1988). Intensity and duration of chronic exposure to solvents were estimated from interviewer-administered questionnaire data. Eight men reported no occupational history of solvent exposure; 13 allied tradesmen (including dry-wall tapers and paperhangers) reported minimal, indirect exposure to solvents and had no history of direct application of solvent-based materials. Cumulative exposure (CEI) to solvents was estimated for the working lifetimes of 85 painters. Fifty cells from each of 91 individuals were scored for sister chromatid exchange; for the remaining 15 persons, a mean of 21.2 cells per individual was examined. Cultures were incubated for 72 h. There was no elevation in the frequency of sister chromatid exchange attributable to cumulative duration of exposure to solvents or to intensity of exposure over the year prior to blood sampling. Smoking was associated with a significant elevation in the level of sister chromatid exchange (6.75 versus 5.73 in nonsmokers).

3.3 Epidemiological studies of carcinogenicity in humans

(a) *Occupational mortality and morbidity statistics*

Detailed data from some of the studies described below are given in Table 23.

(i) *National studies*

The occupations recorded on a 10% sample of death certificates in England and Wales were used to calculate SMRs for deaths occurring around the time of the 1951 Census (Office of Population Censuses and Surveys, 1958), of the 1961 Census (Adelstein, 1972; Office of Population Censuses and Surveys, 1972), of the 1971 Census (Office of Population Censuses and Surveys, 1979) and of the 1981 Census (Office of Population Censuses and Surveys, 1986). The SMRs for various cancer sites among painters and decorators are listed in Table 23. SMRs for all cancers were consistently above the average and those for lung cancer consistently 40% above the national average: 149 (909 deaths) in 1949–53, 143 (1502 deaths) in 1959–63, 139 (847 deaths) in 1970–72 and 142 (803 deaths) in 1979–80, 1982–83. The proportion of current smokers among painters and decorators was reported to be slightly higher than that in the total population (smoking ratio, 110, based on a sample of 7566 men, including 153 painters and decorators; Office of Population Censuses and Surveys, 1979).

Guralnick (1963) divided specific causes of death in the USA in 1950 by occupation and industry as reported on death certificates and compared them with the expected causes of deaths of all working men as reported in the census from the same year. There were 6145 deaths among white male painters and plasterers in the age group 20–64, with a SMR of 114; selected SMRs are: all cancers, 126 (1016 cases); buccal cavity and pharynx, 137 (41); oesophagus, 109 (25); stomach, 127 (130); lung, 155 (248); kidney, 120 (24); bladder, 146 (38); brain, 134 (39); and leukaemia, 117 (41) (see Table 23). Proportionate mortality ratios (PMRs) were used to test for significance; only that for lung cancer was significant.

Dunn and Weir (1965) established in 1954 a fixed cohort of 68 153 working men engaged in occupations suspected of engendering a risk for lung cancer and followed them for mortality through to 1962 for this report. In this group, 12 572 men were painters and decorators. Information on smoking and occupation was gathered for this population until 1957. The number of deaths in the eight–year follow–up was compared with that among men in California, USA, 1959–61. Painters and decorators had an SMR of 129 (91 observed) for lung cancer; adjustment for smoking resulted in a decrease in the SMR to 114. The SMR for all other cancers was 94 (153 observed). [The Working Group noted that, since other cancers were treated as a group, it is impossible to determine the risk for those at specific sites.]

Howe and Lindsay (1983) followed a cohort comprising 415 201 Canadian men with known occupational histories in 1965–69, which represented 10% of the Canadian labour force. Cancer mortality in this cohort was monitored by record linkage with a Canadian mortality data base containing all deaths registered in Canada for the years 1965–73. The only significantly elevated SMR (285; based on five observed cases) was found for cancer of the buccal cavity and pharynx, except lip, in the occupational group of construction and maintenance painters, paperhangers and glaziers as compared to the mortality of the entire cohort.

Table 23. Cancer mortality or incidence in studies of national statistics and of large occupational cohorts of painting trades

Type of neoplasm	National statistics										Occupational cohorts					
	UK, 1949–53 (OPCS, 1958); 'other', painters and decorators; males, 20–64		UK, 1959–63 (OPCS, 1972); painters and decorators; males, 15–64		UK, 1970–72 (OPCS, 1979); painters and decorators; males, 15–64		UK, 1979–80, 1972–83 (OPCS, 1986); painters, decorators and french polishers; males, 20–64		USA, 1950 (Guralnick, 1963); painters and plasterers; white males, 20–64		Sweden, 1958–71 (Englund, 1980; Engholm & Englund, 1982); painters' union		Denmark, 1970–79 (Olsen & Jensen, 1987); painters (construction)		USA, 1975–79 (Matanoski et al., 1986); mixed painters	
	SMR[a]	No.	SMR[a]	No.	SMR[a]	No.	SMR[a]	No.	SMR[b]	No.	SIR[c]	No.	SPIR[d]	No.	SMR[b]	No.
All malignant neoplasms	124	2092**	122	2361**	123	1382**	124	1781**	126	1016**	109	647*	NA		110	927**
Buccal and pharynx	114	16[e]	78	12	138	10	145	40*	137	41	NA		61	5	NA	
Oesophagus	84	31	115	53	130	47	106	57	109	25	215 [195]	17 24]**	148	4	NA	
Stomach	122	360**	120	383**	118	174*	113	132	127	130*	NA [106]	80]	94	11	136	50*
Colon	106	120	101	123	98	78	88	74	112	77	NA		NA		111	93
Rectum	107	103	103	100	101	57	128	82	112	47	NA		NA		NA	
Liver and gall-bladder	65	11	100	26	103 60	9[f] 5[h]	NA		NA		200	12[g]	120 107	3[f] 2[h]	156	20[f]
Larynx	91	21	58	14	127	16	141	21	200	28**	177	14	NA		NA	
Nasal cavity	40	2	120	6	54	2	172	5	NA		NA		125	1	NA	
Lung	149	909**	143	1502**	139	847**	142	803**	155	248**	128 [127]	81* 124]**	149	79**	118	326**
Prostate	105	39	102	43	97	27	109	38	82	28	NA		48	8	98	84
Kidney	86	25	95	39	104	27	NA		120	24	NA		61	4	141	27
Bladder	109	58	118	79	152	66**	116	48	146	38*	NA		112	24	126	40
Non-Hodgkin's lymphoma	109	23	95	38	101	26	NA		NA		NA		56	3	NA	
Hodgkin's disease	113	35	90	37	52	10	144	21	129	22	NA		106	3	NA	
Multiple myeloma	NA		106	19	82	11	NA		NA		NA		NA		NA	
Leukaemia	111	50	98	65	125	43	81	33	117	41	173	13[i]	75	5	116	37

[a] SMR, standardized mortality ratio; expected numbers based on national rates for working men

[b] Expected numbers based on national rates for white males

[c] SIR, standardized incidence ratio; expected numbers based on national rates; in square brackets, SMR

[d] SPIR, standardized proportional incidence ratio; expected numbers based on the proportions of cancers in all persons registered in the Danish Pension Fund

[e] Pharynx

[f] Liver

[g] Intrahepatic bile ducts

[h] Gall-bladder

[i] Lymphatic leukaemia

* Significant at the $p < 0.05$ level

** Significant at the $p < 0.01$ level

NA, not available; OPCS, Office of Population Censuses and Surveys

In cohort studies and national statistics, information on smoking habits is not usually available. A review addressing the effect of smoking as a confounding variable in studies of occupational groups (Simonato *et al.*, 1988) indicates that smoking has a limited effect on the association between lung cancer and occupational exposures: the estimates might be increased by 20–25%. The estimates in the studies described above were usually increased to a greater extent.

(ii) *Other studies*

As part of the US Third National Cancer Survey, both occupation and industry were identified for each subject based on main lifetime employment, recent employment and other jobs held (Williams *et al.*, 1977). Interviews were obtained for a total of 7518 men and women, representing a 57% response rate. The RRs for cancers at particular sites were estimated for specific occupations. Painting (which included painters, construction workers, paper-hangers, and pattern and model makers) was the main lifetime occupation for 27 men and was associated with an excess RR for lung cancer: 4.2, based on 12 cases ($p < 0.01$). There were two cases of leukaemia (RR, 4.0).

In a study carried out in California, USA, using the death certificates of about 200 000 white men during the period 1959–61 (Petersen & Milham, 1980), the cause of death and usual occupation as reported on the death certificate were used to calculate proportionate mortality ratios (PMRs), standardized for age and year of death. The total number of deaths from all causes among painters was 3558. An elevated PMR for lung cancer was reported among painters, but figures were not available to the Working Group.

In a similar study, Milham (1983) analysed the death records of 429 926 men and 25 066 women during the period 1950–79 in Washington State, USA. The PMR for cancers of the lung, bronchus and trachea was significantly elevated among painters (mainly construction and maintenance painters): PMR, 121, $p < 0.01$ for all ages, 251 observed; PMR, 112, not significant for ages 20–64, 103 observed; and among auto painters and body/fender repairmen: PMR, 184, $p < 0.01$ for ages 20–64, 29 observed; PMR, 148, $p < 0.05$ for all ages, 39 observed. The PMR for gastric cancer was also elevated among painters and body/fender repairmen. Among paperhangers and decorators (painters), the PMR for lung cancer was elevated (139; 21 observed), but not significantly for ages 20–64; it was significant for all ages (PMR, 140; $p < 0.05$; 50 observed). In persons with this occupation, cancers of the bladder and other urinary organs (PMR, 179, $p < 0.05$ for all ages, 11 observed; PMR, 186, not significant for ages 20–64, two observed) and reticulosarcoma occurred in excess.

Dubrow and Wegman (1984) examined cancer mortality patterns by occupation for white males over 20 years old in Massachusetts for 1971–73. Using age–standardized mortality ratios, 397 occupational categories defined from information on death certificates were assessed for their association with increased risk for 62 malignancies. Increased risks (at $p < 0.05$) were apparent for stomach cancer (23 deaths; SMR, 158) in construction and maintenance painters; for cancer of the trachea, bronchus and lung in grouped painters (110 deaths; SMR, 131) and in shipyard painters (nine deaths; SMR, 261); and for laryngeal cancer (ten deaths; SMR, 205), skin neoplasms except malignant melanoma (four deaths; SMR, 492) and prostatic cancer (36 deaths; SMR, 146) in grouped painters. Grouped painters aged 55–74 years had a statistically significant increase in risk for buccal cavity and pharynx (14 deaths; SMR, 222); a nonsignificant excess of lymphomas was seen for men in the age group 20–64 years (eight deaths; SMR, 192).

Pearce and Howard (1986) compared cancer deaths among males aged 15–64 years in New Zealand in 1974–78, for whom occupation had been listed on the death certificate, with a 10% sample of census data. The RR for leukaemia was 2.3 in association with the occupation of painting (eight cases; 95% CI, 1.0–4.6). When adjusted for social class, the RR fell to 2.0 (95% CI, 0.86–3.9).

(b)   Cohort studies or studies within a cohort

(i)   Painters

Chiazze et al. (1980) studied workers in ten automobile assembly plants in five large companies in the USA. The plants were selected because of large numbers of employees, similar spray–painting operations, geographic dispersion and adequate records. The study was based on 4760 deaths among active and retired workers from 1970 or 1972 through 1976. A total of 4215 decedents were eligible for study, and employee work records were reviewed; for 253, work histories could not be obtained. The analysis was restricted to white males, who comprised about 80% of the decedents; 226 were spray painters. There was no significant excess proportion of deaths from any cause among spray painters, using either external local deaths or internal non-spray painters deaths. Lung cancer (21 deaths), which was the focus of the study, occurred more frequently among spray painters (PMR, 141) than in the local populations but not more frequently than among other automobile assembly workers (PMR, 108). PMRs greater than unity were noted also for leukaemias and lymphomas and for tumours of the brain, prostate, buccal cavity and pharynx. A nested case–control study covered 263 automotive workers who had died from lung cancer; they were matched by age within two years and by plant of employment with 1001 controls who had died of either cardiovascular disease or accidents. Spray painting was associated with a nonsignificant RR of 1.4 for lung cancer, and there was no indication of a dose–response relationship in association with exposure. The RR for those who had first been exposed at least 15 years prior to death was 1.0. The authors noted that individuals who had worked for only a few years may not have been included among the deaths if they had not been identified by an insurance claim in the company beneficiary file.

Englund (1980) and Engholm and Englund (1982) studied a cohort of 30 580 members of the Swedish painters' union from 1966 to 1974 for mortality and to 1971 for cancer morbidity by matching with national registers. The loss to follow–up was 1%. The SMR for all causes among painters was 102 (2740 cases), and the SIR for cancer was 109 ($p = 0.01$; 647 cases). Excesses were seen for cancers of the oesophagus (17 cases; SIR, 215 [95% CI, 124–340]), liver and bile ducts (12 cases; SIR, 200 [103–349]), lung (81 cases; SIR, 128 [106–152]) and larynx (14 cases; 177 [97–297]) and for lymphatic leukaemia (13 cases; 173 [92–296]). In a study based on population–based registries, about 38 000 painters in the 1960 census were linked to the national cancer registry, 1960–73. Among the 2064 cancers in painters, excesses were seen for cancers of the oesophagus (38 cases; SIR, 148) and of the intrahepatic bile duct (eight cases; SIR, 172). There was also a two–fold excess of pleural tumours based on six cases. The SIR for all cancers was shown to increase with increasing number of years since entry into the union (Engholm & Englund, 1982). The authors suggested in an abstract that smoking habits were no different among painters than among other groups (Engholm *et al.*, 1987).

Dalager *et al.* (1980) examined the risks for cancer among spray painters employed in the aircraft maintenance industry, where there was exposure to zinc chromate primers. Deaths among painters were compared with those expected among US white males using PMRs. The PMRs for all cancers (136, 50 cases) and for lung cancer (184, 21 cases) were significantly raised. The PMRs for cancers at several other sites were increased but not significantly so. The PMR for respiratory cancers increased with duration of employment.

In a study of 2609 male painters belonging to two painters' unions in the Copenhagen area, Mikkelsen (1980) found no increased risk for all cancers combined when the number of cases (82) was compared with those among men in a bricklayers' union (RR, 1.1; 95% CI, 0.8–1.6) or with those among all Copenhagen men (RR, 1.0; 95% CI, 0.8–1.3). Results were not reported for specific sites.

Whorton *et al.* (1983) followed up a group comprising 6424 union members residing in the San Francisco/Oakland Standard Metropolitan Statistical Area, representing six occupations: asbestos workers, bakers, painters, plasterers, plumbers and roofers. Individuals were considered to be members of the cohort if they appeared on union records in July 1976 and 1977. Incident cases of cancer were identified by computer linkage of union rosters to the California Tumor Registry, and the registry's age-, sex- and year–specific incidence rates were used to calculate expected numbers of cancer cases and SIRs. An increased incidence of cancer of the trachea, bronchus, lung and pleura was seen among painters (15 cases; SIR, 199 [95% CI, 112–330]). Relative risks in excess of unity were also observed for leukaemia and for cancers of the prostate and bladder. The authors pointed out that about 15% of all cohort members were of unknown vital status but were assumed to be alive.

In a cohort mortality study of US paint applicators, primarily in new constructions and maintenance, the records of a large international union of painters and allied tradesmen were used (Matanoski *et al.*, 1986). The cohort consisted of 57 175 men who had been born prior to 1940, had had at least one year of union membership, had been members of the union in 1975–79 in four states in different geographical areas, and had died in 1975–79. A total of 1271 (2.2%) individuals were lost to follow–up. Altogether, 5313 deaths occurred (SMR, 88, based on US white male rates). Death certificates were available for all but 288 (5.4%); SMRs were not significantly elevated for cancers at individual sites. Since there was no direct information on individual worker's trades, data from local union chapters were used to define the usual trade of their members; 58% of the cohort belonged to mixed painting locals. Using the US white male population for comparison, significant excess mortality ratios were seen in local chapters for painters for all malignant neoplasms (SMR, 110; 95% CI, 103–117), stomach cancer (136; 101–180) and lung cancer (118; 106–132), and nonsignificant ratios for cancers of the large intestine (111; 90–136), liver (156; 95–241), bladder (126; 90–172) and kidney (141; 93–205) and for leukaemia (116; 82–160) (see also Table 23). When the risks of men in local mixed painting chapters were compared with those of men in specialty locals, the mixed painters had significantly higher mortality from all causes, from malignant neoplasms, from lung cancer, from bladder cancer and from leukaemia. [The Working Group noted that the fact that all painters had to have been active dues–paying members at some time during the follow–up period would tend to have enhanced the 'healthy worker' effect in this population.]

A nested case–control study was conducted of lung cancer incidence in the New York unions included in the study described above (Stockwell & Matanoski, 1985). The 124 male lung cancer cases were identified through the New York State Cancer Registry, and 371 controls without cancer were selected randomly from the union membership and stratified by birth date and geographical location of the unions. Responses to questionnaires on work history, work environment and life–style factors were received from 69 (66%) of the cases and 182 (59%) of the controls; of these, 65 (94%) and 55 (33%) were completed by a proxy for cases and controls, respectively. Painting as the reported usual trade was associated with a high risk (RR, 2.8; 95% CI, 1.5–5.2); high risks were also seen for work in allied trades: painter as a union speciality (RR, 3.2; 95% CI, 1.4–7.1) and ever having worked as a painter (RR, 2.6; 95% CI, 1.3–4.9). In the 57 cases for which the information was available, 53 men were reported to have used spackling compounds (probably containing asbestos), compared with 112 of 161 controls (RR for spackling, 5.2; 95% CI, 1.9–14.5). The authors attempted to adjust for several variables, including asbestos exposure (on the basis of use of spackling compounds). The risk for lung cancer among painters who never wore a respirator remained high (5.4; 95% CI, 1.0–29.3). [The Working Group noted that a high proportion of cases reported using spackling compounds and questioned the accuracy of information obtained from a proxy regarding use of painting materials and of respirators.]

All 93 810 incident cases of cancer recorded in 1970–79 at the Danish Cancer Registry were linked with information on longest employment held submitted by the Supplementary Pension Fund (Olsen & Jensen, 1987). The standardized proportionate incidence ratios (SPIRs) for cancer were reported for each cancer site in each industry and occupation on the basis of the expected proportion of that cancer in all industries. Painters in the construction industry had an increased proportion of lung cancers compared to people in other occupations (SPIR, 149; 95% CI, 119–185; based on 79 cases). Workers in the paint, varnish and lacquer manufacturing industries had an increased proportion of cancers of the nasal cavity and sinus, with a SPIR of 620 (95% CI, 155–2480; based on two cases). In a follow–up study of cases registered through 1984 (Olsen, 1988), the SPIR was reduced to 401 (67–1324) based on two cases of sinonasal cancer. Car painters had a SPIR of 1403 (198–9958) for nasal cavity and sinus cancers based on one case. Several other proportions were above one for these three groups, but the excesses were not significant (Olsen & Jensen, 1987).

(ii) *Paint manufacturers*

Bertazzi *et al.* (1981) followed a small cohort of 427 workers employed in paint manufacturing in Italy. The workers had to have been employed for at least six months at any time from 1946 through 1977 to be eligible for inclusion and were followed for 1954–78. The follow–up was 97.7% complete. There was a significant excess of all cancers in this population (18 cases; SMR, 184; 95% CI, 112–285) when national rates were used as the comparison.

Lung cancers occurred at significant excess when either national (eight cases; SMR, 334; 106–434) or local rates (227; 156–633) were used as a standard, and the risk increased with length of exposure and with latency. These workers were exposed to asbestos as well as to chromate pigments.

A similar study of a larger cohort of 16 243 US male workers in the paint and coating manufacturing industry was reported by Morgan *et al.* (1981). These men had been employed for one year or more after January 1946 in 12 large and 20 medium to small companies and were followed through 31 December 1976. Only plants that retained personnel records for at least 15 years were eligible for the study, and out of 47 eligible plants the 32 largest were finally studied. The overall follow-up rate of the cohort was about 94%. Death certificates could not be obtained for 8.2% decedents. There were 2633 deaths in all (SMR, 86). The cohort was divided into seven subgroups on the basis of their exposures as determined from individual job histories; individuals could appear in multiple exposure groups. Deaths from cancers of the colon and rectum occurred at higher rates in the total population than expected on the basis of numbers among US white males (colon: 65 cases; SMR, 138 [95% CI, 107–176]; rectum: 26 cases; SMR, 139 [91–204]). The risk for respiratory cancers, which was a major focus of the study, was not excessive in this population (SMR, 98; 160 cases); information on smoking habits was not available. Deaths from cancer of the liver and biliary passage occurred more frequently than expected in the subgroups of workers potentially exposed to pigments (seven cases; SMR, 273 [108–555]) and lacquer (five cases; SMR, 255 [81–583]). The SMR for leukaemia was 212 (eight cases [92–418]) in the subgroup of workers exposed to lacquer. A further report on this study (Morgan *et al.*, 1985) provided little additional information.

A small cohort of 416 men who had worked for five years or more in the Swedish paint manufacturing industry during the period 1955–75 were followed for mortality in the years 1961–81 (Lundberg, 1986). Reference numbers were taken from national statistics. Subjects were categorized into lower and higher exposure levels according to duration and intensity of exposure. Overall mortality was low (96 cases; SMR, 88), as was mortality from all cancers (22 cases; SMR, 84; 95% CI, 52–127) and from lung cancer (three cases; SMR, 63; 95% CI, 12–184). The SMR for multiple myeloma was 549 (three cases; 95% CI, 113–1606) and that for cancer of lymphatic and haematopoietic tissues 212 (five cases; 95% CI, 68–496). The three cases of multiple myeloma occurred in workers in the higher exposure category.

(c) *Case–control studies*

(i) *Cancers at multiple sites*

Cancer cases recorded at a cancer centre in New York State, USA, in 1956–65 were compared with all patients with non–neoplastic lesions in regard to occupations related to inhalation of combustion products or chemicals and to personal characteristics (Viadana *et al.*, 1976; Decouflé *et al.*, 1977; Houten *et al.*, 1977). The information was obtained through an inverview at the time of admission for all patients. Each of the 11 591 white male subjects was included for analysis for each occupation held; specific occupations were compared with those of an unexposed clerical group. Painters were analysed as a subgroup of people with chemical exposures and as a subgroup of those with metal–related occupations. Cancer sites for which RRs were increased were: lung (42 cases; RR, 1.7; $p = 0.02$), stomach (eight cases; 2.4; $p = 0.05$), oesophagus (seven cases; 3.0; $p = 0.03$), prostate (nine cases; 1.9), bladder (16 cases; 1.6), kidney (four cases; 2.6) and melanoma (two cases; 3.2). The highest RR was seen in the age group below 60 years for stomach cancer (12.6); for oesophageal cancer, the risk was greater for the age group above 60 years (3.8). These two ratios were even higher among painters with five or more years of exposure (16.6 and 6.9, respectively). For lung and pros- tate cancer, no such dose–response relationship was observed. The elevated lung cancer risk among painters was no longer significant after adjustment for smoking and age (RR, 1.7). The author noted that the risk for stomach cancer was elevated in more than half of the occu- pations, which might be explained by the eastern European origin of the workers. No adjust- ment was made for alcohol drinking (see also Tables 24 and 25).

Coggon *et al.* (1986a,b) identified all cases of cancer in three English counties where chemical, metal and vehicle production industries were situated, using hospital and cancer registration records for the period 1975–80. Males aged 18–54 were included in the study. Occupational and smoking histories were obtained either by mailed questionnaires (re- sponse rate, 52.1%) or from information in hospital records or on death certificates. A total of 2942 cancer cases were identified, and cases of cancer at 15 specific sites were compared with those at all other sites with regard to occupation. Data were corrected for age, resi- dence, source of history and smoking. Laryngeal cancer was more likely to be associated with painting and decorating (RR, 3.4; 95% CI, 1.3–9.0; six cases) than with other occupations; bronchial cancer was also associated with painting, the RR being 1.3 (20 cases; see also Table 24). A borderline significant association was seen for cancer of the stomach (RR, 2.3; 95% CI, 1.0–5.0); other sites for which the RR was above unity were oral cavity (RR, 1.9; five cases), skin (RR, 1.4; four cases), testes (RR, 1.9; nine cases) and malignant melanoma (RR, 1.6; four cases). The authors commented that five patients with testicular cancer had worked as paint sprayers, which results in a RR of 4.9 (95% CI, 1.3–18.2). A nonsignificant RR of 0.7 was found for bladder cancer (see also Table 25).

In the same area of the UK, Magnani *et al.* (1987) examined occupations associated with cancer at five sites - oesophagus, pancreas, melanoma, kidney and brain. Deaths from these cancers in men aged 18–54 for the period 1959–63 and 1965–79 were matched by year of death, age at death and residence to those among four controls who had died from other causes. Occupation and industry were identified from death certificates. No significant risk for any of the cancers was associated with exposure to painting and decorating; however, the RR for oesophageal cancer was 2.0 (95% CI, 0.8–4.9) and that for brain cancer, 1.4 (95% CI, 0.7–2.8). The investigators also described exposures for each occupation, summed these across occupations, and examined the risks of these substances as they relate to the cancers. In this analysis, paints were associated with only a small increase in RR for three cancers - oesophageal and brain cancers and melanoma; none of the associations is significant. The authors noted that only the most recent full-time job was recorded on the death certificate. No adjustment was made for smoking or alcohol drinking.

Table 24. Case–control and other studies of lung cancer among persons exposed in paint manufacture and painting

Reference	Location, time	Type of controls	Source	Exposure	No. of cases (no. of painters)	RR	95% CI	Comments
Case–control studies								
Wynder & Graham (1951)	USA, NG	[Unclear]	Interview		857 (200 fume-exposed; 11 painters)	NG	NG	
Breslow et al. (1954)	USA, 1949–52	Hospital	Interview	Construction and maintenance painters for ≥5 years	518 (22)	1.9	0.93–3.8	Not adjusted
Menck & Henderson (1976)	USA, 1968–70	Estimated population by industry	Death certificates, hospital records	Painter at diagnosis	2161 (45)	SMR, 158	NG	Significant; adjusted for age
Milne et al. (1983)	USA, 1958–62	Deaths from other causes (except pancreas, bladder, nasal, kidney, haematopoietic)	Death certificates, occupation	Painter	925 men (24)	1.8	NG	Significant ($p < 0.01$); adjusted for age
				Paint manufacture	(3)	0.7	NG	Not significant
Kjuus et al. (1986)	Norway, 1979–83	Hospital	Interview and worksite records	Painting and paper-hanging	176 men (5)	1.7	0.4–7.3	Occupation is longest job held; considered exposed if ≥3 years; adjusted for smoking
				Paints, glues, lacquer	17	1.2	0.6–2.6	
Lerchen et al. (1987)	USA, 1980–82	Population and rosters of elderly	Interview	Ever construction painters	333 men (9)	2.7	0.8–8.9	Adjusted for age, ethnicity and smoking
				Asbestos	40	1.1	0.7–1.7	

Table 24 (contd)

Reference	Location, time	Type of controls	Source	Exposure	No. of cases (no. of painters)	RR	95% CI	Comments
Siemiatycki et al. (1987a)	Canada, 1979–85	Other cancers	Interview	Listed as white spirits, but in exposed group construction is 21% of total, mostly painters	857 males	1.1	0.8–1.4	Adjusted for age, socio-economic status, ethnicity, cigarette smoking, blue/white collar; 90% CI
					oat-cell 159 (36)	1.2	1.0–1.5	
					squamous–cell 359 (92)	1.0	0.7–1.3	
					adenocarcinoma 162 (37)	0.8	0.6–1.1	
					other types 177 (32)	1.7	1.2–2.3	
				Long duration, high exposure	44	1.4	NG	
				Construction workers				
Levin et al. (1988)	China, 1984–85	Population	Interview	Ever painter	733 men (15)	1.4	0.5–3.5	Questionable trend; adjusted for age and smoking
Ronco et al. (1988)	Italy, 1976–80	Deaths without smoking-related diseases	Interview	Painter	164 men (5)	1.3	0.43–4.1	Adjusted for age, smoking and other employment in suspect high–risk occupations
Multisite case–control studies								
Viadana et al. (1976); Decouflé et al. (1977); Houten et al. (1977)	USA, 1956–65	Noncancer admissions	Interview at admission	Painter	(42)	1.7	NG	Significant; adjusted for age; non-significant when adjusted for smoking and age
Coggon et al. (1986a)	UK, 1975–80	Other cancers	Interview	Painter	738 men (20)	1.3	NG	Adjusted for age, smoking, residence, respondent

RR, relative risk; CI, confidence interval; NG, not given; SMR, standardized mortality ratio

Table 25. Case–control studies of lower urinary tract cancer among persons exposed in paint manufacture and painting

Reference	Location, time	Type of controls	Source	Exposure	No. of cases (no. of painters)[a]	RR	95% CI	Comments
Wynder et al. (1963)	USA, 1957–61	Hospital, without smoking-related disease	Interview	Ever painter	300 (18)	[2.2]	[1.0–4.5]	No adjustment for smoking
Cole et al. (1972)	USA, 1967–68	General population	Interview	Painter	461 (28 men)	1.2	0.71–1.9	Adjusted for age and smoking
Howe et al. (1980)	Canada, 1974–76	Neighbourhood	Interview	Commercial painting	480 men (≥24)	1.0	0.6–2.3	Unadjusted. After correction for exposure to other suspect 'high–risk' industry, RR for spray painter, 1.0
				Ever spray painting	(≥16)	1.8	0.7–46	
Silverman et al. (1983)	USA, 1977–78	Population	Interview	Ever painter Car painter Paint manufacture	303 men (15) (3) (1)	1.0 0.5 0.2	0.5–2.2 0.1–2.1 0–2.2	Unadjusted
Schoenberg et al. (1984)	USA, 1978–79	Population	Interview	Ever painter Paint exposure	658 men (34) (111)	1.4 1.6	0.85–2.3 1.2–2.1	Adjusted for age, smoking and other employment
Vineis & Magnani (1985)	Italy, 1978–83	Hospital; other urological and surgical	Interview	Painter in building industry Car painter ≥5 years Carpentry painter Spray painter in different industries	512 men (12) (7) (1) (2)	1.0 2.0 0.6 1.2	0.40–2.2 0.60–7.0 0.04–8.4 0.20–5.8	Adjusted for age and smoking
Morrison et al. (1985)	USA, UK, Japan, 1976–78	Population	Interview	Paint and paint manufacture	USA, 430 (35) UK, 399 (23) Japan, 226 (5)	1.5 0.7 0.7	0.9–2.4 0.5–1.2 0.3–1.7	Adjusted for age and smoking; 90% CI
Claude et al. (1988)	FRG, NG	Hospital urological and homes for elderly	Interview	Ever painter Lacquer and paint Spray paints	531 men (15) (78) (52)	1.3 1.5 2.9	0.59–2.7 1.1–2.2 1.7–4.9	Trend, p = 0.04 for exposure to spray paints
Jensen et al. (1987)	Denmark, 1979–81	Population	Interview	Different painting industries Painter 10 years	371 (13)	2.5 1.4	1.1–5.7 1.0–1.9	Adjusted for age, sex and smoking

Table 25 (contd)

Reference	Location, time	Type of controls	Source	Exposure	No. of cases (no. of painters)[a]	RR	95% CI	Comments
Iscovich et al. (1987)	Argentina, 1983–85	Neighbourhood and hospital	Interview	Ever painter	117 (3)	0.55	[0.12–2.5]	Adjusted for age and tobacco smoke, pooling the two control groups
Schifflers et al. (1987)	Belgium, 1984–85	Population	Interview	Painter in high–risk occupation	74 (NG)	NG	NG	No increased risk reported
Risch et al. (1988)	Canada, 1979–82	Population	Interview	Exposed to paints in full–time job at least 6 months, 8–28 years before diagnosis	781 (204 men, 14 women)	1.1 3.9	0.77–1.6 0.9–26.7	Adjusted for smoking
				Commercial painting	(49 men)	0.90	0.39–2.1	
				Spray painting	(67 men)	0.91	0.48–1.7	
Siemiatycki et al. (1987a)	Canada, 1979–85	Other cancers	Interview	Listed as white spirits, but in exposed group construction is 21% of total, mostly painters	486 (91)	1.0	0.8–1.2	Adjusted for age, socioeconomic status, ethnicity, cigarette smoking, blue/white collar work; 90% CI
Multisite studies								
Coggon et al. (1986b)	UK, 1975–80	Other cancers	Interview	Painter	179 (10)	0.7	NG	Adjusted for age, smoking, residence, respondent; bladder and renal pelvis; men aged 18–54 only
Viadana et al. (1976); Decouflé et al. (1977); Houten et al. (1977)	USA, 1956–65	Noncancer admissions	Interview at admission	Painter	(16)	1.6	NG	Not significant

[a]If only discordant pairs noted, no. of painters ≥ number of discordant pairs given

RR, relative risk; CI, confidence interval; NG, not given

[The Working Group noted that the populations studied by Coggon *et al*. (1986a,b) and Magnani *et al*. (1987) may overlap and that only deaths in relatively young men were considered.]

(ii)   *Cancer of the lung*

These studies are summarized in Table 24.

In an early descriptive study, Wynder and Graham (1951) studied a total sample of 857 incident cases of lung carcinoma diagnosed in one hospital in St Louis, MO, USA, over an unspecified period. Of 200 who were 'believed or known to have been exposed to irritative dusts and/or fumes', 11 were painters. [The Working Group found it difficult to clarify the information on the comparison groups.]

Breslow *et al*. (1954) identified 518 cases of lung cancer in 11 Californian hospitals during the period 1949–52. Controls were selected from patients admitted to the same hospital for a condition other than cancer or a chest disease, and matched for age, sex and race. Detailed occupational and smoking histories were obtained by interview. The authors reported that 22 cases had been employed as construction or maintenance painters for at least five years, as had 12 controls [RR, 1.9; 95% CI, 0.93–3.8]. Smoking was not controlled for, although smoking histories had been recorded.

Menck and Henderson (1976) identified deaths from lung cancer for the years 1968–70 (2161 cases) and incident cases of lung cancer for the years 1972–73 (1777 cases) from the Los Angeles County Cancer Surveillance Program. Both were classified by occupation and industry on the basis of either death certificates or hospital records. Of the 3938 subjects, 689 had no reported occupation and 1222 no reported industry of employment. Employment of the population aged 20–64 was estimated from a sample of the population in the 1970 census, and the risk of lung cancer for each occupation was compared to the risk in the total population. The SMR for lung cancer in painters was significantly elevated (45 deaths; SMR, 158; $p < 0.01$; see also Table 24).

Milne *et al*. (1983) compared the occupation and industry of 925 (747 male and 178 female) cases of lung cancer in Alameda County, California, USA, 1958–62, with those of people who had died of other cancers. Usual occupation and industry as stated on the death certificate were coded using the US census classification. When occupations were examined separately, male painters had a significantly increased risk for lung cancer (24 cases) when compared either with all cancer deaths (RR, 1.7; $p < 0.05$) or with those dying of cancers other than of the pancreas, nasal sinus, kidney, bladder, bone and haematopoietic organs (RR, 1.8; $p < 0.01$). There was no increased risk associated with employment in the paint manufacturing industry (RR, 0.7; three cases; see also Table 24).

A study of 176 male incident lung cancer cases, under 80 years of age, admitted in 1979–83 to two hospitals in two neighbouring counties in Norway was conducted by Kjuus *et al*. (1986). Controls were matched on age through admission lists or from the same department records; persons with physical or mental handicaps, general poor health or an admission diagnosis of chronic obstructive pulmonary disease were excluded from the control group. Occupational histories were determined by interview and work site records then coded by job title and separated into three groups according to potential exposure to lung carcinogens, which included painting and paints. Three years was considered to be the minimal exposure classified as positive, and occupation was classified as the longest job held; exposures were included only up to 1970. Within the group, the RR for painting and paperhanging was 1.7 (95% CI, 0.4–7.3; five cases), adjusted for smoking. The RR for lung cancer associated with exposure to paints, glues and lacquer was 1.2 (95% CI, 0.6–2.6; 17 cases), adjusted for smoking, in comparison with all other subjects.

Occupational histories obtained by interview were compared in a case–control study of 506 lung cancer patients (333 men and 173 women) diagnosed in 1980–82, according to the population–based New Mexico Tumor Registry, and 771 controls selected through random telephone numbers or from rosters of elderly (Lerchen *et al.*, 1987). Next–of–kin provided the information for half of cases and 2% of controls. Jobs held by individuals from age 12 years were classified according to an a–priori list of potentially hazardous occupations. Construction workers and painters were included in high–risk occupations; employment for one year or more was classified as ever having been employed in an industry. The RR for lung cancer in men associated with employment as a construction painter was 2.7 (nine cases; 95% CI, 0.8–8.9) compared to never having been employed in that occupation and adjusted for age, ethnicity and smoking.

In the study of Siemiatycki *et al.* (1987a,b), described in detail in the monograph on some petroleum solvents (p. 70), construction workers exposed to white spirits, many of whom were painters, were described as having an excess risk for lung cancer (RR, 1.4 [numbers not given]).

In a cancer registry–based case–control study, Levin *et al.* (1988) identified 833 male lung cancer cases diagnosed between February 1984 and February 1985 in Shanghai, China, and 760 randomly selected male controls from the general urban Shanghai population, frequency matched within five–year age strata. Personal interviews to obtain occupational and smoking histories were obtained for 733 cases and 760 controls. More than 60 industries and occupations were examined; ever *versus* never having worked as a painter was associated with a RR, adjusted for age and smoking, of 1.4 (95% CI, 0.5–3.5). The RR varied according to duration of employment as a painter as follows: < 10 years, 1.9 (seven cases); 10–19 years, 2.8 (two cases); 20–29 years, 2.2 (five cases); ≥30 years, 0.3 (one case; questionable trend). The authors cited multiple comparisons and the use of broad occupational groups as limitations of the study.

Ronco *et al.* (1988) reported a population–based case–control study from two areas in northern Italy which included 164 male lung cancer cases identified from death records during 1976–80 and 492 controls who had died of conditions other than chronic lung disease or smoking–related cancers. Information on smoking and occupation was obtained through interviews of next–of–kin. Many exposures suspected of increasing the risk for lung cancer were evaluated, and individuals who had not held any job in any industry that was associated with exposure to a known or suspected lung carcinogen were classified as nonexposed. The RR for painters, adjusted for age, smoking and employment in other studied exposures, was 1.3 (five cases; 95% CI, 0.43–4.1).

Malker *et al.* (1985) examined the risk for pleural mesothelioma in relation to occupational exposures, including painting. The investigators used the Swedish population–based registries to link incident cancer cases during 1961–79 with 1960 census data on occupation and industry. Altogether, 318 cases of pleural mesothelioma occurred. Standardized incidence ratios (SIR) were calculated for occupations and industrial categories. For workers in the construction industry as a whole, a significant SIR of 1.6 was seen based on 63 cases; painting as a specific industry comported a higher significant SIR (2.9, based on 13 cases); painters and paperhangers as a specific craft showed an SIR of 2.0 (based on 12 cases), which was significant. [The Working Group noted that painters in the construction industry are probably exposed to asbestos.]

(iii) *Cancer of the larynx*

A case–control study of incident laryngeal cancer was carried out by Brown *et al.* (1988) in Texas. Cases consisted of all diagnoses of primary laryngeal cancer among white males aged 30–79 selected from 56 participating hospitals, comprising 220 living cases and 83 dead cases identified during the period 1975–80. Controls consisted of an equal number of white males without respiratory cancer selected from various sources and frequency matched on age, vital status, ethnicity and county of residence. Occupational exposures were examined, controlling for cigarette smoking and alcohol consumption. The RR for painters was elevated (11 cases; RR, 2.3; 95% CI, 0.84–6.3), and a significantly elevated risk was found for workers reportedly exposed to paint (32 cases; RR, 1.8; 1.0–3.2). No clear pattern was evident by duration of exposure.

(iv) *Cancer of the urinary tract*

These studies are summarized in Table 25.

Wynder *et al.* (1963) examined occupational and other risk factors associated with bladder cancer in 300 male patients from seven New York hospitals in 1957–61. Controls consisted of an equal number of male hospital patients who did not have myocardial infarction or cancers of the respiratory system or upper alimentary tract and were matched by age and time of admission. Interviews were conducted directly with the patients. The investigators reported 18 painters among cases and 12 among controls. [The Working Group calculated the RR to be 2.2 (95% CI, 1.0–4.5) for the group that had ever worked as a painter; no adjustment was made for cigarette smoking.]

Cole *et al.* (1972) conducted a case–control study of transitional- or squamous–cell carcinoma of the lower urinary tract in eastern Massachusetts using newly diagnosed cases aged 20–89 during an 18–month period ending 30 June 1968 (Cole *et al.*, 1971). Out of 668 cases ascertained, a random sample of 510 was selected for interview; a usable occupational history was obtained for 461. Controls were selected from the general population of the same area and matched on age and sex. Certain occupations (including painting) were classified as 'suspect'; and each of these groups was compared to nonsuspect industries. The RR for lower urinary tract cancer in male painters, adjusted for age and smoking, was 1.2 (28 cases; 95% CI, 0.71–1.9).

Howe *et al.* (1980) conducted a case–control study of bladder cancer in three areas of Canada; they identified 821 cases through provincial cancer registries in 1974–76 and matched them by age, sex and neighbourhood to 821 controls. Personal interviews were obtained for 632 cases (480 men and 152 women; 77%) and an equal number of controls. Among men, working as a painter was not associated with a risk: the RR for commercial painting was 1.0 (24 cases in discordant pairs; 95% CI, 0.6–2.3); that for spray painting was 1.8 (16 cases in discordant pairs; 0.7–4.6), which was reduced to 1.0 after correction for exposure in other suspect 'high–risk' industries.

As part of the US National Bladder Cancer Study, Silverman *et al*. (1983) conducted a population–based case–control study of bladder cancer in the Detroit, MI, USA, area. They identified 420 male cases diagnosed with transitional- or squamous–cell carcinoma of the lower urinary tract aged 21–84 between 1977–78; interview was obtained for 339 (81%), but the analysis was restricted to 303 white males. Controls were 296 white males stratified for age who were selected from a random digit–dialling survey for those under age 65 and from a random sample of the Health Care Financing Administration lists for those over 65. Employment was measured as 'ever' or 'usual' occupation or industry; 'usually unexposed' were those not employed in the industry of interest. The findings suggest no increased risk for bladder cancer for painters in general (15 cases; RR, 1.0; 95% CI, 0.5–2.2), for painters in the automobile industry (three cases; 0.5; 0.1–2.1) or for paint manufacturers (one case; 0.2; 0–2.2).

A similar case–control study of bladder cancer in 658 white male incident cases aged 21–84 during 1978–79 and of 1258 population controls was conducted in New Jersey, USA, by Schoenberg *et al*. (1984). Controls were selected as by Silverman *et al*. (1983). The RR for bladder cancer in men ever employed as painters, adjusted for age, was 1.4 (34 cases; 95% CI, 0.85–2.3). When occupations were classified by materials used, paint exposure was associated with a risk for bladder cancer (111 cases; RR, 1.6; 95% CI, 1.2–2.1). The risk was higher for those first exposed under age 41 and did not increase with duration of exposure.

A case–control study of bladder cancer in Italy (Vineis & Magnani, 1985) involved 512 male cases aged under 75 between 1978–83 and 596 hospital controls. The controls were matched by age and were subjects with benign urological conditions or surgical conditions. Occupational and smoking histories were obtained by interview. No increased risk was seen for painters in the building industry (RR, 1.0; 95% CI, 0.40–2.2; 12 cases), painters in carpentry (RR, 0.6; 0.04–8.4; one case) or spray painters (RR, 1.2; 0.20–5.8; two cases), but the RR for car painters was 2.0 (95% CI, 0.60–7.0; seven cases).

Morrison *et al*. (1985) examined 15 occupations and the risk for lower urinary tract cancer in Nagoya, Japan (1976–78), Manchester, UK (1976–78), and Boston, USA (1976–77), using incident male cases aged 21–89 and population–based controls. They identified 741 cases in Boston, 577 in Manchester and 348 in Nagoya. Interviews were obtained for 81% of the cases in Boston, 96% in Manchester and 84% in Nagoya; the corresponding figures for the controls were 80%, 90% and 80%. The analysis was limited to 430 cases and 397 controls in Boston, 399 cases and 493 controls in Manchester and 226 cases and 443 controls in Nagoya, for whom smoking histories were known. Occupational exposure to paint or paint manufacture was associated with a risk of bladder cancer only in the Boston population (35 cases; RR, 1.5; 90% CI, 0.9–2.4). This ratio was controlled for age and smoking history. [The

Working Group noted that no specific information was available on how the controls were selected.]

Two publications from the Federal Republic of Germany (Claude et al., 1986, 1988) reported two hospital-based case-control studies of tumours of the lower urinary tract. A total of 340 men and 91 women with such cancer between 1977-82 were matched by age and sex to either hospital patients primarily from urology wards or, for those over 65, to people in homes for the elderly. Subjects were interviewed about occupations, specific exposures and life-style factors. There was no reported excess risk for the occupational category of painting, but the RRs associated with specific exposures suggested a risk of painting in men. Spray painting was associated with an increased risk for cancer of the lower urinary tract (RR, 4.7; 95% CI, 2.1-10.4; 28 cases in discordant pairs), as was exposure to lacquer (RR, 1.6; 95% CI, 0.98-2.5; 45 cases in discordant pairs; Claude et al., 1986). In order to examine occupational risks more extensively, an additional 191 male cases were included, to make a total of 531 (Claude et al., 1988). Painting as an occupation was associated with an increased risk for bladder cancer (RR, 1.3; 95% CI, 0.59-2.7; 15 cases). An examination of the specific exposures indicated significant excess risks for cancer of the lower urinary tract for any exposure to spray paints (RR, 2.9; 95% CI, 1.7-4.9; 52 cases), to lacquer and paints (RR, 1.5; 95% CI, 1.1-2.2; 78 cases) or to chromium/chromate (RR, 2.2; 95% CI, 1.4-3.5). After correction for smoking, a significant trend of increased risk with increasing duration of exposure for individuals exposed to spray paints and chromium/chromate could be seen. [The Working Group questioned the choice of controls and considered that there may have been overlap between the exposure categories.]

Jensen et al. (1987) carried out a case-control study of bladder cancer in Denmark and interviewed 371 patients with invasive and noninvasive lesions diagnosed during 1979-81. The occupations of cases were compared with those of 771 controls selected from residents in the same area. Detailed occupational histories were taken, which included industry, type and place of work and duration; the information was coded according to industry. Significantly more cases than controls were employed in furniture lacquering and painting, industrial painting, sign-post painting, painting firms or car painting (13 cases; RR, 2.5; 95% CI, 1.1-5.7). Employment as a painter for ten years gave a RR of 1.4 (95% CI, 1.0-1.9).

Iscovich et al. (1987) performed a case-control study of 117 bladder cancer cases diagnosed in Argentina in 1983-85 and individually matched on age and sex to one neighbourhood and one hospital control. Hospital controls were selected from the same hospital as the case; about 12% of patients had diseases known to be associated with tobacco smoking. Neighbourhood controls were selected from among persons living in the same street block as the cases. A detailed questionnaire, containing information on smoking, demographic, socioeconomic and medical variables and occupational history for the three occupations of longest duration as well as the most recent one was administered. No increased risk for bladder cancer was observed among painters (three cases; RR, 0.55; [95% CI, 0.12-2.5]).

A pilot case-control study of bladder cancer in Belgium in 1984-85 (Schifflers et al., 1987) included 74 cases and 203 population controls selected from electoral rolls and matched for age and sex. While cases were interviewed by the investigators, most of the controls were interviewed by others. A group of 16 jobs, including painting, were defined as hazardous and associated with a high risk for bladder cancer, but exposure to painting as a specific job did not show a significant excess.

A case–control study from Denmark (Jensen et al., 1988) concentrated on cancers of the renal pelvis and ureter. The 96 cases, aged below 80, were identified from 27 hospitals in 1979–82, and three hospital controls were matched to each case on hospital, age and sex. Patients with urinary tract and smoking–related diseases were not eligible as controls. Questionnaire data on smoking and on occupation and occupational exposures were obtained. An elevated risk for upper urinary tract cancer was associated with occupational exposure as painter or paint manufacturer (RR, 1.8, adjusted for sex and lifetime tobacco consumption; 95% CI, 0.7–4.6; ten cases).

A case–control study of bladder cancer was carried out during the period 1979–82 in Alberta and in Toronto, Ontario (Risch et al., 1988). Cases aged 35–79 were identified through a cancer institute, from a province–wide tumour registry in Alberta, and through review of hospital records in Ontario. Interviews were carried out with 835 (67%) of the cases (826 histologically verified) and 792 (53%) of the controls about jobs in 26 industries that had previously been examined in studies of bladder cancer, and on occupational exposures to fumes, dust, smoke and chemicals. The analysis was carried out on the 781 matched sets for which adequate information was available. Occupational exposure to paints in a full–time job for at least six months, eight to 28 years before diagnosis was not associated with an increased risk for bladder cancer in men (age-adjusted RR, 1.1; 95% CI, 0.77–1.6; 204 cases) but it was for women (RR, 3.9; 0.9–26.7; 14 cases). Little difference in risk was seen between commercial (RR, 0.90; 95% CI, 0.39–2.1; 49 cases) and spray (RR, 0.91; 95% CI, 0.48–1.7; 67 cases) painting in men. The authors noted the problems associated with the very low response rate, the inclusion of cases with borderline malignancies and the potential for recall bias.

In the study of Siemiatycki et al. (1987a,b) (see p. 70), an increased risk for bladder cancer was seen among people exposed to white spirits, 21% of whom worked in construction trades, mostly comprising painting.

(v) Cancer of the biliary tract

Cases of biliary tract cancers were identified from the National Swedish Cancer Registry for the period 1961–79, and the occupations of the patients identified from the 1960 census of occupations (Malker et al.,1986). SIRs were calculated using the incidence rates for the total population and data from the 1960 census with regard to occupation and industrial employment, adjusted by region as well as by age and sex. There were 1304 cases of gall–bladder cancer and 764 cases of other biliary tract cancers in men, and 947 and 346 cases, respectively, in women. Significant SIRs of 1.3 (32 cases) and 1.4 (19 cases) for male painters and paperhangers were reported for gall bladder and other biliary cancers, respectively.

(vi) Cancer of the pancreas

Norell et al. (1986) reported on both a case–control study of pancreatic cancer and a retrospective cohort study of workers based on registry data in Sweden during 1961–79. Information on occupation was obtained through questionnaires. The case–control study included 99 cases of pancreatic cancer (aged 40–79) and 163 hospital controls of the same age and sex with inguinal hernia and 138 population controls of the same age, sex and residence. A significant excess risk was seen for exposure to paint thinners (ten cases; RR *versus* population controls, 2.5; 90% CI, 1.1–5.9; RR *versus* hospital controls, 1.4; 90% CI, 0.7–2.9). In the cohort study, a 20% excess of pancreatic cancer was seen in workers (aged 20–64) in paint and varnish factories (90% CI, 0.7–1.9) and a 30% excess for floor polishing (90% CI, 0.6–2.3).

(vii) *Haematopoietic neoplasms*

Studies on leukaemia are summarized in Table 26.

In a case–control study of leukaemia in three geographical areas of the USA in 1959–62, information was collected on occupations and other subjects by personal interview (Viadana & Bross, 1972). The controls were a random sample from households in the area matched for age and sex. The analysis was limited to 1345 adult leukaemia cases and 1237 adult controls in whites. No association was seen between any occupation and leukaemia in women. The risk for leukaemia in men appeared to be associated with work in the construction industry, and specifically with painting. The risk for painters *versus* nonpainters was 2.8 [1.4–6.0]; that in comparison with clerks was 3.1.

Timonen and Ilvonen (1978) interviewed 45 adults in northern Finland with acute leukaemia or chronic myeloid leukaemia between 1973–77 and a control group of 45 patients from the same hospital about use of drugs and chemicals, including paint. Four cases and four controls had been exposed to paint containing benzene derivatives and lead.

Flodin *et al.* (1986) performed a case–control study on acute myeloid leukaemia on 59 cases aged 20–70 years in 1977–82 from five hospitals in Sweden. Living patients and controls replied to a questionnaire about solvent exposure. Two series of controls were used: 236 matched for sex, age and residence, and 118 selected randomly from the same general population. For 'solvents, all kinds', there were 11 cases exposed and 58 controls (crude rate ratio, 1.2); no case but five controls were classed as 'painters'.

A population–based case–control study of 125 adult leukaemia cases and an equal number of controls matched for age, sex and residence was performed in Sweden in 1980–83 (Lindquist *et al.*, 1987). Information on occupation was obtained by a standardized questionnaire. 'Painters' included spray painters, car painters, machine painters, boat painters, asphalt painters and building painters. Thirteen cases and one control had been painters (RR, 13.0; 95% CI, 2.0–554). The median duration of exposure for painters was 16 years. After exclusion of case–control pairs with a 'painter', 26 patients and seven controls had worked in occupations which also involved exposure to paint and/or solvents and/or glues (RR, 3.7; 95% CI, 1.6–10.1).

Linet *et al.* (1988) linked records for Swedish men by major industry and occupational categories from the 1960 census to cancer registry data for 1961–79 to calculate SIRs for leukaemia subtypes. Expected numbers were based on a 19–year follow–up, taking account of age, region and birth cohort. Among men classified as painters or paperhangers, SIRs were 1.1, 1.0, 1.1 and 0.8 for acute lymphocytic, chronic lymphocytic, acute nonlymphocytic and chronic myelocytic leukaemia, respectively (based on three, 41, 33 and 14 cases, respectively).

Table 26. Case–control and other studies of leukaemia among persons exposed in paint manufacture and painting

Reference	Location, time	Type of controls	Source	Exposure	No. of cases (no. of painters)[a]	RR	95% CI	Comments
Viadana & Bross (1972)	USA, 1959–62	Population	Interview	Painter	845 men (31)	2.8	[1.4–6.0]	In comparison with non–painters
						3.1	NA	In comparison with clerks
Timonen & Ilvonen (1978)	Finland, 1973–77	Hospital	Interview	Paint containing benzene derivatives and lead	45 adults (4)	1.0	–	
Flodin et al. (1986)	Sweden, 1977–82	Population	Interview	Painter	59 adults			
Lindquist et al. (1987)	Sweden, 1980–83	Population	Interview	Painter	125 adults (13)	13	2.0–554	Adjusted for other exposures
				Other professions exposed to paint and/or solvents and/or glues		3.7	1.6–10.1	
				Daily exposure to organic solvents (white spirits) and gasoline		3.0	1.1–9.2	
				Organic solvent		2.0	(significant)	
				Petroleum products		1.4	(significant)	
Linet et al. (1988)	Sweden, 1961–79	Record linkage registry to census (cohort)	1960 Census record	(91) Acute lymphocytic (3)		SIR, 1.1		
				Chronic lymphocytic (41)		SIR, 1.0		
				Acute nonlymphocytic (33)		SIR, 1.1		
				Chronic myelocytic (14)		SIR, 0.8		

A total of 25 cases of Hodgkin's disease in men aged 20–65 was studied in 1978–79 using two controls selected from the Swedish population registry (Olsson & Brandt, 1980). Subjects were asked about occupations, and occupational exposure was defined as handling organic solvents every working day for at least one year within the closest ten–year period. There was a significant association between Hodgkin's disease and exposure to solvents (12 cases; RR, 6.6; 95% CI, 1.8–23.8). Three of the 12 cases and only one of six controls exposed to solvents were painters (RR, 1.7; [0.09–54.6]); the RR for painters among all subjects was 6.7 [0.56–177.0]. [See also the monograph on some petroleum solvents.]

Vianna and Polan (1979) studied mortality in 1950–69 from reticulum–cell sarcoma, lymphosarcoma and Hodgkin's disease among 14 occupational groups considered to be exposed to benzene and/or coal–tar fractions in New York State. The exposed populations were estimated from census data, and deaths were obtained from health department records; mortality, adjusted for age, was presented separately for each cancer site and compared with rates for the state. Among 21 951 painters, the SMR for reticulum–cell sarcoma was 110 (based on nine cases), that for lymphosarcoma, 97 (15 cases) and that for Hodgkin's disease, 135 (21 cases).

Friedman (1986) carried out a case–control study of multiple myeloma among members of the Kaiser Foundation Health Plan in California, USA, and identified 327 cases during the period 1969–82. These were matched by sex, age, race, date of enrollment and residence with 327 controls on the rolls at the time of case diagnosis. Information on occupation was obtained from medical records. Painters as an occupational group occurred more frequently among cases (6) than controls (2). [The Working Group noted that it was not stated how frequently information on occupation was available.]

Morris et al. (1986) conducted a multicentre population–based case–control study in the USA of 698 newly diagnosed cases of multiple myeloma aged under 80 during 1977–81 and 1683 neighbourhood controls matched by age, sex and race. In personal interviews with subjects themselves or with next–of–kin, exposures were ascertained through a question about any exposure to toxic substances. A toxicologist grouped exposures into 20 categories, including 'paints, paint–related products and/or other organic solvents', which resulted in a RR adjusted for age, sex, race and study centre of 1.6 (51 cases exposed to paints and/or solvents; 95% CI, 1.1–2.4); of these cases, 40 had been exposed to paints and paint–related products. This risk showed little variation according to time since first exposure. When only cases who had been interviewed themselves were included, the adjusted RR for paints and/or solvents was 1.8 (39 cases; 95% CI, 1.2–2.7). [The Working Group noted that there may have been bias in the reporting of exposure.]

A case–control study of multiple myeloma in six areas of England and Wales was carried out by Cuzick and De Stavola (1988). A total of 399 cases identified at major regional centres between 1978 and 1984 and 399 age– and sex–matched hospital controls were interviewed about their past occupation and exposure to chemicals and radiation, as well as prior and family history of disease and immunizations. The risk of multiple myeloma in painters, including spray painters, was 1.9 (15 exposed cases; [95% CI, 0.76–4.7]).

Olsson and Brandt (1988) reported a case–control study of 167 male cases of non–Hodgkin's lymphoma aged 20–81 seen in the oncology department of the University Hospital of Lund, Sweden, in 1978–81. Exposure was assessed by interview by one of the authors using a standardized questionnaire, as in the study of Olsson and Brandt (1980; see p. 417). Two control groups comprising a total of 130 men who had been interviewed for two other case–control studies were used to estimate the exposure frequency. The RR for 'organic solvents' was 3.3 (63 exposed cases; 95% CI, 1.9–5.8). The risk for supradiaphragmatic lymphoma was higher (RR, 3.4; 95% CI, 2.3–5.2) than that for lymphomas localized below the diaphragm (RR, 1.4; 95% CI, 1.0–2.0). The risk increased with duration of solvent exposure. Occupational exposure to solvents was associated with employment in machine shops, chemical industry, painting, printing, wood industry and many other types of work; 14% of this population were painters.

(viii) Cancer of the prostate

In a cancer registry-based case–control study in Missouri, USA, conducted by Brownson *et al.* (1988), 1239 cases of histologically confirmed prostatic cancer in white males diagnosed between July 1984 and June 1986 were compared to 3717 white male cancer controls diagnosed in the same time period and frequency-matched by age. Information on occupation, collected routinely using a standardized protocol in all hospitals, was coded at the Registry as usual occupation and industry using the 1980 US census codes. When compared to workers in 'low-risk' industries (wholesale and retail trade, finance, insurance, real estate, business services and professional services), an elevated age–adjusted RR for prostatic cancer was apparent for men whose usual industry was coded as manufacturing of paints and varnishes (five cases; RR, 5.7; 95% CI, 1.4–24.3). However, in the analysis of usual occupation, no risk for exposure to paint was seen. The authors recognized several study limitations, including the use of crude occupational information, multiple comparisons, and use of cancer patients as controls.

(ix) Cancer of the testis

Swerdlow and Skeet (1988) identified 2250 cases of testicular cancer from the South Thames Cancer Registry, UK, for the period 1958–77. The proportion of painters and decorators among cases was compared with that among controls with cancers other than those of the genital system or at an unspecified site and among controls with cancers sampled such that no site represented more than 15% of the cancers in an age group. Occupation was identified from the records for 75% of cases and 73% of controls. The risk for testicular cancer among painters and decorators was about half that in the comparison group of professional, technical workers and artists (RR, 0.45; 15 cases), for both seminoma (RR, 0.44) and teratoma (RR, 0.55).

(x) Cancer of the nasal cavity

Hernberg *et al.* (1983) conducted a case–control study of nasal and sinonasal cancers among cases collected from the cancer registers in Finland and Sweden and from hospitals in Denmark in 1977–80. The 167 cases in live patients who agreed to interview were matched by age, sex and country to controls with colon and rectal cancer. Many of the patients in the subgroup with lesions in the maxillary sinus were not interviewed. Exposures were coded by an industrial hygienist on the basis of intensity, duration and time. Smoking histories were evaluated for the period ten years prior to diagnosis, and smokers were found to be more frequent among cases (54.5%) than among controls (45.5%); the investigators indicated only that snuff use was not an important risk factor. Exposure to paints and lacquers reportedly showed a strong association with nasal cancer, but the investigators indicated that exposure to wood dust was generally a confounding factor. Two cases and no control had been exposed only to lacquers and paints, and both cases had had other potentially carcinogenic exposures.

(d) Cancer in children in relation to parental exposure

Fabia and Thuy (1974) analysed data on paternal occupation for 386 children aged less than five years who had died from malignant disease in the province of Québec, Canada, in 1965–70, and for the 772 control children whose birth registrations immediately preceded and followed those births. Father's occupation at the time of birth, as reported on the birth certificate, was recorded; no specific occupation was given for 30 cases or 56 controls. For ten cases and 11 controls, the paternal occupation at birth was described as painter, dyer and cleaner, excluding other hydrocarbon–related occupations (RR, 2.0; 95% CI, 0.86–4.7) [the Working Group calculated that the RR among those whose father was a painter was 1.2 (eight cases; 95% CI, 0.42–3.6)]. Among the 218 children with leukaemia, five of the fathers were in this occupational group; among the 101 children with central nervous system tumours and the 25 with Wilms' tumour, one (1%) and none (0), respectively, of the fathers were in this occupational group.

Hakulinen et al. (1976) carried out a case–control study of all 1409 children under 15 years of age with cancer reported to the Finnish Cancer Registry during the period 1959–68. After excluding twins and cases for which the father's occupation was unobtainable, 852 cases were available for analysis. The child born immediately before the case in the same maternity welfare district was chosen as a control. Father's occupation recorded at the time of conception was compared for cases and controls. Father's occupation described as 'painter, dyer, printer' was recorded for 12 cases and 15 controls; leukaemia and lymphomas occurred in one case and six controls, brain tumours in five cases and three controls and other tumours in six cases and six controls.

Kwa and Fine (1980) carried out a case–control study of 692 children born in Massachusetts, USA, who died before the age of 15 during 1947–57 and 1963–67. Controls were chosen from among the children whose birth registration immediately preceded and followed that of the case subject, giving a total of 1384 controls. Father's occupation at the time of birth registration was described as 'painter, cleaner, dyer' for 10 cases and 24 controls, comprising seven leukaemias or lymphomas, one neurological cancer and one urinary–tract cancer.

Zack et al. (1980) interviewed the parents of 296 children with cancer attending the Texas Children's Hospital Research Hematology Clinic in Houston, TX, USA, from March 1976 to December 1977. Controls were chosen from among relatives of cases, from among children in the neighbourhoods where the cases lived and from among children who did not have cancer attending the same clinic (33% had haemostatic defects, 24% various anaemias and 23% nonhaematological disorders). Job history from the year before the birth of the child until one year before cancer diagnosis was assessed by personal or telephone interview of a parent. The fathers of none of the cases were described as 'painter, dyer or cleaner'; the corresponding figures for fathers of controls were one for relatives, two for neighbours and one for children attending the same clinic. [The Working Group noted that the selection criteria were given for neither cases nor controls and that it was unclear whether information on exposures was obtained from mothers, from fathers or from both.]

Hemminki *et al.* (1981) described the paternal occupations of 2320 children aged 0–14 with cancer reported to the Finnish Cancer Registry in 1959–75, many of whom had been included in the study of Hakulinen *et al.* (1976). Controls were chosen from among children whose birth had been registered immediately before and immediately after that of the index child. Parental occupation was taken as that in the maternity welfare clinic records at the time of pregnancy. The overall RR for a father's occupation as painter was 1.4 ([95% CI, 0.67–2.9]; based on 40 discordant pairs); the odds ratio for leukaemia was 1.5 ([0.22–10.3]; based on 12 discordant pairs) and that for brain tumours was 2.6 ([0.70–9.6]; based on 14 discordant pairs). The excess of brain tumours was most marked for the more recent study period, 1969–75, in which a significantly elevated RR of 5.0, based on seven discordant pairs, was reported. Maternal occupation was recorded for 2659 children, but no data on mother's exposure to paint was presented. The authors noted that for the earlier period (1959–68) only 63% of the cases had been included in the analysis; but for 1969–75, 86% of cases were included.

In a case–control study (Peters *et al.*, 1981), cases of brain tumours in children under ten years of age at diagnosis in 1972–77 were identified from the Los Angeles County Cancer Surveillance Program. Controls were matched to each case by sex, race and year of birth; matching for social class was attempted by trying to locate the control from among friends of the case or from the same neighbourhood. Mothers of 98 cases (84% of those available) and of 92 controls were interviewed by telephone, and the 92 matched pairs were analysed. Information included working and exposure histories of the mother and father before the pregnancy, during the three trimesters of pregnancy, during nursing and at the time of diagnosis. The authors noted the possibility of biased reporting and recording of exposures. Seven fathers of cases were reported to have had exposure to paints at any time from one year before conception up to the time of diagnosis; the father of one control had been similarly exposed. [The Working Group noted that this study addressed any exposure to paints and not only occupational exposures.]

Sanders *et al.* (1981) studied 6920 children under the age of 15 years who had died of malignant disease in England and Wales in 1959–63 and 1970–72. Father's occupation reported on the child's death certificate was compared with that recorded on the death certificate for a total of 167 646 childhood deaths that had occurred during the same periods. The PMRs for father's occupation described as 'painter or decorator' were 97 (based on 93 cases of cancer) in 1959–63 and 74 (based on 34 cases) in 1970–72. [The Working Group noted that data on specific cancer sites were not given.]

Associations between paternal occupation and childhood leukaemia and brain tumours were investigated in a case–control study in Maryland, USA (Gold *et al.*, 1982). Patients under the age of 20 with leukaemia (1969–74) or brain tumours (1965–74) were ascertained in the Baltimore Standard Metropolitan Statistical Area from death certificates and records from 21 of 23 Baltimore hospitals. Two control groups consisted of children with no malignant disease, selected from birth certificates at the Maryland State Health Department, and of children with malignancies other than leukaemia or brain cancer. Information on occupational exposures of both parents before the birth of the child and between birth and diagnosis was collected by interviewing the mother. A total of 43 children had leukaemia and 70 had brain tumours. Paternal occupational category 'painter' was reported for one case of leukaemia, compared with three normal controls and none of the cancer controls, and no case of brain tumour, compared with one case in normal controls and none in cancer controls.

Wilkins and Sinks (1984) carried out a case–control study of 62 children with Wilms' tumour identified between 1950 and 1981 at the Columbus (Ohio) Children's Hospital Tumor Registry for whom paternal occupation was available from the child's birth certificate. Two groups of controls were chosen from birth certificates, the first matched individually for sex, race and year of birth, and the second for sex, race, year of birth and mother's county of residence at the time of the child's birth. Three of the fathers of cases were reported to be painters compared to one and none in the two sets of controls.

Van Steensel-Moll et al. (1985) carried out a case–control study of 713 children under 15 years of age with leukaemia diagnosed between January 1973 and January 1980 in the Netherlands. Controls were chosen from census records, matched by region, sex and age (to within two months). Information on occupational and other exposures of both parents during pregnancy was obtained by postal survey; the response rate was 88% for parents of cases and 66% for those of controls. The analysis was restricted to 519 patients with acute lymphocytic leukaemia and 507 controls. Twenty-five mothers of children with leukaemia and 11 mothers of controls reported having had occupational exposure to 'paint, petroleum products or other chemicals' during pregnancy (RR, 2.4; 95% CI, 1.2–4.6). These exposures were reported by 140 fathers of children with leukaemia and 113 fathers of controls (1.2; 0.8–1.7). The RR for paternal occupation described as 'painter, cleaner or dyer' was 1.6 (0.5–5.0) for exposures during pregnancy (eight cases) and 1.3 (0.4–4.0) for such exposures one year before the diagnosis of leukaemia (eight cases).

Lowengart et al. (1987) reported a case–control study of 123 children aged ten years or under with leukaemia identified in the Los Angeles County Surveillance Program in 1980–84, representing 57% of eligible cases. Controls were selected from among friends of cases or by random–digit dialling. Interviews were carried out by telephone and included questions on exposure to paints or pigments before, during and after pregnancy and on experiences the children had had from birth to the reference date. The specific types of exposure included in the general category 'paints or pigments' were spray paints, other paints, dyes or pigments, printing inks and lacquers or stains. Excess risks were observed for exposure of fathers to spray paints during the pregnancy (RR, 2.2; [95% CI, 0.91–5.3]) and after the pregnancy (2.0; 0.96–4.4) and for exposure to dyes and pigments during the pregnancy (3.0; [0.41–2.2]) and after the pregnancy (4.5; 0.93–42.8). The RR associated with 'spray paints' or 'dyes or pigments' was higher (RR, 2.5) if the father's exposure had been frequent (≥50 times per year) than if it had been less frequent (< 50 times per year; RR, 1.8) after the birth of the child. Data on maternal occupational exposure were not presented. Use of paints or lacquers in the home by the mother and/or father during the pregnancy and lactation gave a RR of 1.4 (0.79–2.6). [The Working Group noted that the exposure categories overlapped.]

Johnson *et al.* (1987) analysed paternal occupational exposures recorded on the birth certificate of 499 children aged 0–14 who had died of an intracranial or spinal cord tumour in Texas in 1950–79. Children who had been born outside Texas were excluded. Controls were chosen from a 1% sample of live births in Texas during the same period. Maternal occupation could not be assessed. A RR of 1.0 (95% CI, 0.3–3.3) was reported for paternal occupation described as a painter.

4. Summary of Data Reported and Evaluation

4.1 Exposures

Approximately 200 000 workers worldwide are employed in paint manufacture. The total number of painters is probably several millions, a major group being construction painters. Other industries in which large numbers of painters are employed include manufacture of transportation equipment and metal products, automotive and other refinishing operations and furniture manufacture.

Thousands of chemical compounds are used in paint products as pigments, extenders, binders, solvents and additives. Painters are commonly exposed by inhalation to solvents and other volatile paint components; inhalation of less volatile and nonvolatile components is common during spray painting. Dermal contact is the other major source of exposure. Painters may be exposed to other chemical agents that they or their coworkers use.

Painters are commonly exposed to solvents, the main ones being petroleum solvents, toluene, xylene, ketones, alcohols, esters and glycol ethers. Chlorinated hydrocarbons are used in paint strippers and less frequently in paint formulations. Benzene was used as a paint solvent in the past but is currently found in only small amounts in some petroleum solvent–based paints. Titanium dioxide and chromium and iron compounds are used widely as paint pigments, while lead was used commonly in the past. Asbestos has been used as a paint filler and may occur in spackling and taping compounds; painters in the construction industry and shipyards may also be exposed to asbestos. Exposure to silica may occur during the preparation of surfaces in construction and metal painting.

Workers in paint manufacture are potentially exposed to the chemicals that are found in paint products, although the patterns and levels of exposure to individual agents may differ from those of painters. Construction painters may be exposed to dusts and pyrolysis products during the preparation of surfaces and to solvents in paints, although water–based paints have become widely used recently. In metal and automobile painting, metal–based antirust paints and solvent–based paints are often applied by spraying; in addition, newer resin systems, such as epoxy and polyurethane, are commonly used. In contrast to other painting trades, furniture finishing involves the use of more varnishes, which have evolved from cellulose–based to synthetic resin varnishes, including amino resins which may release formaldehyde.

4.2 Human carcinogenicity data

The reports most relevant for assessing the risk for cancer associated with occupational exposures in paint manufacture and painting are three large cohort studies of painters and collections of national statistics on cancer incidence and mortality in which data on cancer at many sites were presented for painters. These show a consistent excess of all cancers, at about 20% above the national average, and a consistent excess of lung cancers, at about 40% above the national average. The available evidence on the prevalence of smoking in painters, although limited, indicates that an excess risk for lung cancer of this magnitude cannot be explained by smoking alone. The risks for cancers of the oesophagus, stomach and bladder were raised in many of the studies, but the excesses were generally smaller and more variable than those for lung cancer. Some of the studies also reported excess risks for leukaemia and for cancers of the buccal cavity and larynx.

Several other small cohort and census–based studies in painters provided estimates of risk for cancer at one or several sites. The risk for lung cancer was reported to be raised in eight, that for stomach cancer in two, that for bladder cancer in two, that for leukaemia in four, that for malignancies of the lymphatic system in three, that for buccal cancer in three, that for laryngeal cancer in one, that for skin cancer in one, and that for prostatic cancer in three. In many studies, risks for cancer were reported only for sites for which the result was statistically significant.

In the three cohort studies of workers involved in the manufacture of paint, two of which were small, there was little to suggest an excess risk of lung cancer or of cancer at any other anatomical site.

Eleven case–control and related studies of lung cancer could be evaluated. All of the studies showed an increased risk for lung cancer among painters. The five studies in which smoking was taken into account showed an increase of 30% or more in risk for lung cancer. Two studies suggested increased risks among painters for laryngeal cancer, and one indicated an increased risk for mesothelioma.

Cancer of the urinary tract has been examined in relation to exposure to paint in 15 case–control and related studies. Eight showed an excess risk for bladder cancer in all painters. In certain studies, specific aspects of exposure to paint were examined: car painters were addressed in two studies, one indicating an excess risk; spray painters were evaluated in three studies, two of which showed an excess risk; and exposure to lacquer and chromium was associated with a risk in one study.

In a study of occupational histories of patients with oesophageal and stomach cancers, high risks were seen for painters. A further study also identified a risk for stomach cancer and another a risk for oesophageal cancer. One study of cancer of the gall–bladder and of the biliary tract showed associations with the occupation of painting. A study of pancreatic cancer reported a high risk for exposure to paint thinners.

Five studies of leukaemia mentioned painters. Two studies showed excess risks. Two small studies of Hodgkin's disease and three studies of multiple myeloma showed increased risks in association with the occupation of painter or with any exposure to paints, paint-related products or organic solvents.

A single study of prostatic cancer showed a significant excess risk for manufacturers of paints and varnishes, and one study reported a high risk for testicular cancer among spray painters.

Twelve studies of childhood cancer mentioned paternal exposure to paint and related substances; four of these also presented data on maternal exposure. Three studies showed an excess of childhood leukaemia in association with paternal exposure and one in association with maternal exposure. Two studies showed an excess risk for brain tumours in the children of male painters. One small study of children with Wilms' tumour showed an excess in those whose fathers were painters. All of these excesses are based on small numbers of children whose parents had been exposed, even in the larger studies. In the other studies, no association was seen between parental exposure to paint and childhood cancers. The type and timing of exposure varied among these studies.

4.3 Other relevant data

Painters may suffer from allergic and nonallergic contact dermatitis, chronic bronchitis and asthma, and adverse effects on the nervous system. There is also some indication of adverse effects in the liver, kidney, blood and blood-forming organs. Many of these effects are also seen in paint production workers.

Of three studies on the fertility of painters, two showed no adverse effect and the third a possible excess frequency of infertility in men. One study reported an excess frequency of spontaneous abortion in female painters, based on self-reported data. Studies of birth weight, perinatal mortality rates and congenital malformations in the offspring of male painters generally showed no adverse effects; few data on female painters were available.

No increase in the frequency of sister chromatid exchange in peripheral lymphocytes was found in one study of painters or in one study of paint manufacturing workers.

4.4 Evaluation[1]

There is *sufficient evidence* for the carcinogenicity of occupational exposure as a painter.

There is *inadequate evidence* for the carcinogenicity of occupational exposure in paint manufacture.

Overall evaluation

Occupational exposure as a painter *is carcinogenic (Group 1)*.

---

[1]For definitions of the italicized terms, see Preamble, pp. 27-30.

Occupational exposure in paint manufacture *is not classifiable as to its carcinogenicity (Group 3)*.

5. References

Adams, R.M. (1983) *Occupational Skin Disease*, New York, Grune and Stratton, pp. 267–278

Adelstein, A.M. (1972) Occupational mortality: cancer. *Ann. occup. Hyg.*, *15*, 53–57

Åkesson, B., Florén, I. & Skerfving, S. (1985) Visual disturbances after experimental human exposure to triethylamine. *Br. J. ind. Med.*, *42*, 848–850

Åkesson, B., Bengtsson, M. & Florén, I. (1986) Visual disturbances after industrial triethylamine exposure. *Int. Arch. occup. environ. Health*, *57*, 297–302

Alexandersson, R. & Hedenstierna, G. (1988) Respiratory hazards associated with exposure to formaldehyde and solvents in acid–curing paints. *Arch. environ. Health*, *43*, 222–227

Alexandersson, R., Plato, N., Kolmodin-Hedman, B. & Hedenstierna, G. (1987) Exposure, lung function, and symptoms in car painters exposed to hexamethylendiisocyanate and biuret modified hexamethylendiisocyanate. *Arch. environ. Health*, *42*, 367–373

Allen, R.A. (1972) Epoxy resins in coatings. In: Madson, W.H., ed., *Federation Series on Coatings Technology*, Unit 20, Philadelphia, PA, Federation of Societies for Paint Technology, pp. 7–61

Allyn, G. (1971) Acrylic resins. In: Madson, W.H., ed., *Federation Series on Coatings Technology*, Unit 17, Philadelphia, PA, Federation of Societies for Paint Technology, pp. 7–34

Angerer, J. & Wulf, H. (1985) Occupational chronic exposure to organic solvents. XI. Alkylbenzene exposure of varnish workers: effects on hematopoietic system. *Int. Arch. occup. environ. Health*, *56*, 307–321

Anshelm Olson, B. (1982) Effects of organic solvents on behavioral performance of workers in the paint industry. *Neurobehav. Toxicol. Teratol.*, *4*, 703–708

Apostoli, P., Brugnone, F., Perbellini, L., Cocheo, V., Bellomo, M.L. & Silvestri, R. (1982) Biomonitoring of occupational toluene exposure. *Int. Arch. occup. environ. Health*, *50*, 153–168

Askergren, A. (1981) Studies on kidney function in subjects exposed to organic solvents. III. Excretion of cells in the urine. *Acta med. scand.*, *210*, 103–106

Askergren, A., Allgén, L.-H. & Bergström, J. (1981a) Studies on kidney function in subjects exposed to organic solvents. II. The effect of desmopression in a concentration test and the effect of exposure to organic solvents on renal concentrating ability. *Acta med. scand.*, *209*, 485–488

Askergren, A., Allgén, L.-H., Karlsson, C., Lundberg, I. & Nyberg, E. (1981b) Studies on kidney function in subjects exposed to organic solvents. I. Excretion of albumin and beta-2-microglobulin in the urine. *Acta med. scand.*, *209*, 479–483

Askergren, A., Brandt, R., Gullquist, R., Silk, B. & Strandell, T. (1981c) Studies on kidney function in subjects exposed to organic solvents. IV. Effect on 51-Cr-EDTA clearance. *Acta med. scand.*, *210*, 373–376

Askergren, A., Beving, H., Hagman, M., Kristensson, J., Linroth, K., Vesterberg, O. & Wennberg, A. (1988) Biological effects of exposure to water–thinned and solvent–thinned paints in house painters (Swed.). *Arb. Hälsa*, *4*, 1–64

Axelson, O., Hane, M. & Hogstedt, C. (1976) A case referent study on neuro–psychiatric disorders among workers exposed to solvents. *Scand. J. Work Environ. Health*, *2*, 14–20

Bäck, B. & Saarinen, L. (1986) Measurement of amines used as curing agents of epoxy resins in work-room air and technical products (Finn.). *Työterveyslaitoksen tutkimuksia, 4*, 31-36

Baker, E.L., Letz, R.E., Eisen, E.A., Pothier, L.J., Plantamura, D.L., Larson, M. & Wolford, R. (1988) Neurobehavioral effects of solvents in construction painters. *J. occup. Med., 30*, 116-123

Barry, T.H. (1939) History of varnish manufacture. In: *Proceedings of a Symposium, Varnish Making, Harrogate, May 1939*, London, Oil & Colour Chemist's Association, pp. 1-15

Beirne, G.J. & Brennan, J.T. (1972) Glomerulonephritis associated with hydrocarbon solvents mediated by antiglomerular basement membrane antibody. *Arch. environ. Health, 25*, 365-369

Belanger, P.L. & Coye, M.J. (1980) *Sinclair Paint Company, Los Angeles, California (Health Hazard Evaluation Report No. 80–68–871)*, Cincinnati, OH, National Institute for Occupational Safety and Health

Bertazzi, P.A., Zocchetti, C., Terzaghi, G.F., Riboldi, L., Guercilena S. & Beretta, F. (1981) Carcinogenicity study of varnish production. A mortality study (Ital.). *Med. Lav., 6*, 465-472

Beving, H., Malmgren, R., Olsson, P., Tornling, G. & Unge, G. (1983) Increased uptake of serotonin in platelets from car painters occupationally exposed to mixtures of solvents and organic isocyanates. *Scand. J. Work Environ. Health, 9*, 253-258

Beving, H., Malmgren, R., Olsson, P. & Unge, G. (1984a) Differences in the respiratory capacity of workers with long-term exposure to vapors from paints free from or containing organic isocyanates. *Scand. J. Work Environ. Health, 10*, 267-268

Beving, H., Kristensson, J., Malmgren, R., Olsson, P. & Unge, G. (1984b) Effects on the uptake kinetics of serotonin (5–hydroxytryptamine) in platelets from workers with long-term exposure to organic solvents. A pilot study. *Scand. J. Work Environ. Health, 10*, 229-234

Beving, H., Malmgren, R. & Olsson, P. (1988) Changed fatty acid composition in platelets from workers with long term exposure to organic solvents. *Br. J. ind. Med., 45*, 565-567

Bjerrehuus, T. & Detlefsen, G.U. (1986) Infertility in Danish painters exposed to organic solvents (Dan.). *Ugeskr. Laeger, 148*, 1105-1106

Bobjer, O. & Knave, B. (1977) Physiological work and exposure to solvents and dust-hazard factors in house painting. In: *Proceedings of an International Symposium on the Control of Air Pollution in the Working Environment, Stockholm, 6–8 September 1977*, Part II, Stockholm, Swedish Work Environment Fund/International Labour Office, pp. 41-61

Bradley, A. & Bodsworth, P.L. (1983) Environmental control of a large paint booth. *Ann. occup. Hyg., 27*, 223-224

Brady, R.E., Jr, Griffith, J.R., Love, K.S. & Field, D.E. (1987) Nontoxic alternatives to antifouling paints. *J. Coatings Technol., 59*, 113-119

Breslow, L., Hoaglin, L., Rasmussen, G. & Abrams, H.K. (1954) Occupations and cigarette smoking as factors in lung cancer. *Am. J. publ. Health, 44*, 171-181

Brewer, G.E.F. (1984) Coatings, electrodeposition. In: Mark, H.F., Bikales, N.M., Overberger, C.G. & Menges, G., eds, *Encyclopedia of Polymer Science and Engineering*, 2nd ed., Vol. 3, New York, John Wiley & Sons, pp. 675-687

Brewer, R.J. & Bogan, R.T. (1984) Cellulose esters, inorganic. In: Mark, H.F., Bikales, N.M., Overgerger, C.G. & Menges, G., eds, *Encyclopedia of Polymer Science and Engineering*, 2nd ed., Vol. 3, New York, John Wiley & Sons, pp. 139-147

Brown, L.M., Mason, T.J., Pickle, L.W., Stewart, P.A., Buffler, P.A., Burau, K., Ziegler, R.G. & Fraumeni, J.F., Jr (1988) Occupational risk factors for laryngeal cancer on the Texas Gulf coast. *Cancer Res., 48*, 1960-1964

Browne, T.D. (1983) Painting and varnishing. In: Parmeggiani, L., ed., *Encyclopedia of Occupational Health and Safety*, 3rd (rev.) ed., Vol. 2, Geneva, International Labour Office, pp. 1583–1587

Brownson, R.C., Chang, J.C., Davis, J.R. & Bagby, J.R., Jr (1988) Occupational risk of prostate cancer: a cancer registry-based study. *J. occup. Med., 30*, 523–526

Brunner, H. (1978) Paint. In: Williams, T.I., ed., *A History of Technology*, Vol. VI, *The Twentieth Century, c. 1900 to c. 1950*, Part I, Oxford, Clarendon Press, pp. 590–606

Cannell, D. (1967) Industrial maintenance paints. In: Standen, A., ed., *Kirk–Othmer Encyclopedia of Chemical Technology*, 2nd ed., Vol. 14, New York, John Wiley & Sons, pp. 474–485

Chapman, E.M. (1941) Observations on the effect of paint on the kidneys with particular reference to the role of turpentine. *J. ind. Hyg. Toxicol., 23*, 277–289

Cherry, N. & Waldron, H.A. (1984) The prevalence of psychiatric morbidity in solvent workers in Britain. *Int. J. Epidemiol., 13*, 197–200

Cherry, N., Hutchins, H., Pace, T. & Waldron, H.A. (1985) Neurobehavioral effects of repeated occupational exposure to toluene and paint solvents. *Br. J. ind. Med., 42*, 291–300

Chiazze, L., Jr, Ference, L.D. & Wolf, P.H. (1980) Mortality among automobile assembly workers. I. Spray painters. *J. occup. Med., 22*, 520–526

Chrostek, W.J. (1980) *Corporation of Veritas, Philadelphia, Pennsylvania (Health Hazard Evaluation Report No. 80–108–705)*, Cincinnati, OH, National Institute for Occupational Safety and Health

Chrostek, W.J. & Levine, M.S. (1981) *Palmer Industrial Coatings Inc., Williamsport, Pennsylvania (Health Hazard Evaluation Report No. 80–153–881)*, Cincinnati, OH, National Institute for Occupational Safety and Health

Claude, J., Kunze, E., Frentzel-Beyme, R., Paczkowski, K., Schneider, J. & Schubert, H. (1986) Lifestyle and occupational risk factors in cancer of the lower urinary tract. *Am. J. Epidemiol., 124*, 578–589

Claude, J., Frentzel-Beyme, R. & Kunze, E. (1988) Occupation and risk of cancer of the lower urinary tract among men. A case–control study. *Int. J. Cancer, 41*, 371–379

Coggon, D., Pannett, B., Osmond, C. & Acheson, E.D. (1986a) A survey of cancer and occupation in young and middle aged men. I. Cancers of the respiratory tract. *Br J. ind. Med., 43*, 332–338

Coggon, D., Pannett, B., Osmond, C. & Acheson, E.D. (1986b) A survey of cancer and occupation in young and middle aged men. II. Non-respiratory cancers. *Br. J. ind. Med., 43*, 381–386

Cohen, S.R. (1974) Occupational health case report – no. 4. Epoxy-type paint. *J. occup. Med., 16*, 201–203

Cole, P., Monson, R.R., Haning, H. & Friedell, G.H. (1971) Smoking and cancer of the lower urinary tract. *New Engl. J. Med., 284*, 129–134

Cole, P., Hoover, R. & Friedell, G.H. (1972) Occupation and cancer of the lower urinary tract. *Cancer, 29*, 1250–1260

Connolly, E.M., Hughes, C.S., Myers, C.P. & Dean, J.C. (1986) *The US Paint Industry: Technology Trends, Markets, Raw Materials*, Menlo Park, CA, SRI International

Cranmer, J.M. & Golberg, L., eds (1986) Proceedings of the workshop on neurobehavioral effects of solvents. *Neurotoxicology, 7*, 1–95

Cuzick, J. & De Stavola, B. (1988) Multiple myeloma. A case–control study. *Br. J. Cancer, 57*, 516–520

Dalager, N.A., Mason, T.J., Fraumeni, J.F., Jr, Hoover, R. & Payne, W.W. (1980) Cancer mortality among workers exposed to zinc chromate paints. *J. occup. Med., 22*, 25–29

Daniell, W.E. & Vaughan, T.L. (1988) Paternal employment in solvent related occupations and adverse pregnancy outcomes. *Br. J. ind. Med., 45*, 193–197

Decouflé, P., Stanislawczyk, K., Houten, L., Bross, I.D.J. & Viadana, E. (1977) *A Retrospective Survey of Cancer in Relation to Occupation (DHEW (NIOSH) Publ. No. 77-178)*, Cincinnati, OH, National Institute for Occupational Safety and Health

Downing, R.S. (1967) Paint and varnish removal. In: Mark, H.F., McKetta, J.J. & Othmer, D.F., eds, *Kirk-Othmer Encyclopedia of Chemical Technology*, 2nd ed., Vol. 14, New York, John Wiley & Sons, pp. 485-493

Drisko, R.W. (1985) Coatings, marine. In: Grayson, M., ed., *Kirk-Othmer Concise Encyclopedia of Chemical Technology*, New York, John Wiley & Sons, p. 295

Dubrow, R. & Wegman, D.H. (1984) Cancer and occupation in Massachusetts: a death certificate study. *Am. J. ind. Med., 6*, 207-230

Dufva, L. (1982) The influence of environmental factors on research and development of paint products. In: Englund, A., Ringen, K. & Mehlman, M.A., eds, *Advances in Modern Environmental Toxicology*, Vol. 2, *Occupational Health Hazards of Solvents*, Princeton, NJ, Princeton Scientific Publishers, pp. 69-74

Dunn, J.E., Jr & Weir, J.M. (1965) Cancer experience of several occupational groups followed prospectively. *Am. J. publ. Health, 55*, 1367-1375

Dupont (1988) *Material Safety Data Sheet – IMRON® Polyurethane Enamel*, No. 7RS, Wilmington, DE

Ehrenreich, T., Yunis, S.L. & Churg, J. (1977) Membranous nephropathy following exposure to volatile hydrocarbons. *Environ. Res., 14*, 35-45

Elofsson, S.-A., Gamberale, F., Hindmarsh, T., Iregren, A., Isaksson, A., Johnsson, I., Knave, B., Lydahl, E., Mindus, P., Persson, H.E., Philipson, B., Steby, M., Struwe, G., Söderman, E., Wennberg, A. & Widén, L. (1980) Exposure to organic solvents. A cross-sectional epidemiologic investigation on occupationally-exposed car and industrial spray painters with special reference to the nervous system. *Scand. J. Work Environ. Health, 6*, 239-273

Engholm, G. & Englund, A. (1982) Cancer incidence and mortality among Swedish painters. In: Englund, A., Ringen, K. & Mehlman, M.A., eds, *Advances in Modern Environmental Toxicology*, Vol. II, *Occupational Health Hazards of Solvents*, Princeton, NJ, Princeton Scientific Publishers, pp. 173-185

Engholm, G., Englund, A. & Löwing, H. (1987) Cancer incidence and mortality in Swedish painters (Abstract). *Scand. J. Work Environ. Health, 13*, 181

Englund, A. (1980) Cancer incidence among painters and some allied trades. *J. Toxicol. environ. Health, 6*, 1267-1273

Fabia, J. & Thuy, T.D. (1974) Occupation of father at time of birth of children dying from malignant diseases. *Br. J. prev. soc. Med., 28*, 98-100

Federation of Societies for Paint Technology (1973) Introduction to coatings technology. In: *Federation Series on Coatings Technology*, Unit 1, Philadelphia, PA, pp. 7-32

Fidler, A.T., Baker, E.L. & Letz, R.E. (1987) Neurobehavioural effects of occupational exposure to organic solvents among construction painters. *Br. J. ind. Med., 44*, 292-308

Finn, R., Fennerty, A.G. & Ahmad, R. (1980) Hydrocarbon exposure and glomerulonephritis. *Clin. Nephrol., 14*, 173-175

Fisher, E.M. (1987) Paints, varnishes and lacquers. In: Johnston, B., ed., *Colliers Encyclopedia*, Vol. 18, London, P.F. Collier, pp. 330-334

Flodin, U., Fredriksson, M., Axelson, O., Persson, B. & Hardell, L. (1986) Background radiation, electrical work, and some other exposures associated with acute myeloid leukemia in a case-referent study. *Arch. environ. Health, 41*, 77-84

Franchini, I., Cavatorta, A., Falzoi, M., Lucertini, S. & Mutti, A. (1983) Early indicators of renal damage in workers exposed to organic solvents. *Int. Arch. occup. environ. Health, 52,* 1-9

Fregert, S. (1981) *Manual of Contact Dermatitis,* 2nd ed., Chicago, IL, Year Book Medical Publishers

Friedman, G.D. (1986) Multiple myeloma: relation to propoxyphene and other drugs, radiation and occupation. *Int. J. Epidemiol., 15,* 424–426

Fytelson, M. (1982) Pigments (organic). In: Mark, H.F., Othmer, D.R., Overberger, C.G., Seaborg, G.T. & Grayson, M., eds, *Kirk-Othmer Encyclopedia of Chemical Technology,* 3rd ed., Vol. 17, New York, John Wiley & Sons, pp. 838-871

Gold, E.B., Diener, M.D. & Szklo, M. (1982) Parental occupations and cancer in children. A case–control study and review of the methodologic issues. *J. occup. Med., 24,* 578-584

Greenburg, L., Mayers, M.R., Heimann, H. & Moskowitz, S. (1942) The effects of exposure to toluene in industry. *J. Am. med. Assoc., 118,* 573-578

Guralnick, L. (1963) *Mortality by Occupation Level and Cause of Death Among Men 20 to 64 Years of Age: USA, 1950 (Vital Statistics Special Reports Vol. 43, No. 5),* Washington DC, US Department of Health, Education, and Welfare

Haglund, U., Lundberg, I. & Zech, L. (1980) Chromosome aberrations and sister chromatid exchanges in Swedish paint industry workers. *Scand. J. Work Environ. Health, 6,* 291-298

Hagmar, L., Nielsen, J. & Skerfving, S. (1987) Clinical features and epidemiology of occupational obstructive respiratory disease caused by small molecular weight organic chemicals. *Monogr. Allergy, 21,* 42-58

Hakulinen, T., Salonen, T. & Teppo, L. (1976) Cancer in the offspring of fathers in hydrocarbon–related occupations. *Br. J. prev. soc. Med., 30,* 138-140

Hamilton, E.C. & Early, L.W. (1972) Nitrocellulose and organosoluble cellulose ethers in coatings. In: Madson, W.H., ed., *Federation Series on Coatings Technology,* Unit 21, Philadelphia, PA, Federation of Societies for Paint Technology, pp. 9-41

Hane, M., Axelson, O., Blume, J., Hogstedt, C., Sundell, L. & Ydreborg, B. (1977) Psychological function changes among house painters. *Scand. J. Work Environ. Health, 3,* 91-99

Hänninen, H., Eskelinen, L., Husman, K. & Nurminen, M. (1976) Behavioral effects of long–term exposure to a mixture of organic solvents. *Scand. J. Work Environ. Health, 4,* 240-255

Hansen, C.M. (1982) Solvent technology in product development. In: Englund, A., Ringen, K. & Mehlman, M.A., eds, *Advances in Modern Environmental Toxicology,* Vol. 2, *Occupational Health Hazards of Solvents,* Princeton, NJ, Princeton Scientific Publishers, pp. 43-52

Hansen, M.K., Larsen, M. & Cohr, K.-H. (1987) Waterborne paints. A review of their chemistry and toxicology and the results of determinations made during their use. *Scand. J. Work Environ. Health, 13,* 473-485

Heaton, N. (1928) *Outlines of Paint Technology,* London, Charles Griffin & Co.

Heidam, L.Z. (1984a) Spontaneous abortions among laboratory workers: a follow up study. *J. Epidemiol. Commun. Health, 38,* 36-41

Heidam, L.Z. (1984b) Spontaneous abortions among dental assistants, factory workers, painters and gardening workers: a follow up study. *J. Epidemiol. Commun. Health, 38,* 149-155

Hellquist, H., Irander, K., Edling, C. & Ödkvist, L.M. (1983) Nasal symptoms and histopathology in a group of spray-painters. *Acta otolaryngol., 96,* 495-500

Hemminki, K., Saloniemi, I., Salonen, T., Partanen, T. & Vainio, H. (1981) Childhood cancer and parental occupation in Finland. *J. Epidemiol. Commun. Health, 35,* 11-15

Hernberg, S., Westerholm, P., Schultz-Larsen, K., Degerth, R., Kuosma, E., Englund, A., Engzell, U., Sand Hansen, H. & Mutanen, P. (1983) Nasal and sinonasal cancer. Connection with occupational exposures in Denmark, Finland and Sweden. *Scand. J. Work Environ. Health, 9*, 315-326

Herrick, R.F., Ellenbecker, M.J. & Smith, T.J. (1988) Measurement of the epoxy content of paint spray aerosol: three case studies. *Appl. ind. Hyg., 3*, 123-128

Hervin, R.L. & Thoburn, T.W. (1975) *Trans World Airlines Main Overhaul Facility, Kansas City, Missouri (Health Hazard Evaluation Report No. 72-96-237)*, Cincinnati, OH, National Institute for Occupational Safety and Health

Högberg, M. & Wahlberg, J.E. (1980) Health screening for occupational dermatoses in house painters. *Contact Derm., 6*, 100-106

Houten, L., Bross, I.D.J., Viadana, E. & Sonnesso, G. (1977) Occupational cancer in men exposed to metals. *Adv. exp. Med. Biol., 91*, 93-102

Howe, G.R. & Lindsay, J.P. (1983) A follow-up study of a ten-percent sample of the Canadian labor force. I. Cancer mortality in males, 1965-73. *J. natl Cancer Inst., 70*, 37-44

Howe, G.R., Burch, J.D., Miller, A.B., Cook, G.M., Estève, J., Morrison, B., Gordon, P., Chambers, L.W., Fodor, G. & Winsor, G.M. (1980) Tobacco use, occupation, coffee, various nutrients, and bladder cancer. *J. natl Cancer Inst., 64*, 701-713

Huré, P. (1986) Application of paints and varnishes inside buildings (Fr.). *Cah. Notes doc., 125*, 587-590

Husman, K. (1980) Symptoms of car painters with long-term exposure to a mixture of organic solvents. *Scand. J. Work Environ. Health, 6*, 19-32

IARC (1972) *IARC Monographs on the Evaluation of Carcinogenic Risk of Chemicals to Man*, Vol. 1, *Some Inorganic Substances, Chlorinated Hydrocarbons, Aromatic Amines, N-Nitroso Compounds, and Natural Products*, Lyon, pp. 29-39

IARC (1974a) *IARC Monographs on the Evaluation of Carcinogenic Risk of Chemicals to Man*, Vol. 4, *Some Aromatic Amines, Hydrazine and Related Substances, N-Nitroso Compounds and Miscellaneous Alkylating Agents*, Lyon, pp. 259-269

IARC (1974b) *IARC Monographs on the Evaluation of Carcinogenic Risk of Chemicals to Man*, Vol. 4, *Some Aromatic Amines, Hydrazine and Related Substances, N-Nitroso Compounds and Miscellaneous Alkylating Agents*, Lyon, pp. 57-64

IARC (1975) *IARC Monographs on the Evaluation of Carcinogenic Risk of Chemicals to Man*, Vol. 8, *Some Aromatic Azo Compounds*, Lyon, pp. 83-89

IARC (1976a) *IARC Monographs on the Evaluation of Carcinogenic Risk of Chemicals to Man*, Vol. 11, *Cadmium, Nickel, Some Epoxides, Miscellaneous Industrial Chemicals, and General Considerations on Volatile Anaesthetics*, Lyon

IARC (1976b) *IARC Monographs on the Evaluation of Carcinogenic Risk of Chemicals to Man*, Vol. 11, *Cadmium, Nickel, Some Epoxides, Miscellaneous Industrial Chemicals, and General Considerations on Volatile Anaesthetics*, Lyon, pp. 131-139

IARC (1976c) *IARC Monographs on the Evaluation of Carcinogenic Risk of Chemicals to Man*, Vol. 11, *Cadmium, Nickel, Some Epoxides, Miscellaneous Industrial Chemicals, and General Considerations on Volatile Anaesthetics*, Lyon, pp. 75-112

IARC (1976d) *IARC Monographs on the Evaluation of Carcinogenic Risk of Chemicals to Man*, Vol. 11, *Cadmium, Nickel, Some Epoxides, Miscellaneous Industrial Chemicals, and General Considerations on Volatile Anaesthetics*, Lyon, pp. 39-74

IARC (1977a) *IARC Monographs on the Evaluation of the Carcinogenic Risk of Chemicals to Man*, Vol. 15, *Some Fumigants, the Herbicides 2,4–D and 2,4,5–T, Chlorinated Dibenzodioxins and Miscellaneous Industrial Chemicals*, Lyon, pp. 155–175

IARC (1977b) *IARC Monographs on the Evaluation of Carcinogenic Risk of Chemicals to Man*, Vol. 14, *Asbestos*, Lyon

IARC (1978a) *IARC Monographs on the Evaluation of the Carcinogenic Risk of Chemicals to Man*, Vol. 16, *Some Aromatic Amines and Related Nitro Compounds – Hair Dyes, Colouring Agents and Miscellaneous Industrial Chemicals*, Lyon, pp. 221–231

IARC (1978b) *IARC Monographs on the Evaluation of the Carcinogenic Risk of Chemicals to Man*, Vol. 16, *Some Aromatic Amines and Related Nitro Compounds – Hair Dyes, Colouring Agents and Miscellaneous Industrial Chemicals*, Lyon, pp. 111–124

IARC (1978c) *IARC Monographs on the Evaluation of the Carcinogenic Risk of Chemicals to Humans*, Vol. 18, *Polychlorinated Biphenyls and Polybrominated Biphenyls*, Lyon, pp. 43–103

IARC (1979a) *IARC Monographs on the Evaluation of the Carcinogenic Risk of Chemicals to Humans*, Vol. 19, *Some Monomers, Plastics and Synthetic Elastomers, and Acrolein*, Lyon, pp. 377–438

IARC (1979b) *IARC Monographs on the Evaluation of the Carcinogenic Risk of Chemicals to Humans*, Vol. 19, *Some Monomers, Plastics and Synthetic Elastomers, and Acrolein*, Lyon, pp. 341–366

IARC (1979c) *IARC Monographs on the Evaluation of the Carcinogenic Risk of Chemicals to Humans*, Vol. 19, *Some Monomers, Plastics and Synthetic Elastomers, and Acrolein*, Lyon, pp. 47–71

IARC (1979d) *IARC Monographs on the Evaluation of the Carcinogenic Risk of Chemicals to Humans*, Vol. 19, *Some Monomers, Plastics and Synthetic Elastomers, and Acrolein*, Lyon, pp. 231–274

IARC (1979e) *IARC Monographs on the Evaluation of the Carcinogenic Risk of Chemicals to Humans*, Vol. 19, *Some Monomers, Plastics and Synthetic Elastomers, and Acrolein*, Lyon, pp. 283–301

IARC (1979f) *IARC Monographs on the Evaluation of the Carcinogenic Risk of Chemicals to Humans*, Vol. 20, *Some Halogenated Hydrocarbons*, Lyon, pp. 429–448

IARC (1979g) *IARC Monographs on the Evaluation of the Carcinogenic Risk of Chemicals to Humans*, Vol. 20, *Some Halogenated Hydrocarbons*, Lyon, pp. 515–531

IARC (1979h) *IARC Monographs on the Evaluation of the Carcinogenic Risk of Chemicals to Humans*, Vol. 20, *Some Halogenated Hydrocarbons*, Lyon, pp. 371–399

IARC (1979i) *IARC Monographs on the Evaluation of the Carcinogenic Risk of Chemicals to Humans*, Vol. 20, *Some Halogenated Hydrocarbons*, Lyon, pp. 545–572

IARC (1979j) *IARC Monographs on the Evaluation of the Carcinogenic Risk of Chemicals to Humans*, Vol. 20, *Some Halogenated Hydrocarbons*, Lyon, pp. 303–325, 349–367

IARC (1979k) *IARC Monographs on the Evaluation of the Carcinogenic Risk of Chemicals to Humans*, Vol. 19, *Some Monomers, Plastics and Synthetic Elastomers, and Acrolein*, Lyon, pp. 320–340

IARC (1980a) *IARC Monographs on the Evaluation of the Carcinogenic Risk of Chemicals to Humans*, Vol. 23, *Some Metals and Metallic Compounds*, Lyon, pp. 325–415

IARC (1980b) *IARC Monographs on the Evaluation of the Carcinogenic Risk of Chemicals to Humans*, Vol. 23, *Some Metals and Metallic Compounds*, Lyon, pp. 205–323

IARC (1980c) *IARC Monographs on the Evaluation of the Carcinogenic Risk of Chemicals to Humans*, Vol. 23, *Some Metals and Metallic Compounds*, Lyon, pp. 39–141

IARC (1981) *IARC Monographs on the Evaluation of the Carcinogenic Risk of Chemicals to Humans*, Vol. 25, *Wood, Leather and Some Associated Industries*, Lyon, pp. 49–197

IARC (1982a) *IARC Monographs on the Evaluation of the Carcinogenic Risk of Chemicals to Humans*, Vol. 29, *Some Industrial Chemicals and Dyestuffs*, Lyon, pp. 345–389

IARC (1982b) *IARC Monographs on the Evaluation of the Carcinogenic Risk of Chemicals to Humans*, Vol. 29, *Some Industrial Chemicals and Dyestuffs*, Lyon, pp. 93-148

IARC (1982c) *IARC Monographs on the Evaluation of the Carcinogenic Risk of Chemicals to Humans*, Vol. 29, *Some Industrial Chemicals and Dyestuffs*, Lyon, pp. 269-294

IARC (1982d) *IARC Monographs on the Evaluation of the Carcinogenic Risk of Chemicals to Humans*, Vol. 29, *Some Industrial Chemicals and Dyestuffs*, Lyon, pp. 213-238

IARC (1982e) *IARC Monographs on the Evaluation of the Carcinogenic Risk of Chemicals to Humans*, Vol. 29, *Some Industrial Chemicals and Dyestuffs*, Lyon, pp. 331-343

IARC (1982f) *IARC Monographs on the Evaluation of the Carcinogenic Risk of Chemicals to Humans*, Vol. 28, *The Rubber Industry*, Lyon

IARC (1983a) *IARC Monographs on the Evaluation of the Carcinogenic Risk of Chemicals to Humans*, Vol. 30, *Miscellaneous Pesticides*, Lyon, pp. 319-328

IARC (1983b) *IARC Monographs on the Evaluation of the Carcinogenic Risk of Chemicals to Humans*, Vol. 32, *Polynuclear Aromatic Compounds, Part 1, Chemical, Environmental and Experimental Data*, Lyon

IARC (1984) *IARC Monographs on the Evaluation of the Carcinogenic Risk of Chemicals to Humans*, Vol. 33, *Polynuclear Aromatic Compounds, Part 2, Carbon Blacks, Mineral Oils and Some Nitroarenes*, Lyon, pp. 35-85

IARC (1985a) *IARC Monographs on the Evaluation of the Carcinogenic Risk of Chemicals to Humans*, Vol. 36, *Allyl Compounds, Aldehydes, Epoxides and Peroxides*, Lyon, pp. 189-226

IARC (1985b) *IARC Monographs on the Evaluation of the Carcinogenic Risk of Chemicals to Humans*, Vol. 35, *Polynuclear Aromatic Compounds, Part 4, Bitumens, Coal-tars and Derived Products, Shale-oils and Soots*, Lyon, pp. 83-159

IARC (1985c) *IARC Monographs on the Evaluation of the Carcinogenic Risk of Chemicals to Humans*, Vol. 36, *Allyl Compounds, Aldehydes, Epoxides and Peroxides*, Lyon, pp. 267-283

IARC (1985d) *IARC Monographs on the Evaluation of the Carcinogenic Risk of Chemicals to Humans*, Vol. 36, *Allyl Compounds, Aldehydes, Epoxides and Peroxides*, Lyon, pp. 245-263

IARC (1986a) *IARC Monographs on the Evaluation of the Carcinogenic Risk of Chemicals to Humans*, Vol. 39, *Some Chemicals Used in Plastics and Elastomers*, Lyon, pp. 113-131

IARC (1986b) *IARC Monographs on the Evaluation of the Carcinogenic Risk of Chemicals to Humans*, Vol. 39, *Some Chemicals Used in Plastics and Elastomers*, Lyon, pp. 287-323

IARC (1986c) *IARC Monographs on the Evaluation of the Carcinogenic Risk of Chemicals to Humans*, Vol. 41, *Some Halogenated Hydrocarbons and Pesticide Exposures*, Lyon, pp. 43-85

IARC (1986d) *IARC Monographs on the Evaluation of the Carcinogenic Risk of Chemicals to Humans*, Vol. 39, *Some Chemicals Used in Plastics and Elastomers*, Lyon, pp. 347-365

IARC (1986e) *IARC Monographs on the Evaluation of the Carcinogenic Risk of Chemicals to Humans*, Vol. 41, *Some Halogenated Hydrocarbons and Pesticide Exposures*, Lyon, pp. 229-235

IARC (1987a) *IARC Monographs on the Evaluation of the Carcinogenic Risk of Chemicals to Humans*, Vol. 42, *Silica and Some Silicates*, Lyon, pp. 39-143

IARC (1987b) *IARC Monographs on the Evaluation of Carcinogenic Risks to Humans*, Suppl. 7, *Overall Evaluations of Carcinogenicity: An Updating of* IARC Monographs *Volumes 1 to 42*, Lyon, pp. 341-343

IARC (1987c) *IARC Monographs on the Evaluation of the Carcinogenic Risk of Chemicals to Humans*, Vol. 42, *Silica and Some Silicates*, Lyon, pp. 185-224

IARC (1987d) *IARC Monographs on the Evaluation of Carcinogenic Risks to Humans*, Suppl. 7, *Overall Evaluations of Carcinogenicity: An Updating of* IARC Monographs *Volumes 1 to 42*, Lyon, pp. 349–350

IARC (1987e) *IARC Monographs on the Evaluation of Carcinogenic Risks to Humans*, Suppl. 7, *Overall Evaluations of Carcinogenicity: An Updating of* IARC Monographs *Volumes 1 to 42*, Lyon, pp. 216–219

IARC (1987f) *IARC Monographs on the Evaluation of Carcinogenic Risks to Humans*, Suppl. 7, *Overall Evaluations of Carcinogenicity: An Updating of* IARC Monographs *Volumes 1 to 42*, Lyon, pp. 211–216

IARC (1987g) *IARC Monographs on the Evaluation of Carcinogenic Risks to Humans*, Suppl. 7, *Overall Evaluations of Carcinogenicity: An Updating of* IARC Monographs *Volumes 1 to 42*, Lyon, pp. 205–207

IARC (1987h) *IARC Monographs on the Evaluation of Carcinogenic Risks to Humans*, Suppl. 7, *Overall Evaluations of Carcinogenicity: An Updating of* IARC Monographs *Volumes 1 to 42*, Lyon, pp. 79–80

IARC (1987i) *IARC Monographs on the Evaluation of Carcinogenic Risks to Humans*, Suppl. 7, *Overall Evaluations of Carcinogenicity: An Updating of* IARC Monographs *Volumes 1 to 42*, Lyon, pp. 202–203

IARC (1987j) *IARC Monographs on the Evaluation of Carcinogenic Risks to Humans*, Suppl. 7, *Overall Evaluations of Carcinogenicity: An Updating of* IARC Monographs *Volumes 1 to 42*, Lyon, pp. 175–176

IARC (1987k) *IARC Monographs on the Evaluation of Carcinogenic Risks to Humans*, Suppl. 7, *Overall Evaluations of Carcinogenicity: An Updating of* IARC Monographs *Volumes 1 to 42*, Lyon, pp. 120–122

IARC (1987l) *IARC Monographs on the Evaluation of the Carcinogenic Risk of Chemicals to Humans*, Vol. 42, *Silica and Some Silicates*, Lyon, pp. 159–173

IARC (1987m) *IARC Monographs on the Evaluation of Carcinogenic Risks to Humans*, Suppl. 7, *Overall Evaluations of Carcinogenicity: An Updating of* IARC Monographs *Volumes 1 to 42*, Lyon, pp. 117

IARC (1987n) *IARC Monographs on the Evaluation of Carcinogenic Risks to Humans*, Suppl. 7, *Overall Evaluations of Carcinogenicity: An Updating of* IARC Monographs *Volumes 1 to 42*, Lyon, pp. 230–232

IARC (1987o) *IARC Monographs on the Evaluation of Carcinogenic Risks to Humans*, Suppl. 7, *Overall Evaluations of Carcinogenicity: An Updating of* IARC Monographs *Volumes 1 to 42*, Lyon, pp. 100–106

IARC (1987p) *IARC Monographs on the Evaluation of Carcinogenic Risks to Humans*, Suppl. 7, *Overall Evaluations of Carcinogenicity: An Updating of* IARC Monographs *Volumes 1 to 42*, Lyon, pp. 264–269

IARC (1987q) *IARC Monographs on the Evaluation of Carcinogenic Risks to Humans*, Suppl. 7, *Overall Evaluations of Carcinogenicity: An Updating of* IARC Monographs *Volumes 1 to 42*, Lyon, pp. 194–195

IARC (1987r) *IARC Monographs on the Evaluation of Carcinogenic Risks to Humans*, Suppl. 7, *Overall Evaluations of Carcinogenicity: An Updating of* IARC Monographs *Volumes 1 to 42*, Lyon, pp. 192–193

IARC (1987s) *IARC Monographs on the Evaluation of Carcinogenic Risks to Humans*, Suppl. 7, *Overall Evaluations of Carcinogenicity: An Updating of* IARC Monographs *Volumes 1 to 42*, Lyon, pp. 131–133

IARC (1989a) *IARC Monographs on the Evaluation of Carcinogenic Risks to Humans*, Vol. 45, *Occupational Exposures in Petroleum Refining, Crude Oil and Major Petroleum Fuels*, Lyon, pp. 159–201

IARC (1989b) *IARC Monographs on the Evaluation of Carcinogenic Risks to Humans*, Vol. 45, *Occupational Exposures in Petroleum Refining, Crude Oil and Major Petroleum Fuels*, Lyon, pp. 203–218

Ikeda, M., Watanabe, T., Kasahara, M., Kamiyama, S., Suzuki, H., Tsunoda, H. & Nakaya, S. (1985) Organic solvent exposure in small scale industries in north-east Japan. *Ind. Health, 23,* 181–189

Iscovich, J., Castelletto, R., Estève, J., Muñoz, N., Colanzi, R., Coronel, A., Deamezola, I., Tassi, V. & Arslan, A. (1987) Tobacco smoking, occupational exposure and bladder cancer in Argentina. *Int. J. Cancer, 40,* 734–740

Jayjock, M.A. & Levin, L. (1984) Health hazards in a small automotive body repair shop. *Ann. occup. Hyg., 28,* 19–29

Jensen, O.M., Wahrendorf, J., Knudsen, J.B. & Sørensen, B.L. (1987) The Copenhagen case-referent study on bladder cancer. Risks among drivers, painters and certain other occupations. *Scand. J. Work Environ. Health, 13,* 129–134

Jensen, O.M., Knudsen, J.B., McLaughlin, J.K. & Sørensen, B.L. (1988) The Copenhagen case-control study of renal pelvis and ureter cancer: role of smoking and occupational exposures. *Int. J. Cancer, 41,* 557–561

Johnson, C.C., Annegers, J.F., Frankowski, R.F., Spitz, M.R. & Buffler, P.A. (1987) Childhood nervous system tumors – an evaluation of the association with paternal occupational exposure to hydrocarbons. *Am. J. Epidemiol., 126,* 605–613

Jones, A. (1938) *Cellulose Lacquers, Finishes and Cements*, Philadelphia, PA, J.B. Lippincott

Kauppinen, T. (1986) Occupational exposure to chemical agents in the plywood industry. *Ann. occup. Hyg., 30,* 19–29

Kelsey, K.T., Wiencke, J.K., Little, F.F., Baker, E.L., Jr & Little, J.B. (1988) Effects of cigarette smoking and solvent exposure on sister chromatid exchange frequency in painters. *Environ. mol. Mutagenesis, 11,* 389–399

Kikukawa, H. (1986) Paint and coatings. *Jpn. chem. Ann., November,* p. 86

Kjuus, H., Skjaerven, R., Langård, S., Lien, J.T. & Aamodt, T. (1986) A case-referent study of lung cancer, occupational exposures and smoking. I. Comparison of title-based and exposure-based occupational information. *Scand. J. Work Environ. Health, 12,* 193–202

Klavis, G. & Drommer, W. (1970) Goodpasture syndrome and the effect of benzene (Ger.). *Arch. Toxikol., 26,* 40–55

Kline, C.H. & Co. (1975) *The Kline Guide to the Paint Industry*, 4th (rev.) ed., Fairfield, NJ

Kominsky, J.R., Rinsky, R. & Stroman, R. (1978) *Goodyear Aerospace Corporation, Akron, Ohio (Health Hazard Evaluation Report No. 77-127-516)*, Cincinnati, OH, National Institute for Occupational Safety and Health

Krivanek, N. (1982) The toxicity of paint pigments, resins, drying oils, and additives. In: Englund, A., Ringen, K. & Mehlman, M.A., eds, *Advances in Modern Environmental Toxicology*, Vol. II, *Occupational Health Hazards of Solvents*, Princeton, NJ, Princeton Scientific Publishers, pp. 1–42

Kurppa, K. & Husman, K. (1982) Car painters' exposure to a mixture of organic solvents. Serum activities of liver enzymes. *Scand. J. Work Environ. Health, 8,* 137–140

Kwa, S.-L. & Fine, L.J. (1980) The association between parental occupation and childhood malignancy. *J. occup. Med.*, 22, 792–794

van der Laan, G. (1980) Chronic glomerulonephritis and organic solvents. A case–control study. *Int. Arch. occup. environ. Health*, 47, 1–8

Lagrue, G., Kamalodine, T., Hirbec, G., Bernaudin, J.-F., Guerrero, J. & Zhepova, F. (1977) The role of inhalation of toxic substances in the development of glomerulonephritis (Fr.). *Nouv. Presse méd.*, 6, 3609–3613

Lam, H.R., Tarding, F., Stokholm, J. & Gyntelberg, F. (1985) Human platelet 5–hydroxytryptamine concentration as a tool in prediction of solvent induced neurotoxic effects. *Acta pharmacol. toxicol.*, 56, 233–238

Landrigan, P.J., Baker, E.L., Jr, Himmelstein, J.S., Stein, G.F., Weddig, J.P. & Straub, W.E. (1982) Exposure to lead from the Mystic River Bridge: the dilemma of deleading. *New Engl. J. Med, 306*, 673–676

Lanson, H.J. (1978) Alkyd resins. In: Grayson, M., ed., *Kirk–Othmer Encyclopedia of Chemical Technology*, 3rd ed., Vol. 2, New York, John Wiley & Sons, pp. 18–50

Larson, B.A. (1978) Occupational exposure to coal tar pitch volatiles at pipeline protective coating operations. *Am. ind. Hyg. Assoc. J.*, 39, 250–255

Lauwerys, R., Bernard, A., Viau, C. & Buchet, J.-P. (1985) Kidney disorders and hematotoxicity of organic solvent exposure. *Scand. J. Work Environ. Health, 11 (Suppl. 1)*, 83–90

Layman, P.L. (1985) Paints and coatings: the global challenge. *Chem. Eng. News, September 30*, 27–68

Lerchen, M.L., Wiggins, C.L. & Samet, J.M. (1987) Lung cancer and occupation in New Mexico. *J. natl Cancer Inst.*, 79, 639–645

Levin, L.I., Zheng, W., Blot, W.J., Gao, Y.-T. & Fraumeni, J.F., Jr (1988) Occupation and lung cancer in Shanghai: a case–control study. *Br. J. ind. Med.*, 45, 450–458

Lind, G. (1939) Significance of haematological changes in lacquer sprayers (Ger.). *Arch. Gewerbepathol. Gewerbehyg.*, 9, 141–166

Lindquist, R., Nilsson, B., Eklund, G. & Gahrton, G. (1987) Increased risk of developing acute leukemia after employment as a painter. *Cancer*, 60, 1378–1384

Lindström, K. & Wickström, G. (1983) Psychological function changes among maintenance house painters exposed to low levels of organic solvent mixtures. *Acta psychiatr. scand.*, 67 *(Suppl. 303)*, 81–91

Lindström, K., Riihimäki, H. & Hänninen, K. (1984) Occupational solvent exposure and neuropsychiatric disorders. *Scand. J. Work Environ. Health*, 10, 321–323

Linet, M.S., Malker, H.S.R., McLaughlin, J.K., Weiner, J.A., Stone, B.J., Blot, W.J., Ericsson, J.L.E. & Fraumeni, J.F., Jr (1988) Leukemias and occupation in Sweden: a registry–based analysis. *Am. J. ind. Med.*, 14, 319–330

Littorin, M., Fehling, C., Attewell, R. & Skerving, C. (1988) Focal epilepsy and exposure to organic solvents: a case–referent study. *J. occup. Med.*, 30, 805–808

Lowell, H.J. (1984) Coating. In: Mark, H.F., Bikales, N.M., Overberger, C.B. & Menges, G., eds, *Encyclopedia of Polymer Science and Engineering*, 2nd ed., Vol. 3, New York, John Wiley & Sons, pp. 615–667

Lowengart, R.A., Peters, J.M., Cicioni, C., Buckley, J., Bernstein, L., Preston-Martin, S. & Rappaport, E. (1987) Childhood leukemia and parents' occupational and home exposures. *J. natl Cancer Inst.*, 79, 39–46

Lundberg, I. (1986) Mortality and cancer incidence among Swedish paint industry workers with long-term exposure to organic solvents. *Scand. J. Work Environ. Health, 12*, 108–113

Lundberg, I. & Håkansson, M. (1985) Normal serum activities of liver enzymes in Swedish paint industry workers with heavy exposure to organic solvents. *Br. J. ind. med., 42*, 596–600

Magnani, C., Coggon, D., Osmond, C. & Acheson, E.D. (1987) Occupation and five cancers: a case–control study using death certificates. *Br. J. ind. Med., 44*, 769–776

Maintz, G. & Werner, L. (1988) Do paints and varnishes damage the respiratory system? (Ger.) *Z. Gesamt. Hyg., 34*, 274–278

Maizlish, N.A., Langolf, G.D., Whitehead, L.W., Fine, L.J., Albers, J.W., Goldberg, J. & Smith, P. (1985) Behavioural evaluation of workers exposed to mixtures of organic solvents. *Br. J. ind. Med., 42*, 579–590

Malker, H.S.R., McLaughlin, J.K., Malker, B.K., Stone, B.J., Weiner, J.A., Erickson J.L.E. & Blot, W.J. (1985) Occupational risks for pleural mesothelioma in Sweden, 1961–79. *J. natl Cancer Inst., 74*, 61–66

Malker, H.S.R., McLaughlin, J.K., Malker, B.K., Stone, B.J., Weiner, J.A., Ericsson, J.L.E. & Blot, W.J. (1986) Biliary tract cancer and occupation in Sweden. *Br. J. ind. Med., 43*, 257–262

Mallov, J.S. (1976) MBK neuropathy among spray painters. *J. Am. med. Assoc., 235*, 1455–1457

Martens, C.R. (1964) *Emulsion and Water-soluble Paints and Coatings*, New York, Reinhold

Matanoski, G.M., Stockwell, H.G., Diamond, E.L., Haring-Sweeney, M., Joffe, R.D., Mele, J.M. & Johnson, M.L. (1986) A cohort mortality study of painters and allied tradesmen. *Scand. J. Work Environ. Health, 12*, 16–21

Matsunaga, J., Une, H., Nakayoshi, N., Momose, Y., Maeda, M., Watanabe, D., Magori, Y., Esaki, H., Kamo, H. & Kuroki, K. (1983) Occupational exposure to organic solvents in the painters of car repair workshops (Jpn.). *Med. Bull. Fukuoka Univ., 10*, 173–178

Matthäus, W. (1964) A contribution to corneal lesions in workers varnishing surfaces in the furniture industry (Ger.). *Klin. Monatsbl. Augenheilkunde, 144*, 713–717

McDowall, M.E. (1985) *Occupational Reproductive Epidemiology: The Use of Routinely Collected Statistics in England and Wales, 1980–82 (Studies on Medical and Population Subjects No. 50)*, London, Her Majesty's Stationery Office, Office of Population Censuses and Surveys

Menck, H.R. & Henderson, B.E. (1976) Occupational differences in rates of lung cancer. *J. occup. Med., 18*, 797–801

Mikkelsen, S. (1980) A cohort study of disability pension and death among painters with special regard to disabling presenile dementia as an occupational disease. *Scand. J. soc. Med., Suppl. 16*, 34–43

Mikkelsen, S., Jørgensen, M., Browne, E. & Gyldensted, C. (1988) Mixed solvent exposure and organic brain damage. A study of painters. *Acta neurol. scand., 78 (Suppl. No. 118)*

Mikulski, P.I., Wiglusz, R., Bublewska, A. & Uselis, J. (1972) Investigation of exposure of ships' painters to organic solvents. *Br. J. ind. Med., 29*, 450–453

Milham, S., Jr (1983) *Occupational Mortality in Washington State 1950–1979 (DHHS (NIOSH) Publication No. 83–116)*, Cincinnati, OH, National Institute for Occupational Safety and Health

Milling Pedersen, L. & Cohr, K.-H. (1984) Biochemical pattern in experimental exposure of humans to white spirit. I. The effects of a 6 hours single dose. *Acta pharmacol. toxicol., 55*, 317–324

Milling Pedersen, L. & Melchior Rasmussen, J. (1982) The haematological and biochemical pattern in occupational organic solvent poisoning and exposure. *Int. Arch. occup. environ. Health, 51*, 113–126

Milling Pedersen, L., Nygaard, E., Nielsen, O.S. & Saltin, B. (1980) Solvent-induced occupational myopathy. *J. occup. Med.*, *22*, 603–606

Milne, K.L., Sandler, D.P., Everson, R.B. & Brown, S.M. (1983) Lung cancer and occupation in Alameda County: a death certificate case–control study. *Am. J. ind. Med.*, *4*, 565–575

Mølhave, L. & Lajer, M. (1976) Organic solvents in the air inspired by house painters (Dan.). *Ugeskr. Laeg.*, *138*, 1230–1237

Morgan, R.W., Kaplan, S.D. & Gaffey, W.R. (1981) A general mortality study of production workers in the paint and coatings manufacturing industry. A preliminary report. *J. occup. Med.*, *23*, 13–21

Morgan, R.W., Claxton, K.W., Kaplan, S.D., Parsons, J.M. & Wong, O. (1985) Mortality of paint and coatings industry workers. A follow-up study. *J. occup. Med.*, *27*, 377–378

Morris, P.D., Koepsell, T.D., Daling, J.R., Taylor, J.W., Lyon, J.L., Swanson, G.M., Child, M. & Weiss, N.S. (1986) Toxic substance exposure and multiple myeloma: a case–control study. *J. natl Cancer Inst.*, *76*, 987–994

Morrison, A.S., Ahlbom, A., Verhoek, W.G., Aoki, K., Leck, I., Ohno, Y. & Obata, K. (1985) Occupation and bladder cancer in Boston, USA, Manchester, UK, and Nagoya, Japan. *J. Epidemiol. Commun. Health*, *39*, 294–300

National Institute for Occupational Safety and Health (1970) *US Census Report*, Cincinnati, OH

National Institute for Occupational Safety and Health (1984) *Recommendations for Control of Occupational Safety and Health Hazards. Manufacture of Paint and Allied Coating Products (DHSS (NIOSH) Publ. No. 84–115)*, Cincinnati, OH

National Institute for Occupational Safety and Health (1987) *Organic Solvent Neurotoxicity (Current Intelligence Bulletin No. 48)*, Cincinnati, OH

Nielsen, J., Sangö, C., Winroth, G., Hallberg, T. & Skerfving, S. (1985) Systemic reactions associated with polyisocyanate exposure. *Scand. J. Work Environ. Health*, *11*, 51–54

Nielsen, J., Welinder, H., Schütz, A. & Skerfving, S. (1988) Specific serum antibodies against phthalic anhydride in occupationally exposed subjects. *J. Allerg. clin. Immunol.*, *82*, 126–133

Norell, S., Ahlbom, A., Olin, R., Erwald, R., Jacobson, G., Lindberg-Navier, I. & Wiechel, K.-L. (1986) Occupational factors and pancreatic cancer. *Br. J. ind. Med.*, *43*, 775–778

O'Brien, D.M. & Hurley, D.E. (1981) *An Evaluation of Engineering Control Technology for Spray Painting (DHHS (NIOSH) Publ. No. 81–121)*, Cincinnati, OH, National Institute for Occupational Safety and Health

Office of Population Censuses and Surveys (1958) *The Registrar General's Decennial Supplement England and Wales 1951, Occupational Mortality*, Part II, Vol. 2, *Tables*, London, Her Majesty's Stationery Office

Office of Population Censuses and Surveys (1970) *Classification of Occupations*, London, Her Majesty's Stationery Office

Office of Population Censuses and Surveys (1972) *The Registrar General's Decennial Supplement, England and Wales 1961, Occupational Mortality Tables*, London, Her Majesty's Stationery Office

Office of Population Censuses and Surveys (1979) *Occupational Mortality 1970–1972, England and Wales, Decennial Supplement*, London, Her Majesty's Stationery Office

Office of Population Censuses and Surveys (1986) *Occupational Mortality 1979–80, 1982–83, Great Britain, Decennial Supplement*, London, Her Majesty's Stationery Office

O'Flynn, R.R., Monkman, S.M. & Waldron, H.A. (1987) Organic solvents and presenile dementia: a case referent study using death certificates. *Br. J. ind. Med.*, *44*, 259–262

Ogata, M., Takatsuka, Y. & Tomokuni, K. (1971) Excretion of hippuric acid and *m*- or *p*- methylhippuric acid in the urine of persons exposed to vapours of toluene and *m*- or *p*-xylene in an exposure chamber and in workshops, with specific reference to repeated exposures. *Br. J. ind. Med.*, *28*, 382–385

Okawa, M.T. & Keith, W. (1977) *United Airlines Maintenance Base, San Francisco International Airport, Burlingame, California* (*Health Hazard Evaluation Report No. 75–195–396*), Cincinnati, OH, National Institute for Occupational Safety and Health

Olsen, J.H. (1983) Risk of exposure to teratogens among laboratory staff and painters. *Dan. med. Bull.*, *30*, 24–28

Olsen, J.H. (1988) Occupational risks of sinonasal cancer in Denmark. *Br. J. ind. Med.*, *45*, 329–335

Olsen, J.H. & Jensen, O.M. (1987) Occupation and risk of cancer in Denmark. An analysis of 93,810 cancer cases, 1970–1979. *Scand. J. Work Environ. Health, 13* (*Suppl. 1*), 1–91

Olsen, J.H. & Rachootin, P. (1983) Organic solvents as possible risk factors of low birthweight (Letter to the Editor). *J. occup. Med.*, *25*, 854–855

Olsen, J. & Sabroe, S. (1980) A case-reference study of neuropyschiatric disorders among workers exposed to solvents in the Danish wood and furniture industry. *Scand. J. Work Environ. Health, 16*, 44–49

Olsson, H. & Brandt, L. (1980) Occupational exposure to organic solvents and Hodgkin's disease in men. A case-referent study. *Scand. J. Work Environ. Health, 6*, 302–305

Olsson, H. & Brandt, L. (1988) Risk of non-Hodgkin's lymphoma among men occupationally exposed to organic solvents. *Scand. J. Work Environ. Health, 14*, 246–251

O'Neill, L.A. (1981) *Health and Safety, Environmental Pollution and the Paint Industry*, Teddington, UK, Paint Research Association

Ørbaek, P., Risberg, J., Rosén, I., Haeger-Aronsen, B., Hagstadius, S., Hjortsberg, U., Regnell, G., Rehnström, S., Svensson, K. & Welinder, H. (1985) Effect of long-term exposure to solvents in the paint industry. A cross-sectional epidemiologic study with clinical and laboratory methods. *Scand. J. Work Environ. Health, 11* (*Suppl. 2*), 1–28

Pearce, N.E. & Howard, J.K. (1986) Occupation, social class and male cancer mortality in New Zealand, 1974–78. *Int. J. Epidemiol.*, *15*, 456–462

Peltonen, K. (1986) Thermal degradation of epoxy powder paint. In: *Proceedings of an International Congress on Industrial Hygiene, Rome, Italy, October 5–9*, Rome, Pontifica Università Urbaniana, pp. 118–119

Peltonen, K., Pfäffli, P., Itkonen, A. & Kalliokoski, P. (1986) Determination of the presence of bisphenol-A and the absence of diglycidyl ether of bisphenol-A in the thermal degradation products of epoxy powder paint. *Am. ind. Hyg. Assoc. J.*, *47*, 399–403

Peters, J.M., Preston-Martin, S. & Yu, M.C. (1981) Brain tumors in children and occupational exposure of parents. *Science, 213*, 235–237

Petersen, G.R. & Milham, S., Jr (1980) *Occupational Mortality in the State of California 1959–61* (*DHEW (NIOSH) Publication No. 80–104*), Cincinnati, OH, National Institute for Occupational Safety and Health

Peterson, J.E. (1984) Painting and coating. In: Cralley, L.J. & Cralley, L.V., eds, *Industrial Hygiene Aspects of Plant Operations*, Vol. 2, *Unit Operations and Product Fabrication*, New York, MacMillan, pp. 222–247

Pham, Q.-T., Mur, J.-M., Teculescu, D., Merou-Poncelet, B., Gaertner, M., Meyer-Bisch, C., Moulin, J.-M. & Massin, N. (1985) Respiratory symptoms and pulmonary function in painters at an industrial vehicle manufacturing plant. Results of a cross-sectional epidemiological study (Fr.). *Arch. Mal. prof., 46*, 31–36

Phillips, L.W. (1976) *Literature Search on Toxic and Carcinogenic Components of Paint (NIOSH Report 210–76–0108; PB 83–117655)*, Cincinnati, OH, National Institute for Occupational Safety and Health

Piper, R. (1965) The hazards of painting and varnishing. *Br. J. ind. Med., 22*, 247–260

Pirilä, V. (1947) On occupational diseases of the skin among paint factory workers, painters, polishers and varnishers in Finland. A clinical and experimental study. *Acta dermatovenerol., 27 (Suppl. 16)*, 1–163

Powell, G.M. (1972) Vinyl resins. In: Madson, W.H., ed., *Federation Series on Coatings Technology*, Unit 19, Philadelphia, PA, Federation of Societies for Paint Technology, pp. 7–55

Priha, E., Riipinen, H. & Korhonen, K. (1986) Exposure to formaldehyde and solvents in Finnish furniture factories in 1975–1984. *Ann. occup. Hyg., 30*, 289–294

Rachootin, P. & Olsen, J. (1983) The risk of infertility and delayed conception associated with exposures in the Danish workforce. *J. occup. Med., 25*, 394–402

Raitta, C., Husman, K. & Tossavainen, A. (1976) Lens changes in car painters exposed to a mixture of organic solvents. *Graefes Arch. klin. exp. Ophthal., 200*, 149–156

Rasmussen, H., Olsen, J. & Lawitsen, J. (1985) Risk of encephalopathia among retired solvent–exposed workers. A case–control study among males applying for nursing home accommodation or other types of social support facilities. *J. occup. Med., 27*, 561–566

Ravnskov, U., Forsberg, B. & Skerfving, S. (1979) Glomerulonephritis and exposure to organic solvents. A case–control study. *Acta med. scand., 205*, 575–579

Reisch, M.S. (1987) Paint sales may top 1986 high. *Chem. Eng. News, September 21*, pp. 51–68

Reyes de la Rocha, S.R., Brown, M.A. & Fortenberry, J.D. (1987) Pulmonary function abnormalities in intentional spray paint inhalation. *Chest, 92*, 100–104

Riala, R. (1982) Chemical hazards in painting in the construction industry. In: Englund, A., Ringen, K. & Mehlman, M.A., eds, *Advances in Modern Environmental Toxicology*, Vol. 2, *Occupational Health Hazards of Solvents*, Princeton, NJ, Princeton Scientific Publishers, pp. 93–95

Riala, R., Kalliokoski, P., Pyy, L. & Wickström, G. (1984) Solvent exposure in construction and maintenance painting. *Scand. J. Work Environ. Health, 10*, 263–266

Ringen, K. (1982) Health hazards among painters. In: Englund, A., Ringen, K. & Mehlman, M.A., eds, *Advances in Modern Environmental Toxicology*, Vol. 2, *Occupational Health Hazards of Solvents*, Princeton, NJ, Princeton Scientific Publishers, pp. 111–138

Risch, H.A., Burch, J.D., Miller, A.B., Hill, G.B., Steele, R. & Howe, G.R. (1988) Occupational factors and the incidence of cancer of the bladder in Canada. *Br. J. ind. Med., 45*, 361–367

Ronco, G., Ciccone, G., Mirabelli, D., Troia, B. & Vineis, P. (1988) Occupation and lung cancer in two industrialized areas of northern Italy. *Int. J. Cancer, 41*, 354–358

Rosen, G. (1953) Occupational health problems of English painters and varnishers in 1825. *Br. J. ind. Med., 10*, 195–199

Rosenberg, C. & Tuomi, T. (1984) Airborne isocyanates in polyurethane spray painting: determination and respirator efficiency. *Am. ind. Hyg. Assoc. J., 45*, 117–121

Rosensteel, R.E. (1974) *Harris Structural Steel Company, Piscataway, New Jersey (Health Hazard Evaluation Report No. 73–99–108)*, Cincinnati, OH, National Institute for Occupational Safety and Health

Sabroe, S. & Olsen, J. (1979) Health complaints and work conditions among lacquerers in the Danish furniture industry. *Scand. J. soc. Med.*, 7, 97–104

Sanders, B.M., White, G.C. & Draper, G.J. (1981) Occupations of fathers of children dying from neoplasms. *J. Epidemiol. Commun. Health*, 35, 245–250

Scheffers, T.M.L., Jongeneelen, F.J. & Bragt, P.C. (1985) Development of effect-specific limit values (ESLVs) for solvent mixtures in painting. *Ann. occup. Hyg.*, 29, 191–199

Schiek, R.C. (1982) Pigments (inorganic). In: Mark, H.F., Othmer, D.F., Overberger, C.G., Seaborg, G.T. & Grayson, M., eds, *Kirk–Othmer Encyclopedia of Chemical Technology*, 3rd ed., Vol. 17, New York, John Wiley & Sons, p. 788

Schifflers, E., Jamart, J. & Renard, V. (1987) Tobacco and occupation as risk factors in bladder cancer: a case–control study in southern Belgium. *Int. J. Cancer*, 39, 287–292

Schoenberg, J.B., Stemhagen, A., Mogielnicki, A.P., Altman, R., Abe, T. & Mason, T.J. (1984) Case–control study of bladder cancer in New Jersey. I. Occupational exposures in white males. *J. natl Cancer Inst.*, 72, 973–981

Schurr, G.G. (1974) Exterior house paints. In: Madison, W.H., ed., *Federation Series on Coatings Technology*, Unit 24, Philadelphia, PA, Federation of Societies for Paint Technology, pp. 5–67

Schurr, G.G. (1981) Paint. In: Mark, H.F., Othmer, D.F., Overberger, C.G., Seaborg, G.T. & Grayson, M., eds, *Kirk–Othmer Encyclopedia of Chemical Technology*, 3rd ed., Vol. 16, New York, John Wiley & Sons, pp. 742–761

Schwartz, D.A. & Baker, E.L. (1988) Respiratory illness in the construction industry. Airflow obstruction among painters. *Chest*, 93, 134–137

Sears, K. (1974) Plasticizers. In: Madison, W.H., ed., *Federation Series on Coatings Technology*, Unit 22, Philadelphia, PA, Federation of Societies for Paint Technology, pp. 5–103

Selikoff, I.J. (1983) *Investigations of Health Hazards in the Painting Trades*, New York, Mount Sinai School of Medicine of the City University of New York

Seppäläinen, A.M. & Lindström, K. (1982) Neurophysiological findings among house painters exposed to solvents. *Scand. J. Work Environ. Health*, 8, 131–135

Seppäläinen, A.M., Husman, K. & Mårtenson, C. (1978) Neurophysiological effects of long-term exposure to a mixture of organic solvents. *Scand. J. Work Environ. Health*, 4, 304–314

Siegel, G.S. (1963) Lead exposure among decorative and house painters. *Arch. environ. Health*, 6, 34–37

Siemiatycki, J., Dewar, R., Nadon, L., Gérin, M., Richardson, L. & Wacholder, S. (1987a) Associations between several sites of cancer and twelve petroleum-derived liquids. Results from a case-referent study in Montreal. *Scand. J. Work Environ. Health*, 13, 493–504

Siemiatycki, J., Wacholder, S., Richardson, L., Dewar, R. & Gérin, M. (1987b) Discovering carcinogens in the occupational environment. Methods of data collection and analysis of a large case-referent monitoring system. *Scand. J. Work Environ. Health*, 13, 486–492

Silverman, D.T., Hoover, R.N., Albert, S. & Graff, K.M. (1983) Occupation and cancer of the lower urinary tract in Detroit. *J. natl Cancer Inst.*, 70, 237–245

Simonato, L., Vineis, P. & Fletcher, A.C. (1988) Estimates of the proportion of lung cancer attributable to occupational exposure. *Carcinogenesis*, 9, 1159–1165

Singer, E. (1957) *Fundamentals of Paint, Varnish and Lacquer Technology*, St Louis, MO, The American Paint Journal Co.

Skerfving, S. (1987) Biological monitoring of exposure to inorganic lead. In: Clarkson, T.W., Friberg, L., Nordberg, G.F. & Sager, P.R., eds, *Biological Monitoring of Toxic Metals*, New York, Plenum, pp. 169–197

Spee, T. & Zwennis, W.C.M. (1987) Lead exposure during demolition of a steel structure coated with lead-based paints. I. Environmental and biological monitoring. *Scand. J. Work Environ. Health, 13*, 52–55

Sterner, J.H. (1941) Study of hazards in spray painting with gasoline as a diluent. *J. ind. Hyg. Toxicol., 23*, 437–448

Stewart, R.D. & Hake, C.L. (1976) Paint remover hazard. *J. Am. med. Assoc., 235*, 398–401

Stockwell, H.G. & Matanoski, G.M. (1985) A case–control study of lung cancer in painters. *J. occup. Med., 27*, 125–126

Struwe, G., Mindus, P. & Jönsson, B. (1980) Psychiatric ratings in occupational health research: a study of mental symptoms in lacquerers. *Am. J. ind. Med., 1*, 23–30

Swedish Work Environment Fund (1987) *Paints, Varnishes, Adhesives*, Stockholm

Swerdlow, A.J. & Skeet, R.G. (1988) Occupational associations of testicular cancer in south east England. *Br. J. ind. Med., 45*, 225–230

Takeuchi, Y., Ono, Y, Hisanaga, N., Iwata, M., Okutani, H., Matsamuto, T., Gotoh, M., Fukaya, Y., Ueno, K., Seki, T. & Mizuno, S. (1982) Environmental and health surveys on car repair workers exposed to organic solvents (Jpn.). *Jpn. J. ind. Health, 24*, 305–313

Timonen, T.T.T. & Ilvonen, M. (1978) Contact with hospital, drugs, and chemicals as aetiological factors in leukaemia. *Lancet, i*, 350–352

Tola, S. & Karskela, V. (1976) Occupational lead exposure in Finland. V. Shipyards and shipbreaking. *Scand. J. Work Environ. Health, 2*, 31–36

Tola, S., Hernberg, S. & Vesanto, R. (1976) Occupational lead exposure in Finland. VI. Final report. *Scand. J. Work Environ. Health, 2*, 115–127

Triebig, D., Claus, D., Csuzda, I., Druschky, K.-F., Holler, P., Kinzel, W., Lehrl, S., Reichwein, P., Weidenhammer, W., Weitbrechst, W.-U., Weltle, D., Schaller, K.H. & Valentin, H. (1988) Cross-sectional epidemiological study on neurotoxicity of solvents in paints and lacquers. *Int. Arch. occup. environ. Health, 60*, 233–241

Ulfvarson, U. (1977) Chemical hazards in the paint industry. In: *Proceedings of an International Symposium on the Control of Air Pollution in the Working Environment, Stockholm, 6–8 September 1977*, Part II, Stockholm, Swedish Work Environment Fund/International Labour Office, pp. 63–75

US Environmental Protection Agency (1979) *Development Document for Effluent Limitations, Guidelines and Standards for the Paint Formulating: Point Source Category (EPA Report 440/1–79/049–b)*, Washington DC

Valciukas, J.A., Lilis, R., Singer, R.M., Glickman, L. & Nicholson, W.J. (1985) Neurobehavioral changes among shipyard painters exposed to solvents. *Arch. environ. Health, 40*, 47–52

Vandervort, R. & Cromer, J. (1975) *Health Hazard Evaluation/Toxicity Determination Report Peabody Galion Corp. (NIOSH–TR 73–47–172; PB 246446)*, Cincinnati, OH, National Institute for Occupational Safety and Health.

Van Steensel-Moll, H.A., Valkenburg, H.A. & Zanen, G.E. (1985) Childhood leukaemia and parental occupation. A register-based case–control study. *Am. J. Epidemiol., 121*, 216–224

Viadana, E. & Bross, I.D.J. (1972) Leukemia and occupations. *Prev. Med., 1*, 513–521

Viadana, E., Bross, I.D.J. & Houten, L. (1976) Cancer experience of men exposed to inhalation of chemicals or to combustion products. *J. occup. Med., 18*, 787–792

Vianna, N.J. & Polan, A. (1979) Lymphomas and occupational benzene exposure. *Lancet, i*, 1394–1395

Vineis, P. & Magnani, C. (1985) Occupation and bladder cancer in males: a case–control study. *Int. J. Cancer, 35*, 599–606 van Vliet, C., Swaen, G.M.F., Slangern, J.J.M., Border, T.D. & Stirman, F. (1987) The organic solvent syndrome. *Int. Arch occup. environ. Health, 59*, 493-501

Volk, O. & Abriss, M. (1976) Interior finishes. In: Madison, W.H., ed., *Federation Series on Coatings Technology*, Unit 23, Philadelphia, PA, Federation of Societies for Paint Technology, pp. 5-21

Welinder, H., Nielsen, J., Bensryd, I. & Skerfving, S. (1988) IgG antibodies against polyisocyanates in car painters. *Clin. Allerg., 18*, 85-93

Wernfors, M., Nielsen, J., Schütz, A. & Skerfving, S. (1986) Phthalic anhydride-induced occupational asthma. *Int. Arch. Allerg. appl. Immunol., 79*, 77-82

White, M.C. & Baker, E.L. (1988) Measurements of respiratory illness among construction painters. *Br. J. ind. Med., 45*, 523-531

Whorton, M.D., Schulman, J., Larson, S.R., Stubbs, H.A. & Austin, D. (1983) Feasibility of identifying high-risk occupations through tumor registries. *J. occup. Med., 25*, 657-660

Wicks, Z.W. (1984) Drying oils. In: Mark, H.F., Bikales, N.M., Overberger, C.G. & Menges, G., eds, *Encyclopedia of Polymer Science and Engineering*, 2nd ed., Vol. 5, New York, John Wiley & Sons, pp. 203-214

Wilkins, J.R., III & Sinks, T.H., Jr (1984) Paternal occupation and Wilms' tumour in offspring. *J. Epidemiol. Commun. Health, 38*, 7-11

Williams, R.A. (1977) Automotive finishes. In: Madson, W.H., ed., *Federation Series on Coatings Technology*, Unit 25, Philadelphia, PA, Federation of Societies for Paint Technology, pp. 7-36

Williams, R.R., Stegens, N.L. & Goldsmith, J.R. (1977) Associations of cancer site and type with occupation and industry from the Third National Cancer Survey interview. *J. natl Cancer Inst., 59*, 1147-1185

Winchester, R.V. & Madjar, V.M. (1986) Solvent effects on workers in the paint, adhesive and printing industries. *Ann. occup. Hyg., 30*, 307-317

World Health Organization (1985) *Chronic Effects of Organic Solvents on the Central Nervous System and Diagnostic Criteria*, Copenhagen

Wynder, E.L. & Graham, E.A. (1951) Etiologic factors in bronchiogenic carcinoma with special reference to industrial exposures. Report of eight hundred fifty-seven proved cases. *Arch. ind. Hyg. occup. Med., 4*, 221-235

Wynder, E.L., Onderdonk, J. & Mantel, N. (1963) An epidemiological investigation of cancer of the bladder. *Cancer, 16*, 1388-1407

Zack, M., Cannon, S., Loyd, D., Heath, C.W., Jr, Falletta, J.M., Jones, B., Housworth, J. & Crowley, S. (1980) Cancer in children of parents exposed to hydrocarbon-related industries and occupations. *Am. J. Epidemiol., 111*, 329-336

Zey, J.N. & Aw, T.-C. (1984) *American Transportation Corporation, Conway, Arkansas (Health Hazard Evaluation Report No. 82-025-1413)*, Cincinnati, OH, National Institute for Occupational Safety and Health

Zimmerman, S.W., Groehler, K. & Beirne, G.J. (1975) Hydrocarbon exposure and chronic glomerulonephritis. *Lancet, ii*, 199-201

SUMMARY OF FINAL EVALUATIONS

Agent	Degree of evidence for carcinogenicity		Overall evaluation
	Humans	Animals	
Petroleum solvents	Inadequate		3
High–boiling aromatic solvents		Inadequate	
Special boiling–range solvents		No data	
White spirits		No data	
Toluene	Inadequate	Inadequate	3
Xylene	Inadequate	Inadequate	3
Cyclohexanone	No data	Inadequate	3
Dimethylformamide	Limited	Inadequate	2B
Morpholine	No data	Inadequate	3
1,2–Epoxybutane	No data	Limited	3
Bis(2,3–epoxycyclopentyl)ether	No data	Limited	3
Glycidyl ethers			
Phenyl glycidyl ether	No data	Sufficient	2B
Bisphenol A diglycidyl ether	No data	Limited	3
Phenol	Inadequate	Inadequate	3
Antimony trioxide	Inadequate	Sufficient	2B
Antimony trisulfide	Inadequate	Limited	3
Titanium dioxide	Inadequate	Limited	3
Paint manufacture and painting (occupational exposures in)			
Painter (occupational exposure as a)	Sufficient		1
Paint manufacture (occupational exposure in)	Inadequate		3

APPENDIX 1

SUMMARY TABLE OF GENETIC AND RELATED EFFECTS

Appendix 1. Summary table of genetic and related effects

	Nonmammalian systems				Mammalian systems			
	Prokaryotes	Lower eukaryotes	Plants	Insects	In vitro — Animal cells	In vitro — Human cells	In vivo — Animals	Humans
	D	G D R	G D A	D G C R	A D G S M C A T	I D G S M C A	D G S M C	D S M C A
Petroleum solvents								
Rubber solvent								
Another special boiling range solvent		–			– –¹			
White spirits		–¹				–¹	–¹	
Toluene	–	–		– –¹ +¹ +¹	–¹ –¹	–¹ ?	+ +	? ?
Xylene	–	+¹ +¹		–¹	–¹	? ?	–¹ +	?
Cyclohexanone	–			–¹	? ?	? ?	–¹	?
Dimethylformamide	–	+¹ –¹	–¹	–¹	– –¹ +¹ +¹ ?	+¹ – – –¹ ?	+¹	
Morpholine	–			–¹ –¹	–¹ –¹	–¹ –¹	–¹ –¹ – –¹	
1,2-Epoxybutane	+ +	+¹ +	+ +¹	+ +¹	–¹ + +¹ +	+¹	–¹ +¹ –¹ +	
Bis(2,3-epoxycyclopentyl)ether	+¹				+¹	+¹	+¹	
Glycidyl ethers								
Allyl (C8–C14) glycidyl ethers	?				–			
Allyl glycidyl ether	+			+¹				

	Nonmammalian systems											Mammalian systems																												
	Prokaryotes		Lower eukaryotes			Plants			Insects			In vitro																		In vivo											
												Animal cells									Human cells								Animals							Humans				
	D	G	D	R	G	A	D	G	C	R	G	A	D	G	S	M	C	A	T	I	D	G	S	M	C	A	T	I	D	G	S	M	C	DL	A	D	S	M	C	A
Bisphenol A diglycidyl ether	+																																							−'
Butane diol glycidyl ether	+'																																							
Butyl glycidyl ether (n = *n*-butyl; t = *tert*-butyl; x, positive *via* intraperitoneal route, negative *via* oral route)	+[4,11]																−1n								x^a									$+1n$						?
tert-Butyl phenyl glycidyl ether	+																																							
Cresyl glycidyl ether	+																																							?
Neopentyl glycol diglycidyl ether	+'																																							
Phenol	−					?				−			−'		+'	+							+					−												
Antimony trioxide	+'																																							
Titanium dioxide	−																			−																				

A, aneuploidy; C, chromosomal aberrations; D, DNA damage; DL, dominant lethal mutation; G, gene mutation; I, inhibition of intercellular communication; M, micronuclei; R, mitotic recombination and gene conversion; S, sister chromatid exchange; T, cell transformation

In completing the tables, the following symbols indicate the consensus of the Working Group with regard to the results for each endpoint:

+   considered to be positive for the specific endpoint and level of biological complexity

+'   considered to be positive, but only one valid study was available to the Working Group.

−   considered to be negative

−'   considered to be negative, but only one valid study was available to the Working Group

?   considered to be equivocal or inconclusive (e.g., there were contradictory results from different laboratories; there were confounding exposures; the results were equivocal)

APPENDIX 2

ACTIVITY PROFILES FOR GENETIC AND RELATED EFFECTS

Methods

The x–axis of the activity profile represents the bioassays in phylogenetic sequence by endpoint, and the values on the y–axis represent the logarithmically transformed lowest effective doses (LED) and highest ineffective doses (HID) tested. The term 'dose', as used in this report, does not take into consideration length of treatment or exposure and may therefore be considered synonymous with concentration. In practice, the concentrations used in all the in–vitro tests were converted to μg/ml, and those for in–vivo tests were expressed as mg/kg bw. Because dose units are plotted on a log scale, differences in molecular weights of compounds do not, in most cases, greatly influence comparisons of their activity profiles. Conventions for dose conversions are given below.

Profile–line height (the magnitude of each bar) is a function of the LED or HID, which is associated with the characteristics of each individual test system – such as population size, cell–cycle kinetics and metabolic competence. Thus, the detection limit of each test system is different, and, across a given activity profile, responses will vary substantially. No attempt is made to adjust or relate responses in one test system to those of another.

Line heights are derived as follows: for negative test results, the highest dose tested without appreciable toxicity is defined as the HID. If there was evidence of extreme toxicity, the next highest dose is used. A single dose tested with a negative result is considered to be equivalent to the HID. Similarly, for positive results, the LED is recorded. If the original data were analysed statistically by the author, the dose recorded is that at which the response was significant ($p < 0.05$). If the available data were not analysed statistically, the dose required to produce an effect is estimated as follows: when a dose–related positive response is observed with two or more doses, the lower of the doses is taken as the LED; a single dose resulting in a positive response is considered to be equivalent to the LED.

In order to accommodate both the wide range of doses encountered and positive and negative responses on a continuous scale, doses are transformed logarithmically, so that effective (LED) and ineffective (HID) doses are represented by positive and negative numbers, respectively. The response, or logarithmic dose unit (LDU_{ij}), for a given test system i and chemical j is represented by the expressions

$$LDU_{ij} = -\log_{10} (\text{dose}), \text{ for HID values; } LDU \leq 0$$

and (1)

$$LDU_{ij} = -\log_{10} (\text{dose} \times 10^{-5}), \text{ for LED values; } LDU \geq 0.$$

–449–

These simple relationships define a dose range of 0 to –5 logarithmic units for ineffective doses (1–100 000 µg/ml or mg/kg bw) and 0 to +8 logarithmic units for effective doses (100 000–0.001 µg/ml or mg/kg bw). A scale illustrating the LDU values is shown in Figure 1. Negative responses at doses less than 1 µg/ml (mg/kg bw) are set equal to 1. Effectively, an LED value ≥100 000 or an HID value ≤1 produces an LDU = 0; no quantitative information is gained from such extreme values. The dotted lines at the levels of log dose units 1 and –1 define a 'zone of uncertainty' in which positive results are reported at such high doses (between 10 000 and 100 000 µg/ml or mg/kg bw) or negative results are reported at such low dose levels (1 to 10 µg/ml or mg/kg bw) as to call into question the adequacy of the test.

Fig. 1. Scale of log dose units used on the y–axis of activity profiles

Positive   Log dose
(µg/ml or mg/kg bw)   units

```
      0.001    .........................................    8     ----
      0.01     .........................................    7     --
      0.1      .........................................    6     --
      1.0      .........................................    5     --
     10        .........................................    4     --
    100        .........................................    3     --
   1000        .........................................    2     --
  10 000       .........................................    1     --
 100 000       .................... 1 ....................   0     ----
               .................... 10 ...................  -1     --
               .................... 100 ..................  -2     --
               ................. 1000 ....................  -3     --
               ............. 10 000 .....................   -4     --
               ............. 100 000 ....................   -5     ----
```

                        Negative
                        (µg/ml or mg/kg bw)

LED and HID are expressed as µg/ml or mg/kg bw.

In practice, an activity profile is computer generated. A data entry programme is used to store abstracted data from published reports. A sequential file (in ASCH) is created for each compound, and a record within that file consists of the name and Chemical Abstracts Service number of the compound, a three–letter code for the test system (see below), the qualitative test result (with and without an exogenous metabolic system), dose (LED or HID), citation number and additional source information. An abbreviated citation for each publication is stored in a segment of a record accessing both the test data file and the citation file. During processing of the data file, an average of the logarithmic values of the data subset is calculated, and the length of the profile line represents this average value. All dose values are plotted for each profile line, regardless of whether results are positive or negative. Results obtained in the absence of an exogenous metabolic system are indicated by a bar (–), and results obtained in the presence of an exogenous metabolic system are indicated by an upward–directed arrow (↑). When all results for a given assay are either positive or negative, the mean of the LDU values is plotted as a solid line; when conflicting data are reported for the same assay (i.e., both positive and negative results), the majority data are shown by a solid line and the minority data by a dashed line (drawn to the extreme conflicting response). In the few cases in which the numbers of positive and negative results are equal, the solid line is drawn in the positive direction and the maximal negative response is indicated with a dashed line.

Profile lines are identified by three–letter code words representing the commonly used tests. Code words for most of the test systems in current use in genetic toxicology were defined for the US Environmental Protection Agency's GENE–TOX Program (Waters, 1979; Waters & Auletta, 1981). For this publication, codes were redefined in a manner that should facilitate inclusion of additional tests in the future. If a test system is not defined precisely, a general code is used that best defines the category of the test. Naming conventions are described below.

Dose conversions for activity profiles

Doses are converted to μg/ml for in–vitro tests and to mg/kg bw per day for in–vivo experiments.

1.   In–vitro test systems

   (a)   Weight/volume converts directly to μg/ml.

   (b)   Molar (M) concentration x molecular weight = mg/ml = 10^3 μg/ml; mM concentration x molecular weight = μg/ml.

   (c)   Soluble solids expressed as % concentration are assumed to be in units of mass per volume (i.e., 1% = 0.01 g/ml = 10 000 μg/ml; also, 1 ppm = 1 μg/ml).

   (d)   Liquids and gases expressed as % concentration are assumed to be given in units of volume per volume. Liquids are converted to weight per volume using the density (D) of the solution (D = g/ml). If the bulk of the solution is water, then D = 1.0 g/ml. Gases are converted from volume to mass using the ideal gas law, PV = nRT. For exposure at 20–37°C at standard atmospheric pressure, 1% (v/v) = 0.4 μg/ml x molecular weight of the gas. Also, 1 ppm (v/v) = 4 x 10^{-5} μg/ml x molecular weight.

   (e)   For microbial plate tests, concentrations reported as weight/plate are divided by top agar volume (if volume is not given, a 2–ml top agar is assumed). For spot tests, in which concentrations are reported as weight or weight/disc, a 1–ml volume is used as a rough approximation.

   (f)   Conversion of asbestos concentrations given in μg/cm^2 are based on the area (A) of the dish and the volume of medium per dish; i.e., for a 100–mm dish: A = πR^2 = π x (5cm)^2 = 78.5 cm^2. If the volume of medium is 10 ml, then 78.5 cm^2 = 10 ml and 1 cm^2 = 0.13 ml.

2.   In–vitro systems using in–vivo activation

   For the body fluid–urine (BF–) test, the concentration used is the dose (in mg.kg bw) of the compound administered to test animals or patients.

3.   In-vivo test systems

(a)   Doses are converted to mg/kg bw per day of exposure, assuming 100% absorption. Standard values are used for each sex and species of rodent, including body weight and average intake per day, as reported by Gold et al. (1984). For example, in a test using male mice fed 50 ppm of the agent in the diet, the standard food intake per day is 12% of body weight, and the conversion is dose = 50 ppm × 12% = 6 mg/kg bw per day.

Standard values used for humans are: weight - males, 70 kg; females, 55 kg; surface area, $1.7 m^2$; inhalation rate, 20 l/min for light work, 30 l/min for mild exercise.

(b)   When reported, the dose at the target site is used. For example, doses given in studies of lymphocytes of humans exposed in vivo are the measured blood concentrations in μg/ml.

Codes for test systems

For specific nonmammalian test systems, the first two letters of the three–symbol code word define the test organism (e.g., SA– for *Salmonella typhimurium*, EC– for *Escherichia coli*). In most cases, the first two letters accurately represent the scientific name of the organism. If the species is not known, the convention used is –S–. The third symbol may be used to define the tester strain (e.g., SA8 for *S. typhimurium* TA1538, ECW for *E. coli* WP2*uvr*A). When strain designation is not indicated, the third letter is used to define the specific genetic endpoint under investigation (e.g., —D for differential toxicity, —F for forward mutation, —G for gene conversion or genetic crossing–over, —N for aneuploidy, —R for reverse mutation, —U for unscheduled DNA synthesis). The third letter may also be used to define the general endpoint under investigation when a more complete definition is not possible or relevant (e.g., —M for mutation, —C for chromosomal aberration).

For mammalian test systems, the first letter of the three–letter code word defines the genetic endpoint under investigation: A— for aneuploidy, B— for binding, C— for chromosomal aberration, D— for DNA strand breaks, G— for gene mutation, I— for inhibition of intercellular communication, M— for micronucleus formation, R— for DNA repair, S— for sister chromatid exchange, T— for cell transformation and U— for unscheduled DNA synthesis.

For animal (i.e., nonhuman) test systems in vitro, when the cell type is not specified, the code letters –IA are used. For such assays in vivo, when the animal species is not specified, the code letters –VA are used. Commonly used animal species are identified by the third letter (e.g., —C for Chinese hamster, —M for mouse, —R for rat, —S for Syrian hamster).

For test systems using human cells in vitro, when the cell type is not specified, the code letters –IH are used. For assays on humans in vivo, when the cell type is not specified, the code letters –VH are used. Otherwise, the second letter specifies the cell type under investigation (e.g., –BH for bone marrow, –LH for lymphocytes).

Some other specific coding conventions used for mammalian systems are as follows: BF– for body fluids, HM– for host-mediated, —L for leucocytes or lymphocytes in vitro

(–AL, animals; –HL, humans), –L– for leucocytes *in vivo* (–LA, animals; –LH, humans), —T for transformed cells.

Note that these are examples of major conventions used to define the assay code words. The alphabetized listing of codes must be examined to confirm a specific code word. As might be expected from the limitation to three symbols, some codes do not fit the naming conventions precisely. In a few cases, test systems are defined by first–letter code words, for example: MST, mouse spot test; SLP, mouse specific locus test, postspermatogonia; SLO, mouse specific locus test, other stages; DLM, dominant lethal test in mice; DLR, dominant lethal test in rats; MHT, mouse heritable translocation test.

The genetic activity profiles and listings that follow were prepared in collaboration with Environmental Health Research and Testing Inc. (EHRT) under contract to the US Environmental Protection Agency; EHRT also determined the doses used. The references cited in each genetic activity profile listing can be found in the list of references in the appropriate monograph.

References

Garrett, N.E., Stack, H.F., Gross, M.R. & Waters, M.D. (1984) An analysis of the spectra of genetic activity produced by known or suspected human carcinogens. *Mutat. Res., 134,* 89–111

Gold, L.S., Sawyer, C.B., Magaw, R., Backman, G.M., de Veciana, M., Levinson, R., Hooper, N.K., Havender, W.R., Bernstein, L., Peto, R., Pike, M.C. & Ames, B.N. (1984) A carcinogenic potency database of the standardized results of animal bioassays. *Environ. Health Perspect., 58,* 9–319

Waters, M.D. (1979) *The GENE–TOX program.* In: Hsie, A.W., O'Neill, J.P. & McElheny, V.K., eds, *Mammalian Cell Mutagenesis: The Maturation of Test Systems (Banbury Report 2),* Cold Spring Harbor, NY, CHS Press, pp. 449–467

Waters, M.D. & Auletta, A. (1981) The GENE–TOX program: genetic activity evaluation. *J. chem. Inf. comput. Sci., 21,* 35–38

TOLUENE

END POINT	TEST CODE	TEST SYSTEM	RESULTS NM	M	DOSE (LED OR HID)	REFERENCE
D	PRB	PROPHAGE, INDUCT/SOS/STRAND BREAKS/X-LINKS	-	-	100.0000	NAKAMURA ET AL., 1987
D	ECL	E. COLI POL A, DIFFERENTIAL TOX (LIQUID)	-	-	400000.0000	MCCARROLL ET AL., 1981b
D	ERD	E. COLI REC, DIFFERENTIAL TOXICITY	-	-	400000.0000	MCCARROLL ET AL., 1981b
D	BSD	B. SUBTILIS REC, DIFFERENTIAL TOXICITY	-	-	127000.0000	MCCARROLL ET AL., 1981a
G	SA0	S. TYPHIMURIUM TA100, REVERSE MUTATION	-	-	2500.0000	SPANGGORD ET AL., 1982
G	SA0	S. TYPHIMURIUM TA100, REVERSE MUTATION	-	-	1000.0000	BOS ET AL., 1981
G	SA0	S. TYPHIMURIUM TA100, REVERSE MUTATION	-	-	2150.0000	NESTMANN ET AL., 1980
G	SA0	S. TYPHIMURIUM TA100, REVERSE MUTATION	-	-	1000.0000	CONNOR ET AL., 1985
G	SA0	S. TYPHIMURIUM TA100, REVERSE MUTATION	-	-	167.0000	HAWORTH ET AL., 1983
G	SA5	S. TYPHIMURIUM TA1535, REVERSE MUTATION	-	-	2500.0000	SPANGGORD ET AL., 1982
G	SA5	S. TYPHIMURIUM TA1535, REVERSE MUTATION	-	-	2150.0000	NESTMANN ET AL., 1980
G	SA5	S. TYPHIMURIUM TA1535, REVERSE MUTATION	-	-	1000.0000	BOS ET AL., 1981
G	SA5	S. TYPHIMURIUM TA1535, REVERSE MUTATION	-	-	167.0000	HAWORTH ET AL., 1983
G	SA7	S. TYPHIMURIUM TA1537, REVERSE MUTATION	-	-	2500.0000	SPANGGORD ET AL., 1982
G	SA7	S. TYPHIMURIUM TA1537, REVERSE MUTATION	-	-	2150.0000	NESTMANN ET AL., 1980
G	SA7	S. TYPHIMURIUM TA1537, REVERSE MUTATION	-	-	1000.0000	BOS ET AL., 1981
G	SA7	S. TYPHIMURIUM TA1537, REVERSE MUTATION	-	-	167.0000	HAWORTH ET AL., 1983
G	SA8	S. TYPHIMURIUM TA1538, REVERSE MUTATION	-	-	2150.0000	NESTMANN ET AL., 1980
G	SA8	S. TYPHIMURIUM TA1538, REVERSE MUTATION	-	-	2500.0000	SPANGGORD ET AL., 1982
G	SA8	S. TYPHIMURIUM TA1538, REVERSE MUTATION	-	-	1000.0000	BOS ET AL., 1981
G	SA9	S. TYPHIMURIUM TA98, REVERSE MUTATION	-	-	1000.0000	BOS ET AL., 1981
G	SA9	S. TYPHIMURIUM TA98, REVERSE MUTATION	-	-	2500.0000	SPANGGORD ET AL., 1982
G	SA9	S. TYPHIMURIUM TA98, REVERSE MUTATION	-	-	2150.0000	NESTMANN ET AL., 1980
G	SA9	S. TYPHIMURIUM TA98, REVERSE MUTATION	-	-	1000.0000	CONNOR ET AL., 1985
G	SA9	S. TYPHIMURIUM TA98, REVERSE MUTATION	-	-	167.0000	HAWORTH ET AL., 1983
G	SAS	S. TYPHIMURIUM (OTHER), REVERSE MUTATION	-	-	1000.0000	CONNOR ET AL., 1985
G	DMX	D. MELANOGASTER, SEX-LINKED RECESSIVES	-	0	13000.0000	RODRIGUEZ ARNAIZ & VILLALOBOS-PIETRINI, 1985b
C	DMH	D. MELANOGASTER, HERITABLE TRANSLOCATIONS	-	0	13000.0000	RODRIGUEZ ARNAIZ & VILLALOBOS-PIETRINI, 1985b
A	DMN	D. MELANOGASTER, ANEUPLOIDY	+	0	8700.0000	RODRIGUEZ ARNAIZ & VILLALOBOS-PIETRINI, 1985a
D	DIA	STRAND BREAKS/X-LINKS, ANIMAL CELLS IN VITRO	+	0	3.0000	SINA ET AL., 1983
T	T7S	CELL TRANSFORMATION, SA7/SHE CELLS	-	0	1000.0000	CASTO, 1981
S	SHL	SCE, HUMAN LYMPHOCYTES IN VITRO	-	0	1500.0000	GERNER-SMIDT & FRIEDRICH, 1978
C	CHL	CHROM ABERR, HUMAN LYMPHOCYTES IN VITRO	-	0	1500.0000	GERNER-SMIDT & FRIEDRICH, 1978
M	MVM	MICRONUCLEUS TEST, MICE IN VIVO	+	0	0.1200	MOHTASHAMIPUR ET AL., 1985
M	MVM	MICRONUCLEUS TEST, MICE IN VIVO	-	0	1700.0000	GAD-EL-KARIM ET AL., 1984
M	MVR	MICRONUCLEUS TEST, RATS IN VIVO	+	0	200.0000	ROH ET AL., 1987
C	CBA	CHROM ABERR, ANIMAL BONE MARROW IN VIVO	+	0	1000.0000	LYAPKALO, 1973
C	CBA	CHROM ABERR, ANIMAL BONE MARROW IN VIVO	+	0	400.0000	ROH ET AL., 1987
C	CBA	CHROM ABERR, ANIMAL BONE MARROW IN VIVO	+	0	800.0000	DOBROKHOTOV, 1972
C	CBA	CHROM ABERR, ANIMAL BONE MARROW IN VIVO	+	0	0.0000	ARISTOV ET AL., 1981
C	CBA	CHROM ABERR, ANIMAL BONE MARROW IN VIVO	-	0	1700.0000	GAD-EL-KARIM ET AL., 1984
S	SLH	SCE, HUMAN LYMPHOCYTES IN VIVO	+	0	75.0000	BAUCHINGER ET AL., 1982
S	SLH	SCE, HUMAN LYMPHOCYTES IN VIVO	-	0	14.0000	HAGLUND ET AL., 1980
S	SLH	SCE, HUMAN LYMPHOCYTES IN VIVO	-	0	40.0000	MAKI-PAAKKANEN ET AL., 198
C	CLH	CHROM ABERR, HUMAN LYMPHOCYTES IN VIVO	+	0	75.0000	BAUCHINGER ET AL., 1982
C	CLH	CHROM ABERR, HUMAN LYMPHOCYTES IN VIVO	-	0	40.0000	MAKI-PAAKKANEN ET AL., 198
C	CLH	CHROM ABERR, HUMAN LYMPHOCYTES IN VIVO	-	0	75.0000	FORNI ET AL., 1971
C	CLH	CHROM ABERR, HUMAN LYMPHOCYTES IN VIVO	-	0	0.0000	PELCLOVA ET AL., 1987
P	SPM	SPERM MORPHOLOGY, MICE	-	0	900.0000	TOPHAM, 1980

TOLUENE
108-88-3

25-OCT-89

LOG DOSE UNITS

XYLENES

END POINT	TEST CODE	TEST SYSTEM	RESULTS NM M	DOSE (LED OR HID)	REFERENCE	
D	PRB	PROPHAGE, INDUCT/SOS/STRAND BREAKS/X-LINKS	- -	36.0000	NAKAMURA ET AL., 1987	(us)
D	ECL	E. COLI POL A, DIFFERENTIAL TOX (LIQUID)	- -	10000.0000	MCCARROLL ET AL., 1981b	(tg)
D	ERD	E. COLI REC, DIFFERENTIAL TOXICITY	- -	10000.0000	MCCARROLL ET AL., 1981b	(tg)
D	BSD	B. SUBTILIS REC, DIFFERENTIAL TOXICITY	- -	100000.0000	MCCARROLL ET AL., 1981a	(tg)
G	SA0	S. TYPHIMURIUM TA100, REVERSE MUTATION	- -	100.0000	ZEIGER ET AL., 1987	(x)
G	SA0	S. TYPHIMURIUM TA100, REVERSE MUTATION	- -	50.0000	SHIMIZU ET AL., 1985	(p)
G	SA0	S. TYPHIMURIUM TA100, REVERSE MUTATION	- -	250.0000	BOS ET AL., 1981	(o,m,p)
G	SA0	S. TYPHIMURIUM TA100, REVERSE MUTATION	- -	166.0000	HAWORTH ET AL., 1983	(o)
G	SA0	S. TYPHIMURIUM TA100, REVERSE MUTATION	- -	16.0000	HAWORTH ET AL., 1983	(m)
G	SA0	S. TYPHIMURIUM TA100, REVERSE MUTATION	- -	100.0000	HAWORTH ET AL., 1983	(p)
G	SA0	S. TYPHIMURIUM TA100, REVERSE MUTATION	- -	500.0000	CONNOR ET AL., 1985	(o,m,p)
G	SA5	S. TYPHIMURIUM TA1535, REVERSE MUTATION	- -	100.0000	ZEIGER ET AL., 1987	(x)
G	SA5	S. TYPHIMURIUM TA1535, REVERSE MUTATION	- -	166.0000	HAWORTH ET AL., 1983	(o)
G	SA5	S. TYPHIMURIUM TA1535, REVERSE MUTATION	- -	16.0000	HAWORTH ET AL., 1983	(m)
G	SA5	S. TYPHIMURIUM TA1535, REVERSE MUTATION	- -	100.0000	HAWORTH ET AL., 1983	(p)
G	SA5	S. TYPHIMURIUM TA1535, REVERSE MUTATION	- -	50.0000	SHIMIZU ET AL., 1985	(p)
G	SA5	S. TYPHIMURIUM TA1535, REVERSE MUTATION	- -	250.0000	BOS ET AL., 1981	(o,m,p)
G	SA7	S. TYPHIMURIUM TA1537, REVERSE MUTATION	- -	50.0000	SHIMIZU ET AL., 1985	(p)
G	SA7	S. TYPHIMURIUM TA1537, REVERSE MUTATION	- -	166.0000	HAWORTH ET AL., 1983	(o)
G	SA7	S. TYPHIMURIUM TA1537, REVERSE MUTATION	- -	16.0000	HAWORTH ET AL., 1983	(m)
G	SA7	S. TYPHIMURIUM TA1537, REVERSE MUTATION	- -	100.0000	HAWORTH ET AL., 1983	(p)
G	SA7	S. TYPHIMURIUM TA1537, REVERSE MUTATION	- -	250.0000	BOS ET AL., 1981	(o,m,p)
G	SA8	S. TYPHIMURIUM TA1538, REVERSE MUTATION	- -	250.0000	BOS ET AL., 1981	(o,m,p)
G	SA9	S. TYPHIMURIUM TA98, REVERSE MUTATION	- -	100.0000	ZEIGER ET AL., 1987	(x)
G	SA9	S. TYPHIMURIUM TA98, REVERSE MUTATION	- -	250.0000	BOS ET AL., 1981	(o,m,p)
G	SA9	S. TYPHIMURIUM TA98, REVERSE MUTATION	- -	50.0000	SHIMIZU ET AL., 1985	(p)
G	SA9	S. TYPHIMURIUM TA98, REVERSE MUTATION	- -	166.0000	HAWORTH ET AL., 1983	(o)
G	SA9	S. TYPHIMURIUM TA98, REVERSE MUTATION	- -	16.0000	HAWORTH ET AL., 1983	(m)
G	SA9	S. TYPHIMURIUM TA98, REVERSE MUTATION	- -	100.0000	HAWORTH ET AL., 1983	(p)
G	SA9	S. TYPHIMURIUM TA98, REVERSE MUTATION	- -	500.0000	CONNOR ET AL., 1985	(o,m,p)
G	SAS	S. TYPHIMURIUM (OTHER), REVERSE MUTATION	- -	100.0000	ZEIGER ET AL., 1987	(x)
G	SAS	S. TYPHIMURIUM (OTHER), REVERSE MUTATION	- -	500.0000	CONNOR ET AL., 1985	(o,m,p)
G	ECW	E. COLI WP2 UVRA, REVERSE MUTATION	- -	50.0000	SHIMIZU ET AL., 1985	(p)
T	T7S	CELL TRANSFORMATION, SA7/SHE CELLS	- 0	1000.0000	CASTO, 1981	(us)
S	SHL	SCE, HUMAN LYMPHOCYTES IN VITRO	- 0	1500.0000	GERNER-SMIDT & FRIEDRICH, 1978	(us)
C	CHL	CHROM ABERR, HUMAN LYMPHOCYTES IN VITRO	- 0	1500.0000	GERNER-SMIDT & FRIEDRICH, 1978	(us)
M	MVM	MICRONUCLEUS TEST, MICE IN VIVO	- 0	400.0000	MOHTASHAMIPUR ET AL., 1985	(o)
M	MVM	MICRONUCLEUS TEST, MICE IN VIVO	- 0	650.0000	MOHTASHAMIPUR ET AL., 1985	(m,p)
S	SLH	SCE, HUMAN LYMPHOCYTES IN VIVO	- 0	10.0000	HAGLUND ET AL., 1980	(us)
S	SLH	SCE, HUMAN LYMPHOCYTES IN VIVO	- 0	55.0000	PAP & VARGA ET AL., 1987	(tg)
P	SPR	SPERM MORPHOLOGY, RATS	(+) 0	400.0000	WASHINGTON ET AL., 1983	(o)

o — ortho
m — meta
p — para
us — unspecified
x — mixture of isomers
tg — technical grade

XYLENES
1330-20-7

25-OCT-89

LOG DOSE UNITS

CYCLOHEXANONE

END POINT	TEST CODE	TEST SYSTEM	RESULTS NM M	DOSE (LED OR HID)	REFERENCE
G	SA0	S. TYPHIMURIUM TA100, REVERSE MUTATION	- -	5000.0000	HAWORTH ET AL., 1983
G	SA5	S. TYPHIMURIUM TA1535, REVERSE MUTATION	- -	5000.0000	HAWORTH ET AL., 1983
G	SA7	S. TYPHIMURIUM TA1537, REVERSE MUTATION	- -	5000.0000	HAWORTH ET AL., 1983
G	SA9	S. TYPHIMURIUM TA98, REVERSE MUTATION	- -	5000.0000	HAWORTH ET AL., 1983
C	CHL	CHROM ABERR, HUMAN LYMPHOCYTES IN VITRO	+ 0	0.0000	COLLIN, 1971
C	CHL	CHROM ABERR, HUMAN LYMPHOCYTES IN VITRO	+ 0	10.0000	LEDERER ET AL., 1971
C	CHL	CHROM ABERR, HUMAN LYMPHOCYTES IN VITRO	+ 0	0.0050	DYSHLOVOI ET AL., 1981
A	AIH	ANEUPLOIDY, HUMAN CELLS IN VITRO	+ 0	0.0050	DYSHLOVOI ET AL., 1981
C	CBA	CHROM ABERR, ANIMAL BONE MARROW IN VIVO	+ 0	100.0000	DE HONDT ET AL., 1983

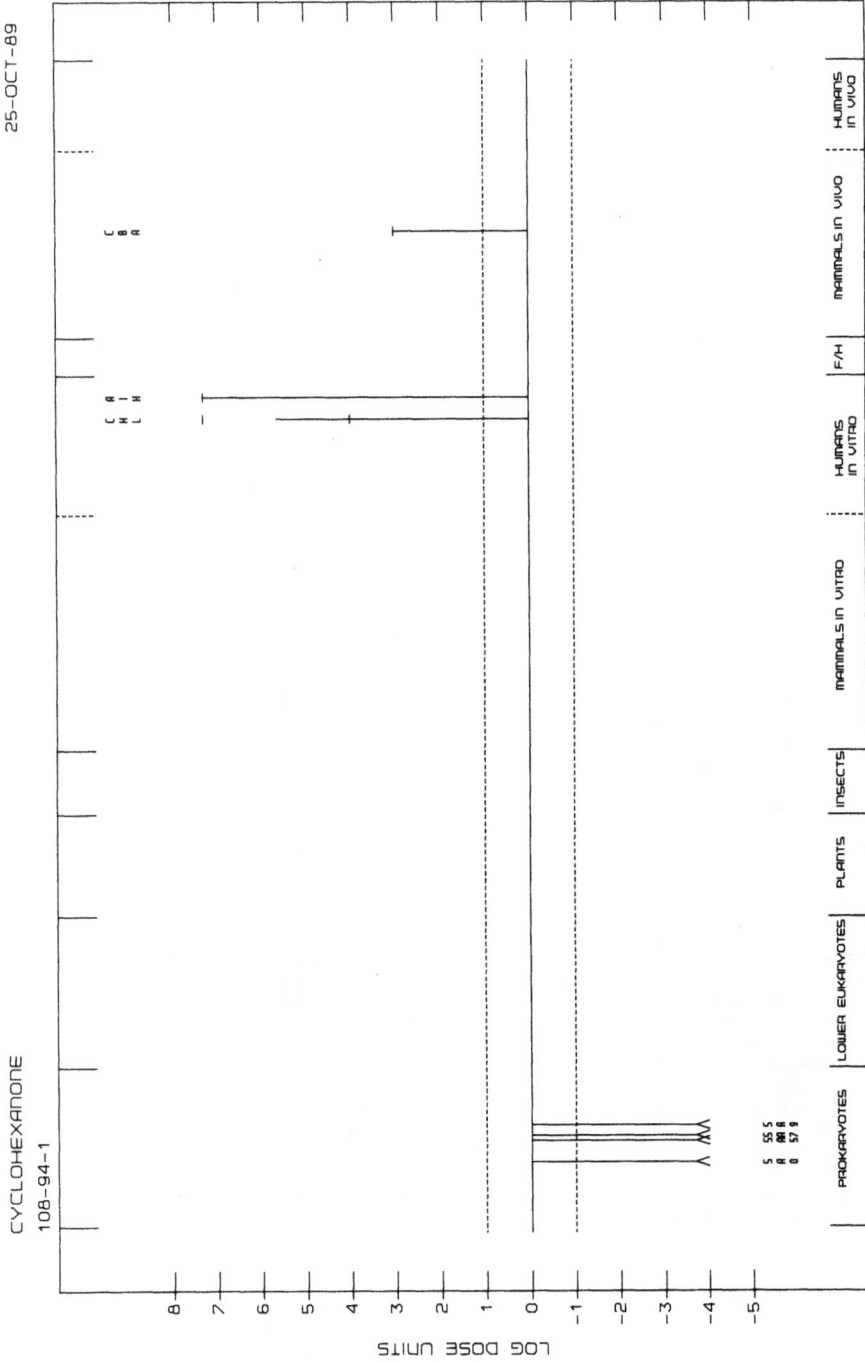

CYCLOHEXANONE
108-94-1

25-OCT-89

DIMETHYLFORMAMIDE

END POINT	TEST CODE	TEST SYSTEM	RESULTS NM M	DOSE (LED OR HID)	REFERENCE
D	ECL	E. COLI POL A, DIFFERENTIAL TOX (LIQUID)	- -	2300.0000	ROSENKRANZ ET AL., 1981
D	ERD	E. COLI REC, DIFFERENTIAL TOXICITY	- -	0.0000	GREEN, 1981
D	ERD	E. COLI REC, DIFFERENTIAL TOXICITY	- 0	0.0000	ICHINOTSUBO ET AL., 1981a
D	ERD	E. COLI REC, DIFFERENTIAL TOXICITY	- -	0.0000	TWEATS, 1981
D	BSD	B. SUBTILIS REC, DIFFERENTIAL TOXICITY	- -	19000.0000	KADA, 1981
G	SAF	S. TYPHIMURIUM, FORWARD MUTATION	0 -	1000.0000	SKOPEK ET AL., 1981
G	SAO	S. TYPHIMURIUM TA100, REVERSE MUTATION	0 -	1250.0000	PURCHASE ET AL., 1978
G	SAO	S. TYPHIMURIUM TA100, REVERSE MUTATION	- -	94.0000	ANTOINE ET AL., 1983
G	SAO	S. TYPHIMURIUM TA100, REVERSE MUTATION	0 -	1000.0000	FALCK ET AL., 1985
G	SAO	S. TYPHIMURIUM TA100, REVERSE MUTATION	- -	5000.0000	MORTELMANS ET AL., 1986
G	SAO	S. TYPHIMURIUM TA100, REVERSE MUTATION	- -	2500.0000	MACDONALD, 1981
G	SAO	S. TYPHIMURIUM TA100, REVERSE MUTATION	- 0	0.0000	ICHINOTSUBO ET AL., 1981b
G	SAO	S. TYPHIMURIUM TA100, REVERSE MUTATION	- -	0.0000	NAGAO & TAKAHASHI, 1981
G	SAO	S. TYPHIMURIUM TA100, REVERSE MUTATION	- -	5000.0000	RICHOLD & JONES, 1981
G	SAO	S. TYPHIMURIUM TA100, REVERSE MUTATION	- -	1000.0000	ROWLAND & SEVERN, 1981
G	SAO	S. TYPHIMURIUM TA100, REVERSE MUTATION	- -	1250.0000	TRUEMAN, 1981
G	SAO	S. TYPHIMURIUM TA100, REVERSE MUTATION	- -	250.0000	VENITT & CROFTON-SLEIGH, 1981
G	SAO	S. TYPHIMURIUM TA100, REVERSE MUTATION	- ?	500.0000	HUBBARD ET AL., 1981
G	SAO	S. TYPHIMURIUM TA100, REVERSE MUTATION	- -	1000.0000	BROOKS & DEAN, 1981
G	SAO	S. TYPHIMURIUM TA100, REVERSE MUTATION	- -	500.0000	BAKER & BONIN, 1981
G	SAO	S. TYPHIMURIUM TA100, REVERSE MUTATION	- -	0.0000	SIMMON & SHEPHERD, 1981
G	SA5	S. TYPHIMURIUM TA1535, REVERSE MUTATION	0 -	1250.0000	PURCHASE ET AL., 1978
G	SA5	S. TYPHIMURIUM TA1535, REVERSE MUTATION	- -	94.0000	ANTOINE ET AL., 1983
G	SA5	S. TYPHIMURIUM TA1535, REVERSE MUTATION	0 -	1000.0000	FALCK ET AL., 1985
G	SA5	S. TYPHIMURIUM TA1535, REVERSE MUTATION	- -	5000.0000	MORTELMANS ET AL., 1986
G	SA5	S. TYPHIMURIUM TA1535, REVERSE MUTATION	- -	5000.0000	RICHOLD & JONES, 1981
G	SA5	S. TYPHIMURIUM TA1535, REVERSE MUTATION	- -	1000.0000	ROWLAND & SEVERN, 1981
G	SA5	S. TYPHIMURIUM TA1535, REVERSE MUTATION	- -	1250.0000	TRUEMAN, 1981
G	SA5	S. TYPHIMURIUM TA1535, REVERSE MUTATION	- -	1000.0000	GATEHOUSE, 1981
G	SA5	S. TYPHIMURIUM TA1535, REVERSE MUTATION	- -	1000.0000	BROOKS & DEAN, 1981
G	SA5	S. TYPHIMURIUM TA1535, REVERSE MUTATION	- -	500.0000	BAKER & BONIN, 1981
G	SA5	S. TYPHIMURIUM TA1535, REVERSE MUTATION	- -	0.0000	SIMMON & SHEPHERD, 1981
G	SA7	S. TYPHIMURIUM TA1537, REVERSE MUTATION	- -	94.0000	ANTOINE ET AL., 1983
G	SA7	S. TYPHIMURIUM TA1537, REVERSE MUTATION	0 -	500.0000	FALCK ET AL., 1985
G	SA7	S. TYPHIMURIUM TA1537, REVERSE MUTATION	- -	5000.0000	MORTELMANS ET AL., 1986
G	SA7	S. TYPHIMURIUM TA1537, REVERSE MUTATION	- -	0.0000	NAGAO & TAKAHASHI, 1981
G	SA7	S. TYPHIMURIUM TA1537, REVERSE MUTATION	- -	5000.0000	RICHOLD & JONES, 1981
G	SA7	S. TYPHIMURIUM TA1537, REVERSE MUTATION	- -	1000.0000	ROWLAND & SEVERN, 1981
G	SA7	S. TYPHIMURIUM TA1537, REVERSE MUTATION	- -	1250.0000	TRUEMAN, 1981
G	SA7	S. TYPHIMURIUM TA1537, REVERSE MUTATION	- -	5000.0000	MACDONALD, 1981
G	SA7	S. TYPHIMURIUM TA1537, REVERSE MUTATION	- -	1000.0000	GATEHOUSE, 1981
G	SA7	S. TYPHIMURIUM TA1537, REVERSE MUTATION	- -	1000.0000	BROOKS & DEAN, 1981
G	SA7	S. TYPHIMURIUM TA1537, REVERSE MUTATION	- -	500.0000	BAKER & BONIN, 1981
G	SA7	S. TYPHIMURIUM TA1537, REVERSE MUTATION	- -	0.0000	SIMMON & SHEPHERD, 1981
G	SA8	S. TYPHIMURIUM TA1538, REVERSE MUTATION	0 -	1250.0000	PURCHASE ET AL., 1978
G	SA8	S. TYPHIMURIUM TA1538, REVERSE MUTATION	- -	94.0000	ANTOINE ET AL., 1983
G	SA8	S. TYPHIMURIUM TA1538, REVERSE MUTATION	0 -	1000.0000	FALCK ET AL., 1985
G	SA8	S. TYPHIMURIUM TA1538, REVERSE MUTATION	- -	5000.0000	RICHOLD & JONES, 1981
G	SA8	S. TYPHIMURIUM TA1538, REVERSE MUTATION	- -	1000.0000	ROWLAND & SEVERN, 1981
G	SA8	S. TYPHIMURIUM TA1538, REVERSE MUTATION	- +	0.0000	TRUEMAN, 1981

DIMETHYLFORMAMIDE

END POINT	TEST CODE	TEST SYSTEM	RESULTS NM M	DOSE (LED OR HID)	REFERENCE
G	SA8	S. TYPHIMURIUM TA1538, REVERSE MUTATION	- -	1000.0000	BROOKS & DEAN, 1981
G	SA8	S. TYPHIMURIUM TA1538, REVERSE MUTATION	- -	500.0000	BAKER & BONIN, 1981
G	SA8	S. TYPHIMURIUM TA1538, REVERSE MUTATION	- -	0.0000	SIMMON & SHEPHERD, 1981
G	SA9	S. TYPHIMURIUM TA98, REVERSE MUTATION	0 -	1250.0000	PURCHASE ET AL., 1978
G	SA9	S. TYPHIMURIUM TA98, REVERSE MUTATION	- -	94.0000	ANTOINE ET AL., 1983
G	SA9	S. TYPHIMURIUM TA98, REVERSE MUTATION	0 -	1000.0000	FALCK ET AL., 1985
G	SA9	S. TYPHIMURIUM TA98, REVERSE MUTATION	- -	5000.0000	MORTELMANS ET AL., 1986
G	SA9	S. TYPHIMURIUM TA98, REVERSE MUTATION	- -	5000.0000	MACDONALD, 1981
G	SA9	S. TYPHIMURIUM TA98, REVERSE MUTATION	- 0	0.0000	ICHINOTSUBO ET AL., 1981b
G	SA9	S. TYPHIMURIUM TA98, REVERSE MUTATION	- -	0.0000	NAGAO & TAKAHASHI, 1981
G	SA9	S. TYPHIMURIUM TA98, REVERSE MUTATION	- -	5000.0000	RICHOLD & JONES, 1981
G	SA9	S. TYPHIMURIUM TA98, REVERSE MUTATION	- -	1000.0000	ROWLAND & SEVERN, 1981
G	SA9	S. TYPHIMURIUM TA98, REVERSE MUTATION	- +	250.0000	TRUEMAN, 1981
G	SA9	S. TYPHIMURIUM TA98, REVERSE MUTATION	- -	250.0000	VENITT & CROFTON-SLEIGH, 1981
G	SA9	S. TYPHIMURIUM TA98, REVERSE MUTATION	- ?	500.0000	HUBBARD ET AL., 1981
G	SA9	S. TYPHIMURIUM TA98, REVERSE MUTATION	- -	1000.0000	GATEHOUSE, 1981
G	SA9	S. TYPHIMURIUM TA98, REVERSE MUTATION	- -	1000.0000	BROOKS & DEAN, 1981
G	SA9	S. TYPHIMURIUM TA98, REVERSE MUTATION	- -	500.0000	BAKER & BONIN, 1981
G	SA9	S. TYPHIMURIUM TA98, REVERSE MUTATION	- -	0.0000	SIMMON & SHEPHERD, 1981
G	SAS	S. TYPHIMURIUM (OTHER), REVERSE MUTATION	0 -	73000.0000	GREEN & SAVAGE, 1978
G	SAS	S. TYPHIMURIUM (OTHER), REVERSE MUTATION	- -	1000.0000	BROOKS & DEAN, 1981
G	ECF	E. COLI (EXCLUDING K12), FORWARD MUTATION	0 -	4000.0000	MOHN ET AL., 1981
G	ECW	E. COLI WP2 UVRA, REVERSE MUTATION	0 -	1000.0000	FALCK ET AL., 1985
G	ECW	E. COLI WP2 UVRA, REVERSE MUTATION	- -	250.0000	VENITT & CROFTON-SLEIGH, 1981
G	ECW	E. COLI WP2 UVRA, REVERSE MUTATION	- -	500.0000	GATEHOUSE, 1981
G	ECW	E. COLI WP2 UVRA, REVERSE MUTATION	- -	0.0000	MATSUSHIMA ET AL., 1981
D	SSB	SACCHAROMYCES, STRAND BREAKS/X-LINKS	- -	1000.0000	KOSSINOVA ET AL., 1981
D	SSD	SACCHAROMYCES, DIFFERENTIAL TOXICITY	+ +	500.0000	SHARP & PARRY, 1981a
R	SCH	S. CEREVISIAE, HOMOZYGOSIS	- -	500.0000	SHARP & PARRY, 1981b
R	SCH	S. CEREVISIAE, HOMOZYGOSIS	+ +	4700.0000	ZIMMERMANN & SCHEEL, 1981
R	SCH	S. CEREVISIAE, HOMOZYGOSIS	- -	167.0000	JAGANNATH ET AL., 1981
R	SCH	S. CEREVISIAE, HOMOZYGOSIS	- -	1000.0000	KASSINOVA ET AL., 1981
G	SCF	S. CEREVISIAE, FORWARD MUTATION	- -	800.0000	MEHTA & VON BORSTEL, 1981
G	SZF	S. POMBE, FORWARD MUTATION	- -	20.0000	LOPRIENO, 1981
A	SCN	S. CEREVISIAE, ANEUPLOIDY	+ +	100.0000	PARRY & SHARP, 1981
G	ASM	ARABIDOPSIS SPECIES, MUTATION	- 0	300000.0000	GICHNER & VELEMINSKY, 1987
G	DMX	D. MELANOGASTER, SEX-LINKED RECESSIVES	- 0	900.0000	WURGLER & GRAF, 1981
D	URP	UDS, RAT PRIMARY HEPATOCYTES	(+) 0	700.0000	WILLIAMS, 1977
D	URP	UDS, RAT PRIMARY HEPATOCYTES	- 0	7300.0000	WILLIAMS & LASPIA, 1979
D	URP	UDS, RAT PRIMARY HEPATOCYTES	- 0	70.0000	ITO, 1982
D	UIA	UDS, OTHER ANIMAL CELLS IN VITRO	- 0	700.0000	MCQUEEN ET AL., 1983
D	UIA	UDS, OTHER ANIMAL CELLS IN VITRO	- 0	70.0000	KLAUNIG ET AL., 1984
G	G5T	MUTATION, L5178Y CELLS, TK LOCUS	(+) -	5000.0000	MCGREGOR ET AL., 1988
G	G5T	MUTATION, L5178Y CELLS, TK LOCUS	- -	4700.0000	MITCHELL ET AL., 1988
G	G5T	MUTATION, L5178Y CELLS, TK LOCUS	(+) -	3000.0000	JOTZ & MITCHELL, 1981
G	G5T	MUTATION, L5178Y CELLS, TK LOCUS	- -	4700.0000	MYHR & CASPARY, 1988
S	SIC	SCE, CHINESE HAMSTER CELLS IN VITRO	- -	100.0000	PERRY & THOMSON, 1981
S	SIC	SCE, CHINESE HAMSTER CELLS IN VITRO	- -	900.0000	EVANS & MITCHELL, 1981
S	SIC	SCE, CHINESE HAMSTER CELLS IN VITRO	- -	6300.0000	NATARAJAN & VAN KESTEREN-VAN LEEUWEN, 198
C	CIC	CHROM ABERR, CHINESE HAMSTER CELLS IN VITRO	- -	6300.0000	NATARAJAN & VAN KESTEREN-VAN LEEUWEN, 198

DIMETHYLFORMAMIDE

END POINT	TEST CODE	TEST SYSTEM	RESULTS NM M	DOSE (LED OR HID)	REFERENCE
C	CIR	CHROM ABERR, RAT CELLS IN VITRO	- 0	300.0000	DEAN, 1981
D	UHF	UDS, HUMAN FIBROBLASTS IN VITRO	- -	90.0000	ROBINSON & MITCHELL, 1981
D	UHF	UDS, HUMAN FIBROBLASTS IN VITRO	0 -	0.0000	AGRELO & AMOS, 1981
D	UHF	UDS, HUMAN FIBROBLASTS IN VITRO	- -	100.0000	MARTIN & MCDERMID, 1981
G	GIH	MUTATION, HUMAN CELLS IN VITRO	- -	500.0000	GUPTA & GOLDSTEIN, 1981
S	SHL	SCE, HUMAN LYMPHOCYTES IN VITRO	- 0	80000.0000	ANTOINE ET AL., 1983
C	CHL	CHROM ABERR, HUMAN LYMPHOCYTES IN VITRO	- 0	80000.0000	ANTOINE ET AL., 1983
C	CHL	CHROM ABERR, HUMAN LYMPHOCYTES IN VITRO	+ 0	0.0070	KOUDELA & SPAZIER ET AL., 1979
S	SVA	SCE, ANIMALS IN VIVO	- 0	2500.0000	PAIKA ET AL., 1981
M	MVM	MICRONUCLEUS TEST, MICE IN VIVO	- 0	2.5000	SALAMONE ET AL., 1981
M	MVM	MICRONUCLEUS TEST, MICE IN VIVO	- 0	1600.0000	KIRKHART, 1981
M	MVM	MICRONUCLEUS TEST, MICE IN VIVO	- 0	1500.0000	TSUCHIMOTO & MATTER, 1981
M	MVM	MICRONUCLEUS TEST, MICE IN VIVO	- 0	2000.0000	ANTOINE ET AL., 1983
M	MVM	MICRONUCLEUS TEST, MICE IN VIVO	+ 0	1.0000	YE ET AL., 1987
C	CLH	CHROM ABERR, HUMAN LYMPHOCYTES IN VIVO	+ 0	0.0000	BERGER ET AL., 1985
C	CLH	CHROM ABERR, HUMAN LYMPHOCYTES IN VIVO	+ 0	0.0000	KOUDELA & SPAZIER ET AL., 1981
P	SPM	SPERM MORPHOLOGY, MICE	- 0	667.0000	ANTOINE ET AL., 1983
P	SPM	SPERM MORPHOLOGY, MICE	- 0	900.0000	TOPHAM, 1980
I	ICR	INHIBIT CELL COMMUNICATION, ANIMAL CELLS	+ 0	3800.0000	CHEN ET AL., 1984

Dimethylformamide was one of the compounds tested in 17 studies on mutation in bacteria in the International Collaborative Program for the Evaluation of Short-term Tests for Carcinogens (de Serres & Ashby, 1981). In two of the studies, results for dimethylformamide were not reported directly, so these are not included in the above listing.

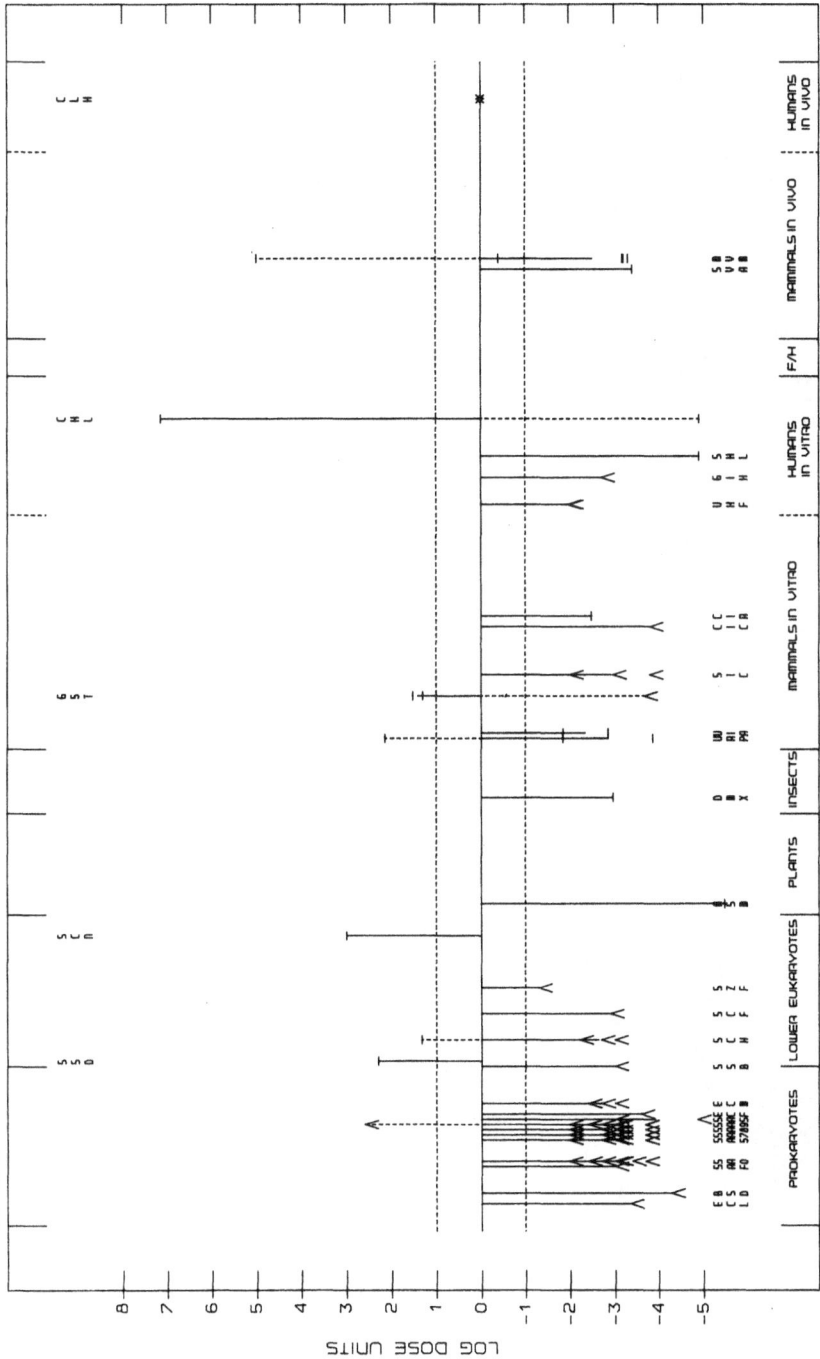

DIMETHYLFORMAMIDE
68-12-2
25-OCT-89

LOG DOSE UNITS

MORPHOLINE

END POINT	TEST CODE	TEST SYSTEM	RESULTS NM	M	DOSE (LED OR HID)	REFERENCE
G	SAO	S. TYPHIMURIUM TA100, REVERSE MUTATION	-	-	5450.0000	HAWORTH ET AL., 1983
G	SAO	S. TYPHIMURIUM TA100, REVERSE MUTATION	-	-	5000.0000	ISHIDATE ET AL., 1984
G	SA5	S. TYPHIMURIUM TA1535, REVERSE MUTATION	-	-	5450.0000	HAWORTH ET AL., 1983
G	SA5	S. TYPHIMURIUM TA1535, REVERSE MUTATION	-	-	5000.0000	ISHIDATE ET AL., 1984
G	SA7	S. TYPHIMURIUM TA1537, REVERSE MUTATION	-	-	5450.0000	HAWORTH ET AL., 1983
G	SA7	S. TYPHIMURIUM TA1537, REVERSE MUTATION	-	-	5000.0000	ISHIDATE ET AL., 1984
G	SA9	S. TYPHIMURIUM TA98, REVERSE MUTATION	-	-	1800.0000	HAWORTH ET AL., 1983
G	SA9	S. TYPHIMURIUM TA98, REVERSE MUTATION	-	-	5000.0000	ISHIDATE ET AL., 1984
G	SAS	S. TYPHIMURIUM (OTHER), REVERSE MUTATION	-	-	5000.0000	ISHIDATE ET AL., 1984
D	URP	UDS, RAT PRIMARY HEPATOCYTES	-	0	100.0000	CONAWAY ET AL., 1984
C	CIC	CHROM ABERR, CHINESE HAMSTER CELLS IN VITRO	-	0	250.0000	ISHIDATE ET AL., 1984
H	HMM	HOST-MEDIATED ASSAY, MICROBIAL CELLS	-	0	250.0000	BRAUN ET AL., 1977
H	HMM	HOST-MEDIATED ASSAY, MICROBIAL CELLS	-	0	40.0000	EDWARDS ET AL., 1979
C	COE	CHROM ABERR, OOCYTES OR EMBRYOS	-	0	500.0000	INUI ET AL., 1979

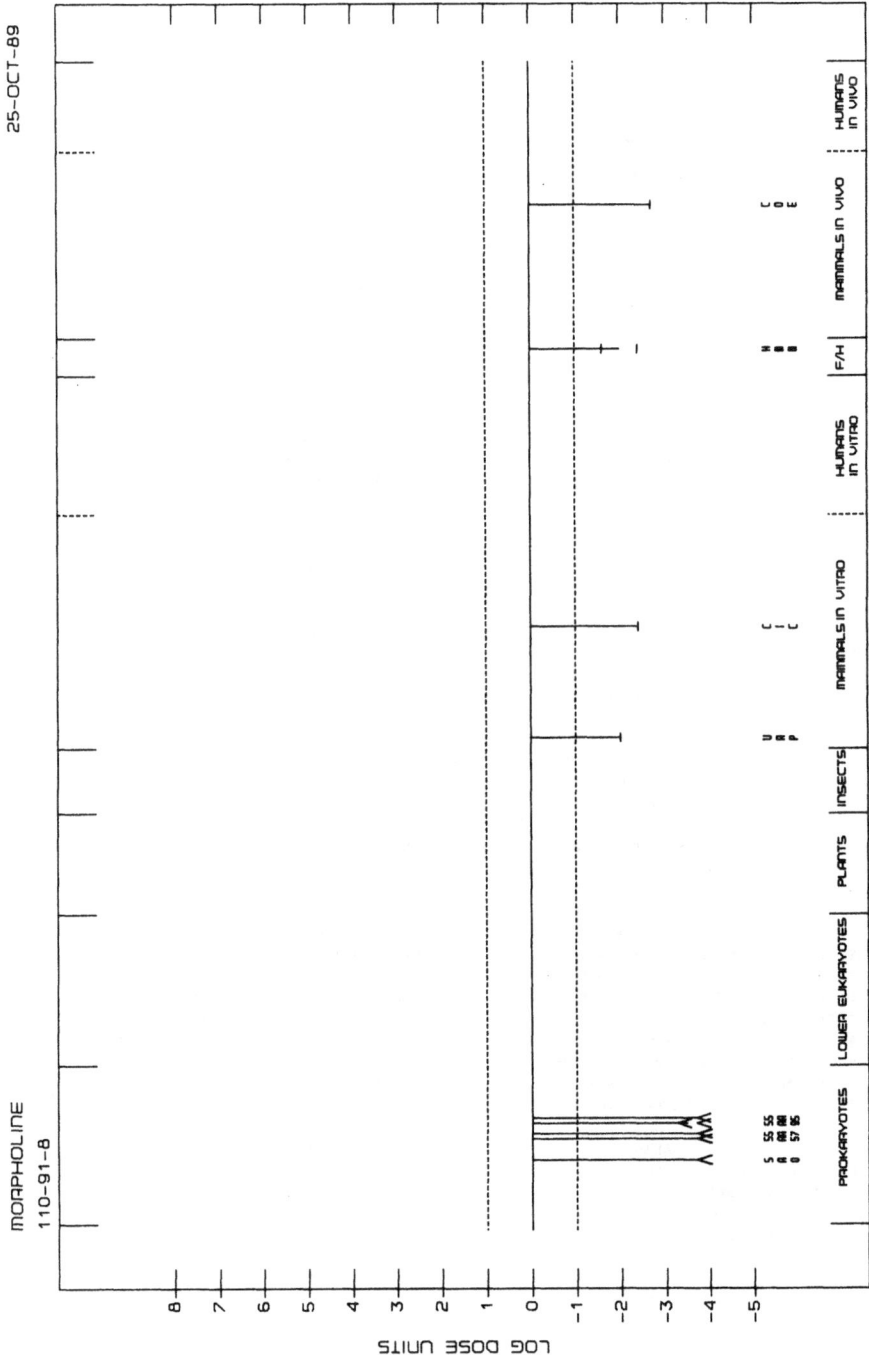

MORPHOLINE
110-91-8

25-OCT-89

1,2-EPOXYBUTANE

END POINT	TEST CODE	TEST SYSTEM	RESULTS NM	M	DOSE (LED OR HID)	REFERENCE
D	PRB	PROPHAGE, INDUCT/SOS/STRAND BREAKS/X-LINKS	+	0	780.0000	NAKAMURA ET AL., 1987
D	ECL	E. COLI POL A, DIFFERENTIAL TOX (LIQUID)	+	0	20000.0000	MCCARROLL ET AL., 1981b
D	ECL	E. COLI POL A, DIFFERENTIAL TOX (LIQUID)	+	0	50.0000	ROSENKRANZ & POIRIER, 1979
D	ERD	E. COLI REC, DIFFERENTIAL TOXICITY	+	0	4300.0000	MCCARROLL ET AL., 1981b
G	SA0	S. TYPHIMURIUM TA100, REVERSE MUTATION	+	0	0.0000	MCMAHON ET AL., 1979
G	SA0	S. TYPHIMURIUM TA100, REVERSE MUTATION	(+)	0	2100.0000	MCCANN ET AL., 1975
G	SA0	S. TYPHIMURIUM TA100, REVERSE MUTATION	+	0	7.0000	SPECK & ROSENKRANZ, 1976
G	SA0	S. TYPHIMURIUM TA100, REVERSE MUTATION	+	0	2100.0000	HENSCHLER ET AL., 1977
G	SA0	S. TYPHIMURIUM TA100, REVERSE MUTATION	-	-	167.0000	DUNKEL ET AL., 1984
G	SA0	S. TYPHIMURIUM TA100, REVERSE MUTATION	-	-	250.0000	SIMMON, 1979a
G	SA0	S. TYPHIMURIUM TA100, REVERSE MUTATION	+	+	2.5000	DE FLORA, 1979
G	SA0	S. TYPHIMURIUM TA100, REVERSE MUTATION	+	+	2000.0000	DE FLORA, 1981
G	SA0	S. TYPHIMURIUM TA100, REVERSE MUTATION	+	0	1100.0000	GERVASI ET AL., 1985
G	SA0	S. TYPHIMURIUM TA100, REVERSE MUTATION	+	+	0.0000	DE FLORA ET AL., 1984
G	SA0	S. TYPHIMURIUM TA100, REVERSE MUTATION	+	+	500.0000	CANTER ET AL., 1986
G	SA0	S. TYPHIMURIUM TA100, REVERSE MUTATION	-	0	360.0000	ROSMAN ET AL., 1987
G	SA0	S. TYPHIMURIUM TA100, REVERSE MUTATION	+	+	500.0000	NTP, 1988
G	SA3	S. TYPHIMURIUM TA1530, REVERSE MUTATION	+	0	17000.0000	CHEN ET AL., 1975
G	SA5	S. TYPHIMURIUM TA1535, REVERSE MUTATION	+	0	2100.0000	MCCANN ET AL., 1975
G	SA5	S. TYPHIMURIUM TA1535, REVERSE MUTATION	+	+	0.0000	DE FLORA ET AL., 1984
G	SA5	S. TYPHIMURIUM TA1535, REVERSE MUTATION	-	-	167.0000	DUNKEL ET AL., 1984
G	SA5	S. TYPHIMURIUM TA1535, REVERSE MUTATION	-	-	250.0000	SIMMON, 1979a
G	SA5	S. TYPHIMURIUM TA1535, REVERSE MUTATION	+	+	42.0000	ROSENKRANZ & POIRIER, 1979
G	SA5	S. TYPHIMURIUM TA1535, REVERSE MUTATION	+	+	1250.0000	WEINSTEIN ET AL., 1981
G	SA5	S. TYPHIMURIUM TA1535, REVERSE MUTATION	+	+	2000.0000	DE FLORA, 1981
G	SA5	S. TYPHIMURIUM TA1535, REVERSE MUTATION	+	+	500.0000	CANTER ET AL., 1986
G	SA5	S. TYPHIMURIUM TA1535, REVERSE MUTATION	+	0	90.0000	ROSMAN ET AL., 1987
G	SA5	S. TYPHIMURIUM TA1535, REVERSE MUTATION	+	+	500.0000	NTP, 1988
G	SA7	S. TYPHIMURIUM TA1537, REVERSE MUTATION	-	-	167.0000	DUNKEL ET AL., 1984
G	SA7	S. TYPHIMURIUM TA1537, REVERSE MUTATION	-	-	250.0000	SIMMON, 1979a
G	SA7	S. TYPHIMURIUM TA1537, REVERSE MUTATION	-	-	20000.0000	DE FLORA, 1981
G	SA7	S. TYPHIMURIUM TA1537, REVERSE MUTATION	-	-	5000.0000	CANTER ET AL., 1986
G	SA7	S. TYPHIMURIUM TA1537, REVERSE MUTATION	-	-	5000.0000	NTP, 1988
G	SA8	S. TYPHIMURIUM TA1538, REVERSE MUTATION	-	-	250.0000	SIMMON, 1979a
G	SA8	S. TYPHIMURIUM TA1538, REVERSE MUTATION	-	-	20000.0000	DE FLORA, 1981
G	SA8	S. TYPHIMURIUM TA1538, REVERSE MUTATION	-	-	167.0000	DUNKEL ET AL., 1984
G	SA8	S. TYPHIMURIUM TA1538, REVERSE MUTATION	-	-	5000.0000	NTP, 1988
G	SA9	S. TYPHIMURIUM TA98, REVERSE MUTATION	-	-	20000.0000	DE FLORA, 1981
G	SA9	S. TYPHIMURIUM TA98, REVERSE MUTATION	-	0	2200.0000	GERVASI ET AL., 1985
G	SA9	S. TYPHIMURIUM TA98, REVERSE MUTATION	-	-	5000.0000	CANTER ET AL., 1986
G	SA9	S. TYPHIMURIUM TA98, REVERSE MUTATION	-	-	250.0000	SIMMON, 1979a
G	SAS	S. TYPHIMURIUM (OTHER), REVERSE MUTATION	+	0	3.5000	ROSENKRANZ & SPECK, 1975
G	SAS	S. TYPHIMURIUM (OTHER), REVERSE MUTATION	-	-	250.0000	SIMMON, 1979a
G	ECW	E. COLI WP2 UVRA, REVERSE MUTATION	+	0	0.0000	MCMAHON ET AL., 1979
G	ECW	E. COLI WP2 UVRA, REVERSE MUTATION	-	-	167.0000	DUNKEL ET AL., 1984
G	KPF	K. PNEUMONIA, FORWARD MUTATION	(+)	+	72.0000	VOOGD ET AL., 1981
G	KPF	K. PNEUMONIA, FORWARD MUTATION	+	0	72.0000	KNAAP ET AL., 1982
R	SCH	S. CEREVISIAE, HOMOZYGOSIS	+	+	5000.0000	SIMMON, 1979b
G	SZF	S. POMBE, FORWARD MUTATION	+	+	29.0000	MIGLIORE ET AL., 1982
G	NCR	N. CRASSA, REVERSE MUTATION	(+)	0	14.0000	KOLMARK & GILES, 1955

1,2-EPOXYBUTANE

END POINT	TEST CODE	TEST SYSTEM	RESULTS NM M		DOSE (LED OR HID)	REFERENCE
G	DMX	D. MELANOGASTER, SEX-LINKED RECESSIVES	+	0	8400.0000	KNAAP ET AL., 1982
G	DMX	D. MELANOGASTER, SEX-LINKED RECESSIVES	+	0	50000.0000	NTP, 1988
C	DMH	D. MELANOGASTER, HERITABLE TRANSLOCATIONS	+	0	50000.0000	NTP, 1988
D	URP	UDS, RAT PRIMARY HEPATOCYTES	-	0	1000.0000	WILLIAMS ET AL., 1982
G	G5T	MUTATION, L5178Y CELLS, TK LOCUS	+	0	63.0000	AMACHER ET AL., 1980
G	G5T	MUTATION, L5178Y CELLS, TK LOCUS	+	+	400.0000	MCGREGOR ET AL., 1987
G	G5T	MUTATION, L5178Y CELLS, TK LOCUS	+	+	55.0000	MITCHELL ET AL., 1988
G	G5T	MUTATION, L5178Y CELLS, TK LOCUS	+	+	50.0000	MYHR & CASPARY, 1985
G	G5T	MUTATION, L5178Y CELLS, TK LOCUS	+	+	50.0000	NTP, 1988
G	G51	MUTATION, L5178Y CELLS, ALL OTHER LOCI	+	0	360.0000	KNAAP ET AL., 1982
S	SIC	SCE, CHINESE HAMSTER CELLS IN VITRO	+	+	16.0000	NTP, 1988
C	CIC	CHROM ABERR, CHINESE HAMSTER CELLS IN VITRO	(+)	(+)	500.0000	NTP, 1988
T	TBM	CELL TRANSFORMATION, BALB/C3T3 CELLS	-	0	50.0000	DUNKEL ET AL., 1981
T	TCS	CELL TRANSFORMATION, SHE, CLONAL ASSAY	+	0	0.0000	PIENTA ET AL., 1981
T	TFS	CELL TRANSFORMATION, SHE, FOCUS ASSAY	(+)	0	50.0000	DUNKEL ET AL., 1981
T	TRR	CELL TRANSFORMATION, RLV/FISCHER RAT	+	0	10.0000	PRICE & MISHRA, 1980
T	TRR	CELL TRANSFORMATION, RLV/FISCHER RAT	+	0	700.0000	DUNKEL ET AL., 1981

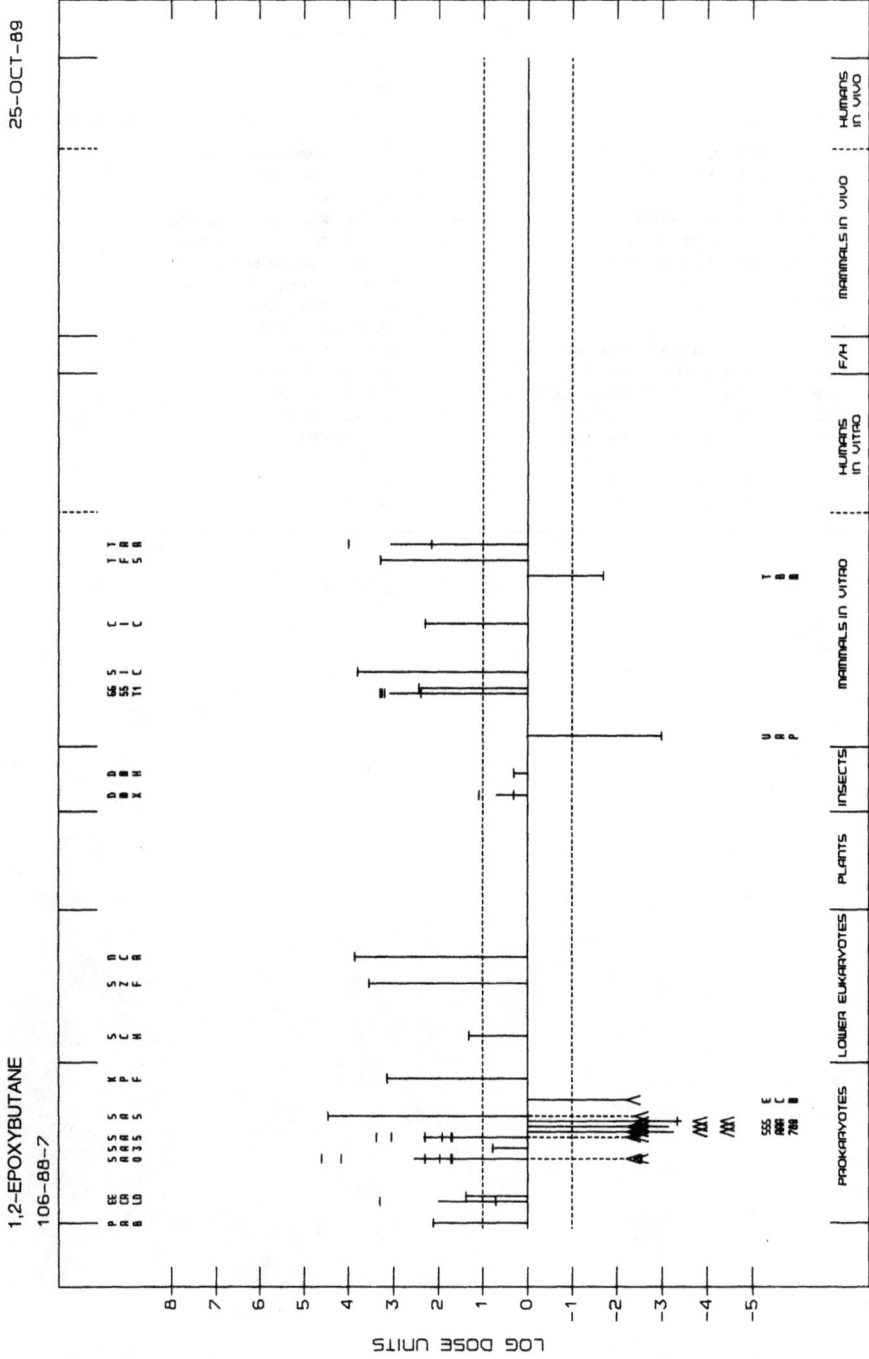

BIS(2,3-EPOXYCYCLOPENTYL)ETHER

END POINT	TEST CODE	TEST SYSTEM	RESULTS NM	M	DOSE (LED OR HID)	REFERENCE
G	SAO	S. TYPHIMURIUM TA100, REVERSE MUTATION	+	+	1425.0000	XIE & DONG, 1984
S	SHL	SCE, HUMAN LYMPHOCYTES IN VITRO	+	0	30.0000	XIE & DONG, 1984
M	MVM	MICRONUCLEUS TEST, MICE IN VIVO	+	0	500.0000	XIE & DONG, 1984

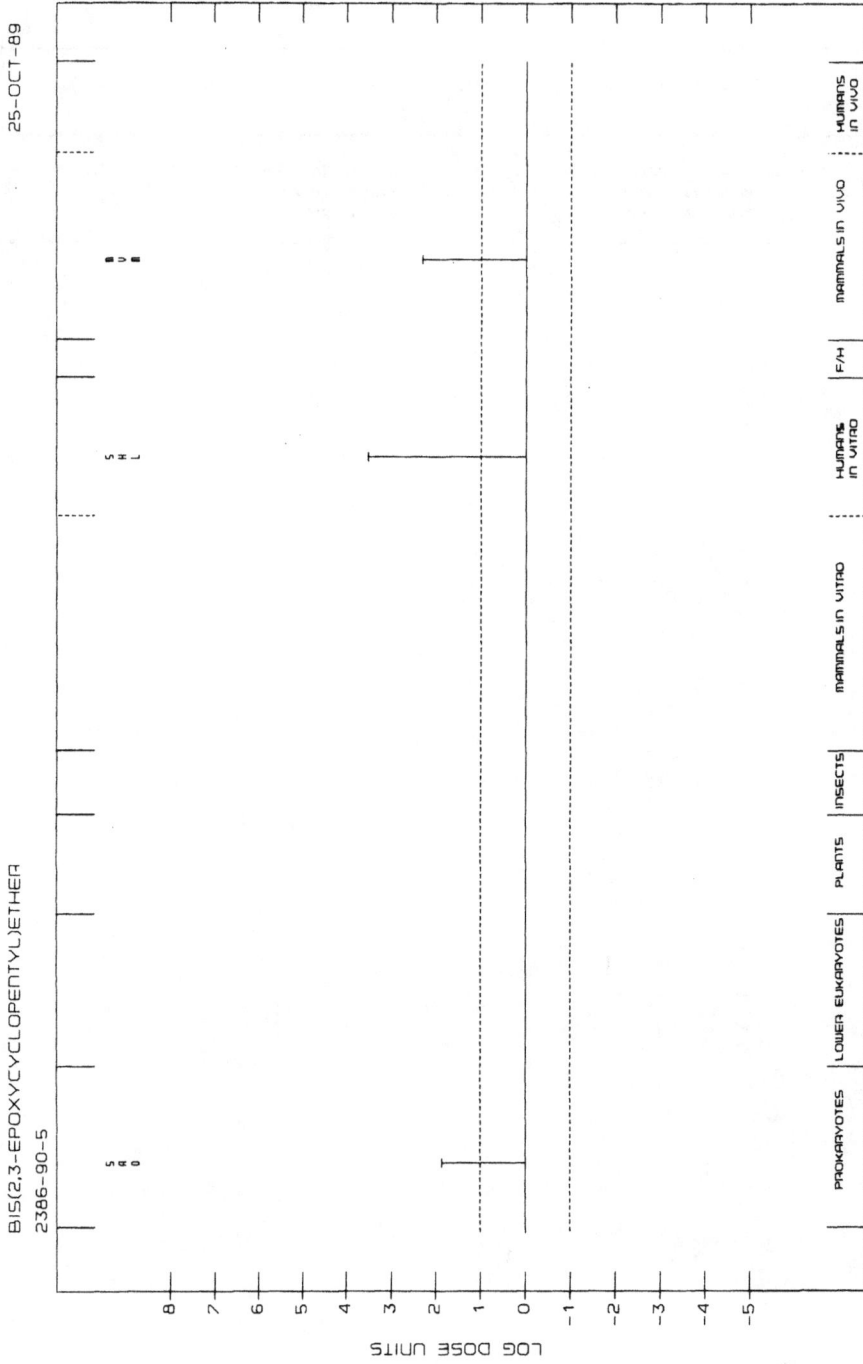

BIS(2,3-EPOXYCYCLOPENTYL)ETHER
2386-90-5

25-OCT-89

LOG DOSE UNITS

PROKARYOTES | LOWER EUKARYOTES | PLANTS | INSECTS | MAMMALS IN VITRO | HUMANS IN VITRO | F/H | MAMMALS IN VIVO | HUMANS IN VIVO

OCTYL GLYCIDYL ETHER

END POINT	TEST CODE	TEST SYSTEM	RESULTS NM M	DOSE (LED OR HID)	REFERENCE
G	SAO	S. TYPHIMURIUM TA100, REVERSE MUTATION	- +	167.0000	THOMPSON ET AL., 1981
G	SA5	S. TYPHIMURIUM TA1535, REVERSE MUTATION	- +	56.0000	THOMPSON ET AL., 1981
G	G5T	MUTATION, L5178Y CELLS, TK LOCUS	- -	225.0000	THOMPSON ET AL., 1981
D	UHF	UDS, HUMAN FIBROBLASTS IN VITRO	- -	1400.0000	THOMPSON ET AL., 1981

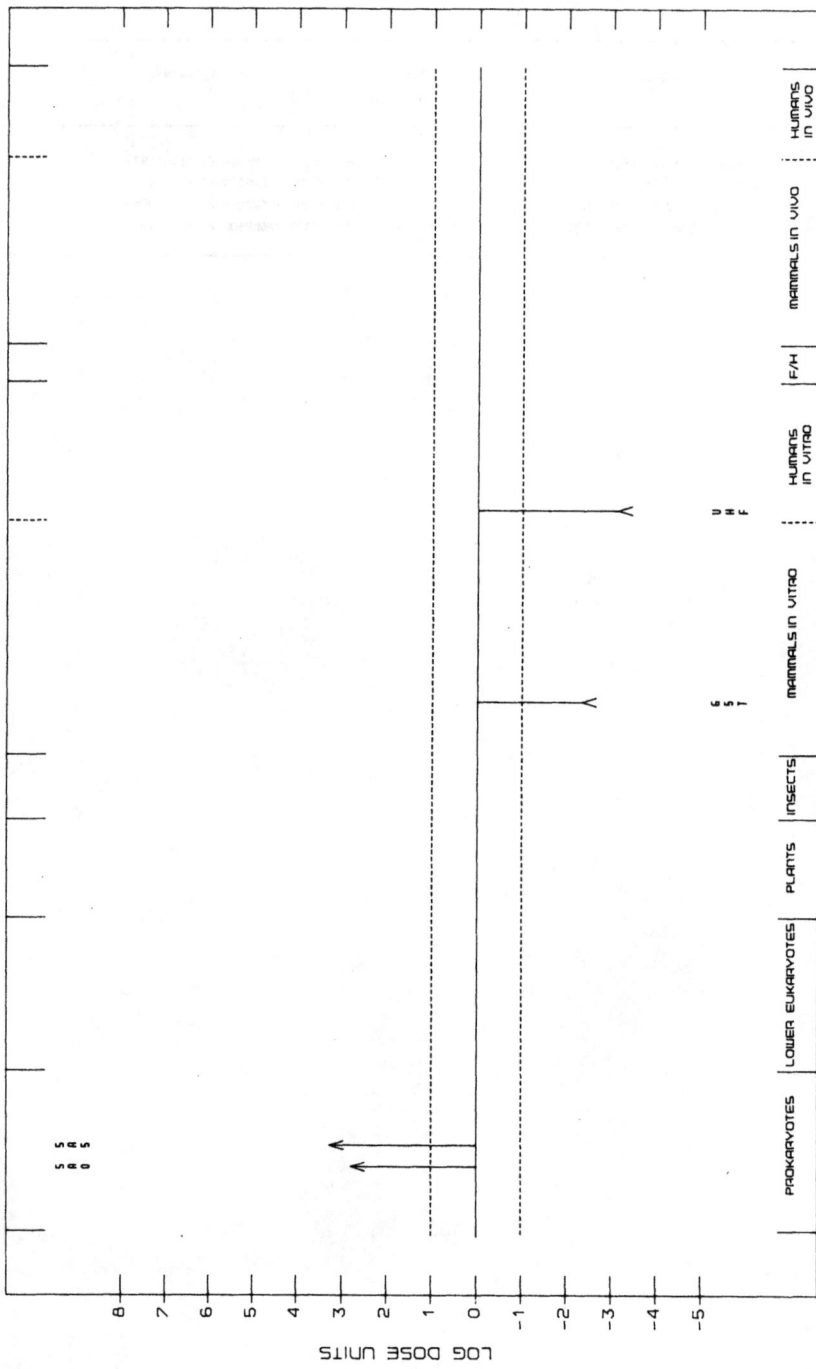

OCTYL GLYCIDYL ETHER
3385-66-8

13-DEC-89

LOG DOSE UNITS

DECYL GLYCIDYL ETHER

END POINT	TEST CODE	TEST SYSTEM	RESULTS NM	M	DOSE (LED OR HID)	REFERENCE
G	SAO	S. TYPHIMURIUM TA100, REVERSE MUTATION	-	+	100.0000	THOMPSON ET AL., 1981
G	SA5	S. TYPHIMURIUM TA1535, REVERSE MUTATION	-	+	100.0000	THOMPSON ET AL., 1981
G	G5T	MUTATION, L5178Y CELLS, TK LOCUS	+	-	80.0000	THOMPSON ET AL., 1981
D	UHF	UDS, HUMAN FIBROBLASTS IN VITRO	-	-	6000.0000	THOMPSON ET AL., 1981

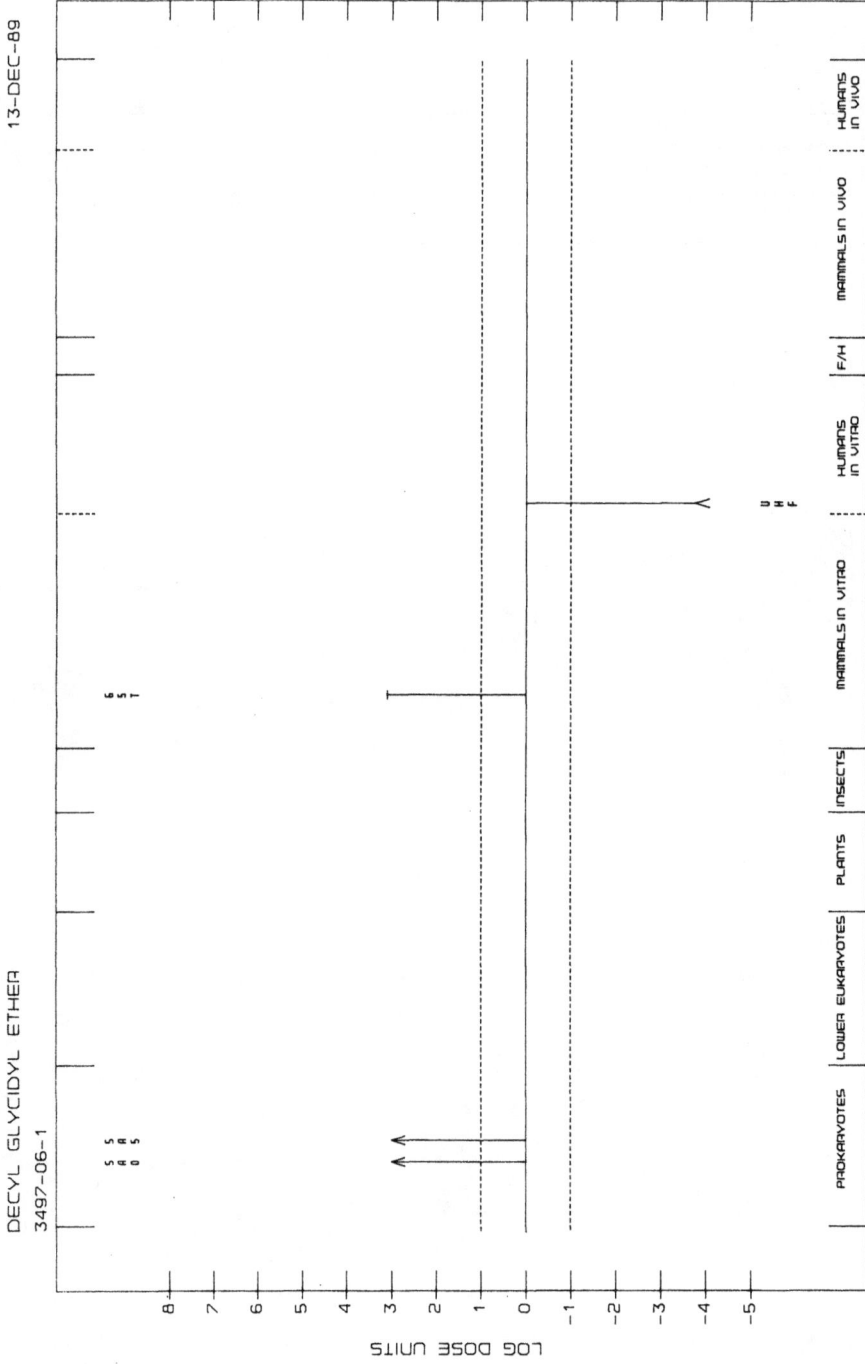

DECYL GLYCIDYL ETHER
3497-06-1

13-DEC-89

LOG DOSE UNITS

PROKARYOTES   LOWER EUKARYOTES   PLANTS   INSECTS   MAMMALS IN VITRO   HUMANS IN VITRO   F/H   MAMMALS IN VIVO   HUMANS IN VIVO

DODECYL GLYCIDYL ETHER

END POINT	TEST CODE	TEST SYSTEM	RESULTS NM M	DOSE (LED OR HID)	REFERENCE
G	SAO	S. TYPHIMURIUM TA100, REVERSE MUTATION	- +	150.0000	THOMPSON ET AL., 1981
G	SA5	S. TYPHIMURIUM TA1535, REVERSE MUTATION	- +	50.0000	THOMPSON ET AL., 1981
G	G5T	MUTATION, L5178Y CELLS, TK LOCUS	- -	225.0000	THOMPSON ET AL., 1981
D	UHF	UDS, HUMAN FIBROBLASTS IN VITRO	- -	6000.0000	THOMPSON ET AL., 1981

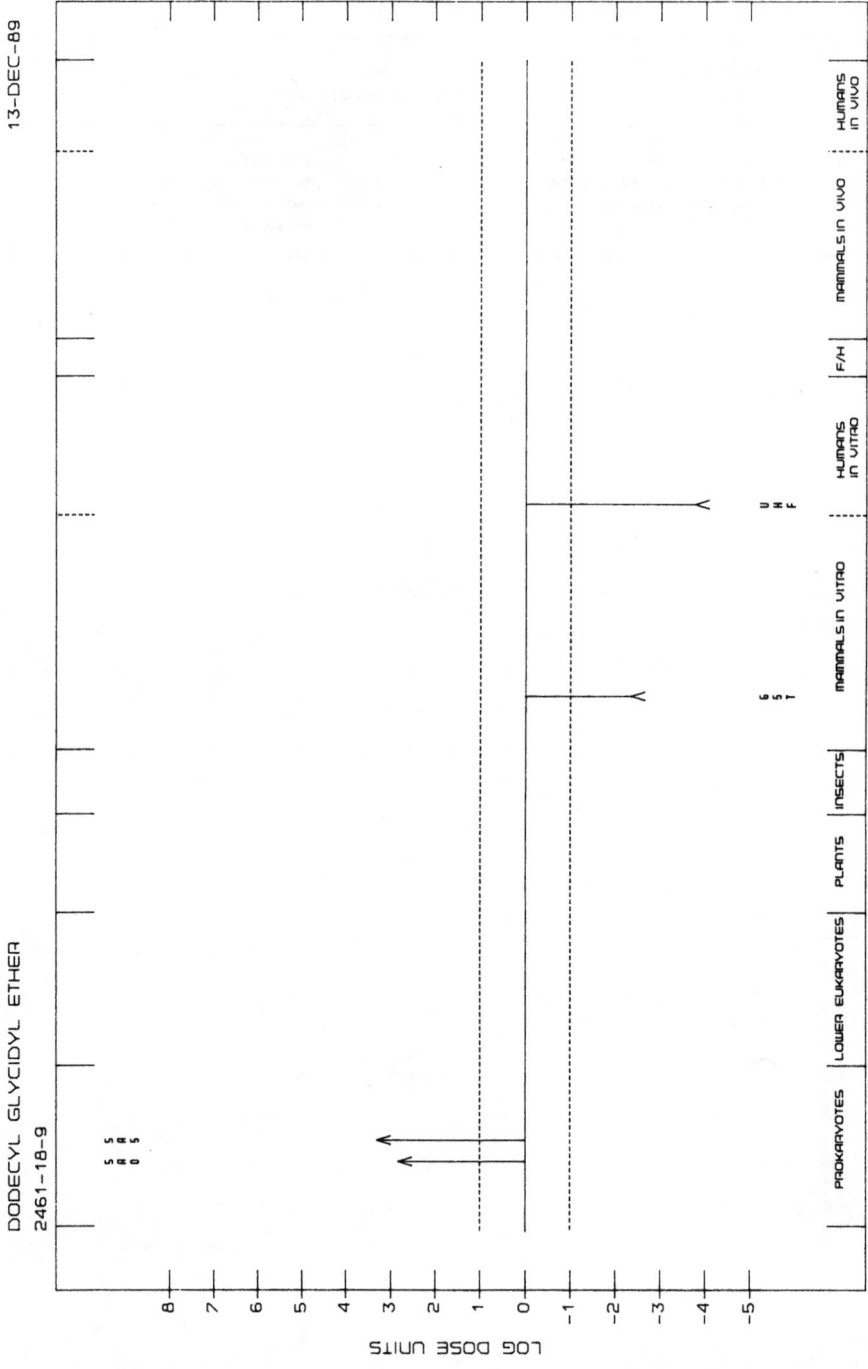

DODECYL GLYCIDYL ETHER
2461-18-9

13-DEC-89

LOG DOSE UNITS

TETRADECYL GLYCIDYL ETHER

END POINT	TEST CODE	TEST SYSTEM	RESULTS NM M	DOSE (LED OR HID)	REFERENCE
G	SA0	S. TYPHIMURIUM TA100, REVERSE MUTATION	- -	2500.0000	THOMPSON ET AL., 1981
G	SA5	S. TYPHIMURIUM TA1535, REVERSE MUTATION	- -	2500.0000	THOMPSON ET AL., 1981
G	G5T	MUTATION, L5178Y CELLS, TK LOCUS	- -	200.0000	THOMPSON ET AL., 1981
D	UHF	UDS, HUMAN FIBROBLASTS IN VITRO	- -	8000.0000	THOMPSON ET AL., 1981

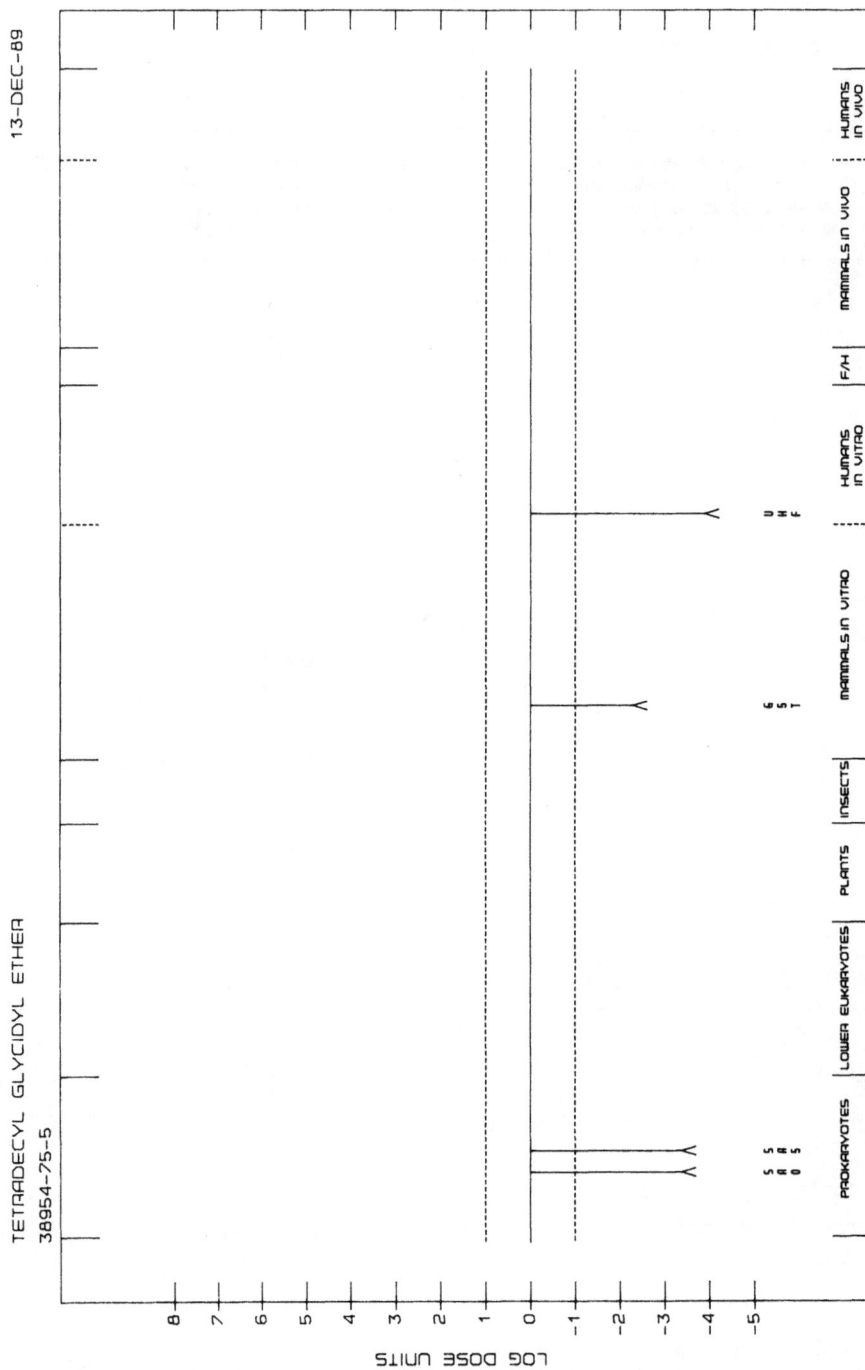

TETRADECYL GLYCIDYL ETHER
38954-75-5

13-DEC-89

LOG DOSE UNITS

PROKARYOTES | LOWER EUKARYOTES | PLANTS | INSECTS | MAMMALS in VITRO | HUMANS in VITRO | F/H | MAMMALS in VIVO | HUMANS in VIVO

ALLYL GLYCIDYL ETHER

END POINT	TEST CODE	TEST SYSTEM	RESULTS NM M	DOSE (LED OR HID)	REFERENCE
G	SA0	S. TYPHIMURIUM TA100, REVERSE MUTATION	+ +	167.0000	CANTER ET AL., 1986
G	SA0	S. TYPHIMURIUM TA100, REVERSE MUTATION	. + +	0.0000	WADE ET AL., 1979
G	SA5	S. TYPHIMURIUM TA1535, REVERSE MUTATION	+ +	50.0000	CANTER ET AL., 1986
G	SA7	S. TYPHIMURIUM TA1537, REVERSE MUTATION	- -	0.0000	CANTER ET AL., 1986
G	SA9	S. TYPHIMURIUM TA98, REVERSE MUTATION	- -	0.0000	CANTER ET AL., 1986
G	SA9	S. TYPHIMURIUM TA98, REVERSE MUTATION	- -	5000.0000	WADE ET AL., 1979
G	ECW	E. COLI WP2 UVRA, REVERSE MUTATION	+ 0	0.0000	HEMMINKI ET AL., 1980
G	DMX	D. MELANOGASTER, SEX-LINKED RECESSIVES	+ 0	5280.0000	YOON ET AL., 1985

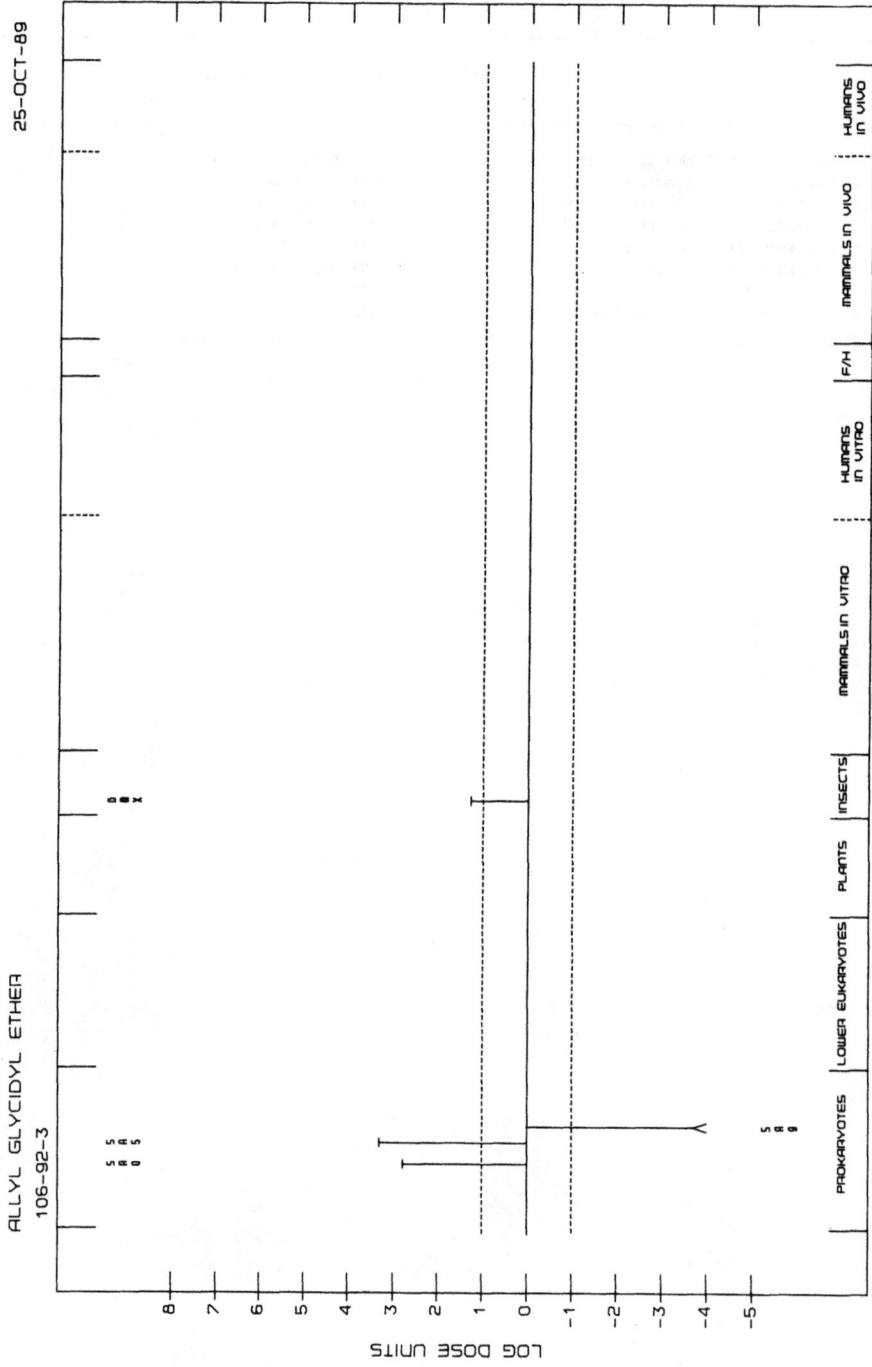

ALLYL GLYCIDYL ETHER
106-92-3

25-OCT-89

LOG DOSE UNITS

PROKARYOTES | LOWER EUKARYOTES | PLANTS | INSECTS | MAMMALS in VITRO | HUMANS in VITRO | F/H | MAMMALS in VIVO | HUMANS in VIVO

BISPHENOL A DIGLYCIDYL ETHER

END POINT	TEST CODE	TEST SYSTEM	RESULTS NM	M	DOSE (LED OR HID)	REFERENCE
G	SA0	S. TYPHIMURIUM TA100, REVERSE MUTATION	-	-	10000.0000	WADE ET AL., 1979
G	SA0	S. TYPHIMURIUM TA100, REVERSE MUTATION	+	0	14.0000	ANDERSEN ET AL., 1978
G	SA0	S. TYPHIMURIUM TA100, REVERSE MUTATION	+	+	16.0000	CANTER ET AL., 1986
G	SA5	S. TYPHIMURIUM TA1535, REVERSE MUTATION	+	+	92.0000	ANDERSEN ET AL., 1978
G	SA5	S. TYPHIMURIUM TA1535, REVERSE MUTATION	+	+	167.0000	CANTER ET AL., 1986
G	SA7	S. TYPHIMURIUM TA1537, REVERSE MUTATION	-	-	0.0000	CANTER ET AL., 1986
G	SA9	S. TYPHIMURIUM TA98, REVERSE MUTATION	-	-	10000.0000	WADE ET AL., 1979
G	SA9	S. TYPHIMURIUM TA98, REVERSE MUTATION	-	-	0.0000	CANTER ET AL., 1986
G	ECW	E. COLI WP2 UVRA, REVERSE MUTATION	+	0	0.0000	HEMMINKI ET AL., 1980
C	CLH	CHROM ABERR, HUMAN LYMPHOCYTES IN VIVO	-	0	0.0000	MITELMAN ET AL., 1980

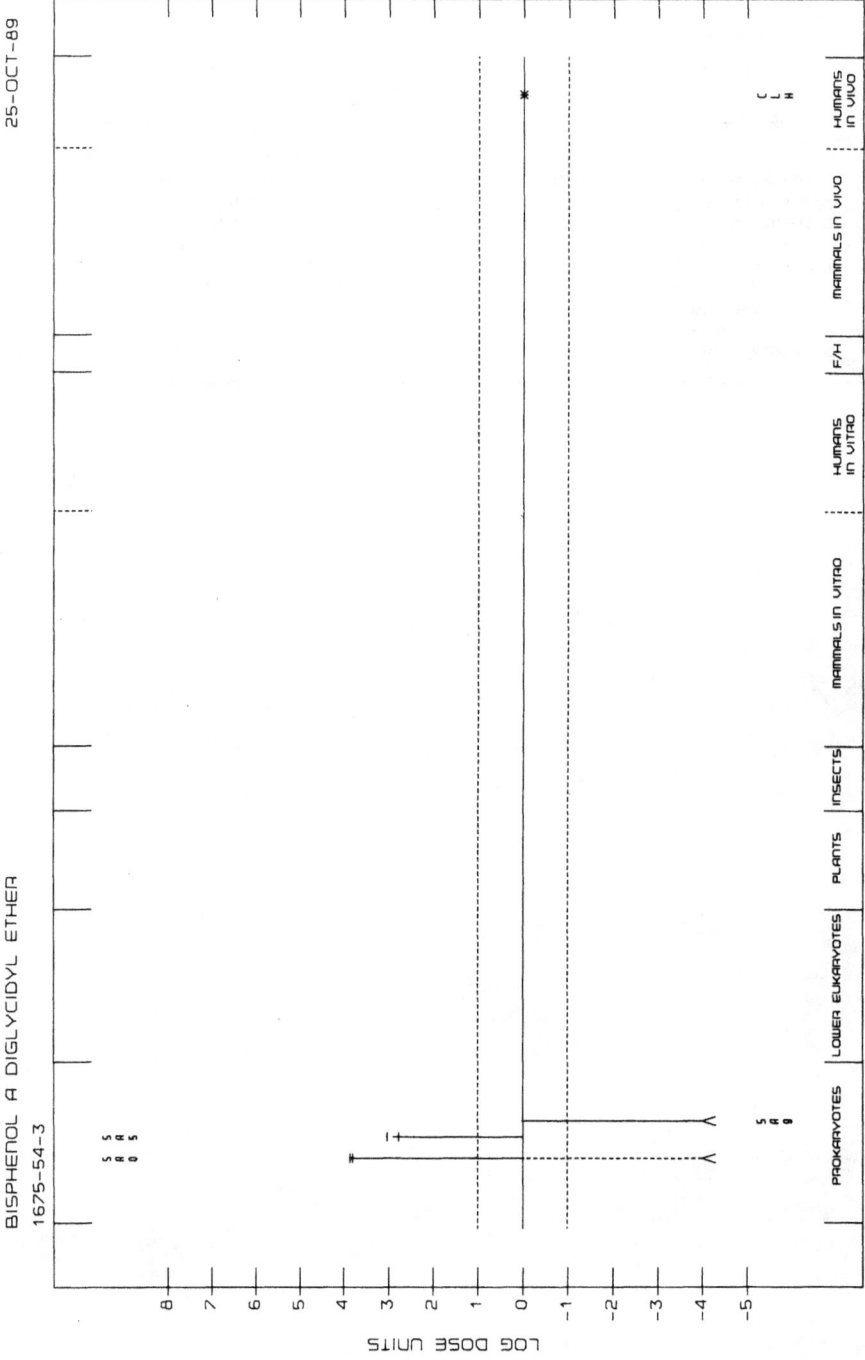

BISPHENOL A DIGLYCIDYL ETHER

1675-54-3

25-OCT-89

LOG DOSE UNITS

1,4-BUTANEDIOL DIGLYCIDYL ETHER

END POINT	TEST CODE	TEST SYSTEM	RESULTS NM	M	DOSE (LED OR HID)	REFERENCE
G	SAO	S. TYPHIMURIUM TA100, REVERSE MUTATION	+	+	167.0000	CANTER ET AL., 1986
G	SA5	S. TYPHIMURIUM TA1535, REVERSE MUTATION	+	+	167.0000	CANTER ET AL., 1986
G	SA7	S. TYPHIMURIUM TA1537, REVERSE MUTATION	?	+	0.0000	CANTER ET AL., 1986
G	SA9	S. TYPHIMURIUM TA98, REVERSE MUTATION	+	+	0.0000	CANTER ET AL., 1986

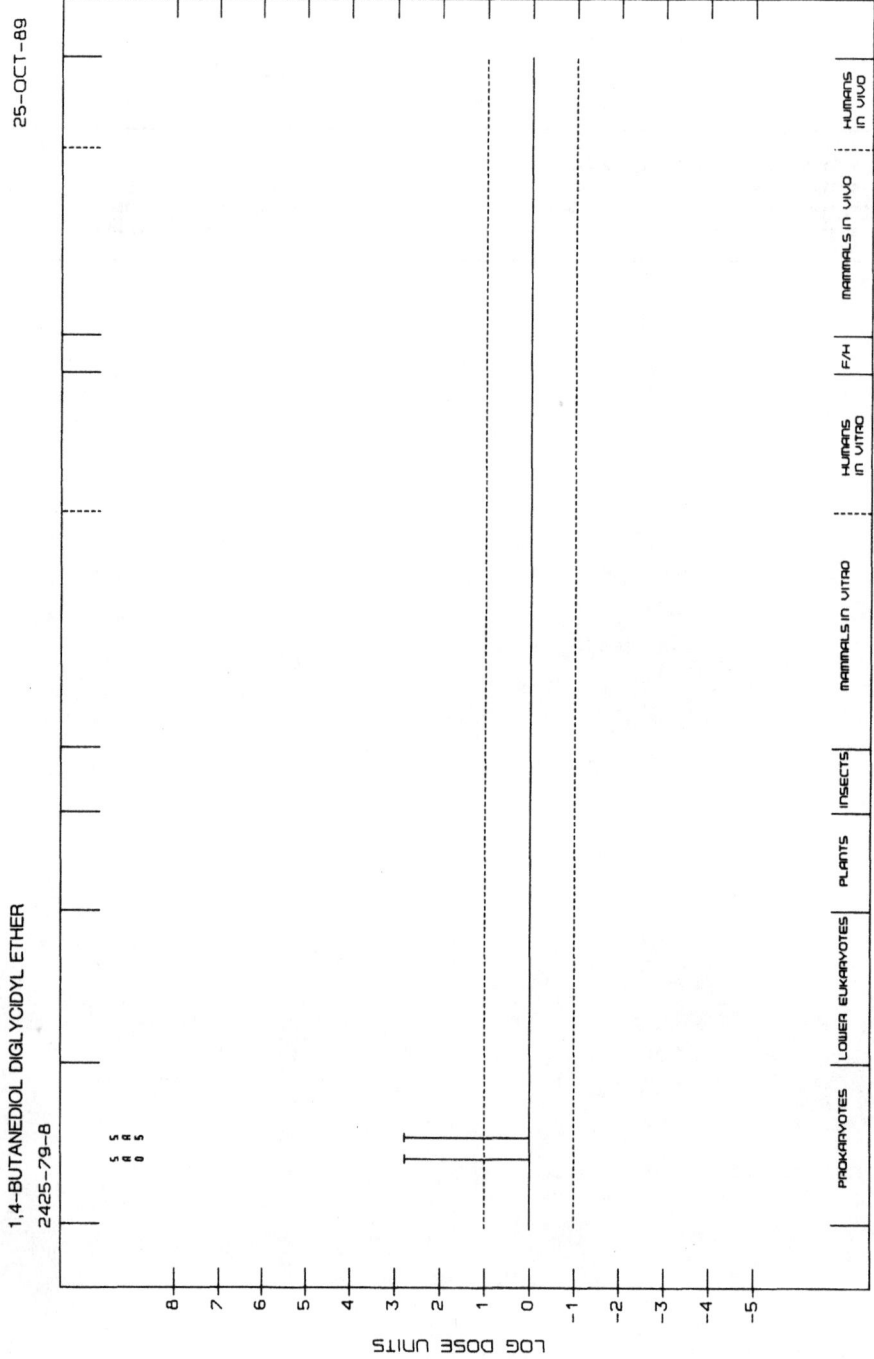

1,4-BUTANEDIOL DIGLYCIDYL ETHER

2425-79-8

25-OCT-89

n-BUTYL GLYCIDYL ETHER

END POINT	TEST CODE	TEST SYSTEM	RESULTS NM	M	DOSE (LED OR HID)	REFERENCE
G	SA0	S. TYPHIMURIUM TA100, REVERSE MUTATION	+	+	10000.0000	WADE ET AL., 1979
G	SA0	S. TYPHIMURIUM TA100, REVERSE MUTATION	+	+	50.0000	CANTER ET AL., 1986
G	SA0	S. TYPHIMURIUM TA100, REVERSE MUTATION	+	0	130.0000	CONNOR ET AL., 1980b
G	SA5	S. TYPHIMURIUM TA1535, REVERSE MUTATION	+	+	167.0000	CANTER ET AL., 1986
G	SA5	S. TYPHIMURIUM TA1535, REVERSE MUTATION	+	0	65.0000	CONNOR ET AL., 1980b
G	SA7	S. TYPHIMURIUM TA1537, REVERSE MUTATION	-	0	130.0000	CONNOR ET AL., 1980b
G	SA8	S. TYPHIMURIUM TA1538, REVERSE MUTATION	-	0	130.0000	CONNOR ET AL., 1980b
G	SA9	S. TYPHIMURIUM TA98, REVERSE MUTATION	-	-	50.0000	WADE ET AL., 1979
G	SA9	S. TYPHIMURIUM TA98, REVERSE MUTATION	-	-	1667.0000	CANTER ET AL., 1986
G	SA9	S. TYPHIMURIUM TA98, REVERSE MUTATION	-	0	130.0000	CONNOR ET AL., 1980b
G	SAS	S. TYPHIMURIUM (OTHER), REVERSE MUTATION	+	+	0.0000	CANTER ET AL., 1986
T	TBM	CELL TRANSFORMATION, BALB/C3T3 CELLS	-	0	670.0000	CONNOR ET AL., 1980b
F	BFA	ANIMAL BODY FLUIDS, MICROBIAL MUTAGENICITY	-	0	200.0000	CONNOR ET AL., 1980b
M	MVM	MICRONUCLEUS TEST, MICE IN VIVO	+	0	675.0000	CONNOR ET AL., 1980b
C	DLM	DOMINANT LETHAL TEST, MICE	(+)	0	1.5000	WHORTON ET AL., 1983

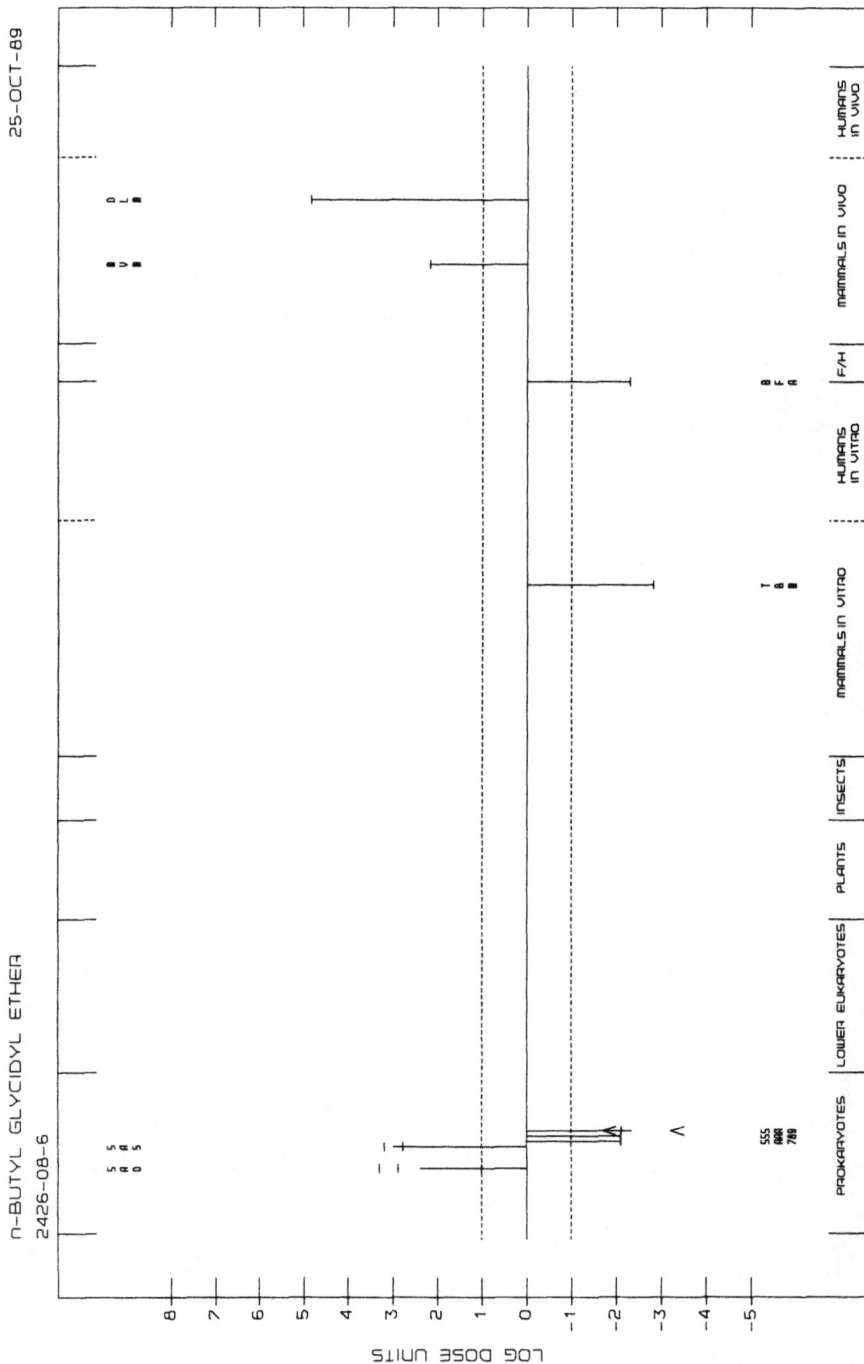

n-BUTYL GLYCIDYL ETHER
2426-08-6

25-OCT-89

t-BUTYL GLYCIDYL ETHER

END POINT	TEST CODE	TEST SYSTEM	RESULTS NM	M	DOSE (LED OR HID)	REFERENCE
G	SA0	S. TYPHIMURIUM TA100, REVERSE MUTATION	+	+	50.0000	CANTER ET AL., 1986
G	SA5	S. TYPHIMURIUM TA1535, REVERSE MUTATION	+	+	167.0000	CANTER ET AL., 1986
G	SA7	S. TYPHIMURIUM TA1537, REVERSE MUTATION	-	(+)	0.0000	CANTER ET AL., 1986
G	SA9	S. TYPHIMURIUM TA98, REVERSE MUTATION	?	(+)	0.0000	CANTER ET AL., 1986

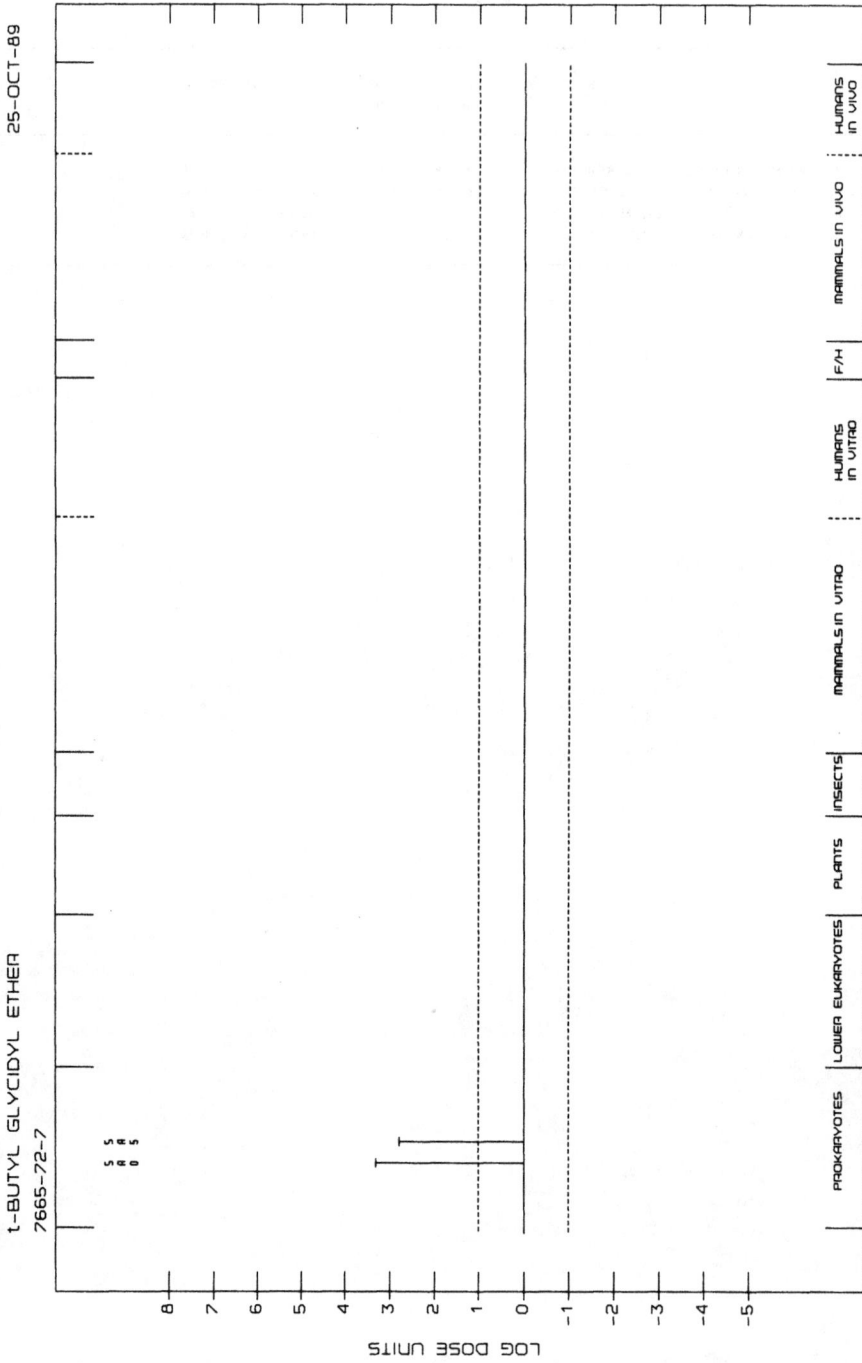

t-BUTYL GLYCIDYL ETHER

7665-72-7

25-OCT-89

t-BUTYLPHENYL GLYCIDYL ETHER

END POINT	TEST CODE	TEST SYSTEM	RESULTS NM M	DOSE (LED OR HID)	REFERENCE
G	SA0	S. TYPHIMURIUM TA100, REVERSE MUTATION	+ -	5.0000	CANTER ET AL., 1986
G	SA0	S. TYPHIMURIUM TA100, REVERSE MUTATION	+ 0	20.0000	NEAU ET AL., 1982
G	SA5	S. TYPHIMURIUM TA1535, REVERSE MUTATION	- -	34.0000	CANTER ET AL., 1986
G	SA5	S. TYPHIMURIUM TA1535, REVERSE MUTATION	- 0	70.0000	NEAU ET AL., 1982
G	SA7	S. TYPHIMURIUM TA1537, REVERSE MUTATION	- -	100.0000	CANTER ET AL., 1986
G	SA9	S. TYPHIMURIUM TA98, REVERSE MUTATION	- -	100.0000	CANTER ET AL., 1986

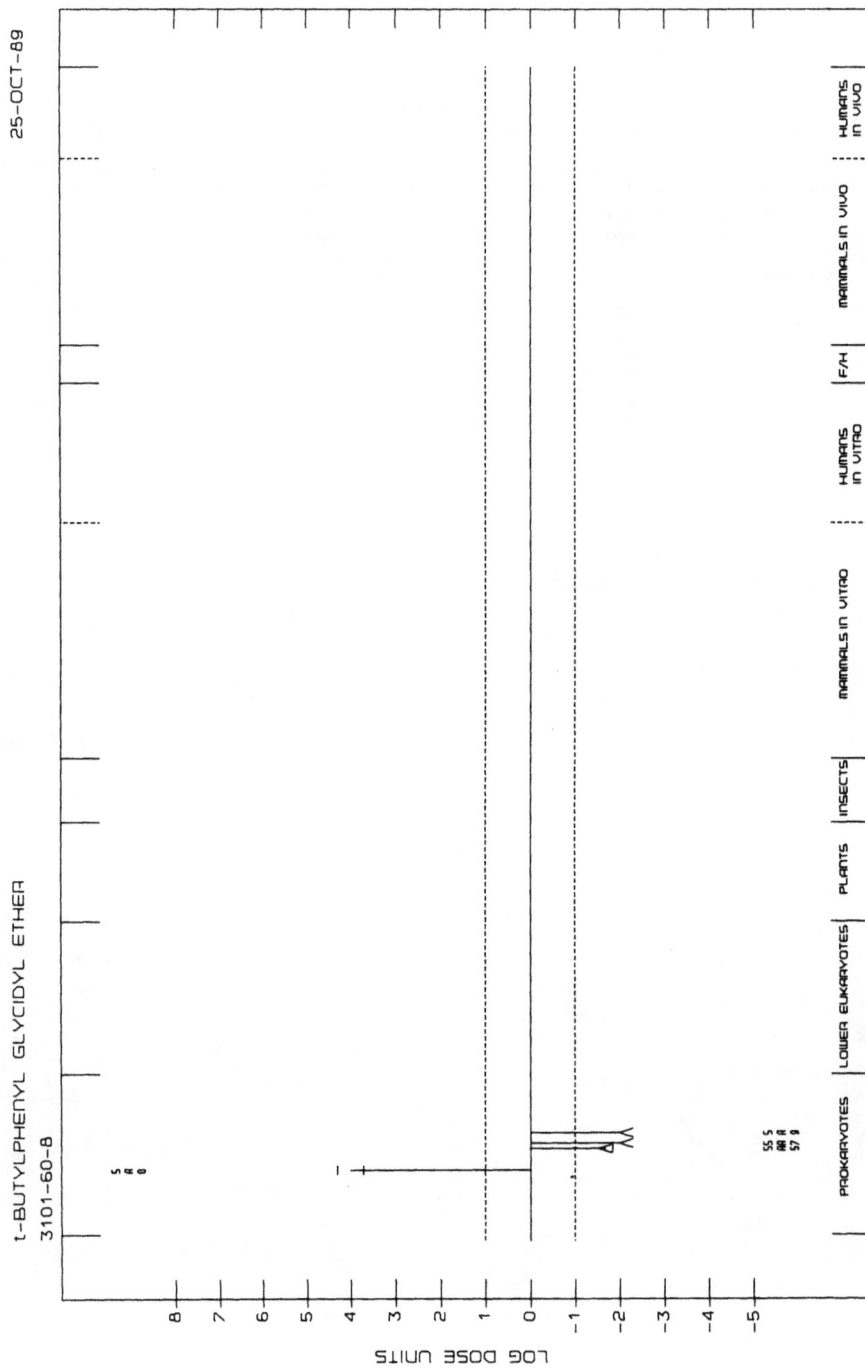

t-BUTYLPHENYL GLYCIDYL ETHER

3101-60-8

25-OCT-89

LOG DOSE UNITS

PROKARYOTES | LOWER EUKARYOTES | PLANTS | INSECTS | MAMMALS IN VITRO | HUMANS IN VITRO | F/H | MAMMALS IN VIVO | HUMANS IN VIVO

O-CRESYL GLYCIDYL ETHER

END POINT	TEST CODE	TEST SYSTEM	RESULTS NM M	DOSE (LED OR HID)	REFERENCE
G	SA0	S. TYPHIMURIUM TA100, REVERSE MUTATION	+ +	17.0000	CANTER ET AL., 1986
G	SA5	S. TYPHIMURIUM TA1535, REVERSE MUTATION	+ -	17.0000	CANTER ET AL., 1986
G	SA7	S. TYPHIMURIUM TA1537, REVERSE MUTATION	- -	0.0000	CANTER ET AL., 1986
G	SA9	S. TYPHIMURIUM TA98, REVERSE MUTATION	- -	0.0000	CANTER ET AL., 1986

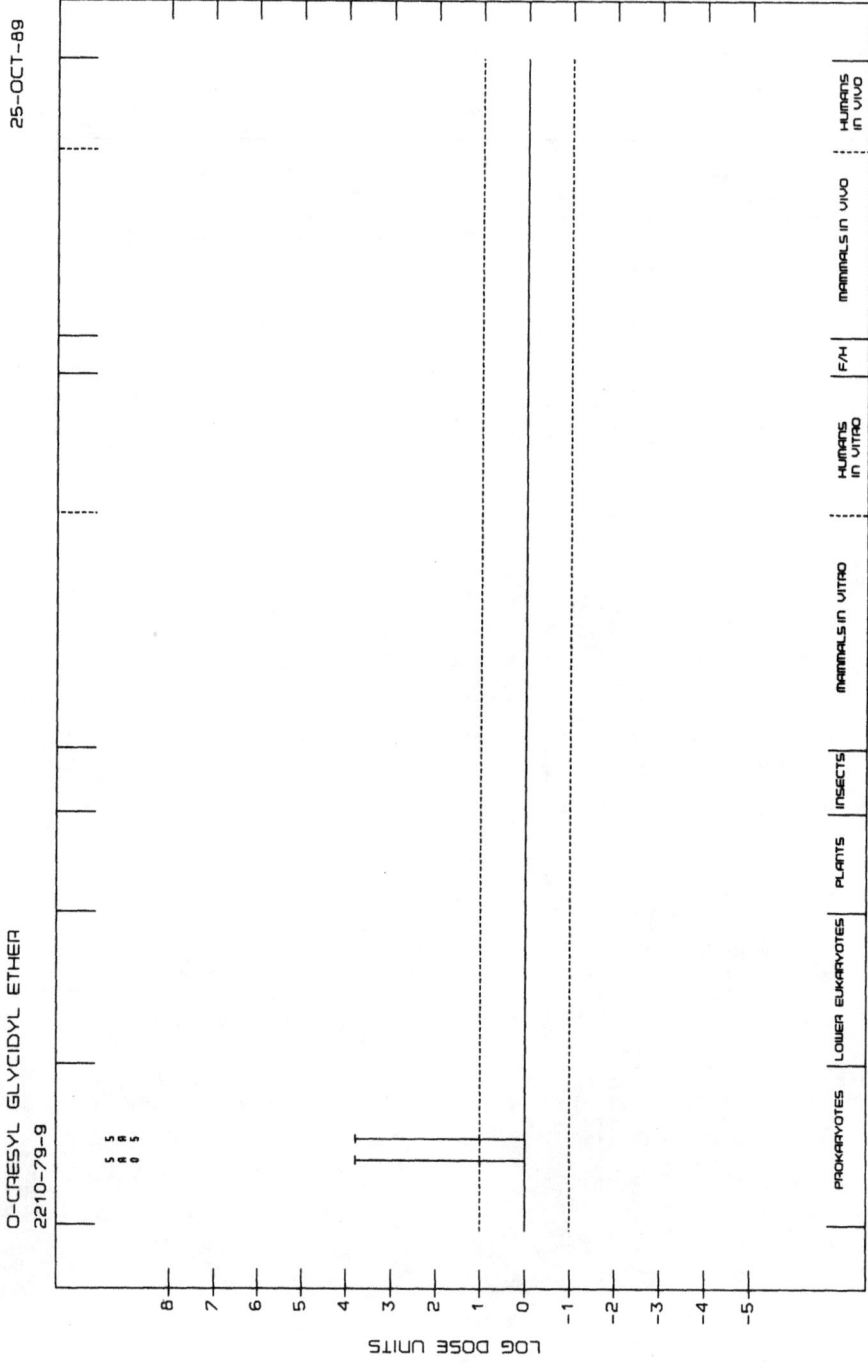

O-CRESYL GLYCIDYL ETHER
2210-79-9

25-OCT-89

LOG DOSE UNITS

PROKARYOTES LOWER EUKARYOTES PLANTS INSECTS MAMMALS IN VITRO HUMANS IN VITRO F/H MAMMALS IN VIVO HUMANS IN VIVO

P-CRESYL GLYCIDYL ETHER

END POINT	TEST CODE	TEST SYSTEM	RESULTS NM	N	DOSE (LED OR HID)	REFERENCE
G	SA0	S. TYPHIMURIUM TA100, REVERSE MUTATION	+	+	17.0000	CANTER ET AL., 1986
G	SA5	S. TYPHIMURIUM TA1535, REVERSE MUTATION	+	+	17.0000	CANTER ET AL., 1986
G	SA7	S. TYPHIMURIUM TA1537, REVERSE MUTATION	-	-	0.0000	CANTER ET AL., 1986
G	SA9	S. TYPHIMURIUM TA98, REVERSE MUTATION	-	-	0.0000	CANTER ET AL., 1986

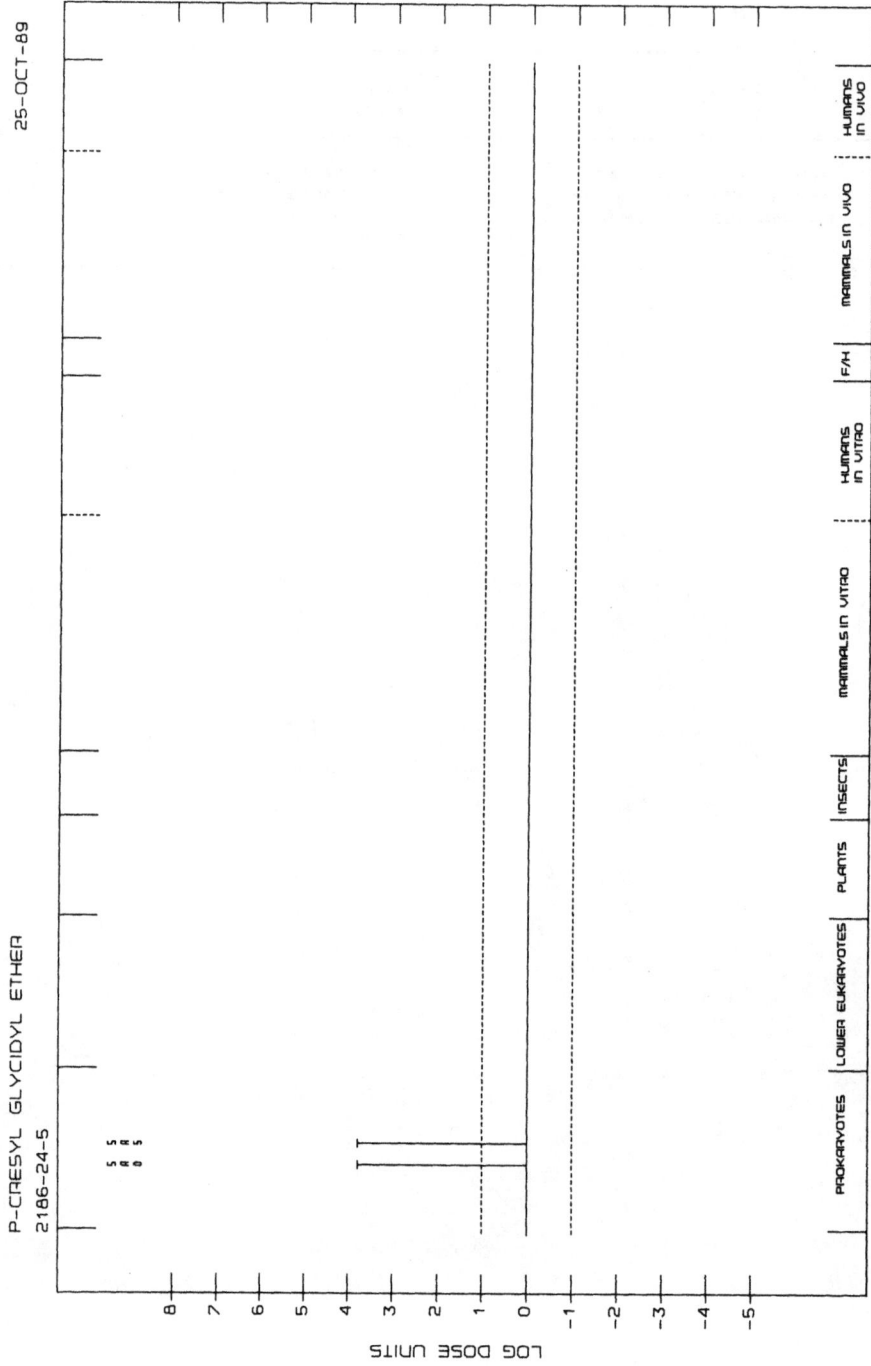

P-CRESYL GLYCIDYL ETHER
2186-24-5

25-OCT-89

LOG DOSE UNITS

PROKARYOTES | LOWER EUKARYOTES | PLANTS | INSECTS | MAMMALS IN VITRO | HUMANS IN VITRO | F/H | MAMMALS IN VIVO | HUMANS IN VIVO

NEOPENTYLGLYCOL DIGLYCIDYL ETHER

END POINT	TEST CODE	TEST SYSTEM	RESULTS NM	M	DOSE (LED OR HID)	REFERENCE
G	SA0	S. TYPHIMURIUM TA100, REVERSE MUTATION	+	+	167.0000	CANTER ET AL., 1986
G	SA5	S. TYPHIMURIUM TA1535, REVERSE MUTATION	+	+	167.0000	CANTER ET AL., 1986
G	SA9	S. TYPHIMURIUM TA98, REVERSE MUTATION	-	-	0.0000	CANTER ET AL., 1986
G	SAS	S. TYPHIMURIUM (OTHER), REVERSE MUTATION	+	+	0.0000	CANTER ET AL., 1986

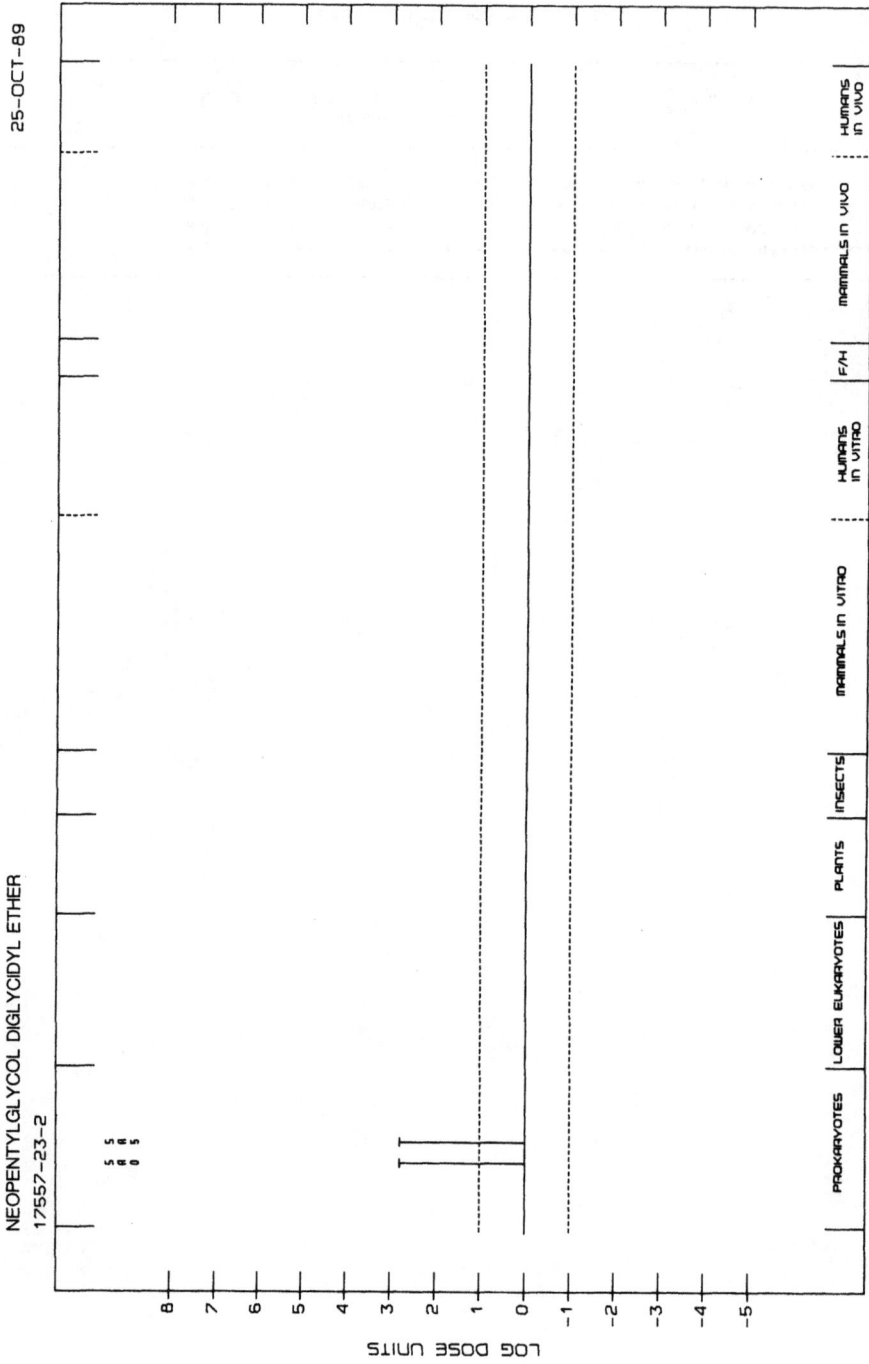

NEOPENTYLGLYCOL DIGLYCIDYL ETHER

17557-23-2

25-OCT-89

LOG DOSE UNITS

PROKARYOTES | LOWER EUKARYOTES | PLANTS | INSECTS | MAMMALS in VITRO | HUMANS in VITRO | F/H | MAMMALS in VIVO | HUMANS in VIVO

PHENYL GLYCIDYL ETHER

END POINT	TEST CODE	TEST SYSTEM	RESULTS NM M		DOSE (LED OR HID)	REFERENCE
G	SAO	S. TYPHIMURIUM TA100, REVERSE MUTATION	+	+	17.0000	CANTER ET AL., 1986
G	SAO	S. TYPHIMURIUM TA100, REVERSE MUTATION	+	0	30.0000	IVIE ET AL., 1980
G	SAO	S. TYPHIMURIUM TA100, REVERSE MUTATION	+	0	25.0000	SEILER, 1984
G	SAO	S. TYPHIMURIUM TA100, REVERSE MUTATION	+	+	25.0000	GREENE ET AL., 1979
G	SA5	S. TYPHIMURIUM TA1535, REVERSE MUTATION	+	+	17.0000	CANTER ET AL., 1986
G	SA5	S. TYPHIMURIUM TA1535, REVERSE MUTATION	+	0	25.0000	IVIE ET AL., 1980
G	SA5	S. TYPHIMURIUM TA1535, REVERSE MUTATION	+	+	2.5000	GREENE ET AL., 1979
G	SA7	S. TYPHIMURIUM TA1537, REVERSE MUTATION	-	0	250.0000	IVIE ET AL., 1980
G	SA7	S. TYPHIMURIUM TA1537, REVERSE MUTATION	-	-	250.0000	GREENE ET AL., 1979
G	SA8	S. TYPHIMURIUM TA1538, REVERSE MUTATION	-	-	250.0000	GREENE ET AL., 1979
G	SA9	S. TYPHIMURIUM TA98, REVERSE MUTATION	-	-	500.0000	CANTER ET AL., 1986
G	SA9	S. TYPHIMURIUM TA98, REVERSE MUTATION	-	0	250.0000	IVIE ET AL., 1980
G	SA9	S. TYPHIMURIUM TA98, REVERSE MUTATION	-	-	250.0000	GREENE ET AL., 1979
G	SAS	S. TYPHIMURIUM (OTHER), REVERSE MUTATION	+	+	0.0000	CANTER ET AL., 1986
G	ECW	E. COLI WP2 UVRA, REVERSE MUTATION	+	0	0.0000	HEMMINKI ET AL., 1980
G	GCO	MUTATION, CHO CELLS IN VITRO	-	0	25.0000	GREENE ET AL., 1979
C	CIC	CHROM ABERR, CHINESE HAMSTER CELLS IN VITRO	-	0	25.0000	SEILER, 1984
T	TCS	CELL TRANSFORMATION, SHE, CLONAL ASSAY	+	0	6.5000	GREENE ET AL., 1979
T	T7S	CELL TRANSFORMATION, SA7/SHE CELLS	+	0	1.6000	GREENE ET AL., 1979
H	HMM	HOST-MEDIATED ASSAY, MICROBIAL CELLS	+	0	2500.0000	GREENE ET AL., 1979
M	MVM	MICRONUCLEUS TEST, MICE IN VIVO	-	0	1000.0000	SEILER, 1984
C	CBA	CHROM ABERR, ANIMAL BONE MARROW IN VIVO	-	0	2.0000	TERRILL ET AL., 1982
C	DLR	DOMINANT LETHAL TEST, RATS	-	0	2.0000	TERRILL ET AL., 1982

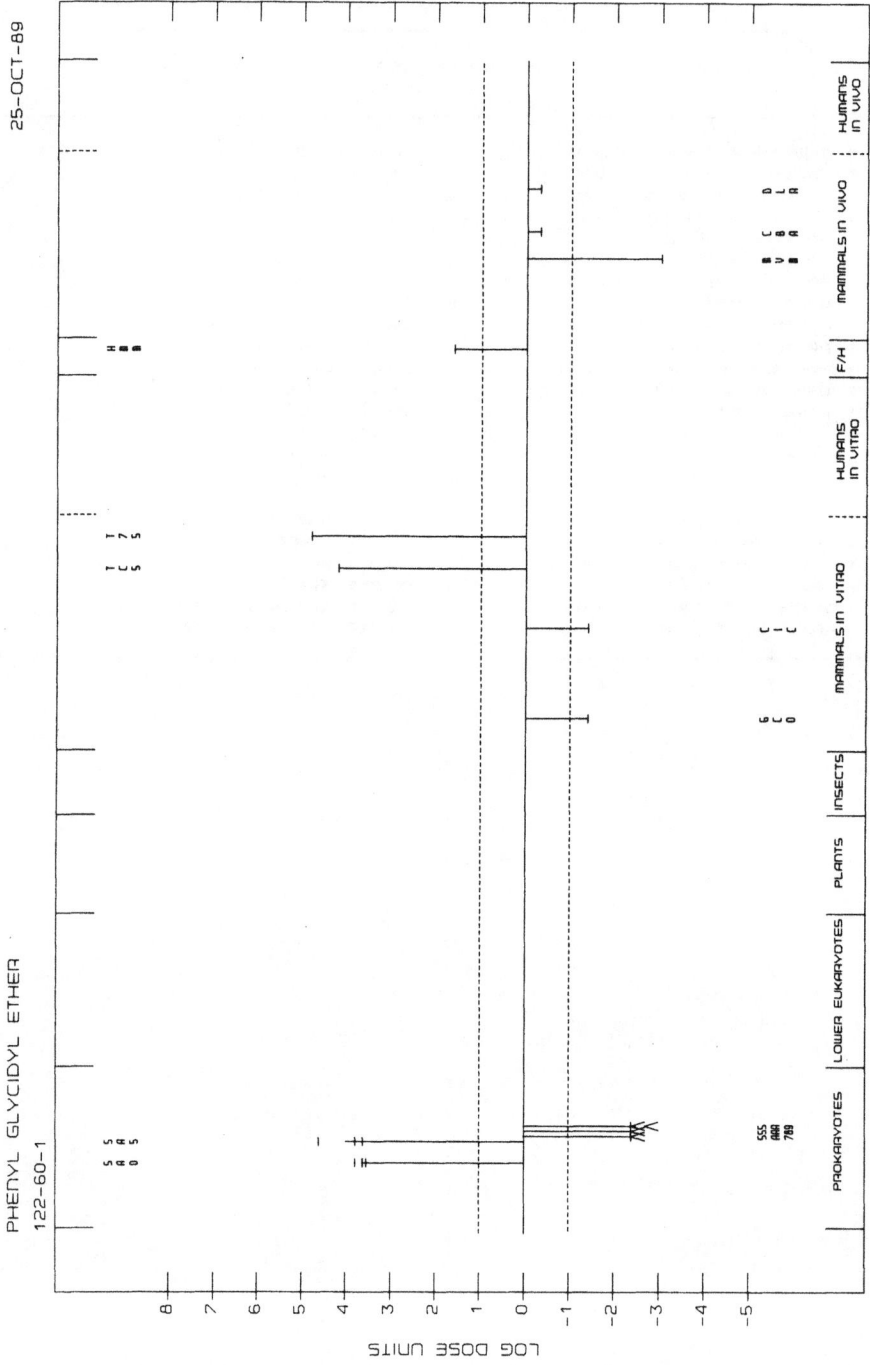

PHENYL GLYCIDYL ETHER

122-60-1

25-OCT-89

LOG DOSE UNITS

PHENOL

END POINT	TEST CODE	TEST SYSTEM	RESULTS NM M	DOSE (LED OR HID)	REFERENCE
G	SA0	S. TYPHIMURIUM TA100, REVERSE MUTATION	- -	250.0000	POOL & LIN, 1982
G	SA0	S. TYPHIMURIUM TA100, REVERSE MUTATION	- -	282.0000	FLORIN ET AL., 1980
G	SA0	S. TYPHIMURIUM TA100, REVERSE MUTATION	- 0	2000.0000	KINOSHITA ET AL., 1981
G	SA0	S. TYPHIMURIUM TA100, REVERSE MUTATION	- -	1500.0000	KAZMER ET AL., 1983
G	SA0	S. TYPHIMURIUM TA100, REVERSE MUTATION	- -	0.0000	COTRUVO ET AL., 1977
G	SA0	S. TYPHIMURIUM TA100, REVERSE MUTATION	- 0	2500.0000	THOMPSON & MELAMPY 453,1981
G	SA0	S. TYPHIMURIUM TA100, REVERSE MUTATION	- -	1250.0000	HAWORTH ET AL., 1983
G	SA5	S. TYPHIMURIUM TA1535, REVERSE MUTATION	- -	250.0000	POOL & LIN, 1982
G	SA5	S. TYPHIMURIUM TA1535, REVERSE MUTATION	- -	282.0000	FLORIN ET AL., 1980
G	SA5	S. TYPHIMURIUM TA1535, REVERSE MUTATION	- 0	50.0000	GILBERT ET AL., 1980
G	SA5	S. TYPHIMURIUM TA1535, REVERSE MUTATION	- -	0.0000	COTRUVO ET AL., 1977
G	SA5	S. TYPHIMURIUM TA1535, REVERSE MUTATION	- -	1250.0000	HAWORTH ET AL., 1983
G	SA7	S. TYPHIMURIUM TA1537, REVERSE MUTATION	- -	250.0000	POOL & LIN, 1982
G	SA7	S. TYPHIMURIUM TA1537, REVERSE MUTATION	- -	282.0000	FLORIN ET AL., 1980
G	SA7	S. TYPHIMURIUM TA1537, REVERSE MUTATION	- -	0.0000	COTRUVO ET AL., 1977
G	SA7	S. TYPHIMURIUM TA1537, REVERSE MUTATION	- -	1250.0000	HAWORTH ET AL., 1983
G	SA8	S. TYPHIMURIUM TA1538, REVERSE MUTATION	- -	250.0000	POOL & LIN, 1982
G	SA8	S. TYPHIMURIUM TA1538, REVERSE MUTATION	- 0	25.0000	GILBERT ET AL., 1980
G	SA8	S. TYPHIMURIUM TA1538, REVERSE MUTATION	- -	0.0000	COTRUVO ET AL., 1977
G	SA8	S. TYPHIMURIUM TA1538, REVERSE MUTATION	- -	1250.0000	HAWORTH ET AL., 1983
G	SA9	S. TYPHIMURIUM TA98, REVERSE MUTATION	- -	282.0000	FLORIN ET AL., 1980
G	SA9	S. TYPHIMURIUM TA98, REVERSE MUTATION	- -	2350.0000	POOL & LIN, 1982
G	SA9	S. TYPHIMURIUM TA98, REVERSE MUTATION	- (+)	2350.0000	GOCKE ET AL., 1981
G	SA9	S. TYPHIMURIUM TA98, REVERSE MUTATION	- -	0.0000	COTRUVO ET AL., 1977
G	SAS	S. TYPHIMURIUM (OTHER), REVERSE MUTATION	- -	0.0000	COTRUVO ET AL., 1977
A	ANN	A. NIDULANS, ANEUPLOIDY	(+) 0	1412.0000	CREBELLI ET AL., 1987
G	DMX	D. MELANOGASTER, SEX-LINKED RECESSIVES	- 0	4700.0000	GOCKE ET AL., 1981
G	DMX	D. MELANOGASTER, SEX-LINKED RECESSIVES	- 0	20000.0000	STURTEVANT, 1952
G	DMX	D. MELANOGASTER, SEX-LINKED RECESSIVES	- 0	5250.0000	WOODRUFF ET AL., 1985
D	DIA	STRAND BREAKS/X-LINKS, ANIMAL CELLS IN VITRO	- 0	94.0000	PELLACK-WALKER & BLUMER, 1986
G	G9H	MUTATION, CHL V79 CELLS, HPRT	+ 0	250.0000	PASCHIN & BAHITOVA, 1982
S	SHL	SCE, HUMAN LYMPHOCYTES IN VITRO	(+) 0	94.0000	MORIMOTO & WOLFF, 1980b
S	SHL	SCE, HUMAN LYMPHOCYTES IN VITRO	- 0	2.0000	JANSSON ET AL., 1986
S	SHL	SCE, HUMAN LYMPHOCYTES IN VITRO	+ 0	0.0050	EREXSON ET AL., 1985a
S	SHL	SCE, HUMAN LYMPHOCYTES IN VITRO	+ +	3.0000	MORIMOTO ET AL., 1983
M	MVM	MICRONUCLEUS TEST, MICE IN VIVO	- 0	250.0000	GAD-EL, SADA
M	MVM	MICRONUCLEUS TEST, MICE IN VIVO	+ 0	265.0000	CIRANNI ET AL., 1988
M	MVM	MICRONUCLEUS TEST, MICE IN VIVO	- 0	188.0000	GOCKE ET AL., 1981
I	ICR	INHIBIT CELL COMMUNICATION, ANIMAL CELLS	- 0	300.0000	MALCOLM ET AL., 1985
I	ICR	INHIBIT CELL COMMUNICATION, ANIMAL CELLS	- 0	0.0000	CHEN ET AL., 1984

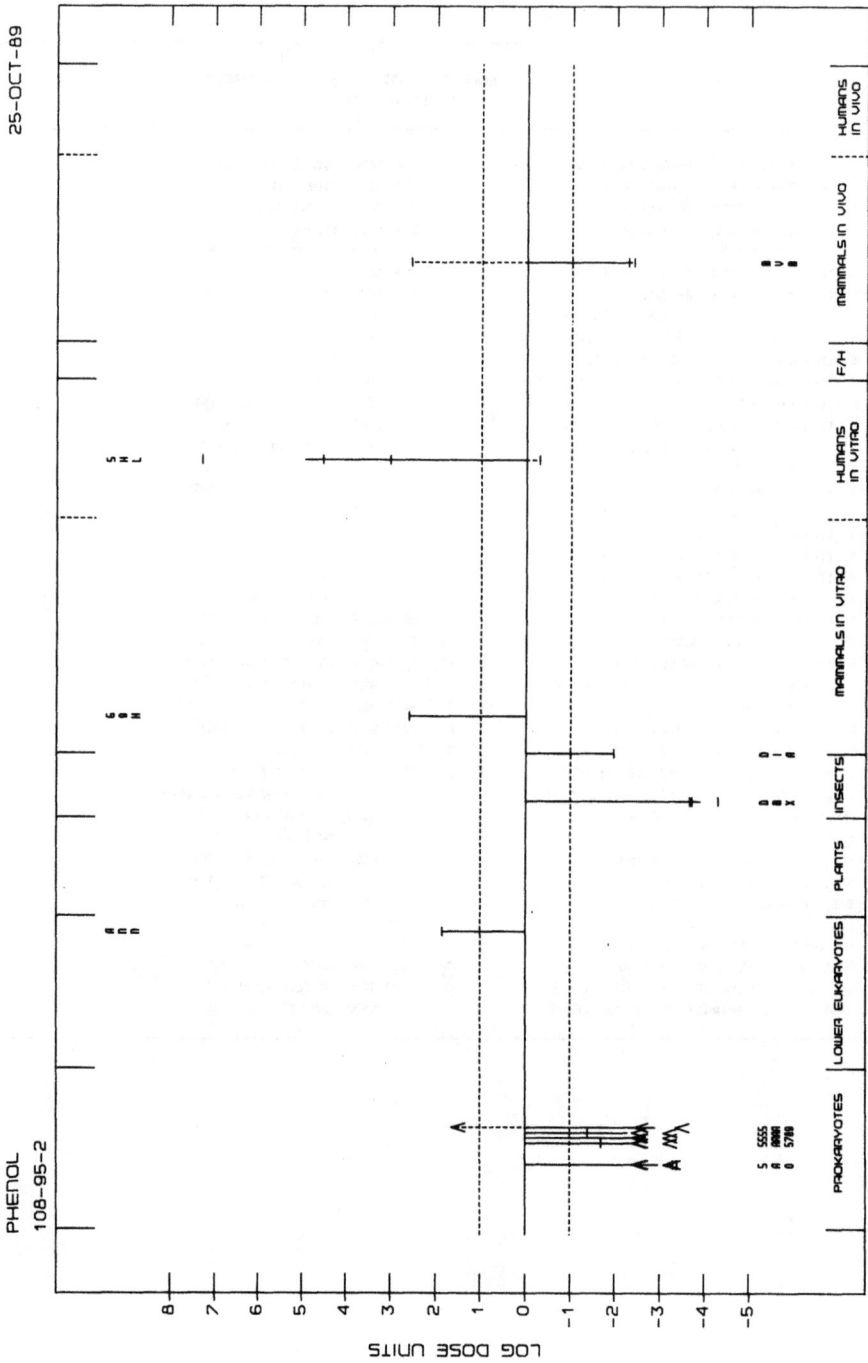

ANTIMONY TRIOXIDE

END POINT	TEST CODE	TEST SYSTEM	RESULTS NM M	DOSE (LED OR HID)	REFERENCE
D	BSD	B. SUBTILIS REC, DIFFERENTIAL TOXICITY	+ 0	720.0000	KANEMATSU ET AL., 1980

ANTIMONY TRIOXIDE
1309-64-4

25-OCT-89

LOG DOSE UNITS

PROKARYOTES   LOWER EUKARYOTES   PLANTS   INSECTS   MAMMALS in VITRO   HUMANS in VITRO   F/H   MAMMALS in VIVO   HUMANS in VIVO

TITANIUM DIOXIDE

END POINT	TEST CODE	TEST SYSTEM	RESULTS NM M	DOSE (LED OR HID)	REFERENCE
G	SA0	S. TYPHIMURIUM TA100, REVERSE MUTATION	- -	167.0000	ZEIGER ET AL., 1988
G	SA0	S. TYPHIMURIUM TA100, REVERSE MUTATION	- -	5000.0000	DUNKEL ET AL., 1985
G	SA5	S. TYPHIMURIUM TA1535, REVERSE MUTATION	- -	5000.0000	DUNKEL ET AL., 1985
G	SA5	S. TYPHIMURIUM TA1535, REVERSE MUTATION	- -	167.0000	ZEIGER ET AL., 1988
G	SA7	S. TYPHIMURIUM TA1537, REVERSE MUTATION	- -	5000.0000	DUNKEL ET AL., 1985
G	SA8	S. TYPHIMURIUM TA1538, REVERSE MUTATION	- -	5000.0000	DUNKEL ET AL., 1985
G	SA9	S. TYPHIMURIUM TA98, REVERSE MUTATION	- -	5000.0000	DUNKEL ET AL., 1985
G	SA9	S. TYPHIMURIUM TA98, REVERSE MUTATION	- -	167.0000	ZEIGER ET AL., 1988
G	SAS	S. TYPHIMURIUM (OTHER), REVERSE MUTATION	- -	167.0000	ZEIGER ET AL., 1988
G	ECW	E. COLI WP2 UVRA, REVERSE MUTATION	- -	5000.0000	DUNKEL ET AL., 1985
T	TCS	CELL TRANSFORMATION, SHE, CLONAL ASSAY	- 0	20.0000	DIPAOLO & CASTO, 1979
T	T7S	CELL TRANSFORMATION, SA7/SHE CELLS	- 0	600.0000	CASTO ET AL., 1979

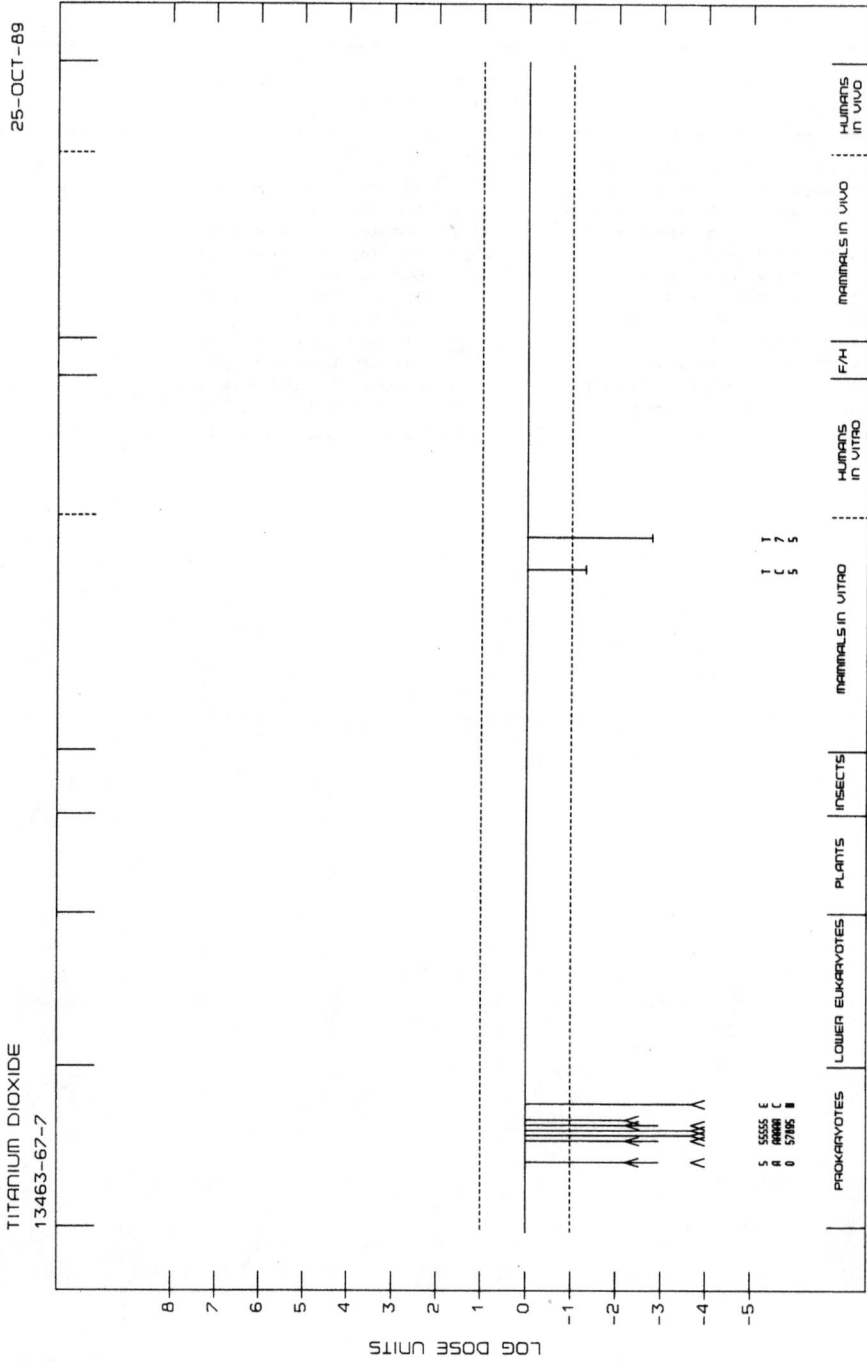

TITANIUM DIOXIDE
13463-67-7

25-OCT-89

LOG DOSE UNITS

PROKARYOTES　LOWER EUKARYOTES　PLANTS　INSECTS　MAMMALS IN VITRO　HUMANS IN VITRO　F/H　MAMMALS IN VIVO　HUMANS IN VIVO

SUPPLEMENTARY CORRIGENDA TO VOLUMES 1–46

Volume 45

p. 163   Table 3   *delete* lines 15, 16 and 17:

Methylcyclohexane	< 1-5 1
+ 1–*cis*–1–Dimethylcyclopentane	
+ 3–Methylhexane[c]	

p. 222   8th column   *replace* Seybolt *by* Saybolt

p. 247   2nd line from bottom   *replace* amidazolines *by* imidazolines

CUMULATIVE CROSS INDEX TO *IARC MONOGRAPHS*
ON THE EVALUATION OF CARCINOGENIC RISKS TO HUMANS

The volume, page and year are given. References to corrigenda are given in parentheses.

A

A-α–C	*40*, 245 (1986); *Suppl. 7*, 56 (1987)
Acetaldehyde	*36*, 101 (1985) (*corr. 42*, 263); *Suppl. 7*, 77 (1987)
Acetaldehyde formylmethylhydrazone (*see* Gyromitrin)	
Acetamide	*7*, 197 (1974); *Suppl. 7*, 389 (1987)
Acridine orange	*16*, 145 (1978); *Suppl. 7*, 56 (1987)
Acriflavinium chloride	*13*, 31 (1977); *Suppl. 7*, 56 (1987)
Acrolein	*19*, 479 (1979); *36,*133 (1985); *Suppl. 7*, 78 (1987);
Acrylamide	*39*, 41 (1986); *Suppl. 7*, 56 (1987)
Acrylic acid	*19*, 47 (1979); *Suppl. 7*, 56 (1987)
Acrylic fibres	*19*, 86 (1979); *Suppl. 7*, 56 (1987)
Acrylonitrile	*19*, 73 (1979); *Suppl. 7*, 79 (1987)
Acrylonitrile-butadiene-styrene copolymers	*19*, 91 (1979); *Suppl. 7*, 56 (1987)
Actinolite (*see* Asbestos)	
Actinomycins	*10*, 29 (1976) (*corr. 42*, 255); *Suppl. 7*, 80 (1987)
Adriamycin	*10*, 43 (1976); *Suppl. 7*, 82 (1987)
AF-2	*31*, 47 (1983); *Suppl. 7*, 56 (1987)
Aflatoxins	*1*, 145 (1972) (*corr. 42*, 251); *10*, 51 (1976); *Suppl. 7*, 83 (1987)
Aflatoxin B_1 (*see* Aflatoxins)	
Aflatoxin B_2 (*see* Aflatoxins)	
Aflatoxin G_1 (*see* Aflatoxins)	
Aflatoxin G_2 (*see* Aflatoxins)	
Aflatoxin M_1 (*see* Aflatoxins)	
Agaritine	*31*, 63 (1983); *Suppl. 7*, 56 (1987)
Alcohol drinking	*44*
Aldrin	*5*, 25 (1974); *Suppl. 7*, 88 (1987)
Allyl chloride	*36*, 39 (1985); *Suppl. 7*, 56 (1987)

Allyl isothiocyanate   *36*, 55 (1985); *Suppl. 7*, 56 (1987)
Allyl isovalerate   *36*, 69 (1985); *Suppl. 7*, 56 (1987)
Aluminium production   *34*, 37 (1984); *Suppl. 7*, 89 (1987)
Amaranth   *8*, 41 (1975); *Suppl. 7*, 56 (1987)
5-Aminoacenaphthene   *16*, 243 (1978); *Suppl. 7*, 56 (1987)
2-Aminoanthraquinone   *27*, 191 (1982); *Suppl. 7*, 56 (1987)
para-Aminoazobenzene   *8*, 53 (1975); *Suppl. 7*, 390 (1987)
ortho-Aminoazotoluene   *8*, 61 (1975) (*corr. 42*, 254); Suppl..
                                                             7, 56 (1987)

para-Aminobenzoic acid   *16*, 249 (1978); *Suppl. 7*, 56 (1987)
4-Aminobiphenyl   *1*, 74 (1972) (*corr. 42*, 251); Suppl.
                                                             7, 91 (1987)

2-Amino-3,4-dimethylimidazo[4,5-*f*]quinoline (*see* MeIQ)
2-Amino-3,8-dimethylimidazo[4,5-*f*]quinoxaline (*see* MeIQx)
3-Amino-1,4-dimethyl-5*H*-pyrido[4,3-*b*]indole (*see* Trp-P-1)
2-Aminodipyrido[1,2-*a*:3′,2′-*d*]imidazole (*see* Glu-P-2)
1-Amino-2-methylanthraquinone   *27*, 199 (1982); *Suppl. 7*, 57 (1987)
2-Amino-3-methylimidazo[4,5-*f*]quinoline (*see* IQ)
2-Amino-6-methyldipyrido[1,2-*a*:3′,2′-*d*]-imidazole (*see* Glu-P-1)
2-Amino-3-methyl-9*H*-pyrido[2,3-*b*]indole (*see* MeA-α-C)
3-Amino-1-methyl-5*H*-pyrido[4,3-*b*]indole (*see* Trp-P-2)
2-Amino-5-(5-nitro-2-furyl)-1,3,4-thiadiazole   *7*, 143 (1974); *Suppl. 7*, 57 (1987)
4-Amino-2-nitrophenol   *16*, 43 (1978); *Suppl.7*, 57 (1987)
2-Amino-5-nitrothiazole   *31*, 71 (1983); *Suppl. 7*, 57 (1987)
2-Amino-9*H*-pyrido[2,3-*b*]indole [*see* A-α-C]
11-Aminoundecanoic acid   *39*, 239 (1986); *Suppl. 7*, 57 (1987)
Amitrole   *7*, 31 (1974); *41*, 293 (1986)
                                                             Suppl. 7, 92 (1987)

Ammonium potassium selenide (*see* Selenium and selenium
  compounds)
Amorphous silica (*see also* Silica)   *Suppl. 7*, 341 (1987)
Amosite (*see* Asbestos)
Anabolic steroids (*see* Androgenic (anabolic) steroids)
Anaesthetics, volatile   *11*, 285 (1976); *Suppl. 7*, 93 (1987)
Analgesic mixtures containing phenacetin (*see also* Phenacetin)   *Suppl. 7*, 310 (1987)
Androgenic (anabolic) steroids   *Suppl. 7*, 96 (1987)
Angelicin and some synthetic derivatives (*see also* Angelicins)   *40*, 291 (1986)
Angelicin plus ultraviolet radiation (*see also* Angelicin and some   *Suppl. 7*, 57 (1987)
  synthetic derivatives)

Angelicins   *Suppl. 7*, 57 (1987)
Aniline   *4*, 27 (1974) (*corr. 42*, 252); *27*, 39
                                                             (1982); *Suppl. 7*, 99 (1987)

ortho-Anisidine   *27*, 63 (1982); *Suppl. 7*, 57 (1987)
para-Anisidine   *27*, 65 (1982); *Suppl. 7*, 57 (1987)
Anthanthrene   *32*, 95 (1983); *Suppl. 7*, 57 (1987)

B

Benzene	7, 203 (1974) (corr. 42, 254); 29, 93, 391 (1982); Suppl. 7, 120 (1987)
Benzidine	1, 80 (1972); 29, 149, 391 (1982); Suppl. 7, 123 (1987)
Benzidine–based dyes	Suppl. 7, 125 (1987)
Benzo[b]fluoranthene	3, 69 (1973); 32, 147 (1983); Suppl. 7, 58 (1987)
Benzo[j]fluoranthene	3, 82 (1973); 32, 155 (1983); Suppl. 7, 58 (1987)
Benzo[k]fluoranthene	32, 163 (1983); Suppl. 7, 58 (1987)
Benzo[ghi]fluoranthene	32, 171 (1983); Suppl. 7, 58 (1987)
Benzo[a]fluorene	32, 177 (1983); Suppl. 7, 58 (1987)
Benzo[b]fluorene	32, 183 (1983); Suppl. 7, 58 (1987)
Benzo[c]fluorene	32, 189 (1983); Suppl. 7, 58 (1987)
Benzo[ghi]perylene	32, 195 (1983); Suppl. 7, 58 (1987)
Benzo[c]phenanthrene	32, 205 (1983); Suppl. 7, 58 (1987)
Benzo[a]pyrene	3, 91 (1973); 32, 211 (1983); Suppl. 7, 58 (1987)
Benzo[e]pyrene	3, 137 (1973); 32, 225 (1983); Suppl. 7, 58 (1987)
para-Benzoquinone dioxime	29, 185 (1982); Suppl. 7, 58 (1987)
Benzotrichloride (see also α–Chlorinated toluenes)	29, 73 (1982); Suppl. 7, 148 (1987)
Benzoyl chloride	29, 83 (1982) (corr. 42, 261); Suppl. 7, 126 (1987)
Benzoyl peroxide	36, 267 (1985); Suppl. 7, 58 (1987)
Benzyl acetate	40, 109 (1986); Suppl. 7, 58 (1987)
Benzyl chloride (see also α–Chlorinated toluenes)	11, 217 (1976) (corr. 42, 256); 29, 49 (1982); Suppl. 7, 148 (1987)
Benzyl violet 4B	16, 153 (1978); Suppl. 7, 58 (1987)
Bertrandite (see Beryllium and beryllium compounds)	
Beryllium and beryllium compounds	1, 17 (1972); 23, 143 (1980) (corr. 42, 260); Suppl. 7, 127 (1987)

Beryllium acetate (see Beryllium and beryllium compounds)
Beryllium acetate, basic (see Beryllium and beryllium compounds)
Beryllium–aluminium alloy (see Beryllium and beryllium compounds)
Beryllium carbonate (see Beryllium and beryllium compounds)
Beryllium chloride (see Beryllium and beryllium compounds)
Beryllium–copper alloy (see Beryllium and beryllium compounds)
Beryllium–copper–cobalt alloy (see Beryllium and beryllium compounds)
Beryllium fluoride (see Beryllium and beryllium compounds)
Beryllium hydroxide (see Beryllium and beryllium compounds)
Beryllium–nickel alloy (see Beryllium and beryllium compounds)
Beryllium oxide (see Beryllium and beryllium compounds)
Beryllium phosphate (see Beryllium and beryllium compounds)

C

Cadmium sulphate (*see* Cadmium and cadmium compounds)
Cadmium sulphide (*see* Cadmium and cadmium compounds)
Calcium arsenate (*see* Arsenic and arsenic compounds)
Calcium chromate (*see* Chromium and chromium compounds)
Calcium cyclamate (*see* Cyclamates)
Calcium saccharin (*see* Saccharin)

Cantharidin	*10*, 79 (1976); *Suppl. 7*, 59 (1987)
Caprolactam	*19*, 115 (1979) (*corr. 42*, 258); *39*, 247 (1986) (*corr. 42*, 264); *Suppl. 7*, 390 (1987)
Captan	*30*, 295 (1983); *Suppl. 7*, 59 (1987)
Carbaryl	*12*, 37 (1976); *Suppl. 7*, 59 (1987)
Carbazole	*32*, 239 (1983); *Suppl. 7*, 59 (1987)
3-Carbethoxypsoralen	*40*, 317 (1986); *Suppl. 7*, 59 (1987)
Carbon blacks	*3*, 22 (1973); *33*, 35 (1984); *Suppl. 7*, 142 (1987)
Carbon tetrachloride	*1*, 53 (1972); *20*, 371 (1979); *Suppl. 7*, 143 (1987)
Carmoisine	*8*, 83 (1975); *Suppl. 7*, 59 (1987)
Carpentry and joinery	*25*, 139 (1981); *Suppl. 7*, 378 (1987)
Carrageenan	*10*, 181 (1976) (*corr. 42*, 255); *31*, 79 (1983); *Suppl. 7*, 59 (1987)
Catechol	*15*, 155 (1977); *Suppl. 7*, 59 (1987)

CCNU (*see* 1-(2-Chloroethyl)-3-cyclohexyl-1-nitrosourea)
Ceramic fibres (*see* Man-made mineral fibres)
Chemotherapy, combined, including alkylating agents
   (*see* MOPP and other combined chemotherapy including
   alkylating agents)

Chlorambucil	*9*, 125 (1975); *26*, 115 (1981); *Suppl. 7*, 144 (1987)
Chloramphenicol	*10*, 85 (1976); *Suppl. 7*, 145 (1987)
Chlordane (*see also* Chlordane/Heptachlor)	*20*, 45 (1979) (*corr. 42*, 258)
Chlordane/Heptachlor	*Suppl. 7*, 146 (1987)
Chlordecone	*20*, 67 (1979); *Suppl. 7*, 59 (1987)
Chlordimeform	*30*, 61 (1983); *Suppl. 7*, 59 (1987)
Chlorinated dibenzodioxins (other than TCDD)	*15*, 41 (1977); *Suppl. 7*, 59 (1987)
α-Chlorinated toluenes	*Suppl. 7*, 148 (1987)
Chlormadinone acetate (*see also* Progestins; Combined oral contraceptives)	*6*, 149 (1974); *21*, 365 (1979)

Chlornaphazine (*see* N,N-Bis(2-chloroethyl)-2-naphthylamine)

Chlorobenzilate	*5*, 75 (1974); *30*, 73 (1983); *Suppl. 7*, 60 (1987)
Chlorodifluoromethane	*41*, 237 (1986); *Suppl. 7*, 149 (1987)
1-(2-Chloroethyl)-3-cyclohexyl-1-nitrosourea (*see also* Chloroethyl nitrosoureas)	*26*, 137 (1981) (*corr. 42*, 260); *Suppl. 7*, 150 (1987)

Citrinin   *40*, 67 (1986); *Suppl. 7*, 60 (1987)
Citrus Red No. 2   *8*, 101 (1975) (*corr. 42*, 254);
                                                                     Suppl. 7, 60 (1987)
Clofibrate   *24*, 39 (1980); *Suppl. 7*, 171 (1987)
Clomiphene citrate   *21*, 551 (1979); *Suppl. 7*, 172 (1987)
Coal gasification   *34*, 65 (1984); *Suppl. 7*, 173 (1987)
Coal–tar pitches (*see also* Coal–tars)   *Suppl. 7*, 174 (1987)
Coal–tars   *35*, 83 (1985); *Suppl. 7*, 175 (1987)
Cobalt–chromium alloy (*see* Chromium and chromium
    compounds)
Coke production   *34*, 101 (1984); *Suppl. 7*, 176 (1987)
Combined oral contraceptives (*see also* Oestrogens, progestins   *Suppl. 7*, 297 (1987)
    and combinations)
Conjugated oestrogens (*see also* Steroidal oestrogens)   *21*, 147 (1979)
Contraceptives, oral (*see* Combined oral contraceptives;
    Sequential oral contraceptives)
Copper 8–hydroxyquinoline   *15*, 103 (1977); *Suppl. 7*, 61 (1987)
Coronene   *32*, 263 (1983); *Suppl. 7*, 61 (1987)
Coumarin   *10*, 113 (1976); *Suppl. 7*, 61 (1987)
Creosotes (*see also* Coal–tars)   *Suppl. 7*, 177 (1987)
meta–Cresidine   *27*, 91 (1982); *Suppl. 7*, 61 (1987)
para–Cresidine   *27*, 92 (1982); *Suppl. 7*, 61 (1987)
Crocidolite (*see* Asbestos)
Crude oil   *45*, 119 (1989)
Crystalline silica (*see also* Silica)   *Suppl. 7*, 341 (1987)
Cycasin   *1*, 157 (1972) (*corr. 42*, 251); *10*,
                                                                     121 (1976); *Suppl. 7*, 61 (1987)
Cyclamates   *22*, 55 (1980); *Suppl. 7*, 178 (1987)
Cyclamic acid (*see* Cyclamates)
Cyclochlorotine   *10*, 139 (1976); *Suppl. 7*, 61 (1987)
Cyclohexanone   *47*, 157 (1989)
Cyclohexylamine (*see* Cyclamates)
Cyclopenta[*cd*]pyrene   *32*, 269 (1983); *Suppl. 7*, 61 (1987)
Cyclopropane (*see* Anaesthetics, volatile)
Cyclophosphamide   *9*, 135 (1975); *26*, 165 (1981);
                                                                     Suppl. 7, 182 (1987)

D

2,4–D (*see also* Chlorophenoxy herbicides; Chlorophenoxy   *15*, 111 (1977)
    herbicides, occupational exposures to)
Dacarbazine   *26*, 203 (1981); *Suppl. 7*, 184 (1987)
D & C Red No. 9   *8*, 107 (1975); *Suppl. 7*, 61 (1987)
Dapsone   *24*, 59 (1980); *Suppl. 7*, 185 (1987)
Daunomycin   *10*, 145 (1976); *Suppl. 7*, 61 (1987)

1,2–Dibromo–3–chloropropane 15, 139 (1977); 20, 83 (1979); Suppl. 7, 191 (1987)

Dichloroacetylene 39, 369 (1986); Suppl. 7, 62 (1987)

ortho–Dichlorobenzene 7, 231 (1974); 29, 213 (1982); Suppl. 7, 192 (1987)

para–Dichlorobenzene 7, 231 (1974); 29, 215 (1982); Suppl. 7, 192 (1987)

3,3′–Dichlorobenzidine 4, 49 (1974); 29, 239 (1982); Suppl. 7, 193 (1987)

trans–1,4–Dichlorobutene 15, 149 (1977); Suppl. 7, 62 (1987)

3,3′–Dichloro–4,4′–diaminodiphenyl ether 16, 309 (1978); Suppl. 7, 62 (1987)

1,2–Dichloroethane 20, 429 (1979); Suppl. 7, 62 (1987)

Dichloromethane 20, 449 (1979); 41, 43 (1986); Suppl. 7, 194 (1987)

2,4–Dichlorophenol (see Chlorophenols; Chlorophenols, occupational exposures to)

(2,4–Dichlorophenoxy)acetic acid (see 2,4–D)

2,6–Dichloro–para–phenylenediamine 39, 325 (1986); Suppl. 7, 62 (1987)

1,2–Dichloropropane 41, 131 (1986); Suppl. 7, 62 (1987)

1,3–Dichloropropene (technical–grade) 41, 113 (1986); Suppl. 7, 195 (1987)

Dichlorvos 20, 97 (1979); Suppl. 7, 62 (1987)

Dicofol 30, 87 (1983); Suppl. 7, 62 (1987)

Dicyclohexylamine (see Cyclamates)

Dieldrin 5, 125 (1974); Suppl. 7, 196 (1987)

Dienoestrol (see also Nonsteroidal oestrogens) 21, 161 (1979)

Diepoxybutane 11, 115 (1976) (corr. 42, 255); Suppl. 7, 62 (1987)

Diesel and gasoline engine exhausts 46, 41 (1989)

Diesel fuels 45, 219 (1989) (corr. 47, 505)

Diethyl ether (see Anaesthetics, volatile)

Di(2–ethylhexyl)adipate 29, 257 (1982); Suppl. 7, 62 (1987)

Di(2–ethylhexyl)phthalate 29, 269 (1982) (corr. 42, 261); Suppl. 7, 62 (1987)

1,2–Diethylhydrazine 4, 153 (1974); Suppl. 7, 62 (1987)

Diethylstilboestrol 6, 55 (1974); 21, 173 (1979) (corr. 42, 259); Suppl. 7, 273 (1987)

Diethylstilboestrol dipropionate (see Diethylstilboestrol)

Diethyl sulphate 4, 277 (1974); Suppl. 7, 198 (1987)

Diglycidyl resorcinol ether 11, 125 (1976); 36, 181 (1985); Suppl. 7, 62 (1987)

Dihydrosafrole 1, 170 (1972); 10, 233 (1976); Suppl. 7, 62 (1987)

Dihydroxybenzenes (see Catechol; Hydroquinone; Resorcinol)

Dihydroxymethylfuratrizine 24, 77 (1980); Suppl. 7, 62 (1987)

E

Fuel oils (heating oils)	*45*, 239 (1989) (*corr. 47*, 505)
Furazolidone	*31*, 141 (1983); *Suppl. 7*, 63 (1987)
Furniture and cabinet-making	*25*, 99 (1981); *Suppl. 7*, 380 (1987)
2-(2-Furyl)-3-(5-nitro-2-furyl)acrylamide (*see* AF-2)	
Fusarenon-X	*11*, 169 (1976); *31*, 153 (1983); *Suppl. 7*, 64 (1987)

G

Gasoline	*45*, 159 (1989) (*corr. 47*, 505)
Gasoline engine exhaust (*see* Diesel and gasoline engine exhausts)	
Glass fibres (*see* Man-made mineral fibres)	
Glasswool (*see* Man-made mineral fibres)	
Glas filaments (*see* Man-made mineral fibres)	
Glu-P-1	*40*, 223 (1986); *Suppl. 7*, 64 (1987)
Glu-P-2	*40*, 235 (1986); *Suppl. 7*, 64 (1987)
L-Glutamic acid, 5-[2-(4-hydroxymethyl)phenylhydrazide] (*see* Agaratine)	
Glycidaldehyde	*11*, 175 (1976); *Suppl. 7*, 64 (1987)
Some glycidyl ethers	*47*, 237 (1989)
Glycidyl oleate	11, 183 (1976); *Suppl. 7*, 64 (1987)
Glycidyl stearate	*11*, 187 (1976); *Suppl. 7*, 64 (1987)
Griseofulvin	*10*, 153 (1976); *Suppl. 7*, 391 (1987)
Guinea Green B	*16*, 199 (1978); *Suppl. 7*, 64 (1987)
Gyromitrin	*31*, 163 (1983); *Suppl. 7*, 391 (1987)

H

Haematite	*1*, 29 (1972); *Suppl. 7*, 216 (1987)
Haematite and ferric oxide	*Suppl. 7*, 216 (1987)
Haematite mining, underground, with exposure to radon	*1*, 29 (1972); *Suppl. 7*, 216 (1987)
Hair dyes, epidemiology of	*16*, 29 (1978); *27*, 307 (1982)
Halothane (*see* Anaesthetics, volatile)	
α-HCH (*see* Hexachlorocyclohexanes)	
β-HCH (*see* Hexachlorocyclohexanes)	
γ-HCH (*see* Hexachlorocyclohexanes)	
Heating oils (*see* Fuel oils)	
Heptachlor (*see also* Chlordane/Heptachlor)	*5*, 173 (1974); *20*, 129 (1979)
Hexachlorobenzene	*20*, 155 (1979); *Suppl. 7*, 219 (1987)
Hexachlorobutadiene	*20*, 179 (1979); *Suppl. 7*, 64 (1987)
Hexachlorocyclohexanes	*5*, 47 (1974); *20*, 195 (1979) (*corr. 42*, 258); *Suppl. 7*, 220 (1987)
Hexachlorocyclohexane, technical-grade (*see* Hexachlorocyclohexanes)	

Hexachloroethane   *20*, 467 (1979); *Suppl. 7*, 64 (1987)
Hexachlorophene   *20*, 241 (1979); *Suppl. 7*, 64 (1987)
Hexamethylphosphoramide   *15*, 211 (1977); *Suppl. 7*, 64 (1987)
Hexoestrol (*see* Nonsteroidal oestrogens)
Hycanthone mesylate   *13*, 91 (1977); *Suppl. 7*, 64 (1987)
Hydralazine   *24*, 85 (1980); *Suppl. 7*, 222 (1987)
Hydrazine   *4*, 127 (1974); *Suppl. 7*, 223 (1987)
Hydrogen peroxide   *36*, 285 (1985); *Suppl. 7*, 64 (1987)
Hydroquinone   *15*, 155 (1977); *Suppl. 7*, 64 (1987)
4-Hydroxyazobenzene   *8*, 157 (1975); *Suppl. 7*, 64 (1987)
17α-Hydroxyprogesterone caproate (*see also* Progestins) *21*, 399 (1979) (*corr. 42*, 259)
8-Hydroxyquinoline   *13*, 101 (1977); *Suppl. 7*, 64 (1987)
8-Hydroxysenkirkine   *10*, 265 (1976); *Suppl. 7*, 64 (1987)

I

Indeno[1,2,3-*cd*]pyrene   *3*, 229 (1973); *32*, 373 (1983);
                                                          Suppl. 7, 64 (1987)

IQ   *40*, 261 (1986); *Suppl. 7*, 64 (1987)
Iron and steel founding   *34*, 133 (1984); *Suppl. 7*, 224 (1987)
Iron–dextran complex   *2*, 161 (1973); *Suppl. 7*, 226 (1987)
Iron–dextrin complex   *2*, 161 (1973) (*corr. 42*, 252);
                                                          Suppl. 7, 64 (1987)

Iron oxide (*see* Ferric oxide)
Iron oxide, saccharated (*see* Saccharated iron oxide)
Iron sorbitol–citric acid complex   *2*, 161 (1973); *Suppl. 7*, 64 (1987)
Isatidine   *10*, 269 (1976); *Suppl. 7*, 65 (1987)
Isoflurane (*see* Anaesthetics, volatile)
Isoniazid (*see* Isonicotinic acid hydrazide)
Isonicotinic acid hydrazide   *4*, 159 (1974); *Suppl. 7*, 227 (1987)
Isophosphamide   *26*, 237 (1981); *Suppl. 7*, 65 (1987)
Isopropyl alcohol   *15*, 223 (1977); *Suppl. 7*, 229 (1987)
Isopropyl alcohol manufacture (strong–acid process)   *Suppl. 7*, 229 (1987)
   (*see also* Isopropyl alcohol)
Isopropyl oils   *15*, 223 (1977); *Suppl. 7*, 229 (1987)
Isosafrole   *1*, 169 (1972); *10*, 232 (1976);
                                                          Suppl. 7, 65 (1987)

J

Jacobine   *10*, 275 (1976); *Suppl. 7*, 65 (1987)
Jet fuel   *45*, 203 (1989)
Joinery (*see* Carpentry and joinery)

K

L

M

Man–made mineral fibres   *43*, 39 (1988)
Mannomustine   *9*, 157 (1975); *Suppl. 7*, 65 (1987)
MCPA (*see also* Chlorophenoxy herbicides; Chlorophenoxy   *30*, 255 (1983)
  herbicides, occupational exposures to)
MeA–α–C   *40*, 253 (1986); *Suppl. 7*, 65 (1987)
Medphalan   *9*, 168 (1975); *Suppl. 7*, 65 (1987)
Medroxyprogesterone acetate   *6*, 157 (1974); *21*, 417 (1979)
                                                  (*corr. 42*, 259); *Suppl. 7*, 289 (1987)

Megestrol acetate (*see* also Progestins; Combined oral
  contraceptives)
MeIQ   *40*, 275 (1986); *Suppl. 7*, 65 (1987)
MeIQx   *40*, 283 (1986); *Suppl. 7*, 65 (1987)
Melamine   *39*, 333 (1986); *Suppl. 7*, 65 (1987)
Melphalan   *9*, 167 (1975); *Suppl. 7*, 239 (1987)
6–Mercaptopurine   *26*, 249 (1981); *Suppl. 7*, 240 (1987)
Merphalan   *9*, 169 (1975); *Suppl. 7*, 65 (1987)
Mestranol (*see also* Steroidal oestrogens)   *6*, 87 (1974); *21*, 257 (1979)
                                                  (*corr. 42*, 259)

Methanearsonic acid, disodium salt (*see* Arsenic and arsenic
  compounds)
Methanearsonic acid, monosodium salt (*see* Arsenic and arsenic
  compounds
Methotrexate   *26*, 267 (1981); *Suppl. 7*, 241 (1987)
Methoxsalen (*see* 8–Methoxypsoralen)
Methoxychlor   *5*, 193 (1974); *20*, 259 (1979);
                                                  Suppl. 7, 66 (1987)

Methoxyflurane (*see* Anaesthetics, volatile)
5–Methoxypsoralen   *40*, 327 (1986); *Suppl. 7*, 242 (1987)
8–Methoxypsoralen (*see also* 8–Methoxypsoralen plus ultraviolet   *24*, 101 (1980)
  radiation)
8–Methoxypsoralen plus ultraviolet radiation   *Suppl. 7*, 243 (1987)
Methyl acrylate   *19*, 52 (1979); *39*, 99 (1986);
                                                  Suppl. 7, 66 (1987)

5–Methylangelicin plus ultraviolet radiation (*see also* Angelicin
  and some synthetic derivatives)   *Suppl. 7*, 57 (1987)
2–Methylaziridine   *9*, 61 (1975); *Suppl. 7*, 66 (1987)
Methylazoxymethanol acetate   *1*, 164 (1972); *10*, 131 (1976);
                                                  Suppl. 7, 66 (1987)

Methyl bromide   *41*, 187 (1986) (*corr. 45*, 283);
                                                  Suppl. 7, 245 (1987)

Methyl carbamate   *12*, 151 (1976); *Suppl. 7*, 66 (1987)
Methyl–CCNU [*see* 1–(2–Chloroethyl)–3–(4–methylcyclohexyl)–
  1–nitrosourea]
Methyl chloride   *41*, 161 (1986); *Suppl. 7*, 246 (1987)
1–, 2–, 3–, 4–, 5– and 6–Methylchrysenes   *32*, 379 (1983); *Suppl. 7*, 66 (1987)

MOPP and other combined chemotherapy including alkylating agents — *Suppl. 7*, 254 (1987)

Morpholine — *47*, 199 (1989)

5-(Morpholinomethyl)-3-[(5-nitrofurfurylidene)amino]-2-oxazolidinone — *7*, 161 (1974); *Suppl. 7*, 67 (1987)

Mustard gas — *9*, 181 (1975) (*corr. 42*, 254); *Suppl. 7*, 259 (1987)

Myleran (*see* 1,4-Butanediol dimethanesulphonate)

N

Nafenopin — *24*, 125 (1980); *Suppl. 7*, 67 (1987)

1,5-Naphthalenediamine — *27*, 127 (1982); *Suppl. 7*, 67 (1987)

1,5-Naphthalene diisocyanate — *19*, 311 (1979); *Suppl. 7*, 67 (1987)

1-Naphthylamine — *4*, 87 (1974) (*corr. 42*, 253); *Suppl. 7*, 260 (1987)

2-Naphthylamine — *4*, 97 (1974); *Suppl. 7*, 261 (1987)

1-Naphthylthiourea — *30*, 347 (1983); *Suppl. 7*, 263 (1987)

Nickel acetate (*see* Nickel and nickel compounds)

Nickel ammonium sulphate (*see* Nickel and nickel compounds)

Nickel and nickel compounds — *2*, 126 (1973) (*corr. 42*, 252); *11*, 75 (1976); *Suppl. 7*, 264 (1987) (*corr. 45, 283*)

Nickel carbonate (*see* Nickel and nickel compounds)

Nickel carbonyl (*see* Nickel and nickel compounds)

Nickel chloride (*see* Nickel and nickel compounds)

Nickel–gallium alloy (*see* Nickel and nickel compounds)

Nickel hydroxide (*see* Nickel and nickel compounds)

Nickelocene (*see* Nickel and nickel compounds)

Nickel oxide (*see* Nickel and nickel compounds)

Nickel subsulphide (*see* Nickel and nickel compounds)

Nickel sulphate (*see* Nickel and nickel compounds)

Niridazole — *13*, 123 (1977); *Suppl. 7*, 67 (1987)

Nithiazide — *31*, 179 (1983); *Suppl. 7*, 67 (1987)

5-Nitroacenaphthene — *16*, 319 (1978); *Suppl. 7*, 67 (1987)

5-Nitro-*ortho*-anisidine — *27*, 133 (1982); *Suppl. 7*, 67 (1987)

9-Nitroanthracene — *33*, 179 (1984); *Suppl. 7*, 67 (1987)

7-Nitrobenz[*a*]anthracene — *46*, 247 (1989)

6-Nitrobenzo[*a*]pyrene — *33*, 187 (1984); *Suppl. 7*, 67 (1987); *46*, 255 (1989)

4-Nitrobiphenyl — *4*, 113 (1974); *Suppl. 7*, 67 (1987)

6-Nitrochrysene — *33*, 195 (1984); *Suppl. 7*, 67 (1987); *46*, 267 (1989)

Nitrofen (technical-grade) — *30*, 271 (1983); *Suppl. 7*, 67 (1987)

3-Nitrofluoranthene — *33*, 201 (1984); *Suppl. 7*, 67 (1987)

Phenylbutazone	*13*, 183 (1977); *Suppl. 7*, 316 (1987)
meta-Phenylenediamine	*16*, 111 (1978); *Suppl. 7*, 70 (1987)
para-Phenylenediamine	*16*, 125 (1978); *Suppl. 7*, 70 (1987)
N-Phenyl-2–naphthylamine	*16*, 325 (1978) (*corr. 42*, 257); *Suppl. 7*, 318 (1987)
ortho-Phenylphenol	*30*, 329 (1983); *Suppl. 7*, 70 (1987)
Phenytoin	*13*, 201 (1977); *Suppl. 7*, 319 (1987)
Piperazine oestrone sulphate (*see* Conjugated oestrogens)	
Piperonyl butoxide	*30*, 183 (1983); *Suppl. 7*, 70 (1987)
Pitches, coal–tar (*see* Coal-tar pitches)	
Polyacrylic acid	*19*, 62 (1979); *Suppl. 7*, 70 (1987)
Polybrominated biphenyls	*18*, 107 (1978); *41*, 261 (1986); *Suppl. 7*, 321 (1987)
Polychlorinated biphenyls	*7*, 261 (1974); *18*, 43 (1978) (*corr. 42*, 258); *Suppl. 7*, 322 (1987)
Polychlorinated camphenes (*see* Toxaphene)	
Polychloroprene	*19*, 141 (1979); *Suppl. 7*, 70 (1987)
Polyethylene	*19*, 164 (1979); *Suppl. 7*, 70 (1987)
Polymethylene polyphenyl isocyanate	*19*, 314 (1979); *Suppl. 7*, 70 (1987)
Polymethyl methacrylate	*19*, 195 (1979); *Suppl. 7*, 70 (1987)
Polyoestradiol phosphate (*see* Oestradiol–17β)	
Polypropylene	*19*, 218 (1979); *Suppl. 7*, 70 (1987)
Polystyrene	*19*, 245 (1979); *Suppl. 7*, 70 (1987)
Polytetrafluoroethylene	*19*, 288 (1979); *Suppl. 7*, 70 (1987)
Polyurethane foams	*19*, 320 (1979); *Suppl. 7*, 70 (1987)
Polyvinyl acetate	*19*, 346 (1979); *Suppl. 7*, 70 (1987)
Polyvinyl alcohol	*19*, 351 (1979); *Suppl. 7*, 70 (1987)
Polyvinyl chloride	*7*, 306 (1974); *19*, 402 (1979); *Suppl. 7*, 70 (1987)
Polyvinyl pyrrolidone	*19*, 463 (1979); *Suppl. 7*, 70 (1987)
Ponceau MX	*8*, 189 (1975); *Suppl. 7*, 70 (1987)
Ponceau 3R	*8*, 199 (1975); *Suppl. 7*, 70 (1987)
Ponceau SX	*8*, 207 (1975); *Suppl. 7*, 70 (1987)
Potassium arsenate (*see* Arsenic and arsenic compounds)	
Potassium arsenite (*see* Arsenic and arsenic compounds)	
Potassium bis(2-hydroxyethyl)dithiocarbamate	*12*, 183 (1976); *Suppl. 7*, 70 (1987)
Potassium bromate	*40*, 207 (1986); *Suppl. 7*, 70 (1987)
Potassium chromate (*see* Chromium and chromium compounds)	
Potassium dichromate (*see* Chromium and chromium compounds)	
Prednisone	*26*, 293 (1981); *Suppl. 7*, 326 (1987)
Procarbazine hydrochloride	*26*, 311 (1981); *Suppl. 7*, 327 (1987)
Proflavine salts	*24*, 195 (1980); *Suppl. 7*, 70 (1987)
Progesterone (*see also* Progestins; Combined oral contraceptives)	*6*, 135 (1974); *21*, 491 (1979) (*corr. 42*, 259)

Q

R

S

Saccharated iron oxide	2, 161 (1973); *Suppl. 7*, 71 (1987)
Saccharin	22, 111 (1980) (*corr. 42*, 259); *Suppl. 7*, 334 (1987)
Safrole	1, 169 (1972); 10, 231 (1976); *Suppl. 7*, 71 (1987)
The sawmill industry (including logging) (*see* The lumber and sawmill industry (including logging))	
Scarlet Red	8, 217 (1975); *Suppl. 7*, 71 (1987)
Selenium and selenium compounds	9, 245 (1975) (*corr. 42*, 255); *Suppl. 7*, 71 (1987)
Selenium dioxide (*see* Selenium and selenium compounds)	
Selenium oxide (*see* Selenium and selenium compounds)	
Semicarbazide hydrochloride	12, 209 (1976) (*corr. 42*, 256); *Suppl. 7*, 71 (1987)
Senecio jacobaea L. (*see* Pyrrolizidine alkaloids)	
Senecio longilobus (*see* Pyrrolizidine alkaloids)	
Seneciphylline	10, 319, 335 (1976); *Suppl. 7*, 71 (1987)
Senkirkine	10, 327 (1976); 31, 231 (1983); *Suppl. 7*, 71 (1987)
Sepiolite	42, 175 (1987); *Suppl. 7*, 71 (1987)
Sequential oral contraceptives (*see also* Oestrogens, progestins and combinations)	*Suppl. 7*, 296 (1987)
Shale-oils	35, 161 (1985); *Suppl. 7*, 339 (1987)
Shikimic acid (*see also* Bracken fern)	40, 55 (1986); *Suppl. 7*, 71 (1987)
Shoe manufacture and repair (*see* Boot and shoe manufacture and repair)	
Silica (*see also* Amorphous silica; Crystalline silica)	42, 39 (1987)
Slagwool (*see* Man-made mineral fibres)	
Sodium arsenate (*see* Arsenic and arsenic compounds)	
Sodium arsenite (*see* Arsenic and arsenic compounds)	
Sodium cacodylate (*see* Arsenic and arsenic compounds)	
Sodium chromate (*see* Chromium and chromium compounds)	
Sodium cyclamate (*see* Cyclamates)	
Sodium dichromate (*see* Chromium and chromium compounds)	
Sodium diethyldithiocarbamate	12, 217 (1976); *Suppl. 7*, 71 (1987)
Sodium equilin sulphate (*see* Conjugated oestrogens)	
Sodium fluoride (*see* Fluorides)	
Sodium monofluorophosphate (*see* Fluorides)	
Sodium oestrone sulphate (*see* Conjugated oestrogens)	
Sodium *ortho*-phenylphenate (*see also* *ortho*-Phenylphenol)	30, 329 (1983); *Suppl. 7*, 392 (1987)
Sodium saccharin (*see* Saccharin)	
Sodium selenate (*see* Selenium and selenium compounds)	

Sodium selenite (*see* Selenium and selenium compounds)
Sodium silicofluoride (*see* Fluorides)
Soots   *3*, 22 (1973); *35*, 219 (1985);
                                                      Suppl. 7, 343 (1987)
Spironolactone   *24*, 259 (1980); *Suppl. 7*, 344 (1987)
Stannous fluoride (*see* Fluorides)
Steel founding (*see* Iron and steel founding)
Sterigmatocystin   *1*, 175 (1972); *10*, 245 (1976);
                                                      Suppl. 7, 72 (1987)
Steroidal oestrogens (*see also* Oestrogens, progestins and   *Suppl. 7*, 280 (1987)
    combinations)
Streptozotocin   *4*, 221 (1974); *17*, 337 (1978);
                                                      Suppl. 7, 72 (1987)

Strobane® (*see* Terpene polychlorinates)
Strontium chromate (*see* Chromium and chromium compounds)
Styrene   *19*, 231 (1979) (*corr. 42*, 258);
                                                      Suppl. 7, 345 (1987)
Styrene–acrylonitrile copolymers   *19*, 97 (1979); *Suppl. 7*, 72 (1987)
Styrene–butadiene copolymers   *19*, 252 (1979); *Suppl. 7*, 72 (1987)
Styrene oxide   *11*, 201 (1976); *19*, 275 (1979);
                                                      36, 245 (1985); *Suppl. 7*, 72 (1987)
Succinic anhydride   *15*, 265 (1977); *Suppl. 7*, 72 (1987)
Sudan I   *8*, 225 (1975); *Suppl. 7*, 72 (1987)
Sudan II   *8*, 233 (1975); *Suppl. 7*, 72 (1987)
Sudan III   *8*, 241 (1975); *Suppl. 7*, 72 (1987)
Sudan Brown RR   *8*, 249 (1975); *Suppl. 7*, 72 (1987)
Sudan Red 7B   *8*, 253 (1975); *Suppl. 7*, 72 (1987)
Sulfafurazole   *24*, 275 (1980); *Suppl. 7*, 347 (1987)
Sulfallate   *30*, 283 (1983); *Suppl. 7*, 72 (1987)
Sulfamethoxazole   *24*, 285 (1980); *Suppl. 7*, 348 (1987)
Sulphisoxazole (*see* Sulfafurazole)
Sulphur mustard (*see* Mustard gas)
Sunset Yellow FCF   *8*, 257 (1975); *Suppl. 7*, 72 (1987)
Symphytine   *31*, 239 (1983); *Suppl. 7*, 72 (1987)

T

2,4,5-T (*see also* Chlorophenoxy herbicides; Chlorophenoxy
    herbicides, occupational exposures to)   *15*, 273 (1977)
Talc   *42*, 185 (1987); *Suppl. 7*, 349 (1987)
Tannic acid   *10*, 253 (1976) (*corr. 42*, 255);
                                                      Suppl. 7, 72 (1987)
Tannins (*see also* Tannic acid)   *10*, 254 (1976); *Suppl. 7*, 72 (1987)
TCDD (*see* 2,3,7,8-Tetrachlorodibenzo–*para*-dioxin)
TDE (*see* DDT)

Terpene polychlorinates	*5*, 219 (1974); *Suppl. 7*, 72 (1987)
Testosterone (*see also* Androgenic (anabolic) steroids)	*6*, 209 (1974); *21*, 519 (1979)
Testosterone oenanthate (*see* Testosterone)	
Testosterone propionate (*see* Testosterone)	
2,2',5,5'-Tetrachlorobenzidine	*27*, 141 (1982); *Suppl. 7*, 72 (1987)
2,3,7,8-Tetrachlorodibenzo-*para*-dioxin	*15*, 41 (1977); *Suppl. 7*, 350 (1987)
1,1,1,2-Tetrachloroethane	*41*, 87 (1986); *Suppl. 7*, 72 (1987)
1,1,2,2-Tetrachloroethane	*20*, 477 (1979); *Suppl. 7*, 354 (1987)
Tetrachloroethylene	*20*, 491 (1979); *Suppl. 7*, 355 (1987)
2,3,4,6-Tetrachlorophenol (*see* Chlorophenols; Chlorophenols, occupational exposures to)	
Tetrachlorvinphos	*30*, 197 (1983); *Suppl. 7*, 72 (1987)
Tetraethyllead (*see* Lead and lead compounds)	
Tetrafluoroethylene	*19*, 285 (1979); *Suppl. 7*, 72 (1987)
Tetramethyllead (*see* Lead and lead compounds)	
Thioacetamide	*7*, 77 (1974); *Suppl. 7*, 72 (1987)
4,4'-Thiodianiline	*16*, 343 (1978); *27*, 147 (1982); *Suppl. 7*, 72 (1987)
Thiotepa (*see* Tris(1-aziridinyl)phosphine sulphide)	
Thiouracil	*7*, 85 (1974); *Suppl. 7*, 72 (1987)
Thiourea	*7*, 95 (1974); *Suppl. 7*, 72 (1987)
Thiram	*12*, 225 (1976); *Suppl. 7*, 72 (1987)
Titanium dioxide	*47*, 307 (1989)
Tobacco habits other than smoking (*see* Tobacco products, smokeless)	
Tobacco products, smokeless	*37* (1985) (*corr. 42*, 263); *Suppl. 7*, 357 (1987)
Tobacco smoke	*38* (1986) (*corr. 42*, 263); *Suppl. 7*, 357 (1987)
Tobacco smoking (*see* Tobacco smoke)	
ortho-Tolidine (*see* 3,3'-Dimethylbenzidine)	
2,4-Toluene diisocyanate (*see also* Toluene diisocyanates)	*19*, 303 (1979); *39*, 287 (1986)
2,6-Toluene diisocyanate (*see also* Toluene diisocyanates)	*19*, 303 (1979); *39*, 289 (1986)
Toluene	*47*, 79 (1989)
Toluene diisocyanates	*39*, 287 (1986) (*corr. 42*, 264); *Suppl. 7*, 72 (1987)
Toluenes, α-chlorinated (*see* α-Chlorinated toluenes)	
ortho-Toluenesulphonamide (*see* Saccharin)	
ortho-Toluidine	*16*, 349 (1978); *27*, 155 (1982); *Suppl. 7*, 362 (1987)
Toxaphene	*20*, 327 (1979); *Suppl. 7*, 72 (1987)
Tremolite (*see* Asbestos)	
Treosulphan	*26*, 341 (1981); *Suppl. 7*, 363 (1987)
Triaziquone (*see* Tris(aziridinyl)-*para*-benzoquinone)	
Trichlorfon	*30*, 207 (1983); *Suppl. 7*, 73 (1987)

1,1,1-Trichloroethane   *20*, 515 (1979); *Suppl. 7*, 73 (1987)
1,1,2-Trichloroethane   *20*, 533 (1979); *Suppl. 7*, 73 (1987)
Trichloroethylene   *11*, 263 (1976); *20*, 545 (1979);
                                                                   Suppl. 7, 364 (1987)

2,4,5-Trichlorophenol (*see also* Chlorophenols; Chlorophenols   *20*, 349 (1979)
    occupational exposures to)
2,4,6-Trichlorophenol (*see also* Chlorophenols; Chlorophenols,   *20*, 349 (1979)
    occupational exposures to)
(2,4,5-Trichlorophenoxy)acetic acid (*see* 2,4,5-T)
Trichlorotriethylamine hydrochloride   *9*, 229 (1975); *Suppl. 7*, 73 (1987)
T$_2$-Trichothecene   *31*, 265 (1983); *Suppl. 7*, 73 (1987)
Triethylene glycol diglycidyl ether   *11*, 209 (1976); *Suppl. 7*, 73 (1987)
4,4',6-Trimethylangelicin plus ultraviolet radiation (*see also*   *Suppl. 7*, 57 (1987)
    Angelicin and some synthetic derivatives)
2,4,5-Trimethylaniline   *27*, 177 (1982); *Suppl. 7*, 73 (1987)
2,4,6-Trimethylaniline   *27*, 178 (1982); *Suppl. 7*, 73 (1'987)
4,5',8-Trimethylpsoralen   *40*, 357 (1986); *Suppl. 7*, 366 (1987)
Triphenylene   *32*, 447 (1983); *Suppl. 7*, 73 (1987)
Tris(aziridinyl)-*para*-benzoquinone   *9*, 67 (1975); *Suppl. 7*, 367 (1987)
Tris(1-aziridinyl)phosphine oxide   *9*, 75 (1975); *Suppl. 7*, 73 (1987)
Tris(1-aziridinyl)phosphine sulphide   *9*, 85 (1975); *Suppl. 7*, 368 (1987)
2,4,6-Tris(1-aziridinyl)-*s*-triazine   *9*, 95 (1975); *Suppl. 7*, 73 (1987)
1,2,3-Tris(chloromethoxy)propane   *15*, 301 (1977); *Suppl. 7*, 73 (1987)
Tris(2,3-dibromopropyl)phosphate   *20*, 575 (1979); *Suppl. 7*, 369 (1987)
Tris(2-methyl-1-aziridinyl)phosphine oxide   *9*, 107 (1975); *Suppl. 7*, 73 (1987)
Trp-P-1   *31*, 247 (1983); *Suppl. 7*, 73 (1987)
Trp-P-2   *31*, 255 (1983); *Suppl. 7*, 73 (1987)
Trypan blue   *8*, 267 (1975); *Suppl. 7*, 73 (1987)
Tussilago farfara L. (*see* Pyrrolizidine alkaloids)

U

Ultraviolet radiation   *40*, 379 (1986)
Underground haematite mining with exposure to radon   *1*, 29 (1972); *Suppl. 7*, 216 (1987)
Uracil mustard   *9*, 235 (1975); *Suppl. 7*, 370 (1987)
Urethane   *7*, 111 (1974); *Suppl. 7*, 73 (1987)

V

Vinblastine sulphate   *26*, 349 (1981) (*corr. 42*, 261);
                                                                   Suppl. 7, 371 (1987)
Vincristine sulphate   *26*, 365 (1981); *Suppl. 7*, 372 (1987)
Vinyl acetate   *19*, 341 (1979); *39*, 113 (1986);
                                                                   Suppl. 7, 73 (1987)

W

X

Y

Z

Ziram   *12*, 259 (1976); *Suppl. 7*, 74 (1987)

IARC MONOGRAPHS ON THE EVALUATION OF THE CARCINOGENIC RISK OF CHEMICALS TO HUMANS
(English editions only)

(Available from booksellers through the network of WHO Sales Agents*)

*A list of these Agents may be obtained by writing to the World Health Organization, Distribution and Sales Service, 1211 Geneva 27, Switzerland

IARC MONOGRAPHS SERIES

Volume 27
Some aromatic amines, anthraquinones and nitroso compounds, and inorganic fluorides used in drinking-water and dental preparations
1982; 341 pages; Sw. fr. 40.-

Volume 28
The rubber industry
1982; 486 pages; Sw. fr. 70.-

Volume 29
Some industrial chemicals and dyestuffs
1982; 416 pages; Sw. fr. 60.-

Volume 30
Miscellaneous pesticides
1983; 424 pages; Sw. fr. 60.-

Volume 31
Some food additives, feed additives and naturally occurring substances
1983; 14 pages; Sw. fr. 60.-

Volume 32
Polynuclear aromatic compounds, Part 1, Chemical, environmental and experimental data
1984; 477 pages; Sw. fr. 60.-

Volume 33
Polynuclear aromatic compounds, Part 2, Carbon blacks, mineral oils and some nitroarenes
1984; 245 pages; Sw. fr. 50.-

Volume 34
Polynuclear aromatic compounds, Part 3, Industrial exposures in aluminium production, coal gasification, coke production, and iron and steel founding
1984; 219 pages; Sw. fr. 48.-

Volume 35
Polynuclear aromatic compounds, Part 4, Bitumens, coal-tars and derived products, shale-oils and soots
1985; 271 pages; Sw. fr.70.-

Volume 36
Allyl compounds, aldehydes, epoxides and peroxides
1985; 369 pages; Sw. fr. 70.-

Volume 37
Tobacco habits other than smoking; betel-quid and areca-nut chewing; and some related nitrosamines
1985; 291 pages; Sw. fr. 70.-

Volume 38
Tobacco smoking
1986; 421 pages; Sw. fr. 75.-

Volume 39
Some chemicals used in plastics and elastomers
1986; 403 pages; Sw. fr. 60.-

Volume 40
Some naturally occurring and synthetic food components, furocoumarins and ultraviolet radiation
1986; 444 pages; Sw. fr. 65.-

Volume 41
Some halogenated hydrocarbons and pesticide exposures
1986; 434 pages; Sw. fr. 65.-

Volume 42
Silica and some silicates
1987; 289 pages; Sw. fr. 65.-

*Volume 43
Man-made mineral fibres and radon
1988; 300 pages; Sw. fr. 65.-

Volume 44
Alcohol and alcoholic beverages
1988; 416 pages; Sw. fr. 65.-

Volume 45
Occupational exposures in petroleum refining; crude oil and major petroleum fuels
1989; 322 pages; Sw. fr. 65.-

Volume 46
Diesel and gasoline engine exhausts and some nitroarenes
1989; 458 pages; Sw. fr. 65.-

Supplement No. 1
Chemicals and industrial processes associated with cancer in humans (IARC Monographs, Volumes 1 to 20)
1979; 71 pages; out of print

Supplement No. 2
Long-term and short-term screening assays for carcinogens: a critical appraisal
1980; 426 pages; Sw. fr. 40.-

Supplement No. 3
Cross index of synonyms and trade names in Volumes 1 to 26
1982; 199 pages; Out of print

Supplement No. 4
Chemicals, industrial processes and industries associated with cancer in humans (IARC Monographs, Volumes 1 to 29)
1982; 292 pages; Out of print

Supplement No. 5
Cross index of synonyms and trade names in Volumes 1 to 36
1985; 259 pages; Sw. fr. 60.-

*Supplement No. 6
Genetic and related effects: An updating of selected IARC Monographs from Volumes 1-42
1987; 730 pages; Sw. fr. 80.-

Supplement No. 7
Overall evaluations of carcinogenicity: An updating of IARC Monographs Volumes 1-42
1987; 440 pages; Sw. fr. 65.-

*From Volume 43 and Supplement No. 6 onwards, the series title has been changed to IARC MONOGRAPHS ON THE EVALUATION OF CARCINOGENIC RISKS TO HUMANS from IARC MONOGRAPHS ON THE EVALUATION OF THE CARCINOGENIC RISK OF CHEMICALS TO HUMANS

PUBLICATIONS OF THE INTERNATIONAL
AGENCY FOR RESEARCH ON CANCER
SCIENTIFIC PUBLICATIONS SERIES

(Available from Oxford University Press) through local bookshops

No. 1 LIVER CANCER
1971; 176 pages; out of print

No. 2 ONCOGENESIS AND HERPESVIRUSES
Edited by P.M. Biggs, G. de-Thé & L.N. Payne
1972; 515 pages; out of print

No. 3 N-NITROSO COMPOUNDS: ANALYSIS
AND FORMATION
Edited by P. Bogovski, R. Preussmann & E. A. Walker
1972; 140 pages; out of print

No. 4 TRANSPLACENTAL CARCINOGENESIS
Edited by L. Tomatis & U. Mohr
1973; 181 pages; out of print

*No. 5 PATHOLOGY OF TUMOURS IN
LABORATORY ANIMALS. VOLUME 1.
TUMOURS OF THE RAT. PART 1
Editor-in-Chief V.S. Turusov
1973; 214 pages

*No. 6 PATHOLOGY OF TUMOURS IN
LABORATORY ANIMALS. VOLUME 1.
TUMOURS OF THE RAT. PART 2
Editor-in-Chief V.S. Turusov
1976; 319 pages
*reprinted in one volume, Price £50.00

No. 7 HOST ENVIRONMENT INTERACTIONS IN
THE ETIOLOGY OF CANCER IN MAN
Edited by R. Doll & I. Vodopija
1973; 464 pages; £32.50

No. 8 BIOLOGICAL EFFECTS OF ASBESTOS
Edited by P. Bogovski, J.C. Gilson, V. Timbrell
& J.C. Wagner
1973; 346 pages; out of print

No. 9 N-NITROSO COMPOUNDS IN THE
ENVIRONMENT
Edited by P. Bogovski & E. A. Walker
1974; 243 pages; £16.50

No. 10 CHEMICAL CARCINOGENESIS ESSAYS
Edited by R. Montesano & L. Tomatis
1974; 230 pages; out of print

No. 11 ONCOGENESIS AND HERPESVIRUSES II
Edited by G. de-Thé, M.A. Epstein & H. zur Hausen
1975; Part 1, 511 pages; Part 2, 403 pages; £65.-

No. 12 SCREENING TESTS IN CHEMICAL
CARCINOGENESIS
Edited by R. Montesano, H. Bartsch & L. Tomatis
1976; 666 pages; £12.-

No. 13 ENVIRONMENTAL POLLUTION AND
CARCINOGENIC RISKS
Edited by C. Rosenfeld & W. Davis
1976; 454 pages; out of print

No. 14 ENVIRONMENTAL N-NITROSO
COMPOUNDS: ANALYSIS AND FORMATION
Edited by E.A. Walker, P. Bogovski & L. Griciute
1976; 512 pages; £37.50

No. 15 CANCER INCIDENCE IN FIVE
CONTINENTS. VOLUME III
Edited by J. Waterhouse, C. Muir, P. Correa
& J. Powell
1976; 584 pages; out of print

No. 16 AIR POLLUTION AND CANCER IN MAN
Edited by U. Mohr, D. Schmähl & L. Tomatis
1977; 311 pages; out of print

No. 17 DIRECTORY OF ON-GOING RESEARCH
IN CANCER EPIDEMIOLOGY 1977
Edited by C.S. Muir & G. Wagner
1977; 599 pages; out of print

No. 18 ENVIRONMENTAL CARCINOGENS:
SELECTED METHODS OF ANALYSIS
Edited-in-Chief H. Egan
VOLUME 1. ANALYSIS OF VOLATILE
NITROSAMINES IN FOOD
Edited by R. Preussmann, M. Castegnaro, E.A. Walker
& A.E. Wassermann
1978; 212 pages; out of print

No. 19 ENVIRONMENTAL ASPECTS OF
N-NITROSO COMPOUNDS
Edited by E.A. Walker, M. Castegnaro, L. Griciute
& R.E. Lyle
1978; 566 pages; out of print

No. 20 NASOPHARYNGEAL CARCINOMA:
ETIOLOGY AND CONTROL
Edited by G. de-Thé & Y. Ito
1978; 610 pages; out of print

No. 21 CANCER REGISTRATION AND ITS
TECHNIQUES
Edited by R. MacLennan, C. Muir, R. Steinitz
& A. Winkler
1978; 235 pages; £35.-

SCIENTIFIC PUBLICATIONS SERIES

SCIENTIFIC PUBLICATIONS SERIES

No. 43 LABORATORY DECONTAMINATION
AND DESTRUCTION OF CARCINOGENS IN
LABORATORY WASTES: SOME N-NITROSAMINES
Edited by M. Castegnaro, G. Eisenbrand, G. Ellen,
L. Keefer, D. Klein, E.B. Sansone, D. Spincer,
G. Telling & K. Webb
1982; 73 pages; £7.50

No. 44 ENVIRONMENTAL CARCINOGENS:.
SELECTED METHODS OF ANALYSIS
Editor-in-Chief H. Egan
VOLUME 5. SOME MYCOTOXINS
Edited by L. Stoloff, M. Castegnaro, P. Scott,
I.K. O'Neill & H. Bartsch
1983; 455 pages; £22.50

No. 45 ENVIRONMENTAL CARCINOGENS:
SELECTED METHODS OF ANALYSIS
Editor-in-Chief H. Egan
VOLUME 6. N-NITROSO COMPOUNDS
Edited by R. Preussmann, I.K. O'Neill, G. Eisenbrand,
B. Spiegelhalder & H. Bartsch
1983; 508 pages; £22.50

No. 46 DIRECTORY OF ON-GOING RESEARCH
IN CANCER EPIDEMIOLOGY 1982
Edited by C.S. Muir & G. Wagner
1982; 722 pages; out of print

No. 47 CANCER INCIDENCE IN SINGAPORE
1968-1977
Edited by K. Shanmugaratnam, H.P. Lee & N.E. Day
1982; 171 pages; out of print

No. 48 CANCER INCIDENCE IN THE USSR
Second Revised Edition
Edited by N.P. Napalkov, G.F. Tserkovny,
V.M. Merabishvili, D.M. Parkin, M. Smans & C.S. Muir,
1983; 75 pages; £12.-

No. 49 LABORATORY DECONTAMINATION AND
DESTRUCTION OF CARCINOGENS IN
LABORATORY WASTES: SOME POLYCYCLIC
AROMATIC HYDROCARBONS
Edited by M. Castegnaro, G. Grimmer, O. Hutzinger,
W. Karcher, H. Kunte, M. Lafontaine, E.B. Sansone,
G. Telling & S.P. Tucker
1983; 81 pages; £9.-

No. 50 DIRECTORY OF ON-GOING RESEARCH
IN CANCER EPIDEMIOLOGY 1983
Edited by C.S. Muir & G. Wagner
1983; 740 pages; out of print

No. 51 MODULATORS OF EXPERIMENTAL
CARCINOGENESIS
Edited by V. Turusov & R. Montesano
1983; 307 pages; £22.50

No. 52 SECOND CANCER IN RELATION TO
RADIATION TREATMENT FOR CERVICAL
CANCER
Edited by N.E. Day & J.D. Boice, Jr
1984; 207 pages; £20.-

No. 53 NICKEL IN THE HUMAN ENVIRONMENT
Editor-in-Chief F.W. Sunderman, Jr
1984: 530 pages; £32.50

No. 54 LABORATORY DECONTAMINATION
AND DESTRUCTION OF CARCINOGENS IN
LABORATORY WASTES: SOME HYDRAZINES
Edited by M. Castegnaro, G. Ellen, M. Lafontaine,
H.C. van der Plas, E.B. Sansone & S.P. Tucker
1983; 87 pages; £9.-

No. 55 LABORATORY DECONTAMINATION
AND DESTRUCTION OF CARCINOGENS IN
LABORATORY WASTES: SOME N-NITROSAMIDES
Edited by M. Castegnaro, M. Benard,
L.W. van Broekhoven, D. Fine, R. Massey,
E.B. Sansone, P.L.R. Smith, B. Spiegelhalder,
A. Stacchini, G. Telling & J.J. Vallon
1984; 65 pages; £7.50

No. 56 MODELS, MECHANISMS AND ETIOLOGY
OF TUMOUR PROMOTION
Edited by M. Börszönyi, N.E. Day, K. Lapis
& H. Yamasaki
1984; 532 pages; £32.50

No. 57 N-NITROSO COMPOUNDS:
OCCURRENCE, BIOLOGICAL EFFECTS
AND RELEVANCE TO HUMAN CANCER
Edited by I.K. O'Neill, R.C. von Borstel, C.T. Miller,
J. Long & H. Bartsch
1984; 1011 pages; £80.-

No. 58 AGE-RELATED FACTORS IN
CARCINOGENESIS
Edited by A. Likhachev, V. Anisimov & R. Montesano
1985; 288 pages; £20.-

No. 59 MONITORING HUMAN EXPOSURE TO
CARCINOGENIC AND MUTAGENIC AGENTS
Edited by A. Berlin, M. Draper, K. Hemminki
& H. Vainio
1984; 457 pages; £27.50

No. 60 BURKITT'S LYMPHOMA: A HUMAN
CANCER MODEL
Edited by G. Lenoir, G. O'Conor & C.L.M. Olweny
1985; 484 pages; £22.50

No. 61 LABORATORY DECONTAMINATION
AND DESTRUCTION OF CARCINOGENS IN
LABORATORY WASTES: SOME HALOETHERS
Edited by M. Castegnaro, M. Alvarez, M. Iovu,
E.B. Sansone, G.M. Telling & D.T. Williams
1984; 53 pages; £7.50

No. 62 DIRECTORY OF ON-GOING RESEARCH
IN CANCER EPIDEMIOLOGY 1984
Edited by C.S. Muir & G.Wagner
1984; 728 pages; £26.-

No. 63 VIRUS-ASSOCIATED CANCERS IN AFRICA
Edited by A.O. Williams, G.T. O'Conor, G.B. de-Thé
& C.A. Johnson
1984; 774 pages; £22.-

SCIENTIFIC PUBLICATIONS SERIES

SCIENTIFIC PUBLICATIONS SERIES

No. 85 ENVIRONMENTAL CARCINOGENS:
METHODS OF ANALYSIS AND EXPOSURE
MEASUREMENT. VOLUME 10. BENZENE
AND ALKYLATED BENZENES
Edited by L. Fishbein & I.K. O'Neill
1988; 318 pages; £35.-

No. 86 DIRECTORY OF ON-GOING RESEARCH
IN CANCER EPIDEMIOLOGY 1987
Edited by D.M. Parkin & J. Wahrendorf
1987; 685 pages; £22.-

No. 87 INTERNATIONAL INCIDENCE OF
CHILDHOOD CANCER
Edited by D.M. Parkin, C.A. Stiller, G.J. Draper,
C.A. Bieber, B. Terracini & J.L. Young
1988; 402 pages; £35.-

No. 88 CANCER INCIDENCE IN FIVE
CONTINENTS. VOLUME V
Edited by C. Muir, J. Waterhouse, T. Mack,
J. Powell & S. Whelan
1988; 1004 pages; £50.-

No. 89 METHODS FOR DETECTING DNA
DAMAGING AGENTS IN HUMANS:
APPLICATIONS IN CANCER EPIDEMIOLOGY
AND PREVENTION
Edited by H. Bartsch, K. Hemminki & I.K. O'Neill
1988; 518 pages; £45.-

No. 90 NON-OCCUPATIONAL EXPOSURE TO
MINERAL FIBRES
Edited by J. Bignon, J. Peto & R. Saracci
1988; 530 pages; £45.-

No. 91 TRENDS IN CANCER INCIDENCE IN
SINGAPORE 1968-1982
Edited by H.P. Lee, N.E. Day &
K. Shanmugaratnam
1988; 160 pages; £25.-

No. 92 CELL DIFFERENTIATION, GENES
AND CANCER
Edited by T. Kakunaga, T. Sugimura,
L. Tomatis and H. Yamasaki
1988; 204 pages; £25.-

No. 93 DIRECTORY OF ON-GOING RESEARCH
IN CANCER EPIDEMIOLOGY 1988
Edited by M. Coleman & J. Wahrendorf
1988; 662 pages; £26.-

No. 94 HUMAN PAPILLOMAVIRUS AND CERVICAL
CANCER
Edited by N. Muñoz, F.X Bosch & O.M. Jensen.
1989; 154 pages; £18.-

No. 95 CANCER REGISTRATION: PRINCIPLES
AND METHODS
Edited by O.M. Jensen, D.M. Parkin, R.
MacLennan, C.S. Muir and R. Skeet.

No. 96 PERINATAL AND MULTIGENERATION
CARCINOGENESIS
Edited by N.P. Napalkov, J.M. Rice, L. Tomatis & H.
Yamasaki
1989; 436 pages; £56.-

No. 97 OCCUPATIONAL EXPOSURE TO SILICA
AND CANCER RISK
Edited by L. Simonato, A.C. Fletcher, R. Saracci & T.
Thomas
c. 160 pages (in press)

No. 98 CANCER INCIDENCE IN JEWISH
MIGRANTS TO ISRAEL, 1961-1981
Edited by R. Steinitz, D.M. Parkin, J.L. Young,
C.A. Bieber & L. Katz
c. 300 pages (in press)

No. 99 PATHOLOGY OF TUMOURS IN
LABORATORY ANIMALS, VOL. I.
TUMOURS OF THE RAT (2nd edition).
Edited by V.S. Turusov & U. Mohr
c. 700 pages (in press)

INFORMATION BULLETINS ON THE
SURVEY OF CHEMICALS BEING
TESTED FOR CARCINOGENICITY*

No. 8 (1979)
Edited by M.-J. Ghess, H. Bartsch
& L. Tomatis
604 pages; Sw. fr. 40.-

No. 9 (1981)
Edited by M.-J. Ghess, J.D. Wilbourn,
H. Bartsch & L. Tomatis
294 pages; Sw. fr. 41.-

No. 10 (1982)
Edited by M.-J. Ghess, J.D. Wilbourn
& H. Bartsch
362 pages; Sw. fr. 42.-

No. 11 (1984)
Edited by M.-J. Ghess, J.D. Wilbourn,
H. Vainio & H. Bartsch
362 pages; Sw. fr. 50.-

No. 12 (1986)
Edited by M.-J. Ghess, J.D. Wilbourn,
A. Tossavainen & H. Vainio
385 pages; Sw. fr. 50.-

No. 13 (1988)
Edited by M.-J. Ghess, J.D. Wilbourn
& A. Aitio
404 pages; Sw. fr. 43.-

NON-SERIAL PUBLICATIONS

(Available from IARC)

ALCOOL ET CANCER
By A. Tuyns (in French only)
1978; 42 pages; Fr. fr. 35.-

CANCER MORBIDITY AND CAUSES OF
DEATH AMONG DANISH BREWERY
WORKERS
By O.M. Jensen
1980; 143 pages; Fr. fr. 75.-

DIRECTORY OF COMPUTER SYSTEMS
USED IN CANCER REGISTRIES
By H.R. Menck & D.M. Parkin
1986; 236 pages; Fr. fr. 50.-

*Available from IARC; or the World Health Organization Distribution and Sales Services, 1211 Geneva 27, Switzerland or WHO Sales Agents.

www.ingramcontent.com/pod-product-compliance
Lightning Source LLC
Chambersburg PA
CBHW081757200326
41597CB00023B/4058